Additive Manufacturing Applications for Metals and Composites

K.R. Balasubramanian
National Institute of Technology, Tiruchirappalli, India

V. Senthilkumar
National Institute of Technology, Trichy, India

IGI Global
PUBLISHER of TIMELY KNOWLEDGE

A volume in the Advances in Civil and Industrial Engineering (ACIE) Book Series

Published in the United States of America by
IGI Global
Engineering Science Reference (an imprint of IGI Global)
701 E. Chocolate Avenue
Hershey PA, USA 17033
Tel: 717-533-8845
Fax: 717-533-8661
E-mail: cust@igi-global.com
Web site: http://www.igi-global.com

Library of Congress Cataloging-in-Publication Data

Names: Balasubramanian, K. R., 1976- editor. | Senthilkumar, V., editor.
Title: Additive manufacturing applications for metals and composites / K.R. Balasubramanian, V. Senthilkumar, editors.
Description: Hershey, PA : Engineering Science Reference, 2020. | Includes bibliographical references and index. | Summary: "This book provides research on advancing methods and technological developments within additive manufacturing practices. Special attention is paid to the material design of additive manufacturing of parts, the choice of feedstock materials, the metallurgical behavior and synthesis principle during the manufacturing process, and the resulted microstructures and properties, as well as the relationship between these factors"-- Provided by publisher.
Identifiers: LCCN 2020000302 (print) | LCCN 2020000303 (ebook) | ISBN 9781799840541 (h/c) | ISBN 9781799852438 (s/c) | ISBN 9781799840558 (eISBN)
Subjects: LCSH: Three-dimensional printing. | Metals. | Composite materials.
Classification: LCC TS171.95 .A325 2020 (print) | LCC TS171.95 (ebook) | DDC 621.9/88--dc23
LC record available at https://lccn.loc.gov/2020000302
LC ebook record available at https://lccn.loc.gov/2020000303

This book is published in the IGI Global book series Advances in Civil and Industrial Engineering (ACIE) (ISSN: 2326-6139; eISSN: 2326-6155)

British Cataloguing in Publication Data
A Cataloguing in Publication record for this book is available from the British Library.

For electronic access to this publication, please contact: eresources@igi-global.com.

Advances in Civil and Industrial Engineering (ACIE) Book Series

Ioan Constantin Dima
University Valahia of Târgovişte, Romania

ISSN:2326-6139
EISSN:2326-6155

Mission

Private and public sector infrastructures begin to age, or require change in the face of developing technologies, the fields of civil and industrial engineering have become increasingly important as a method to mitigate and manage these changes. As governments and the public at large begin to grapple with climate change and growing populations, civil engineering has become more interdisciplinary and the need for publications that discuss the rapid changes and advancements in the field have become more in-demand. Additionally, private corporations and companies are facing similar changes and challenges, with the pressure for new and innovative methods being placed on those involved in industrial engineering.

The **Advances in Civil and Industrial Engineering (ACIE) Book Series** aims to present research and methodology that will provide solutions and discussions to meet such needs. The latest methodologies, applications, tools, and analysis will be published through the books included in **ACIE** in order to keep the available research in civil and industrial engineering as current and timely as possible.

Coverage

- Structural Engineering
- Coastal Engineering
- Productivity
- Production Planning and Control
- Optimization Techniques
- Urban Engineering
- Operations Research
- Materials Management
- Construction Engineering
- Engineering Economics

IGI Global is currently accepting manuscripts for publication within this series. To submit a proposal for a volume in this series, please contact our Acquisition Editors at Acquisitions@igi-global.com or visit: http://www.igi-global.com/publish/.

The Advances in Civil and Industrial Engineering (ACIE) Book Series (ISSN 2326-6139) is published by IGI Global, 701 E. Chocolate Avenue, Hershey, PA 17033-1240, USA, www.igi-global.com. This series is composed of titles available for purchase individually; each title is edited to be contextually exclusive from any other title within the series. For pricing and ordering information please visit http://www.igi-global.com/book-series/advances-civil-industrial-engineering/73673. Postmaster: Send all address changes to above address.

Titles in this Series

For a list of additional titles in this series, please visit:
http://www.igi-global.com/book-series/advances-civil-industrial-engineering/73673

Reconstructing Urban Ambiance in Smart Public Places
Hisham Abusaada (Housing and Building National Research Center (HBRC), Egypt) Ashraf M. Salama (University of Strathclyde, Glasgow, UK) and Abeer Elshater (Ain Shams Universit, Egypt)
Engineering Science Reference • © 2020 • 300pp • H/C (ISBN: 9781799838562) • US $225.00

Multi-Objective Optimization of Industrial Power Generation Systems Emerging Research and Opportunities
Timothy Ganesan (Royal Bank of Canada, Canada)
Engineering Science Reference • © 2020 • 233pp • H/C (ISBN: 9781799817109) • US $195.00

Impact of Industry 4.0 on Architecture and Cultural Heritage
Cecilia Maria Bolognesi (Politecnico of Milano, Italy) and Cettina Santagati (Università di Catania, Italy)
Engineering Science Reference • © 2020 • 422pp • H/C (ISBN: 9781799812340) • US $225.00

Handbook of Research on Urban-Rural Synergy Development Through Housing, Landscape, and Tourism
Aleksandra Krstić-Furundžić (University of Belgrade, Serbia) and Aleksandra Djukić (University of Belgrade, Serbia)
Engineering Science Reference • © 2020 • 437pp • H/C (ISBN: 9781522599326) • US $275.00

Re-Coding Homes Through Flexible Interiors Emerging Research and Opportunities
Nilüfer Saglar Onay (Politecnico di Torino, Italy) S. Banu Garip (Istanbul Technical University, Turkey) and Ervin Garip (Istanbul Technical University, Turkey)
Engineering Science Reference • © 2020 • 165pp • H/C (ISBN: 9781522589587) • US $175.00

Green Building Management and Smart Automation
Arun Solanki (Gautam Buddha University, India) and Anand Nayyar (Duy Tan University, Vietnam)
Engineering Science Reference • © 2020 • 312pp • H/C (ISBN: 9781522597544) • US $215.00

Handbook of Research on Implementation and Deployment of IoT Projects in Smart Cities
Krishnan Saravanan (Anna University Chennai – Regional Office Tirunelveli, India) Golden Julie (Anna University, India) and Harold Robinson (SCAD College of Engineering and Technology, India)
Engineering Science Reference • © 2019 • 415pp • H/C (ISBN: 9781522591993) • US $295.00

Handbook of Research on Digital Research Methods and Architectural Tools in Urban Planning and Design
Hisham Abusaada (Housing and Building National Research Center, Egypt) Carsten Vellguth (German Academic Exchange Service, Germany) and Abeer Elshater (Ain Shams University, Egypt)
Engineering Science Reference • © 2019 • 445pp • H/C (ISBN: 9781522592389) • US $265.00

IGI Global
PUBLISHER of TIMELY KNOWLEDGE

701 East Chocolate Avenue, Hershey, PA 17033, USA
Tel: 717-533-8845 x100 • Fax: 717-533-8661
E-Mail: cust@igi-global.com • www.igi-global.com

Table of Contents

Detailed Table of Contents

Additive manufacturing (AM) is also referred to as 3D printing, rapid prototyping, solid freeform fabrication, rapid manufacturing, desktop manufacturing, direct digital manufacturing, layered manufacturing, generative manufacturing, layered manufacturing, solid free-form fabrication, rapid prototype, tool-less model making, etc. It is emerging as an important manufacturing technology. It is the process of building up of layer-by-layer by depositing a material to make a component using the digital 3D model data. The main advantages of AM are mass customization, minimisation of waste, freedom of designing complex structures, and ability to print large structures. AM is broadly applicable to all classes of materials including metals, ceramics, polymers, composites, and biological systems. The AM methods used for producing complex geometrical shapes are classified based either on energy source (laser, electron beam) used or the material feed stock (powder feed, wire feed).

Additive manufacturing (AM) is going to cover all the segments of industries from missile industry to biomedical industry. This marked change of technology is due to the distinctive potential of AM to fabricate the parts with intricate designs and reduce fabrication expenditure (free from machining, waste generation, assembly of various parts) with small production runs and short turnaround times. This chapter extensively discussed industrially practiced AM technology. In this chapter, all additive manufacturing materials like metal, alloys, polymer, ceramics, composite, etc. have been given focus for various applications. Additive manufacturing technology is cost effective: no loss of metal and easy to fabricate both larger and intricate shapes. This technology already has taken a primary position in aerospace industries as well as the medical and household industries.

is provided in two-fold perspective: (1) the advancement of the arc welding process as an additive manufacturing technology and (2) the progress in the production of metal/alloys and composites through these technologies. The chapter will provide important insights into the application of arc welding technology in additive manufacturing of metals and composites for advanced applications in the era of Industry 4.0.

Wire + Arc additive manufacturing (WAAM) processes have become popular because of their proven capabilities to produce large metallic components with high deposition rates (promoted by arc-based processes) compared to conventional additive manufacturing processes such as powder bed fusion, binder jetting, direct energy deposition, etc. The applications of WAAM processes were constantly increasing in the manufacturing sector, which necessitates an understanding of the process capability to various metals. This chapter outlines the significant outcomes of the WAAM process for most of the engineering metals in terms of microstructure and mechanical properties. Discussion on various defects associated with the processed components is also presented. Potential application of WAAM for different metals such as aluminum and its alloys, titanium, and steels was discussed. The research indicates that the components manufactured by the WAAM process have significant microstructural changes and improved mechanical properties.

Additive manufacturing (AM) technology can be employed to produce multimaterial parts. In this approach, multiple types of materials are used for the fabrication of a single part. Custom-built functionally graded, heterogeneous, or porous structures and composite materials can be fabricated thorough this process. In this method, metals, plastics, and ceramics have been used with suitable AM methods to obtain multi-material products depending on functional requirements. The process of making composite materials by AM can either be performed during the material deposition process or by a hybrid process in which the combination of different materials can be performed before or after AM as a previous or subsequent stage of production of a component. Composite processes can be employed to produce functionally graded materials (FGM).

Additive manufacturing (AM) opens up the possibility of a direct build-up of components with sophisticated internal features or overhangs that are difficult to manufacture by a single conventional method. As a cost-efficient, tool-free, and digital approach to manufacturing components with complex geometries, AM of metals offers many critical benefits to various sectors such as aerospace, medical, automotive, and energy compared to conventional manufacturing processes. Direct laser fabrication (DLF) uses pre-alloyed powder mix or in-situ alloying of the elemental powders for metal additive manufacturing with excellent chemical homogeneity. It, therefore, shows great promise to enable the production of complex engineering components. This technique allows the highest build rates of the AM techniques with no restrictions on deposit size/shape and the fabrication of graded and hybrid materials by simultaneously feeding different filler materials. The advantages and disadvantages of DLF on the fabrication of compositionally complex metallic alloys are discussed in the chapter.

In order to use 3D printing technology as a sanction, it is necessary to optimize topology, component unification, and reduce weight need for advanced manufacturing design. In the case of metal 3D printing, it is necessary to manage deformation and defects in the process cause of using laser, and support generation and design optimization must be accompanied for efficiency. Currently, design progresses through simulation before actual production in AM field. This chapter explores design in additive manufacturing.

Additive manufacturing (AM) has developed and gained popularity across the globe into a multi-billion-dollar industry that involves many materials and techniques. AM has created itself as a technology for the manufacturing of metallic parts with enhanced mechanical characteristics that are scientifically sound and commercially feasible. However, there are various challenges, from business point of view, like high machine and material costs. Considering the complexities involved, sustainable manufacturing, optimization tools, and simulation models are necessary in order to save time and costly trial and errors. Topology optimization and simulation of AM processes are commercially available and are receiving attention from scientists and industry. Thus, this chapter is designed to provide readers with a brief introduction to AM technologies with typical applications. The main objective of this chapter is to provide the current trends and innovations in the field of design for additive manufacturing (DFAM), topology optimization, and simulation technologies.

For 3D printing technology to be used at the manufacturing site, excellent 3D printers, materials, and software are essential. Moreover, in the additive manufacturing (AM) process, software simulation is becoming more important as materials are diversified, and output shapes are more complicated and larger. The goal of the AM process simulation is to prevent build-up failures by predicting the macroscopic distortion and stress of the part. In the AM process simulation, structural deflection or thermal deformation easily occurs in the case where the shape of the additive manufacturing products is large and complex. So, it is necessary to provide more optimized parameters for the build-up process and more precise production of supporters. This chapter is an example of applying AM process simulation to industrial and medical parts to produce excellent products.

Additive manufacturing (AM) has emerged as the most versatile process in the manufacturing sector. The advantages of AM such as applicability in a wide range of industries, ease of manufacturing, and reduction in waste production have increased its demand over the past decades. Out of the many techniques under AM, direct metal laser sintering (DMLS) is one of the most efficient manufacturing techniques that uses a high-powered laser beam to sinter metal powders in a layer-by-layer fashion. With the current usage of computational modeling, the prediction of microstructure evolution and other thermo-mechanical properties of different materials have been of great advantage to researchers. Along with a detailed classification of AM techniques, this chapter focuses on the use of continuum, phase field, and atomistic modeling under the DMLS process. The results show that multiscale modeling can be advantageous in gaining deeper insight into various phenomena like diffusion and sintering.

Emerging additive manufacturing technologies have been gaining interest from different industries and widened their fields of application among aerospace and defense. The introduction of powder bed fusion processes was one of the significant developments in terms of direct metal part manufacturing of different materials and complex geometries, presenting good properties, and decreasing the need for tooling to allow fast product development as well as small-volume production. In this respect, nickel-based superalloys are one of the most employed material groups for aerospace and defense applications due to their mechanical strength, creep, wear, and oxidation resistance at both ambient and elevated

Preface

Additive manufacturing also referred as 3D printing has come forth as a transformative approach in industrial production for the creation of lighter, more potent and flexible components. In the past few years, metal additive manufacturing technologies have revolutionized the manufacturing industry due to its potential to produce complex geometries with special connections in contrast to the traditional production techniques. In particular, components that are multi-functional can be modelled to provide solutions to structural, protective engineering and insulation problems at the same time. Expiration of earlier patents on 3D printing technologies, cost reduction of 3D printers, continuous encouragement from the government and the flexibility in mass manufacturing of customized parts are key factors in driving the AM industry so fast. A recent survey by Wohlers Associates indicates about 50% of 3D printing will evolve around the manufacturing of commercial products in 2020 and an increase of the AM market to 2% in the world manufacturing economy of 12 trillion US dollars in the year 2030. (Ngo et al., 2018)

The demand for AM technologies in aerospace, automotive and medical fields is due to limitations of traditional processes such as long lead time, lack of flexibility and expensive tooling.(Culmone et al., 2019; Fasel et al., 2020; Gisario et al., 2019; Ngo et al., 2018) Among various AM technologies for producing metal parts, processes such as powder bed fusion (PBF) and direct energy depositions (DED) techniques are finding their place in industrial applications (Adeyemi et al., 2018). High-quality metal powder and wire feedstock are critical to the successful building of parts in AM. A number of new and dissimilar metals in powdered form to suit exact process requirements that are included in library of materials are aluminium, steel, copper, titanium, stainless steel and copper, cobalt chrome, nickel and titanium-based alloys in addition to precious metals like platinum, amber, silver and palladium. A broad range of wire feedstock options is also available like those of steel and stainless-steel alloys as well as pure metals like aluminium, niobium, titanium, molybdenum and tungsten are available as wire feedstock (Ngo et al., 2018).

Due to the emergence of additive manufacturing, designers were able to come up with complicated design like never before and manufacturers were able to produce components at ease. Additive manufacturing is desirable when the variation in production is high and design tends to be more complicated like those of aerospace and naval application. Additive techniques render huge benefits in case of new invention where it calls for a deal of money due to the need of preparation of moulds in case of traditional procedures, therefore enabling the producers to be amply competent in the securities industry. Based on the place of usage the additive techniques like those of Selective Laser Melting/ Selective Laser Sintering (SLM/SLS), binder jetting, Direct Metal Deposition (DMD), wire arc additive manufacturing (WAAM), powder based and wire -based Direct Energy Deposition (DED), Laser Engineering Net Shaping (LENS) and Laser Metal Deposition (LMD) are used. (Azarniya et al., 2019; Froes & Dutta, 2014)

Composite materials have been used widely due to their inheriting superior mechanical strength and high performance in various applications. Metal matrix composites offer production of light weight component which make them as a better alternative to their metal counterparts. Application of AM techniques that are used for the production of metals with minor modification make them suitable for producing composites; some commonly used techniques are Selective Laser Melting/ Selective Laser Sintering (SLM/SLS), powder based and wire -based Direct Energy Deposition (DED), Laser Metal Deposition (LMD), Direct Metal Deposition (DMD), Laser Engineering Net Shaping (LENS), Plasma Deposition Manufacturing (PDM), and sheet lamination (Fasel et al., 2020; Froes & Dutta, 2014; Kirka et al., 2018)

Though parts can be produced in AM in various sizes from micro and macro scale, constraints on anisotropic behaviour, residual stress built in and its associated residual stress, post processing required for altering microstructure and poor fatigue life are challenged in the field (Liu et al., 2018; Munford et al., 2020). This book will bea unique guide to additive manufacturing/ three-dimensional (3D) printing of metals and composites. Examining a range of manufacturing technologies and their capability to manufacture both functional live scale products as wells as prototypes, the chapters that are discussed in the book address metal components and explore some of the main research issues associated with the use of these technologies. The newer technologies that are currently under development are also explored here. The book also presents unique real-life case studies from industry, the contents specified here offers the perspective of engineers who are associated in the fields of aerospace and transportation systems; it is also intent to help engineers to design components and manufacturing networks. Written by the leading university and industry experts, the book is aimed at providing a comprehensive insight for students and practitioners in the field.

The book consists of 14 chapters formulated by experts and the following sections briefly outline various chapters enclosed in the book.

Chapter 1 represents an introduction to metal AM techniques and its benefits in the fabrication process over conventional methods. It also explains about the conventional CNC and AM processes which are detailed. Chapter 2 throws light o nthree AM processes, namely, Powder Bed Fusion (PBF), Electron Beam Melting (EBM), Binder Jetting (BJ) which are majorly used for commercial purposes and their advantage and disadvantages are also detailed. The contents also explain about the role of 3D printing in future manufacturing industries.

Chapter 3 discusses briefly about the advancements in additive manufacturing of Ni-based superalloys which are used in aero-engines. The influence of pre-alloyed powder, process parameters, and scanning strategy on the microstructure and mechanical properties in as-built condition are also explained. Chapter 4 summarises the process-structure-property relationships in laser powder-bed fusion of Invar 36. The process parameters for L-PBF of Invar 36 to obtain optimum properties were also detailed. Chapter 5 details the fundamental concepts of the three arc welding technologies, namely metal arc, tungsten arc and plasma arc welding methods which are explained in detail, the chapter also provides information about the various modifications that can be introduced into these methods to create their capabilities as additive manufacturing processes.

Chapter 6 elucidates the production of complex shapes using WAAM technique. The technique of WAAM automation and modelling advancements has known to improve the process penetration into the market for producing a wide range of products. It also deciphers the reason why CMT has become a popular heat source for WAAM processes, which is followed by Chapter7 that expounds the development of products with multi-material and composite structures using different additive manufacturing techniques. The mechanical, metallurgical and functional properties enhanced through various AM processes are discussed in brief with their results. The chapter also details the benefits of functionally graded materials developed through hybrid AM process rather than the existing additive manufacturing systems. Chapter 8 briefs about the Direct Laser Fabrication (DLF) which provides benefits to the healthcare sector for the development towards the creation of high-quality prosthetic devices/implants. This technique can be used to repair/remanufacture damaged components within an assembly such as tool/dies and turbine blades and thus avoiding the necessity of the fabrication of new components. The future perspectives of DLF as a tool for the fabrication of high-quality, high-end components, including the development of efficient laser sources, improved powder utilization during the deposition process, the design of flexible/protective process chambers for DLF platforms and effective powder recycling strategies to reduce the environmental impacts of the process are detailed. Chapter 9 discusses the current status of the AM process simulation software, and the AM process simulation. It also states that the AM process simulation must be performed when the shape of the stacked output is large and complex. The chapter ends with a marvellous postulation claiming that despite the promising results reported, one must keep in mind the key challenges in additive manufacturing technology. The topology optimization and cutting-edge AM simulation technologies is presented in Chapter 10. Applications in these two directions were cited with case study examples. Despite the promising results reported, one must keep in mind of the key challenges in additive manufacturing technology. Chapter 11 describes about a compensation that can minimize deformation and predict the deformation due to high thermal source (Laser) in manufacturing cranial implants with a 3D printer. Chapter 12 focuses on the importance of multiscale modelling in AM processes. It states that using phase field and atomistic modelling, simulations are performed to understand the necking behaviour of AlSi10Mg in one of the AM processes (DMLS). The chapter ends by giving out a few areas where additional scope of research is required proceeded by Chapter 13 that briefs on PBF technology, which brings to light about the fact that it is important to note that due to the inherent nature of the processes, additive manufacturing may lead to new process and supply chains as well as new manufacturability problems related to high residual stresses and leading deformations, inferior surface quality and dimensional accuracy in contrast to machining, anisotropic material properties, etc. Chapter 14 compiles the evolution and global status of LAM technology, highlighting its advantages and freedoms for various industrial applications. It discusses how LAM is contributing to Industry 4.0 for the fabrication of tailored parts for engineering and prosthetic applications through case studies. It compiles research, development and deployment scenarios of this new technology in developing economies along with the future scope of the technology.

A lot of people rendered their help in the preparation of this book directly and indirectly. Though it is very difficult to name everyone at this point, few of them whose immense contribution for the fruition of the effort must be remembered. The authors of this book have a long association with each other, both working at National Institute of Institute of Technology, Tiruchirappalli (NITT) for over a decade. Both authors have been teaching Manufacturing Technology, Welding Technology and Machine Design, Metal forming and Machining for the past several years. Further they have supervised several master and doctoral students in the area of laser processing, metal additive manufacturing and process simulation. Interaction with the graduate, doctoral students and industry personnel that are working in the Additive Manufacturing field are the driving force for taking the assignment of editing of this book that covers the recent advances in additive manufacturing of metals and composites. Our heartfelt thanks to our Institute Director Dr Mini Shaji Thomas for her support in our efforts for taking up of the project and its timely completion. Our friends and colleagues have helped tremendously in the completion of the work and we owe a lot for their constant encouragement and criticism during the editing of work. We thank all our reviewers, especially, Dr. Özgür Poyraz, Dr. Mostafa Yakout, Fredrick Madaraka, Dr. Anagh Deshpande, Dr. Jithin Joseph, Dr. Evren Yasa who have spent time to review the individual chapters and making comments and suggestions for the improvement of the chapter. Special thanks to our PhD students Mr. S. Divakar and Mr. C.T. Justus Panicker for helping us in our editorial work.

REFERENCES

Adeyemi, A., Akinlabi, E. T., & Mahamood, R. M. (2018). Powder bed based laser additive manufacturing process of stainless steel: A review. *Materials Today: Proceedings*, *5*(9), 18510–18517. doi:10.1016/j.matpr.2018.06.193

Azarniya, A., Colera, X. G., Mirzaali, M. J., Sovizi, S., Bartolomeu, F., St Weglowski, M., Wits, W. W., Yap, C. Y., Ahn, J., Miranda, G., Silva, F. S., Madaah Hosseini, H. R., Ramakrishna, S., & Zadpoor, A. A. (2019). Additive manufacturing of Ti–6Al–4V parts through laser metal deposition (LMD): Process, microstructure, and mechanical properties. *Journal of Alloys and Compounds*, *804*, 163–191. doi:10.1016/j.jallcom.2019.04.255

Culmone, C., Smit, G., & Breedveld, P. (2019). Additive manufacturing of medical instruments: A state-of-the-art review. *Additive Manufacturing*, *27*(April), 461–473. doi:10.1016/j.addma.2019.03.015

Fasel, U., Keidel, D., Baumann, L., Cavolina, G., Eichenhofer, M., & Ermanni, P. (2020). Composite additive manufacturing of morphing aerospace structures. In Manufacturing Letters (Vol. 23, pp. 85–88). doi:10.1016/j.mfglet.2019.12.004

Froes, F. H., & Dutta, B. (2014). The additive manufacturing (AM) of titanium alloys. *Advanced Materials Research, 1019*, 19–25. . doi:10.4028/www.scientific.net/AMR.1019.19

Gisario, A., Kazarian, M., Martina, F., & Mehrpouya, M. (2019). Metal additive manufacturing in the commercial aviation industry: A review. *Journal of Manufacturing Systems*, *53*(June), 124–149. doi:10.1016/j.jmsy.2019.08.005

Kirka, M. M., Plotkowski, A., Nandwana, P., Chaudhary, A., Babu, S. S., & Dehoff, R. R. (2018). Progress in the processing and understanding of alloy 718 fabricated through powder bed additive manufacturing processes. In Minerals, Metals and Materials Series. doi:10.1007/978-3-319-89480-5_4

Liu, Z. Y., Stief, P., Dantan, J., Etienne, A., & Siadat, A. (2018). Residual Stress in Metal Additive Manufacturing. *4th CIRP Conference on Surface Integrity (CSI 2018).* 10.1016/j.procir.2018.05.039

Munford, M., Hossain, U., Ghouse, S., & Jeffers, J. R. T. (2020). Prediction of anisotropic mechanical properties for lattice structures. *Additive Manufacturing, 32*(December), 101041. doi:10.1016/j.addma.2020.101041

Ngo, T. D., Kashani, A., Imbalzano, G., Nguyen, K. T. Q., & Hui, D. (2018). Additive manufacturing (3D printing): A review of materials, methods, applications and challenges. *Composites. Part B, Engineering*, *143*, 172–196. Advance online publication. doi:10.1016/j.compositesb.2018.02.012

Chapter 1
Introduction to Additive Manufacturing

K. R. Balasubramanian
https://orcid.org/0000-0001-7179-5566
National Institute of Technology, Tiruchirappalli, India

V. Senthilkumar
National Institute of Technology, Trichy, India

Divakar Senthilvel
National Institute of Technology, Tiruchirappalli, India

ABSTRACT

Additive manufacturing (AM) is also referred to as 3D printing, rapid prototyping, solid freeform fabrication, rapid manufacturing, desktop manufacturing, direct digital manufacturing, layered manufacturing, generative manufacturing, layered manufacturing, solid free-form fabrication, rapid prototype, tool-less model making, etc. It is emerging as an important manufacturing technology. It is the process of building up of layer-by-layer by depositing a material to make a component using the digital 3D model data. The main advantages of AM are mass customization, minimisation of waste, freedom of designing complex structures, and ability to print large structures. AM is broadly applicable to all classes of materials including metals, ceramics, polymers, composites, and biological systems. The AM methods used for producing complex geometrical shapes are classified based either on energy source (laser, electron beam) used or the material feed stock (powder feed, wire feed).

INTRODUCTION

A world-wide contention, product development based on client requirement and reduced product development time have quickened the need for advanced technologies. AM is one such process that significantly reduces the time taken for component design and development. The product development in AM takes place in two stages namely, virtual and physical. In virtual stage, a CAD model of the product is devel-

DOI: 10.4018/978-1-7998-4054-1.ch001

oped and analyzed using 3D modelling and analysis software. In physical stage, a complete product is built directly from the CAD data without the use of any tool / work holding devices.

AM process provides simplified product development of complex 3D products from any CAD design. To develop a product using AM process, one needs dimensions of the product, basic knowledge on the working and materials which are compatible for AM machines. However, any conventional manufacturing process needs detailed study regarding various parameters like feature fabrication order, type of tool to be used in order to obtain a complete product. In AM, each part is made by thin layer by layer by the addition of materials of fixed thickness. As long as each layer has minimum thickness, the final product will be close to the CAD data. Most of the commercially available AM machine are similar in building the product by layer-based manner, whereas they differ in material used, generation and bonding manner of the layers which directly impacts the accuracy and properties of the final product.

This chapter discuss briefly about the historical perspective, various metal additive manufacturing process, their benefits, drawbacks and a few limitations for each process. As AM has improved the overall product development process, the chapter also includes few advanced processes like Hybrid Layered Manufacturing (HLM), and Electron Beam Free Form Fabrication (EBF3) which will be predominant manufacturing processes in the near future (Asiabanpour, Mokhtar, & Houshmand, 2008; Gibson, Rosen, & Stucker, 2015).

HISTORICAL PERSPECTIVE

The origin for the additive manufacturing can be traced to the development of two major fields namely, Topography and Photo sculpture.

1. **Topography:** Topography is defined as a layered method to produce moulds for topographical relief maps in the early 1900's by Blanther. Both, positive and negative 3D exteriors were collected from several wax plates that were cut through the topographical outline.
2. **Photo sculpture:** Photo sculpture is a technique which is used to create models of 3D objects. The technique includes capturing photographs at once using 24 cameras at equal spaces in an annular space followed by capturing the silhouette of the snap to shape $1/24^{th}$ of the cylindrical portion of an item (Asiabanpour et al., 2008).

PRELIMINARY WORKS IN ADDITIVE MANUFACTURING

Development in AM process sustained over 1960's and 1970's and quite a few patents were made on various devices. A few of them are

1. A method to fabricate substances from powdered supplies by warming particles close by and joining it by means of power sources such as laser beam, plasma or even electron beam [Ciraud, 1972].
2. Method to create plastic designs by 3D selective polymerization on a photosensitive polymer material at the junction of dual laser beams [Swainson, 1977].
3. A novel kind of photopolymer AM process to make an object in layers. A disguise regulates the contact of the UV rays while making a cross-section of the product. [Kodama, 1981].

4. Mirror systems were used to guide the laser beam to polymer over an x-y plotter [Herbert, 1982].

EVOLUTION OF AM

It is the novel name which was formerly called as Rapid prototyping (RP) and widely called as 3D-Printing. RP is made use in several companies to define the method used for producing a component (prototype) or a part even before its entry into the market. Whereas, now the trend has been changed, as the final products are manufactured directly and introduced into the market. Hence the term "Prototyping" is no longer appropriate. Also, the products are produced in an additive approach. Hence a committee under ASTM has agreed to use the novel term ***"Additive Manufacturing"*** (Gibson et al., 2015).

PROCESS FLOW

AM comprises a sequence of stages which initiate from a CAD model and ends up in creating a physical feature. The AM process differs from product to product. For example, some simple products may require additive manufacturing for the visualization only. Whereas, certain complex geometries may require careful cleaning and post-processing. The following steps are carried out in any AM process.

Creating a CAD Model

Developing a software model is the foremost step in any AM process. Any commercially available CAD modeling software which provides output in a 3D solid representation may be used to achieve this. Reverse Engineering (also called as scanning) may also be adapted to create a CAD model.

Converting CAD Model to STL File

STL refers to (Stereolithography or Standard Triangle Language or Standard Tessellation Language). The .STL file format explains it as representing a 3D object in its triangular form. STL files are unique from CAD drawing in the sense that, a circle in any CAD model would look like a triangle in the case of an STL file which is shown in figure 1. Some of the few advanced modeling software can even create .STL files with some add-on tools.

Transfer and File Modification

The .STL file created should be moved to any AM machine. At this stage, some modifications to the file may be made so that, the product is produced within the machine's workspace.

Setting the Machine

The machine setup should be made before the commencement of the build process. At this stage, the process parameters are set as per the required product specifications.

Figure 1. Difference between STL and CAD model
(Adapted from (McCue, 2019))

Product Building

Product building is generally an automated process. Certain factors has to be considered during building to ensure proper power backup and continuous supply of the material. More attention should be given if a multi-colored building is opted.

Product Removal

Subsequently the built product will be detached from the machine. Utmost care must be taken while removing the product from the workspace.

Post-Processing

Some products may be weak since the processing is just complete. Hence additional care must be taken while handling it. Support structures need to be removed at this stage and proper finishing of the product is needed. In the case of FDM products, acetone fumes should be introduced in order to attain a good surface texture

Application

Now the products are ready to use, they may also require some mechanical or electrical assemblies to be used as a final product. Most of the AM machines use fragile components that need utmost care and proper conditioning (Gibson et al., 2015).

AM METHODS

AM process can be sorted based on the medium chosen to prepare the samples as Liquid, powder, and solid (wire-based) as shown in figure 2. The liquid-based systems include Stereolithography (SLA), Digital Light Processing (DLP) and Inkjet 3D Printing. The solid-based systems include Fused Deposition Modelling (FDM), Wire Arc Additive Manufacturing (WAAM) and Electron beam Freeform (EBF3). The powder-based systems include Selective Laser Sintering (SLS), Electron Beam Melting (EBM),

Selective Laser Melting (SLM), Direct Metal Laser Sintering (DMLS) (Paolini, Kollmannsberger, & Rank, 2019).

Even though, 3D Printing and Additive Manufacturing terms are used interchangeably, they have distinct features in their mode of layer by layer addition. Hence, the American Society for Testing and Materials (ASTM) group 'ASTM F42 Additive Manufacturing' has formulated a standard to classify the AM process into 7 broad categories in the year 2012 (Loughborough University, n.d.)

The categories are as follows:

VAT Photopolymerization

This technique makes use of a liquid photopolymer resin stored in a vat, from which the product is made in a layer-by layer. Ultraviolet rays are used to cure the resin at required spots. A moving platform travels down after each layer is cured. The products do not get any structural support as it gets solidified from a liquid resin, hence it needs additional support in the form of unbonded material. Plastics and UV curable photopolymer resin can be printed using this technique.

Material Jetting

In this method, the object is created similar to a 2-D printer. The material is forced towards the build platform through a nozzle. A DOD (drop on demand) or a continuous approach is followed for material delivery. A thermal or piezoelectric method is followed to deposit the material over the build plate. Additional post-processing technique may be required to finish the product. Liquid photopolymer and casting wax can be printed using material jetting.

Binder Jetting

Two distinct materials are used in binder jetting process. They are, powder-based material (for product building) and a liquid binder (to bind the powder material). Structural parts cannot be printed using binder jetting process, since the print head deposit alternate surfaces of powder and binder for each layer. Similar to other powder-based AM process, the support structures are built along with the product which needs additional post processing technique to remove these structures. Full colour sand stone, sintered tungsten carbide, Bronze infiltrated stainless steel, Sintered Inconel alloy and Sintered stainless steel can be printed using binder jetting.

Material Extrusion

It is a common AM process wherein the material gets heated and is drawn out from the nozzle. The nozzle moves in horizontal direction, whereas the platform moves in the vertical direction to build a complete product. The pressure and heating should always be constant in order to obtain the desired output. A spool is used to supply material. It is one of the most commercial and affordable AM process. ABS (Acrylonitrile butadiene styrene), Polylactic Acid (PLA), (Polyamide) PA, (Polyetheretherketone) PEEK, (High Impact Polystyrene) HIPS, (Polyetherimide) PEI, and (Thermoplastic Polyurethane) TPU can be printed using material extrusion.

Powder Bed Fusion (PBF)

PBF process comprises of DMLS, EBM, SLS, SLM, SHS. All the processes have some similar working principles like blade, roller or a hopper to provide a fresh layer of powdered material. Laser beam or an Electron beam is used as a power source in the powder bed fusion process. The only difference between the DMLS and SLS process is the kind of material being used. The former sinters metal powder, whereas the latter sinters plastics. The power source selectively melts the powder layer by layer to form a complete product. EBM has an exception that it requires vacuum and can be employed for production of functional metal products.

Sheet Lamination

It has two distinct process.

1. **Ultrasonic Additive Manufacturing (UAM):** Sheets of metal are welded using Ultrasonic Welding. Additional machining process is needed to remove unwanted materials formed as a result of welding. Since no melting is involved, only less heat is generated. Hence this process is apt for Al, Cu, Ti and SS.
2. **Laminated Object Manufacturing (LOM):** Sheets of paper are bonded using adhesives. Printing is done in a cross-hatching manner making it convenient for removing undesirable objects. LOM products cannot be used in structural applications, but they are most suited for printing artistic and pictorial prototypes.

Direct Energy Deposition (DED)

DED is a kind of complicated process which is often used in repairing worn-out components or to fill additional substances over an actual product. The material is directly deposited over the platform using a nozzle that has 4-5 axis movement based on the requirement. Followed by a power source to melt the definite surface and then solidification occurs. Commonly utilized power sources include Laser/Plasma or Electron beam.

After successful completion of the first layer, the same steps (different for each process as explained above) are done over and over again until the entire product is built as per the requirements. A few processes may demand post-processing to obtain the final object (usually a model / prototype / structure).

METAL AM METHODS

Metal AM process is a kind of manufacturing process through which metal components are manufactured from the raw material. Metal AM process offers high level of design variations and broad variety of materials. The need for metal AM products rises as simple and complex parts can be produced, and the cost of production per component is constant even for low or high quantity. AM process has few limitations and its process uniqueness make it possible to manufacture products that are not possible to fabricate using conventional methods.

Figure 2. AM process

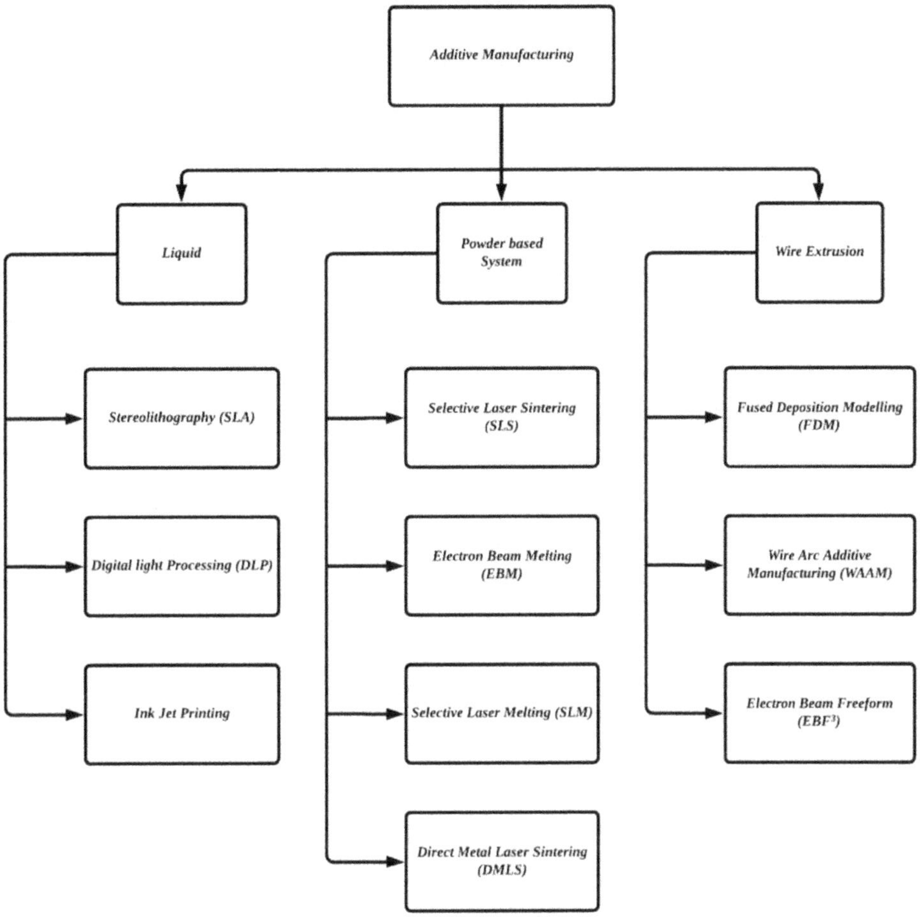

Two types of metal AM process are in practice based on the mode of supply of powder material as shown in figure 3, which are discussed below

1. **Powder bed Fusion (PBF):** PBF is one among the swift and popular metal AM process to create metal products. Compared to traditional machining processes, AM makes use of CAD model to build the 3-D object. Most of the AM techniques are done in a bottom-up building manner making use of material agglomeration rather than material subtraction (Bassoli, Sola, Celesti, Calcagnile, & Cavallini, 2018). A subset of AM process in which the heat source joins the powdered material to make a complete product. The powder materials are kept in a platform also called as powder bed which moves down after the completion of each layer until the entire 3D - object is complete (Goodridge & Ziegelmeier, 2017).
2. **Direct Energy Deposition (DED):** This process is also referred as, Laser cladding or Laser Metal Deposition. This process also uses the same powdered material. The only difference is the mode of supply of the powder. The powder is fed directly through the heat source and is directed to the melt pool. These processes are more accurate and follows a computerized system for material deposition having a thickness in the range of 0.1 mm to few centimetres.

Figure 3. Types of metal AM process

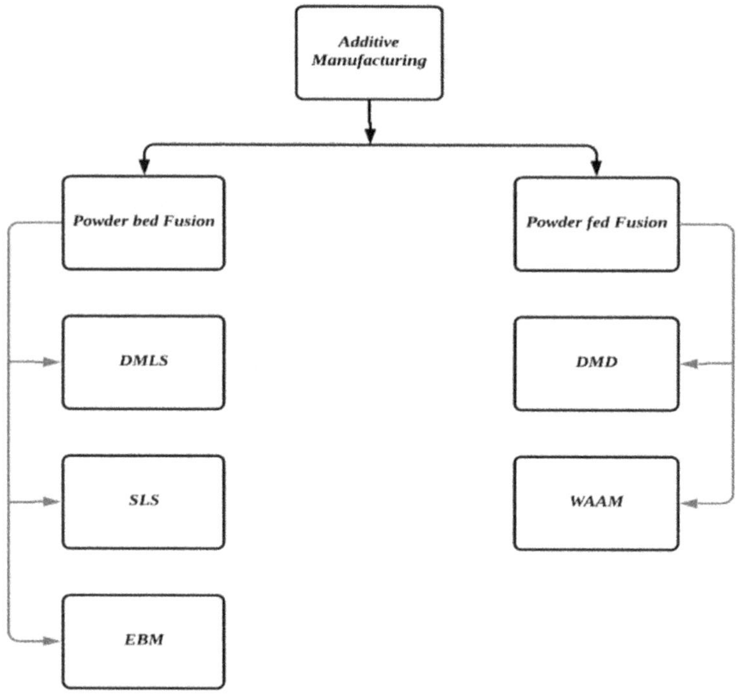

A recent development in this process is termed as LENS (Laser Engineered Net Shaping). A special process to add material over any object that helps in repairing defective, valuable metal parts such as turbine blades and tool inserts.

1. Salient Features of Direct Energy Deposition.

    a. The cladding material gets well adhered in terms of metallurgical aspects
    b. Undercut is usually not present (Inovar Communications Ltd, n.d.).

POWDER-BED SYSTEMS

Selective Laser Sintering (SLS)

SLS is a kind of metal AM technique that falls under powder-based systems. SLS uses miniscule particles of ceramics, plastic, glass or even metal (DMLS) that gets fused to form a solid object by a high-power laser beam. SLS was established and patented by Deckard and Beaman in the early 1990s.

Similar to other 3D printing methods, a CAD file is a base for SLS. Later these CAD files are changed into .STL files to be fed to a 3D printing machine.

1. Steps involved in SLS

a. Conversion of CAD files to .STL files.
b. The powder materials (polymer and metals) are spread over the build plate.
c. A computer-controlled laser pulses over the build plate, tracing the object above it.
d. The intensity of the laser causes the powder material to get sintered and fuses it.
e. As the first layer is finished, the bed moves down (less than 0.1 mm) and introduces a fresh layer of materials. And this step gets repeated until the whole object is printed as shown in figure 4.
f. Finally, the object is left in the machine to get cooled down and then removed (Palermo, 2013).

In most of the cases, additional tooling (sanding and support removal) is not needed in SLS. Prototypes and end-use products can be printed using SLS. The complex structural objects generally contain the initiating materials and porosity that needs additional sintering in order to enhance its density. The sintering process for ceramics has both the solid and liquid phase. This is explained by taking SiC as an example. The efficiency of SiC prepared by SLS is attributed to the type of sintering followed. As sintering takes place, grain growth occurs and reduction in pore size followed by reduced volume and surface energy. Minimum self-diffusion coefficient and stable covalent bond makes it hard to obtain dense SiC object. In most of the cases, a dense SiC object may be prepared when sintering is done at the elevated temperature and pressure but results in higher production cost. Nevertheless, the ceramics obtained has less flexural strength and higher silicon volume fraction resulting in poor mechanical properties and efficiency deterioration of the product (Song et al., 2019).

SLS is suitable in regions where high-quality materials in minimum quantity are required. Aerospace industries make use of SLS to make prototypes for their parts. This is because producing a physical mould that remains idle for a longer period is not cost-effective. However, storing the .STL files are easier and design modifications are even possible if required.

Electron Beam Melting (EBM)

Like other metal AM processes, EBM (founded by Arcam AB) also builds the product from a CAD file as shown in figure 5. EBM finds its application in the energy sector, automotive, aerospace and biomedical industries. It falls under the category of powder bed fusion process wherein an electron beam melts the metal powder selectively to produce the complex-shaped parts. EBM also aids in producing meshes and internal assembly which is not possible in other traditional techniques (Chern et al., 2019). This technique allows to produce parts without any residual stress and the vacuum environment retains the chemical specification of the product (“GE”, 2018).

Since the process finds its root from electric charges, only conductive materials can be printed. If the material is a non-conductor, there won’t be any interaction between the powder and beam. Hence polymer / ceramic printing is technically not feasible. Only metal parts can be printed. In recent days, only Ti, and Co-Cr alloys are used. Ti alloys are preferred due to their biocompatibility, less weight, and high strength. EBM is used extensively to manufacture turbine blades and engine components (V, 2019).

The EBM process is handy in control, it has improved reaction rate and reliability. The vacuum can aid in eliminating the creation of thick oxide film over the Ti products resulting in improved mechanical properties. Lack of fusion is almost removed in EBM due to increased energy usage and high penetration. The metallurgical combining ability of the melt puddle and layers is also increased due to higher wettability. The process parameters to produce a complete lamellar structure was also experimented. A

Figure 4. Schematic of SLS, SLM, and DMLS
(Adapted from (Palermo, 2013))

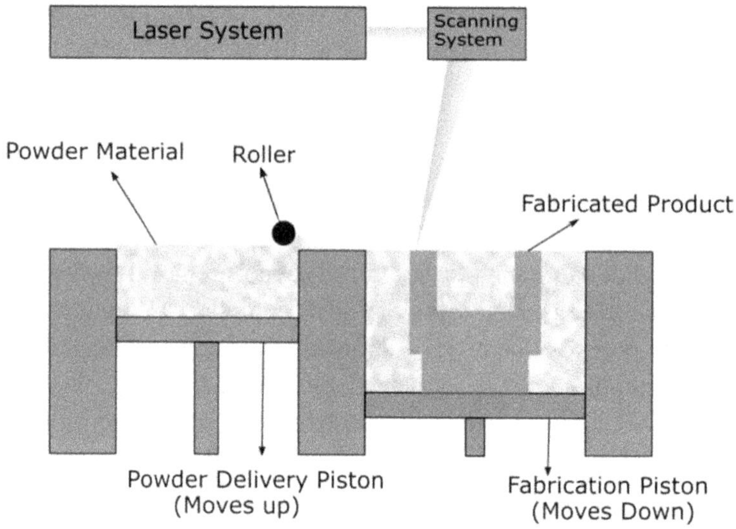

lot of researchers have reported that EBM is an appropriate technique to produce Ti alloy components. Still, few problems like the relationship between the forming process conditions and mechanical properties needs a deep research (Materials et al., 2019).

Selective Laser Melting (SLM)

SLM is a generative manufacturing process specifically for metals. The process steps are similar to SLS. The thickness of each layer varies between 15 – 500 μm. SLM process can print near-complete dense products from a CAD model. Since Ti and its alloys are difficult to produce from conventional methods, SLM is best suited for these materials. Current researches are carried out with Al alloys and composites (Sercombe, Li, Sercombe, & Li, 2016). SLM can produce dense objects than conventional powder metallurgy process (Zhenlu Zhou, Zhen Tan, Dingyong He, Zheng Zhou, Li Cui, Yiming Wang, Wei Shao, 2020). Process parameters related to Laser power, Scan speed, Powder flow, and temperature affect the quality of print directly. Unlike SLS, SLM needs support structure for overhanging parts that need to be removed at the final stages. A band saw is used to remove the product from the build plate. Process parameters such as laser power and beam travel speed has greater tendency to influence the property of product by creating crack, void and balling effect (KurianAntony, N.Arivazhagan, 2014).

The time taken for printing is an addition of Primary and auxiliary time. The primary time represents the time needed to scan the powder and the auxiliary time represents the time taken for powder deposition and lowering the platform (Aboulkhair et al., 2019).

1. SLM is preferred due to the following reasons
    a. A wide variety of materials are available
    b. Complex shapes can be made
    c. Lead time is reduced since tooling is not required

Figure 5. Schematic of electron beam melting
(Adapted from (Suard, 2014))

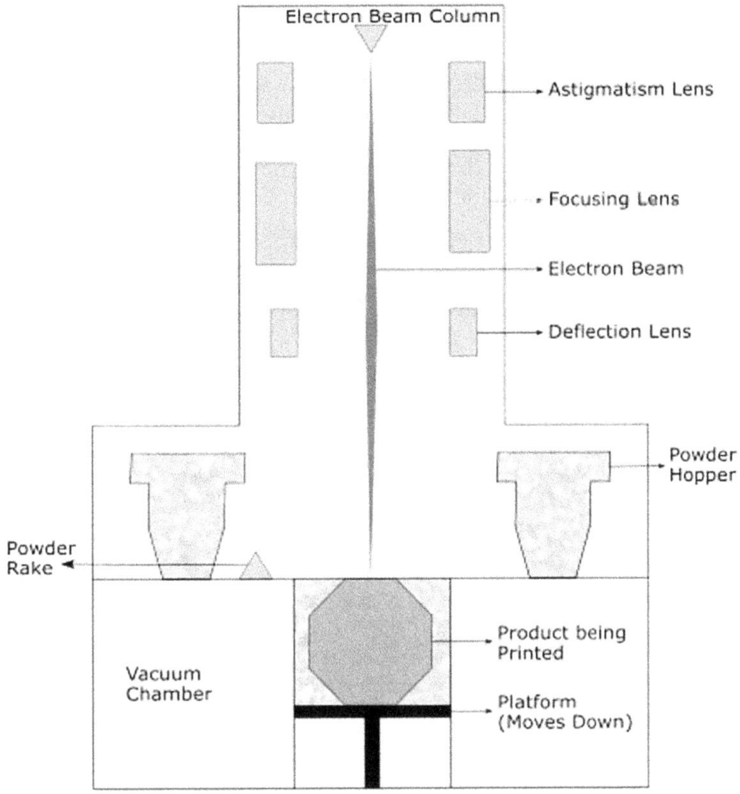

   d. Multiple parts can be produced at a time
   e. High level of degree of freedom
2. A few limitations of SLM are
   a. Production cost is high
   b. Specialized design, skill and a high degree of knowledge is required
   c. Surface finish is rough
   d. Post-processing is time-consuming (Murphy, 2019).
   e. Growth of tensile residual stress that leads to cracking (Kalentics et al., 2019).

Direct Metal Laser Sintering (DMLS)

Like SLS, the parts will be built layer by layer based on the 3D CAD design. The significant difference is the temperature at which the material is sintered. In SLS polyamides are sintered at a temperature range of 160 – 200 °C, whereas in DMLS the metal gets sintered at a temperature range of about 1510 – 1600 °C which means high power laser is required. DMLS is opted for producing metal parts as a prototype or low volume production by omitting the time required for the tooling process (Rafieazad, Mohammadi, & Nasiri, 2019). Even high detailed products can be built using DMLS rather than the conventional process (Sculpteo, n.d.). Table 2 shows the key difference between SLS, SLM and DMLS process.

The principle is the use of a high-power laser beam to strengthen the material using solid or liquid phase sintering through a predefined path thereby creating a required object. A few authors have stated that the sintering process causes a change in density, microstructure and properties of the product (Nandy, Yedla, Gupta, Sarangi, & Sahoo, 2019). The process parameters like fast solidification, and rapid cooling of tiny melt puddle that creates improved microstructure, but residual thermal stress is also higher inside the product. Therefore, annealing is done to relieve internally developed stress (Rafieazad et al., 2019). Additional processing is required to improve the surface finish of the component (Atzeni et al., 2016).

1. Some of the limitations in DMLS are

    a. Slow process and low build volume
    b. Porosity cannot be eliminated in this process (Jones, 2019).
    c. Poor surface finish (Atzeni et al., 2016).

DIRECT ENERGY DEPOSITION

Table 1. Differences between SLS, SLM, and DMLS

	SLS	DMLS	SLM
Process involved	Sintering	Sintering	Melting
Temperature	160 – 200 °C	1510 - 1600 °C	1000 – 4000 °C
Laser used	CO_2	Precise, High - power	High - power
Material Used (Powder)	Plastics, Ceramics, and Nylon	Metal and Metal Alloy	Metal

(Lasers, n.d.), (Maria CristinaTanzi, SilviaFarè, 2019), (Hooper, 2018)

Direct Metal Deposition (DMD)

DMD is an advanced metal AM process that holds good for components with complex geometry using minimal energy and material. It produces completely dense and working products (B. Chen et al., 2018). DMD can be used in repairing and restoring worn out surface over expensive material surfaces (Govekara, Kuznetsova, Kotara, & Masaki Kondo, 2018). Functionally Graded Materials (FGM) are mostly produced using the DMD process (Akbari & Kovacevic, 2018). FGM are composites having groundbreaking form and function that can perform well even in adverse conditions (B. Chen, Su, Xie, Tan, & Feng, 2020). Heat generation is made possible from sources such as Laser, Electron beam, Plasma arc or TIG welding. A tiny melt pool is created by the heat source, followed by the addition of the materials to the melt pool by a concentrated stream or a wire feed system in order to produce an elevated region of the product. A complete geometry can be printed by moving the base using an automated system. The heat source reaches the substrate along with the gas and powder material. Since the powder materials are fed through the nozzle, the positional flexibility is higher compared to powder bed systems. Dissimilar metal wall can be prepared between Zirconium and Stainless Steel using DMD process with Copper as

Figure 6. Schematic of DMD

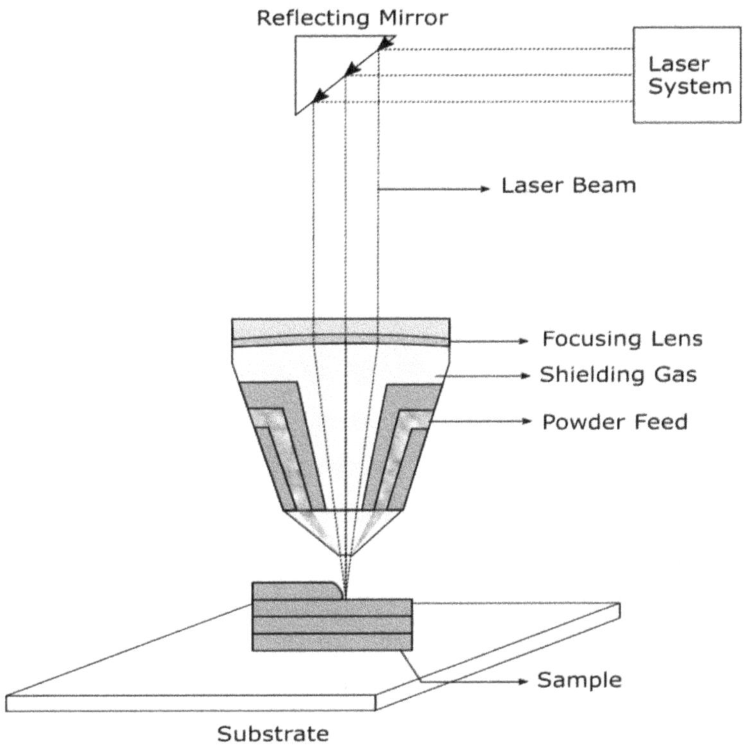

an interlayer (Khodabakhshi, Farshidianfar, Bakhshivash, Gerlich, & Khajepour, 2019). The schematic of DMD is shown in figure 6.

1. DMD is preferred due to the following reasons (Caiazzo, 2018)
    a. Low Distortion of the sample being produced
    b. HAZ (Heat affected zone) is less
    c. The surface finish of the component is good
    d. Degrees of freedom is high
    e. High deposition rate
2. Limitations of DMD
    a. Laser source needs gas shielding
    b. In the case of electron beam source, the vacuum is needed to reduce attenuation
    c. Support structures are needed. (Creative Commons, 2018)

Wire Arc Additive Manufacturing (WAAM)

WAAM was initially proposed by Baker in the year 1925. The process involves the use of an electric arc as a heat supply to melt and deposit the filler wire and build the required product. In the current scenario, WAAM is one of the promising manufacturing processes for a variety of materials like Ti, Al, Ni alloy

Table 2. Features of different WAAM techniques

Type of WAAM Process	Source of Energy	Characteristics
GTAW	GTAW	Non – Consumable wire Electrode
		Individual wire feed
		Deposition rate- 1-2 Kg/ hr
GMAW	GMAW	Consumable Electrode
		Deposition rate 3-4 Kg/ hr
	CMT	Reciprocating electrode
		Low heat and high tolerance
		Deposition rate 2-3 Kg/hr
PAW	Plasma	Non consumable electrode
		Deposition rate 2-4 Kg/hr

(Wu et al., 2018)

and steel. Fabrication time for WAAM is about 40 – 60% less than that of the traditional machining process. Moreover, the post-processing time also gets reduced by 15-20% (Dhinakaran et al., 2019). Energy consumption and lead time is less than that of the Laser and electron beam process. It has improved material deposition and usage. The products exhibit good internal stability and improved mechanical properties. Al alloys are used due to their excellent physical properties, good thermal conductivity, better corrosion resistance and less dense (Su, Chen, Gao, & Wang, 2019). The basic components required to make any WAAM system are wire feed system, a robotic system, welding torch and energy supply (Dinovitzer, Chen, Laliberte, Huang, & Frei, 2019), which makes it easily available at lower price. The schematic of WAAM is shown in figure 7.

Table 4 shows the recent applications of different alloys manufactured using WAAM.

Based on the type of heat source, WAAM is classified into three types, as shown in Table 3.

1. Plasma Arc welding (PAW) based
2. Gas Metal Arc welding (GMAW) based
3. Gas Tungsten Arc welding (GTAW) based (Dinovitzer et al., 2019).

Two different design systems are available in WAAM robotic process

System 1 – Inert gas environment is maintained by a closed chamber
System 2 – A local shielding mechanism moves over the build area aiding in the fabrication of huge structures.

The part fabrication involves process planning (pre-processing), deposition, post-processing. The robot movement, welding parameters are done by computer programming for the given CAD model (Wu et al., 2018).

Some commonly found defects in WAAM products are porosity, residual stress, delamination, oxidation, cracking, and deformation. Some of these can be minimized by suitable post-processing techniques.

A few issues faced when wire is selected:

Table 3. Application of WAAM in different alloys

	Alloy				
	Bi-metal	**Al**	**Ni**	**Ti**	**Steel**
Aerospace	✓	✓	✓	✓	
Automobile	✓	✓			✓
Marine				✓	✓
Corrosion resistant	✓		✓	✓	
Elevated Temperature	✓		✓	✓	
Tools and Mould					✓

(Wu et al., 2018)

1. Residual stress and distortion caused due to heat generated while welding.
2. Poor Accuracy.
3. The printed parts have poor surface quality (Horgar et al., 2018).

Electron Beam Freeform Fabrication (EBF^3)

EBF^3 falls under the category of solid extrusion system which uses a thin filament material to print the product. EBF^3 is one of the recent developments in AM. Part construction using a layer additive approach is defined as a Solid Freeform Fabrication method (SFF). SFF methods offer minimal production time, cost and lead time. SFF aid in controlling the composition and microstructure at a higher level. Other SFF processes like SLS and EBM needs additional post-processing to achieve functional products/prototypes. Some most common issues faced in any Laser-assisted system arc the limitation in build area, operating at a low rate of deposition and a few applications on huge products (Taminger & Hafley, 2019). EBF^3 is a novel technique for making metal structural products.

Figure 7. Schematic of WAAM

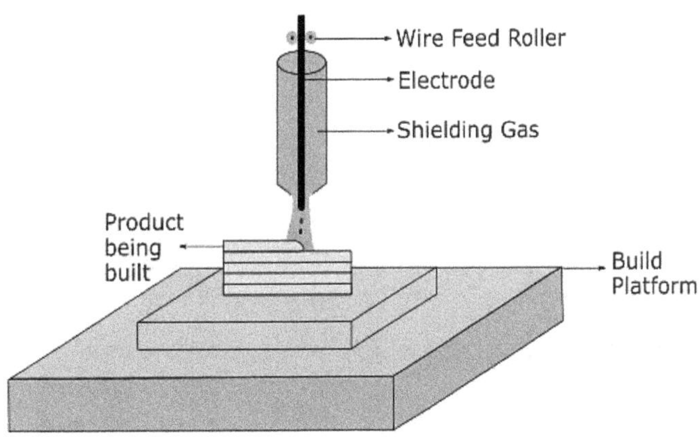

The schematic of the EBF^3 components is shown in figure 8. EBF^3 process makes use of a focused high-power beam of electrons under the presence of a vacuum. The wire is passed through the beam in order to create a pool of molten metal. Solidification of material occurs instantly and is capable of self-supporting (support structures are needless) (Tucker, 2017). The same step is repeated until the entire product is complete (Taminger & Hafley, 2019) (Seufzer & Taminger, n.d.).

1. Merits of EBF3 over other AM process
   a. Full-dense products can be built.
   b. Stacking of multiple parts inside the build area is possible.
   c. The only process that can be carried out even at zero gravity conditions (Tucker, 2017).
   d. Provides the feasible solution to problems related to deposition rate, material compatibility and overall process efficiency (Taminger & Hafley, 2019).
   e. Highly accurate dimensions can be maintained.
2. De-Merits of EBF3
   a. EBF3 has the highest equipment cost compared to similar techniques, due to its high accuracy.
   b. Part manufacturing is limited to homogenous materials only (Tucker, 2017).
   c. Distortion as well as residual stresses are high (Z. Chen, Ye, & Xu, 2018).
3. Future scope

Product development for non-weldable alloys is still under research. Producing complex geometrical parts with reduced residual stress and good surface finish is on high demand (Taminger & Hafley, 2019) (Seufzer & Taminger, n.d.).

Figure 8. Schematic of EBF^3
(Adapted from (Seufzer & Taminger, n.d.))

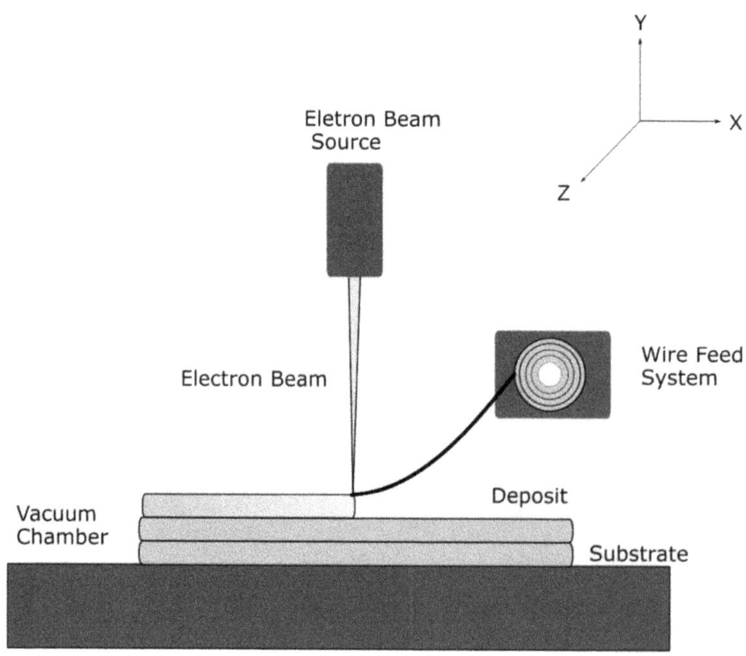

LENS

The LENS process uses the prevailing methods of AM to create completely dense metal components from a CAD model. This process finds its root from laser-based systems. Nd: YAG laser is used to join metallic materials as they are projected over the deposition area. The constant development in modern computer-controlled systems have paved way for faster process improvements and creating complex metal structures directly from a CAD model. The LENS process was initially focussed on research purposes, but recent studies have made it possible for direct product development. The product fabricated by LENS has an upper hand over traditionally manufactured product. Earlier, researches were focused on enhancing the surface finish and to identify the effect of process parameter on the final product.

In the same manner as other AM techniques, a .STL file is created from a CAD model. The .STL files are sliced into layers of definite thickness and algorithm for motion control to feed material at the required region using a present-day slicing software. A secondary software is used to commute the sliced file into a motion control. Thus G-codes are generated for the sliced file, that shows the precise areas where the material is to be deposited. It comprises the information about shutter reaction, dwell series and location. Hence these files are fairly large.

After the completion of motion control, the files are sent to LENS data-processing system. The components in LENS consists of a steady Nd: YAG laser system, glove box, powder feeding system and movement controlling system. The laser system is fixed on the peak position in a wagon. The emitted laser beam is focused to the glove box using traditional optical system which is usually a plano-convex lens.

Next comes the powder feed system. This part is a critical system in LENS. Past researches were not appropriate due to the unsteady and discontinuous powder feed. In order to supply unvarying quantity of powder regardless of the direction of travel, a powder delivery arrangement is created. Powdered material is focussed through the nozzle towards the beam focus region. It gets used up in creating new layers. A substrate acts as the base for building new product. The beam gets directed towards the parent material and generate a melt pool wherein the powdered material is focussed and later reaches to a molten state. The melt pool is moved through the parent material or the former layer. The same process is repeated to obtain a complete product. The process needs a previously deposited layer to create the following layers. Support structures are not needed for the limited geometries (Keichep & Smugeresky, n.d.).

HYBRID LAYERED MANUFACTURING (HLM)

Hybrid Layered Manufacturing combines the benefits of additive manufacturing with that of conventional machining processes such as CNC or subtractive manufacturing simultaneously in an effective manner. Both these processess are focussed independently, hence they are rapid in action (Suryakumar, Karunakaran, Bernard, Chandrasekhar, & Raghavender, 2011). The need for adopting the hybrid manufacturing process is that the present AM process does not meet the requirements of advanced products in terms of precision. Even metal parts need supplementary post-processing in order to achieve a final product. Such a hybrid process eliminates the shortcomings of a single process and combines its benefits. The HLM process can be integrated with more than one process or even follow a particular sequence to obtain a complete product (Manogharan, Wysk, Harrysson, & Aman, 2015). At the end of the deposition (near net shape), the conventional machining process is carried out to obtain a complete product. Hybrid Layered

Manufacturing combined with MIG/TIG/Laser cladding/DED may be opted for low-cost production (Kapil, Legesse, Kumar, & Karunakaran, 2017).

Part customization is made possible which serves as the key to choose HLM process. Products for a particular application is made using HLM by making a rough part in AM followed by machining process. Recent studies focus on combining AM in production line to check the viablity of HLM process (Ambrogio, Gagliardi, Muzzupappa, & Filice, 2019).

As this process uses GMAW to deposit matrial, it offers high material deposit rate, safety and reduced price. More than one torch can also be assembled to an existing CNC spindle head. If a homogenous product is desired, filler wires of different diameters can be used in such a way that dense wire can be used to fill the inner regions of the product and slender wires for the periphery of a product. Likewise, composite products can also be built using the same method. HLM has deposition rate of about 50-100 g/min, whereas a laser or EBM offers only 2-10 g/min. Precise geometry and metallurgical ascpects depend directly on size of the droplet, flux etc. which can be very well altered in HLM. Complete automation for both addititve and subractive manufacturing is feasible in HLM (Karunakaran, Suryakumar, Pushpa, & Akula, 2010). Table 5 shows the materials that can be used in different AM processes.

Its technical market practability is justified for injection molding techniques followed by improved product development also. Since, efficiency and in-depth penetration are the maanjor concerns in bonding processess, HLM also offers low energy input and better heat dissipation. In addition, HLM offers improved advantages when complexity of the part material cost is high (Suryakumar et al., 2011).

Applications

1. HLM process is efficient in producing turbine blades rather than the traditional manufacturing process.
2. Aluminium, Mild Steel, and tool steel are currently produced using the HLM process (Kapil et al., 2017).
3. Titanium and Inconel alloys are currently under research.
4. Near net shapes can be developed by HLM (Karunakaran et al., 2010).

CONCLUSION

This chapter presents an introduction to metal AM techniques and its benefits on the fabrication process over conventional methods. Metal AM process is currently used in aero, auto and pharmaceutical industries in an increased rate to produce fully functional products that are difficult to produce using conventional techniques. The high sales volume of marketable metal AM products shows an in-depth need of metal AM products in near future. The sales growth is due to the interaction of several aspects like high end simulation software's, inexpensive laser systems, improved knowledge in welding and metallurgical aspects of a material. Nevertheless, metal AM materials exhibit a few defects such as distortion, poor surface finish, presence of voids, and distortion which must be taken care to improve the market share of AM products.

Despite the fact that the properties of metal AM products are mostly similar to traditionally manufactured parts, the AM process parameters can influence the mechanical properties to a larger extent.

Table 5. Materials compatible for different metal AM processes.

Process	Materials
SLS	Polyamide 12 (PA 12), Polyamide 11 (PA 11), Aluminium filled Nylon (Alumide), Gas filled Nylon (PA-GF), Carbon fiber filled Nylon (PA-CF)
SLM and DMLS	Aluminum Alloys, Stainless steel and tool steel, Titanium alloys, Cobalt Chromium super alloys, Nickel based super Alloys (Inconel), Precious Metals
EBM	Titanium alloys and Chromium-Cobalt Alloys
DMD	Steel, Nickel based alloys, Titanium, Cobalt, Copper
WAAM	Stainless steel, Carbon steel, Low Alloy steel, Aluminum Alloys, Titanium alloys, Nickel based alloys. Most of the weldable metals can be used in WAAM method, provided they are in wire form.
EBF^3	Weldable alloys. Current research is based on Aluminium and Titanium alloys. (Taminger & Hafley, 2019)
LENS	Titanium, Inconel and Stainless Steel. High performance metals.
HLM	Aluminium, Mild steel, Tool steel, Titanium and Inconel Super Alloys. (Kapil et al., 2017)

Hence a thorough understanding of the effect of process parameters can yield better products. Some common metallurgical defects can be eliminated by proper understanding of the microstructure evolution due to thermal cycles. Higher standardization of process parameters can give rise to repeatability in process and in turn consistent products with improved mechanical properties. Researches are underway for better correlation between method, morphology, mechanical properties, and performance of parts produced by AM.

Huge components are difficult to produce using the accessible equipment's like Laser- PBF due to its high investment cost, hindering the productivity of the firm. A complete insight and innovation on such processes are required for technological advancements in AM. At present, most of the software's required for AM are proprietary. Higher accessibility of open source software's can aid a huge number of researchers in AM process development. Rapid part certification can help in increased commercialization of AM products.

The basic differences between a conventional CNC and AM process are explained in the early parts of the chapter. Metal printing is made possible using the above methods with its own pros and cons for each method. EBF^3 has the highest investment cost among all the other metal AM processes due to its accuracy and minimum residual stress. Producing better and defect free products at a faster rate in a cost-effective way stands as the foremost goal. Reaching the milestone with the available facilities may need persistent R&D for a prolonged time.

FUTURE SCOPE

Research is still being carried out to enhance the mechanical properties and to reduce the production time. Optimization of process parameters based on application. Computational analysis aid in predicting the defects during the product development stage.

RESEARCH GRANT

This research received no specific grant from any funding agency in the public, commercial, or not-for-profit sectors.

REFERENCES

Aboulkhair, N. T., Simonelli, M., Parry, L., Ashcroft, I., Tuck, C., & Hague, R. (2019). Progress in Materials Science 3D printing of Aluminium alloys : Additive Manufacturing of Aluminium alloys using selective laser melting. *Progress in Materials Science, 106*(July), 100578. doi:10.1016/j.pmatsci.2019.100578

Akbari, M., & Kovacevic, R. (2018). An investigation on mechanical and microstructural properties of 316LSi parts fabricated by a robotized laser / wire direct metal deposition system. *Additive Manufacturing, 23*(August), 487–497. doi:10.1016/j.addma.2018.08.031

Ambrogio, G., Gagliardi, F., Muzzupappa, M., & Filice, L. (2019). Additive-incremental forming hybrid manufacturing technique to improve customised part performance. *Journal of Manufacturing Processes, 37*(November), 386–391. doi:10.1016/j.jmapro.2018.12.008

Asiabanpour, B., Mokhtar, A., & Houshmand, M. (2008). Rapid manufacturing. In *Collaborative Engineering*. Theory and Practice. doi:10.1007/978-0-387-47321-5_7

Atzeni, E., Barletta, M., Calignano, F., Iuliano, L., Rubino, G., & Tagliaferri, V. (2016). Abrasive Fluidized Bed (AFB) finishing of AlSi10Mg substrates manufactured by Direct Metal Laser Sintering (DMLS). *Additive Manufacturing, 10*, 15–23. doi:10.1016/j.addma.2016.01.005

Bassoli, E., Sola, A., Celesti, M., Calcagnile, S., & Cavallini, C. (2018). *Development of Laser-Based Powder Bed Fusion Process Parameters and Scanning Strategy for New Metal Alloy Grades : A Holistic Method Formulation*. doi:10.3390/ma11122356

Caiazzo, F. (2018). Laser-aided Directed Metal Deposition of Ni-based superalloy powder. *Optics & Laser Technology, 103*, 193–198. doi:10.1016/j.optlastec.2018.01.042

Chen, B., Su, Y., Xie, Z., Tan, C., & Feng, J. (2020). *Development and characterization of 316L / Inconel625 functionally graded material fabricated by laser direct metal deposition*. doi:10.1016/j.optlastec.2019.105916

Chen, B., Yao, Y., Song, X., Tan, C., Cao, L., & Feng, J. (2018). *Microstructure and mechanical properties of additive manufacturing AlSi10Mg alloy using direct metal deposition*. doi:10.1080/00150193.2018.1392147

Chen, Z., Ye, H., & Xu, H. (2018). Distortion control in a wire-fed electron-beam thin-walled Ti-6Al-4V freeform. *Journal of Materials Processing Technology, 258*(March), 286–295. doi:10.1016/j.jmatprotec.2018.04.008

Chern, A. H., Nandwana, P., Yuan, T., Kirka, M. M., Dehoff, R. R., Liaw, P. K., & Duty, C. E. (2019). A review on the fatigue behavior of Ti-6Al-4V fabricated by electron beam melting additive manufacturing. *International Journal of Fatigue, 119*(August), 173–184. doi:10.1016/j.ijfatigue.2018.09.022

Creative Commons, B.-N.-S. (2018). *Direct metal deposition (DMD) – OpenLearn*. Open University.

Dhinakaran, V., Ajith, J., Yasin, A. F., Jagadeesha, T., Sathish, T., & Stalin, B. (2019). Materials Today : Proceedings Wire Arc Additive Manufacturing (WAAM) process of nickel based superalloys – A review. *Materials Today: Proceedings*. doi:10.1016/j.matpr.2019.08.159

Dinovitzer, M., Chen, X., Laliberte, J., Huang, X., & Frei, H. (2019). Effect of wire and arc additive manufacturing (WAAM) process parameters on bead geometry and microstructure. *Additive Manufacturing, 26*(October), 138–146. doi:10.1016/j.addma.2018.12.013

GE. (2018). *Electron Beam Melting (EBM)*. GE Additive.

Gibson, I., Rosen, D., & Stucker, B. (2015). Additive manufacturing technologies: 3D printing, rapid prototyping, and direct digital manufacturing. In *Additive Manufacturing Technologies: 3D* (2nd ed.). Printing, Rapid Prototyping, and Direct Digital Manufacturing. doi:10.1007/978-1-4939-2113-3

Goodridge, R., & Ziegelmeier, S. (2017). 7 - Powder bed fusion of polymers. Laser Additive Manufacturing. doi:10.1016/B978-0-08-100433-3.00007-5

Govekara, E., Kuznetsova, A., Kotara, M., & Kondo, M. (2018). Annular laser beam based direct metal deposition A new methodology to analyze the functional and physical architecture of assembly oriented product family Japan identification. *Procedia CIRP, 74*, 222–227. doi:10.1016/j.procir.2018.08.099

Hooper, P. A. (2018). Melt pool temperature and cooling rates in laser powder bed fusion. *Additive Manufacturing*, *22*(May), 548–559. doi:10.1016/j.addma.2018.05.032

Horgar, A., Fostervoll, H., Nyhus, B., Ren, X., Eriksson, M., & Akselsen, O. M. (2018). Additive manufacturing using WAAM with AA5183 wire. *Journal of Materials Processing Technology*, *259*(March), 68–74. doi:10.1016/j.jmatprotec.2018.04.014

Inovar Communications Ltd. (n.d.). *Metal Additive Manufacturing processes*. Author.

Jones, G. (2019). *Direct Metal Laser Sintering (DMLS) – Simply Explained*. All3DP.

Kalentics, N., Sohrabi, N., Tabasi, H. G., Griffiths, S., Jhabvala, J., Leinenbach, C., Burn, A., & Logé, R. E. (2019). Healing cracks in selective laser melting by 3D laser shock peening. *Additive Manufacturing, 30*(April), 100881. doi:10.1016/j.addma.2019.100881

Kapil, S., Legesse, F., Kumar, R., & Karunakaran, K. P. (2017). ScienceDirect Hybrid Layered Manufacturing of Turbine Blades. *Materials Today: Proceedings*, *4*(8), 8837–8847. doi:10.1016/j.matpr.2017.07.234

Karunakaran, K. P., Suryakumar, S., Pushpa, V., & Akula, S. (2010). Robotics and Computer-Integrated Manufacturing Low cost integration of additive and subtractive processes for hybrid layered manufacturing. *Robotics and Computer-integrated Manufacturing*, *26*(5), 490–499. doi:10.1016/j.rcim.2010.03.008

Keichep, D. M., & Smugeresky, J. E. (n.d.). *Using the Laser Engineered Net Shaping (LENS) Process to Produce Complex Components from a CAD Solid Model*. Academic Press.

Khodabakhshi, F., Farshidianfar, M. H., Bakhshivash, S., Gerlich, A. P., & Khajepour, A. (2019). Dissimilar metals deposition by directed energy based on powder-fed laser additive manufacturing. *Journal of Manufacturing Processes*, *43*(May), 83–97. doi:10.1016/j.jmapro.2019.05.018

Kurian & Arivazhagan. (2014). *Numerical and experimental investigations on laser melting of stainless steel 316L metal powders _ Elsevier Enhanced Reader.pdf*. Elsevier.

Lasers, S. (n.d.). *Direct Metal Laser Sintering | 3D Selective Laser Melting*. Academic Press.

Loughborough University. (n.d.). *The 7 categories of Additive Manufacturing*. Additive Manufacturing Research Group | Loughborough University.

Manogharan, G., Wysk, R., Harrysson, O., & Aman, R. (2015). AIMS- a Metal Additive-Hybrid Manufacturing System : System Architecture and Attributes. *Procedia Manufacturing*, *1*, 273–286. doi:10.1016/j.promfg.2015.09.021

Materials, H., Konyashin, I., Hinners, H., Ries, B., Kirchner, A., Klöden, B., ... Sidorenko, D. (2019). International Journal of Refractory Metals Additive manufacturing of WC-13% Co by selective electron beam melting : Achievements and challenges. *International Journal of Refractory Metals & Hard Materials*, *84*(June), 105028. doi:10.1016/j.ijrmhm.2019.105028

McCue, T. (2019). *STL Files: What They Are and How to Use Them*. Academic Press.

Murphy, J. (2019). *Selective Laser Melting (SLM) – 3D Printing Simply Explained*. All3DP.

Nandy, J., Yedla, N., Gupta, P., Sarangi, H., & Sahoo, S. (2019). Sintering of AlSi10Mg particles in direct metal laser sintering process: A molecular dynamics simulation study. *Materials Chemistry and Physics*, *236*(March), 121803. doi:10.1016/j.matchemphys.2019.121803

Palermo, E. (2013). *What is Selective Laser Sintering?* Live Science.

Paolini, A., Kollmannsberger, S., & Rank, E. (2019). Additive manufacturing in construction: A review on processes, applications, and digital planning methods. *Additive Manufacturing*, *30*(October), 100894. doi:10.1016/j.addma.2019.100894

Rafieazad, M., Mohammadi, M., & Nasiri, A. M. (2019). On microstructure and early stage corrosion performance of heat treated direct metal laser sintered AlSi10Mg. *Additive Manufacturing*, *28*(April), 107–119. doi:10.1016/j.addma.2019.04.023

Sculpteo. (n.d.). *DMLS Metal Technology for 3D Printing*. Sculpteo.

Sercombe, T. B., Li, X., Sercombe, T. B., & Li, X. (2016). *Selective laser melting of aluminium and aluminium metal matrix composites : review*. doi:10.1080/10667857.2016.1161147

Seufzer, W. J., & Taminger, K. M. (n.d.). *Control Methods for the Electron Beam Free Form Fabrication Process*. Academic Press.

Song, S., Gao, Z., Lu, B., Bao, C., Zheng, B., & Wang, L. (2019). Performance optimization of complicated structural SiC/Si composite ceramics prepared by selective laser sintering. *Ceramics International*, *46*(1), 568–575. doi:10.1016/j.ceramint.2019.09.004

Su, C., Chen, X., Gao, C., & Wang, Y. (2019). Applied Surface Science E ff ect of heat input on microstructure and mechanical properties of Al-Mg alloys fabricated by WAAM. *Applied Surface Science*, *486*(December), 431–440. doi:10.1016/j.apsusc.2019.04.255

Suard, M. (2014). *Scheme of the electron beam melting process.* Download Scientific Diagram.

Suryakumar, S., Karunakaran, K. P., Bernard, A., Chandrasekhar, U., Raghavender, N., & Sharma, D. (2011). Computer-Aided Design Weld bead modeling and process optimization in Hybrid Layered Manufacturing. *Computer Aided Design, 43*(4), 331–344. doi:10.1016/j.cad.2011.01.006

Taminger, K. M. B., & Hafley, R. A. (2019). *Electron beam freeform fabrication : A rapid metal deposition process.* Academic Press.

Tanzi & Farè. (2019). *Manufacturing Technologies 3.1.* doi:10.1016/B978-0-08-101034-1.00003-7

Tucker, D. (2017). *Additive Manufacturing: Electron Beam Freeform Fabrication.* Tata Technologies PLM Solutions.

V, C. (2019). *The Complete Guide to Electron Beam Melting (EBM) in 3D Printing.* 3Dnatives.

Wu, B., Pan, Z., Ding, D., Cuiuri, D., Li, H., Xu, J., & Norrish, J. (2018). A review of the wire arc additive manufacturing of metals: Properties, defects and quality improvement. *Journal of Manufacturing Processes, 35*(August), 127–139. doi:10.1016/j.jmapro.2018.08.001

Zhou, Z., Tan, Z., He, D., Zhou, Z., Cui, L., Wang, Y., & Shao, W. G. W. (2020). *Fabrication of three-dimensional connected W-Cu10Sn composites by selective laser melting _ Elsevier Enhanced Reader. pdf.* Elsevier.

ADDITIONAL READING

Behera, M. P., Dougherty, T., & Singamneni, S. (2019). Conventional and additive manufacturing with metal matrix composites: A perspective. *Procedia Manufacturing, 30*, 159–166. doi:10.1016/j.promfg.2019.02.023

Dutta, B., Babu, S., & Jared, B. (2019). Raw materials for metal additive manufacturing. *Science, Technology and Applications of Metals in Additive Manufacturing*, 77–91. doi:10.1016/b978-0-12-816634-5.00004-2

Friel, R. J., & Harris, R. A. (2013). Ultrasonic additive manufacturing A hybrid production process for novel functional products. *Procedia CIRP, 6*(1), 35–40. doi:10.1016/j.procir.2013.03.004

Graziosi, S., Rosa, F., Casati, R., Solarino, P., Vedani, M., & Bordegoni, M. (2017). Designing for Metal Additive Manufacturing: A Case Study in the Professional Sports Equipment Field. *Procedia Manufacturing, 11*(June), 1544–1551. doi:10.1016/j.promfg.2017.07.288

Herzog, D., Seyda, V., Wycisk, E., & Emmelmann, C. (2016). Additive manufacturing of metals. *Acta Materialia, 117*, 371–392. doi:10.1016/j.actamat.2016.07.019

Ji, X., Mirkoohi, E., Ning, J., & Liang, S. Y. (2020). Analytical modeling of post-printing grain size in metal additive manufacturing. *Optics and Lasers in Engineering, 124*(May 2019), 105805. doi:10.1016/j.optlaseng.2019.105805

Kerbrat, O., Mognol, P., & Hascoët, J. Y. (2011). A new DFM approach to combine machining and additive manufacturing. *Computers in Industry*, *62*(7), 684–692. doi:10.1016/j.compind.2011.04.003

Oliveira, J. P., & LaLonde, A. J. M. (2020). Processing parameters in laser powder bed fusion metal additive manufacturing. *Materials & Design*.

Vanmeensel, K., Lietaert, K., Vrancken, B., Dadbakhsh, S., Li, X., & Kruth, J. P. … Van Humbeeck, J. (2018). Additively manufactured metals for medical applications. In Additive Manufacturing: Materials, Processes, Quantifications and Applications. doi:10.1016/B978-0-12-812155-9.00008-6

KEY TERMS AND DEFINITIONS

Cladding: A thin layer of coating applied over a material or structure.

Distortion: The act of twisting or altering something out of its true state.

Miniscule: Tiny particles of material to be printed.

Polymerization: Process of reacting monomer molecules together in a chemical reaction to form polymer chains or three-dimensional networks.

Post-Process: The final step in the additive manufacturing process, where products receive finishing touches such as smoothing and painting.

Prototype: A first or preliminary version of a product from which other forms are developed.

Residual Stress: A kind of stress that persist in a solid material even after the source of the stresses has been removed. Residual stress can be desirable or undesirable.

Chapter 2
Processes and Application in Additive Manufacturing:
Practices in Aerospace, Automobile, Medical, and Electronic Industries

Ajit Behera
https://orcid.org/0000-0001-5357-7733
National Institute of Technology, Rourkela, India

ABSTRACT

Additive manufacturing (AM) is going to cover all the segments of industries from missile industry to biomedical industry. This marked change of technology is due to the distinctive potential of AM to fabricate the parts with intricate designs and reduce fabrication expenditure (free from machining, waste generation, assembly of various parts) with small production runs and short turnaround times. This chapter extensively discussed industrially practiced AM technology. In this chapter, all additive manufacturing materials like metal, alloys, polymer, ceramics, composite, etc. have been given focus for various applications. Additive manufacturing technology is cost effective: no loss of metal and easy to fabricate both larger and intricate shapes. This technology already has taken a primary position in aerospace industries as well as the medical and household industries.

INTRODUCTION

The cutting edge technology of advanced manufacturing area is the Additive manufacturing (AM) in this current era. According to the ASTM standard, AM can be defined as "the process of joining materials to make the structure from 3D model data, usually layer upon layer, as opposed to subtractive manufacturing technologies" (Guo & Leu, 2013; Herzog, Seyda, Wycisk, & Emmelmann, 2016). This manufacturing process allows the direct conversion of analog to digital processes and designs the construction files into fully functional objects that is why called 3D-printing and also computer aided design (CAD) technology. In the 1980s, the rapid prototyping (RP) was first introduced to make a 3D

DOI: 10.4018/978-1-7998-4054-1.ch002

prototype layer-by-layer from a CAD (Frazier, 2014; Rengier et al., 2010). In this technology, we can use metals, polymer, ceramics, glass or other materials that build up layer-by-layer. Depending on the type of materials, the layers are interconnected by specific binder system or by the help of laser/electron beam energy (Tofail et al., 2018).

The additive manufacturing technology stated in 1980s in Japan. After a couple of year stereolithography was invented in 1983 before the development of the rapid prototyping (Kruth, 1991). In 1986 at University of Texas, Austin introduced the first additional manufacturing (AM) technology on stage due to the development of RP technology and the need for high performance manufacturing with the ability to produce complex parts (Gross, Erkal, Lockwood, Chen, & Spence, 2014). Hideo Kodama from Nagoya Municipal Industrial Research Institute constructed the earliest equipment of additive manufacturing and he found out two methodology of additive manufacturing. Since then, there are usually dozens of different technologies invented in 3D printing. Three years later, a new technology was invented: Select Laser Sintering (SLS). First commercial SLS was found in 1990. At latest stage of 20thcentury, the bio-printer was fabricated which was able to 3D print the first kidney (Y. Huang, Leu, Mazumder, & Donmez, 2015; Irwin, Reutzel, Michaleris, Keist, & Nassar, 2016; Zhai, Lados, & LaGoy, 2014). Around ten years later, first 3D printing kit was launched to the market for commercial production for different purposes. Now, we are using larger scale printers to print large 3D structures such as large frame and equipment of cars. 3D printing provides the direction for printing everything everywhere. Different techniques of post AM processes adopted in some materials such as heat treatment of additive manufactured green product of superalloy (Yadollahi, Shamsaei, Thompson, Elwany, & Bian, 2017).

Recently, aero-industries like Boeing, have installed more than tens of thousands AM facilities on both military and commercial aircraft. By AM, costlier titanium parts of aerocraft can be fabricated, expecting the reduction of around 3,000,000 USD per aircraft. GE Aviation is using metal based 3D printer to fabricates more than thousands of fuel nozzles per year for the use of LEAP engine ("Additive Manufacturing Industry overview | GE Additive," n.d.). Airbus industry is producing the metal based 3D printed brackets and bleed pipes in aircraft. The demand of 3D printing in aerospace industries are rapid prototyping, direct digital manufacturing (DDM) and rapid tooling and repairing of parts made of the metals, alloys, polymers, ceramics, and composites. For metal design, the primary3D printings in aero-industries are powder bed fusion and directed energy deposition. For non-metallic design, the primary3D printings are material jetting, material extrusion, and vat photopolymerization (Zenou & Grainger, 2018).

STEPS OF MATERIAL FABRICATION BY MANUFACTURING

AM technology consists of five basic steps as follow (figure 1):

Step I: **Development of computerized 3D solid model:** Generally, 3D printing work uses two digital imaging techniques: computed tomography (CT) and magnetic resonance imaging (MRI). The data collected from those two scans is transferred to a 3D model by the software. Then, the model is changed to Standard Triangulation Language (.STL) file format and imported into the additive manufacturing setup (Qin, Qi, Scott, & Jiang, 2019; Velu, Calais, Jayakumar, & Raspall, 2019). Other way is to design the structure by the help of softwere.

Step II: **Choice of AM type**: The choice of the specific additive manufacturing technique is highly essential with respect to the materials to be used, economic point of view, geometric and material property, and time.

Step III: **Choice of raw materials:** according to the specific technology, powder/wire will be chosen. Dimension of the raw materials is more important to gain 100% of energy.

Step IV: **Optimization of process parameter:** Important process parameters to be focused such as (i) the source power (laser or e-beam power), power density, focus, (ii) Feed rate of the powder or wire, (iii) traverse velocity, (iv) Hatch (x-y) spacing, and (v) z-range increment.

Step V: **Layer by layer build up:** The structure is made layer-by-layer on the 3D printing machine. Different 3D printing technology build the layer-by-layer set up in specific way using equipped source such as laser or electron beam. Other processes use to spray binder/solvent onto powdered ceramic or polymer to bind the materials in layer and in between the layer (Jiménez, Romero, Domínguez, Espinosa, & Domínguez, 2019; LaMonica, 2013).

After the completion of manufacturing, several post treatment also followed as per requirement (optional) like sand blasting (for surface smoothness), heat treatment (for releasing stresses) etc.

Figure 1. Steps of additive manufacturing

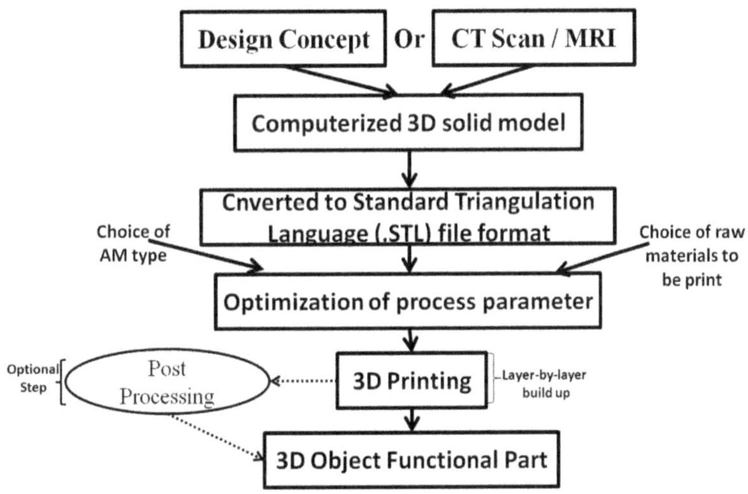

ADVANTAGE OF ADDITIVE MANUFACTURING PROCESS

AM provides a novel platform among the world class technologies that going to obsolete all the multi-step traditional manufacturing processes. AM offers the following advantages:

- **Faster Production:** The manufacture speed is much better than the traditional manufacturing methods. One step overall shorter production process can easily facilitate the market requirement.

- **Free from assembly:** This process avoids the assembly of various part for a complex structural element.
- **Design freedom for complicated shape**: lay-by-layer development of shape gives fully freedom for any shapes and geometry according to the design customization in the computer.
- **Enhancement of the reliability:** Single accurate final piece results a better reliability. As comparison to the traditional manufacturing, AM provides the product with more structural efficiency.
- **Cost effectiveness:** Higher structural efficiency requires some materials that are able to minimize the overall costs for certain components
- **Added Flexibility:** 3D printers in industrial bulk design can able to provide multiple options and iterations at the fabrication stages, results an optimum and efficient automobile design. Able to customize the designs as per customers demand and for on-site spares manufacturing.
- **Green Production:** AM produces zero waste, which can be consider as a green production.
- **To achieve light-weight:** AM able to manufacture the ultra-light materials with optimised strength. In automobile this light weight structure proved to provide the higher fuel efficiency. Aerogel and microlattice can be made by AM

DISADVANTAGE OF ADDITIVE MANUFACTURING PROCESS

Beside the above uncompetitive advantages, there are some limitations found in AM are discussed below:

- **Limitation of size:** There is a limitation to manufacture the higher dimension objects.
- **More time consuming:** By increasing the dimension, there is increasing in production time in that single manufacturing layer-by-layer step. For example, a 10x10x10 cm^3 cubical object takes approximately 120 hr manufacture time.
- **Imperfections in parts:** Parts produced by AM sometime rough and ribbed surface. This is due to plastic beads or the presence of coarse sized particles that are stacked on top of each other. Therefore, there should be a additional step for post processing.
- **Cost:** AM equipment are considered very expensive for investment. The metal printing SLM machine will be around 4 crore in entry level, without any additional module of the machine. Also, AM cost of production for smaller size are cheap as comparison with the larger size.
- **Energy consumption:** a continuous power supply is required to manufacture the object. More is the manufacturing time, more is the power consumption.

A number of investigations have been conducted to overcome the above mentioned AM drawbacks. It is expected that at the middle of this 20th century, AM will be the only replacement for all the conventional processes like liquid/powder metallurgy synthesis route.

VARIOUS POPULAR INDUSTRIAL PRACTICE OF ADDITIVE MANUFACTURING

Generally, Additive manufacturing can be classified into three categories. First one is sintering in which material is heated less than the melting temperature. The second one is to melt the materials that are melts layer-by-layer of metal powder using laser source or electron beam source. The third one is stereolithography (photopolymerisation process), in which the ultraviolet laser source is given to vat of photopolymer resin to form torque-resistant ceramic structure that can withstand at high temperatures(S. H. Huang, Liu, Mokasdar, & Hou, 2013; Tuomi et al., 2014). All the above printing processes simultaneously require software, hardware and the materials to work together.

All the AM techniques processed layer-by-layer fashion of 3D structure, where as a number of AM techniques are there such as direct metal laser melting (DMLM)(Lawrence E. Murr et al., 2012), selective laser melting (SLM)(Spears & Gold, 2016), selective laser sintering (SLS) (Shahzad, Deckers, Zhang, Kruth, & Vleugels, 2014),selective heat sintering (SHS) (Rahman et al., 2018), Electron beam melting (EBM)(L.E. Murr et al., 2010),Material jetting process (MJP) (Jared et al., 2017), precise extrusion manufacturing (PEM) (Guo & Leu, 2013), precision extrusion deposition (PED)(Snyder, Rin Son, Hamid, & Sun, 2016), Laminated object manufacture (LOM) (Klosterman, Chartoff, Osborne, & Graves, n.d.), Shape deposition manufacturing (SDM)(A. . Cooper et al., 1999), Laser engineered net shaping (LENS) (Yuzhou Li, Hu, Cong, Zhi, & Guo, 2017), Binder jetting (Gaytan et al., 2015), Powder bed fusion (Khairallah, Anderson, Rubenchik, & King, 2016), Electron beam powder bed fusion (EPBF) (Gonzalez et al., 2019), photopolymer vat processes (PVP) (Wilts et al., 2019), stereolithography (SLA) (H. Wu et al., 2017) etc.GE is currently offering solutions for four of the several processes recognised by the American Society for Testing and Materials (ASTM)("Additive Manufacturing & 3D Printing Processes | GE Additive," n.d.; Tofail et al., 2018), which are discussed below:

- Powder Bed Fusion (PBF)
- Direct Metal Laser Melting (DMLM)
- Electron Beam Melting (EBM)
- Binder Jetting process (BJP)

Powder Bed Fusion Process

The PBF is a3D printing technique involve in particle fusion process in the powder bed. By different source like laser or electron beam, melted/semi-melted (fusible) particles are obtained in the powder bed. The ultrathin layer of powder is spread by the help of a spread guide or roller over the successive layer. After completion of the required shape, the unfused powder is taken away for re-use in the next fabrication process. The Airbus A350 XWB wing brackets have been printed using this PBF technique (Williams, Mistree, & Rosen, 2011). Figure 2 is showing the schematics of the PBF process.

Direct Metal Laser Melting (DMLM) Process

The DMLM is another form of PBF techniques, involve with the complete melting of metal powder into liquid pools. Like PBF technique, the printing processed layer-by-layer forming a micro-thin layers of fully melted metal powders such as Ti, Cr, Co, Al, and various alloys etc. The excess powder is removed after the printing, leaving a smooth surface with dense and homogeneous volume of the structural part that usually requires negligible post-processing. The full melting of the consecutive powder bed results around 100 percent densities. DMLM is used to manufacture jet engines and other parts of airplane. GE is also engaging the DMLM technology for printing the fuel nozzles for its LEAP engines that made down the 20 traditionally fabricated parts to just one (Manfredi et al., 2014; Vayre, Vignat, & Villeneuve, 2012). Figure 3 shows the schematics of the DMLM process.

Figure 2. Schematics of powder bed fusion AM machine: selective laser melting

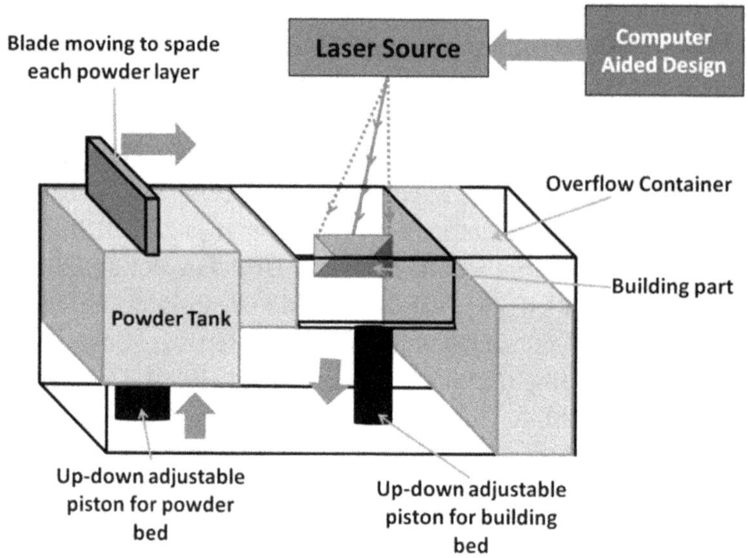

Electron Beam Melting (EBM) Process

In EBM process, the powder fusion will be done by high-power electron beam source in the vacuum chamber. Here the beam of electrons concentrically focused by the help of electromagnetic coils to melt the metal powder. This hot process favour to produce the structure without inherent residual stress. The process in the vacuum gives a clean and no reaction environment. EBM is used to manufacture highly précised components that highly demands in aerospace and medical industries. Orthopaedic implants are fabricated by EBM process, yields a slightly rough surface that is suitable for bone adhesion in implant. EBM made equipment are used in new jet engine structural part and in rocket engine prototypes ("3D

Figure 3. Schematics of direct metal laser melting process

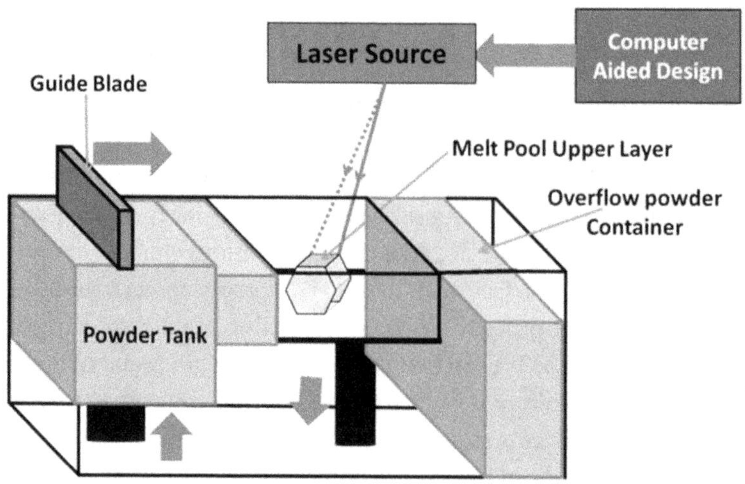

Printing in Aerospace & Aviation | GE Additive," n.d.). Generally EBM is a faster AM technique than DMLM, but the deposited layers are thicker and the surface is rougher. EBM works with a wide range of metals including stainless steel, Ti, Cu, Co, Cr etc. (Gu, 2015). Figure 4 depicts the schematics of EBM process.

Figure 4. Schematics of EBM process

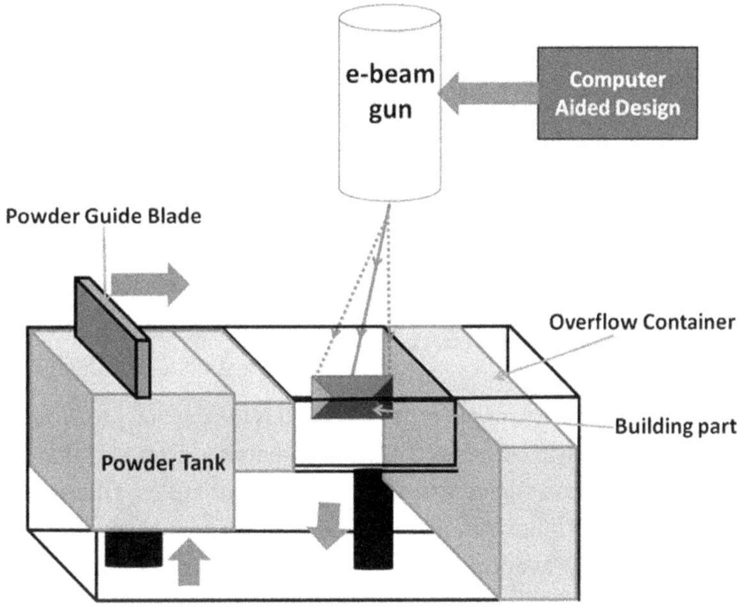

Binder Jetting Process (BJP)

BJP is a 3D printing technique in which the droplet of the liquids acts as a binding agent to construct the structure lay-by-layer in the powder bed as shown in figure 5. The print head strategically spray the binder into the powder and successively the part constructed through the layering of powder and binder ("ExOne | Binder Jetting Technology," n.d.). A holder containing a section of liquid binder moves across the bed of the powder and selectively deposit to bind an area together one layer at a time. In this process, the common materials are metals, ceramics, and sand. After printing the whole structure, the binder can be removed from the solid metal part by the help of debind and sintering processes. The liquid binder printing head as well as powder depositor layer move on x, y and z axes alternatively to finish the structure (Liu, Wang, Sparks, Liou, & Newkirk, 2017). Successive powder layer will be spread by the spread-guide for further liquid binding process. Gradually the structure is made and the loose powder has been collected for recycling purpose. This is a green process. Further treatment is required for the development of pore in the product.

Figure 5. Schematic of BJP 3D printing process

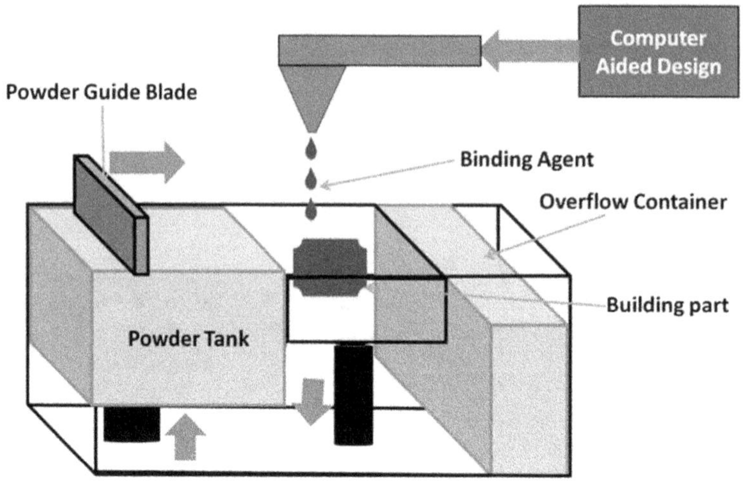

INDUSTRIAL 3D PRINTING MATERIALS

Different kinds of materials are utilized for 3D printing.AM has the capacity to build any type of complex shape with the help of metals, alloys, ceramics, polymer, bioactive glass and their combinations. Various textured materials are developed such as powder, filament, granules, fillets, resin etc. for different printing technologies. Also, at the industrial practice there are very specific materials uses for specific purpose such as pre-alloyed NiTi powder is use to develop the smart materials in single cartridge.

Metal-based Materials

Metals are used to construct the high strength and durable structure. The use of metal in 3D printing is growing in wide variety of sector such as high strength metal manufacturer industries to corrosion protection metal manufacturing industry. Metal powder can make complex, bespoke parts with geometries that conventional technique are unable to manufacture. Metal AM structures can be topologically optimized to maximize their performance while minimizing their weight and the total number of components in an assembly. By metal powder, the light metal and superalloys also manufactured. The metal powder AM processes are associated with high cost and high temperature processing so that not easy for manufacturing as that of conventional processes. Different 3D printing can produce structures from a large range of metals or alloys including stainless steel, Al, Ti, Ni, Cr, Fe, Cr, Au, Ag, Pt, Pd and inconel (Kok et al., 2018). Various 3D printing materials has been discussed below:

- **Stainless steel:** This is an Ideal printing materials for utensils, cookware and other items. Stainless Steel is the cheapest form of metal printing, very strong and suitable for very large objects. Stainless Steel can be printed using different metal 3D printing processes: DMLS and Binder Jetting. This material is good to print small parts, resistant against corrosion and high temperatures. Stainless Steel in powder form is used in sintering/melting/fusion/EBM processes. Stainless steel can be heat treated for the high strength applications. It provides strong resistance against corrosion (A. S. Wu, Brown, Kumar, Gallegos, & King, 2014).

- **Aluminum:** Aluminum is the ideal material for thin metal objects. This metal is printed with the SLM technology. The Al is commonly used in foundries for fine objects and complex geometries, and also to create precise projects. The second advantage of Al is its very low weight among the metal material. These properties make Al effective in areas where a strength/mass ratio, as well as good thermal properties, are required (Yali Li & Gu, 2014).
- **Titanium**- Ti is known as a strongest and lightest metal in AM industrial applications. It is used in the process called DMLS. The preferred choice for strong, solid fixtures structure made up of Titanium powder generally composed of Ti (88-90%), Al (5.50-6.5%) and V (3.50-4.50%). Ti 3D printed structures are mainly used in high-tech industries like aero-industries. Also used in medical industries due to its biocompatible and resists corrosion properties (Dutta & (Sam) Froes, 2015).
- **Gold and Silver**: The best way to print the jeweller is the AM. Currently 3D printed gold and silver rings, earrings, bracelets and necklaces are available in the market. These metals use the DMLS or SLM process for printing. Both Au and Ag are not easy to make the structure with lasers due to their high thermal conductivity and high reflectivity properties. So higher temperature is required to print these materials (Khan & Dickens, 2010; Seifert et al., 2015).
- **Nitinol:** From 19th century, NiTi alloy has got a high commercial demand to manufacture all the type of smart structure due to its two peculiar property like shape memory effect and superelasticity. Nitinol is also known as the shape memory ally that can regain its shape after deformation by the help of external stimuli like thermal energy, magnetic energy and stress. Though it is a high temperature materials, DMLM and SLM is suitable method in AM (Walker, Andani, Haberland, & Elahinia, 2014).
- **Other Metals:** Other than above discussed materials, so many materials can be used as per their specific properties and application. E.g. maraging steel, Inconel-In718, Inconel-In625, cobalt, chromium, bronze, copper etc. are used in 3D printing (Bourell et al., 2017).

Carbon-based Materials

Graphene is widely used for 3D printing of functional materials because of its higher strength and conductivity. This material is the best, for some of the devices that need to be flexible, such as touch screens. Compounds such as carbon fiber are used as the top layer of plastic materials to make the plastic stronger. In the 3D printing industry, carbon fibers can be combined with plastics to replace metals quickly and conveniently. Conductive carbomorphs allow manufacturers to reduce the number of steps needed to assemble an electromechanical device. Graphene is also used in solar panels and component parts (Chang et al., 2018; Compton & Nguyen, 2010). Graphene is better at electricity conductivity, stronger, and lighter than other conductors in today's market. The semiconductor industry is interested in producing large amounts of graphene. For example, IBM recently adopted the grapheme use in LED lighting. The ability to print 3D materials for use in LEDs significantly reduces the cost of lighting production (Willis, Brockmeyer, Hudson, & Poupyrev, 2012).

Polymer-based Materials

Currently polymer occupies a large sector of area in structural application. Compared with other application of 3D materials, resin has limited flexibility and strength. Made of liquid polymers, the resin reaches its final state upon exposure to ultraviolet light (Chen et al., 2017). It is mainly used for technologies such as SLA, DLP, and FDM (Fused Deposition Modeling) technologies. Several types of resins that can be used for 3D printing (such as moldable resins, resistive resins, flexible resins, etc.) resin materials have excellent chemical resistance. Exposure to heat can cause premature polymerization. Polymer is one of the most versatile materials for 3D printed toys and household accessories such as vases, utensils, flooring and designs. Plastic materials are available in powder and filament forms. Powdered plastics are used in the sintering process (mostly SLS), and filament forms are used in the FDM process. Few polymers are used in additive manufacturing are discussed below:

- **Polylactic acid (PLA):** PLA is made from renewable resources such as sugar cane or corn starch. Also known as "green plastic". One of the most environmentally friendly options for 3D printers is biodegradation, since polysaccharide acids are obtained from natural products such as sugar cane and corn starch. PLA comes in a soft and solid form and can able to print on cold surfaces (Spoerk, Arbeiter, Cajner, Sapkota, & Holzer, 2017).
- **Acrylonitrile butadiene styrene (ABS):** ABS is a thermoplastic 3D printing filaments commonly used for personal or home 3D printing objects, and a reference material. Availability of ABS with variety of colors are high in market. Used only for manufacturers and engineers who want to produce high quality prototypes. A hot bed is required for printing. Because ABS materials have a high melting point, they tend to deform when cooled during printing. Another disadvantage of ABS filament is its non-biodegradable toxic nature that emits toxic gases with a terrible smell at high temperatures (Quan et al., 2016).
- **Polyvinyl Alcohol Plastic (PVA):**PVA is a plastic used for low-cost domestic supporting materials. PVA is not suitable for products that require high strength but may be a low cost option for temporary items (Chaudhuri, Mondal, Ray, & Sarkar, 2016).
- **Polycarbonate (PC):** Less frequent than polycarbonate types mentioned above, polycarbonate features a nozzle design and only works with 3D printers operating at high temperatures. Best of all, polycarbonate is used to make plastic fasteners and cheap molding trays (Mohamed, Masood, & Bhowmik, 2017).
- **Polyethylene terephthalate (PET):** Like Nylon, PET is one of the most used plastics. This material is used in the thermoforming process. It can also be combined with other materials such as fiber glass to make engineering resins. PETG is a modified version of PET that means "modified with glycol". As a result, filaments that are weaker, less transparent and easier to use than PET are formed. It combines ABS (temperature resistance, strong) and PLA (easy printing) functions. UV rays can weaken the material and have less scratch resistance (Zander, 2019).
- **High Impact Polystyrene (HIPS):** HIPS is a plastic filament used for supporting structures of FDM printers. It can be compared with ABS in terms of ease of use, but the only difference is the HIP's high resistance to dissolve in water and its low cost produces strong steam. Therefore, it is recommended to use in a well-ventilated place. Without a constant heat flux, this material can clog the nozzles and the printer feed tube (Shi, Wang, Chen, & Huang, 2008).

- **Nylon:** Nylon (called polyamide) is a synthetic thermoplastic linear polyamide. It is a known 3D printing filament for its flexibility, durability, low friction and corrosion resistance. Nylon is generally used in clothing and accessories with the ability to creating complex and delicate shapes. Mainly used as filament in FDM or FFF (Fused Filament Fabrication) 3D printer. This material is recognized as one of the most economical and durable plastic materials (Fathi & Dickens, 2013).

Ceramic-based Materials

Ceramic is one of the hard and brittle materials used in 3D printing system. Other than high temperature resistance and high strength properties, this type of material does not corrode or wear easily like metal and plastic. This material is commonly used for binder jets, steoliolithography (SLA) and digital light processing (DLP) technologies. Ceramics produce high precision parts with a smooth and shiny surface. It also has resistance to acids, heat and bleach. Because of their fragility, there are limits to printing objects with closed and woven parts. Ceramic is an excellent material for tableware such as cups, bowls, egg cups and coasters. Also suitable for candle sticks, tiles, vases, arts and other items (Travitzky et al., 2014).

Other Materials

If there is an aspect that distinguishes 3D printing from other forms of manufacturing, then it is the variety of materials that can be printed. A single 3D printer can print on multiple materials to create different types of objects. In traditional manufacturing, different machines are required for each type of material, but not for 3D printing. Recently, the materials available for 3D printing have evolved significantly.

- **Food:** With the launch of the pasta extruder, printing with 3D foods has never been easier. Chocolate has become the most common food item that fabricated using the 3D printing. But there are also specific printers that can use sugar, pasta, meat and some new food ingredients continue to evolve over time (Lipton, Cutler, Nigl, Cohen, & Lipson, 2015).
- **Biological tissue Materials:** Much research is being done on the potential of 3D printing tissue in many medical applications. Additive manufactured kidneys, ear tissues and all these products are successfully designed. The idea is to print most of the human organs in 3D printing and replace body parts for transplantation in any complex cases. It will be revolutionized the entire medical industry in next few year (Studart, 2016).
- **Building:** Now peoples are thinking for larger structural part like buildings. In recent year, Dubai is trying to install a large 3D printer to develop the cluster of building (Delgado Camacho et al., 2018).

APPLICATION OF ADDITIVE MANUFACTURING PRODUCT

A wide variety of industries preferentially adopted additive manufacturing due to ultimate inherent advantages of the process. Some of the industrial high end products are given below.

In Aerospace Industries

Aerospace industries act as a national security sector, access to air transportation to space activities. In aerospace industries, the material has been chosen mostly according to the reliability at continuous adverse condition. The consulting firm of aerospace sector announces recently on the investigation about the growth rate of additive manufacturing of 23.01% between 2017 and 2021("Additive manufacturing in aerospace is growing-3Dnatives," n.d.; "Engineering and manufacturing - great.gov.uk international," n.d.). Additive manufacturing fulfil the production of fairly large piece without assemble of many parts. For example the fuselage panel designed by STELIA Aerospace.GE companies expanded their production on turbine blades, fuel systems and guide vanes or even parts which carry oil and water can be made using 3D metal printing. Spirit aerosystems uses additive manufacturing recently to installing the Boeing 787's first Ti-components. Figure 6(a) shows a solid metal part manufactured by additive manufacturing with a motor inside. The engine, sealed inside the part, adjusts the rocket's fuel mixture while the rocket is flying. In case of the nozzle (figure 6(b)), 3D printing allowed to reduce the weight by 25 percent and the number of the parts used to create the nozzle reduced from 18 to 1, and more complex cooling pathways and supports, giving the nozzle a five-fold enhanced life. In April 2016, GE sent the first two LEAP-1A production engines with 19 3D print nozzles to Airbus ("Airbus Gets 1st Production LEAP-1A Jet Engines - GE Reports," n.d.). LEAP engines using these nozzles are about 15% more fuel efficient than other nozzles("Additive Manufacturing in the Aerospace Industry > ENGINEERING.com," n.d.).

Figure 6. (a) Adjustive part of the rocket's fuel mixture while the rocket is in flight, (b) CFM International's 3D-printed fuel nozzle, and (c) MTU Aero Engines
("GE Aviation celebrates 30,000th 3D printed fuel nozzle - 3D Printing Industry," n.d.; "Transformation In 3D: How A Walnut-Sized Part Changed The Way GE Aviation Builds Jet Engines - GE Reports," n.d.)

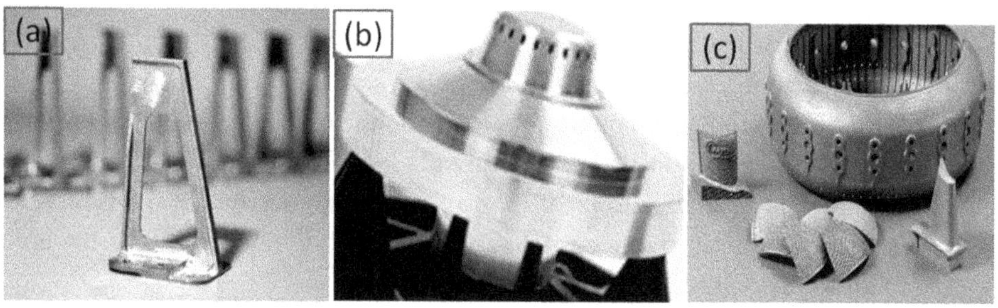

Figure 6(c) showing the MTU aero engine. NASA already declared the achievement of an important milestone by manufacturing the whole 3D-printed rocket engine. NASA's largely 3D printed rocket engine prototype successfully tested a series of 12 test firings at NASA's Marshall Space Flight Centre in Huntsville, Alabama. The fuel injector, fuel turbopump, valves and many major components of the engine were all 3D printed. But the project of creating a fully printed 3D engine was not completed, and the main combustion chamber is the main obstacle. In the 2015, NASA developed an engine prototype 75 percent of which was made with 3D printed parts with the first full-scale copper rocket engine part("NASA 3-D Prints First Full-Scale Copper Rocket Engine Part | NASA," n.d.; "NASA develops first 3-D printed copper rocket engine part - The Economic Times," n.d.). During the test, the combustion chamber was eroded within 10 second firing time, results the increase in internal pressures of the engine. While the new 3D printed combustion chamber in more recent tests, the engine was able to run for 30 seconds, with the ability to go higher life period("Students Aim for Space with 3D-Printed Rocket Engine | Space,"). The table 1 depicted the different additive manufactured part of aircraft using specific materials and technique.

Table 1. Various 3D printed parts of the aircraft

Part of the Aircraft	Materials Used	Process Adopted
Suspension wishbone & GE Jet Engine	Ti and Al	DMLS/SLM
Tarmac nozzle bezel (Engine compartment)	Glass-filled Nylon	SLS
Air flow ducting	Nylon 12	SLS
Seat backs & entry doors	Standard Resin	SLA
Brackets and door handles	Castable Resin or Wax	SLA & Material Jetting
Console control part (Cabin accessories)	Standard Resin	SLA
Dashboard interface	Digital ABS	Material Jetting
Headlight prototypes	Transparent Resin	Material Jetting & SLA

(*Additive Manufacturing for the Aerospace Industry*, 2019; "Direct Metal Laser Melting (DMLM) | GE Additive"; *3DIndustrial Printing with Polymers*, 2018)

Specific brackets are the dominant part in any satellites that assemble the consecutive structure with reflectors and feeder facilities mounted at each end. Main objectives of the bracket (figure 7) are to affix the constituents securely to the satellite's body and to act as insulator from the vast temperature fluctuations (from -170 °C to +100 °C), experienced outside of earth's atmosphere. The stress fluctuation is also extremely high with respect to the temperature. Titanium can fulfil these criteria and by 3D printing it is possible to reduce the cost as well zeroed down the waste. It is estimated that the production were more than 20% with decreasing the weight nearly one-kilogram reduction per satellite.

Automobile Industries

To reduce the assemble pats and joints, 3D printing of single larger part is really preferable by automotive industries now-a-days. Additive manufacturing has already been adopted in almost all the automotive industries for vehicle parts. There are a number of industries that already continuing the production of the parts given in the table 2 produced by different 3D printing processes.

Medical Applications

The medical industry is one of the biggest users of additive manufacturing. 3D printed surgical cutting, dental brackets, drill guides, skull plate, etc. are the growing application within the medical sector. All the bone implants are promoted by 3D printing due to better surface finish of the product. Low volume implant, require porosity as per requirement is the focus of interest on this technology. Even the drugs

Figure 7. 3D printed Ti frames of a satellite
("Aerospace 3D printing Applications | 3D Hubs")

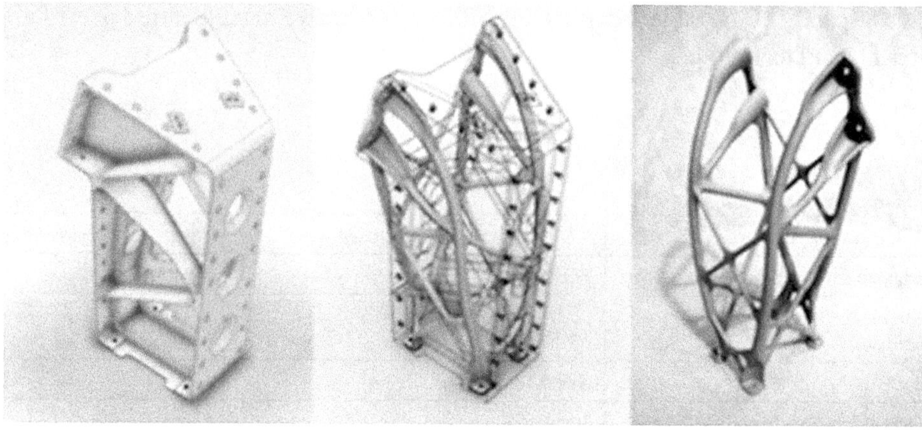

are designed by the additive manufacturing technique, which proved an efficient delivery inside the body which are specific to the body and body needs. Currently, advanced techniques like DMLS, SLS, SLM, and EBM are being used to manufacture implants with sufficient dense in metallic form. Knee joint and hip joint (figure 8a) are patient-specific structural systems that permits perfect fit and higher load bearing capacity. For these implant the traditional processes are adopted like CNC machining, investment casting or plastic molding, which takes much time and also they need specific tool's design, development and production for shaping the implant. But the 3D printing technique solved this big problem to develop the perfect fit implant and able to distribute higher load uniformly (Barazanchi, Li, Al-Amleh, Lyons, & Waddell, 2017). 3D printed plate (figure 8b) is also reliable structure to support at the broken bone. Pharmaceutical companies are developing molecule level printers that can print drugs on demand.

Electronics Industries

Miniaturization of the devices in electronic industries is a focus of interest now-a-days. Too much precision with many sophisticated equipment are required to develop the electronic circuits in very small spaces. By using additive manufacturing technique, the above difficulties successively eliminated. 3D printing made possible to manufacture the mobile phones. Without the need to include separate circuit boards, the assembly process is greatly reduced and the total weight is also reduced. Additive processing machines can produce conductive, resistive, dielectric and semiconductor inks that can be processed to make active, inactive components. Shielding, antennas and sensors are just some of the types of electronic components you can make. The world of electronics is still young to do 3D printing, from drones and satellites to laptops and smart phones. Electronic devices play an increasingly important role in our lives. However, these devices rely on electronic components such as printed circuit boards (PCBs), antennas and sensors to work (Flowers, Reyes, Ye, Kim, & Wiley, 2017; Ota et al., 2016). 3D printing redefines the way traditionally partially designed these components by providing faster product development and greater design complexity, especially in the nonplanar (nonplanar) geometry area. An electromagnet designed for additive manufacturing by solid works using the nano dimension Add-in is given in figure 9(a). Figure 9(b) showing the 3D printed UAV antenna.

Table 2. various 3D printed automobiles parts

3D Printed Parts	Materials Used	Process Adopted
Exhausts and emissions (cooling vents)	Aluminium alloys	SLM
Pumps and valves (within the fluid handling system)	aluminium alloys	SLM and EBM
wind breakers and bumpers	polymers	SLM
Hot work steels and polymers Manufacturing process	steels and polymers	SLS, SLM, and fused deposition modelling for prototypes
Interior and seating (*dashboards and seat frames*)	Polymers	stereo-lithography and SLM
Suspension springs, tyres and hubcaps	Aluminium alloys and polymers	SLS, SLM and Inkjet technology
Electronics (*such as sensors, and single part control panels*)	polymers	SLS
Framework and doors	aluminium alloys	SLM
Engine components	titanium and aluminium	SLM, EBM
OEM components	steel or aluminium alloys	SLM, EBM
Under the hood (**Heat resistant functional parts**)	Nylon	SLS
Interior accessories	Resin	SLA
Air ducts (**Air conditioning ducting**)	Nylon	SLS
Full scale panels	Resin	SLA
Metal parts made from 3d printed patterns	Wax	SLA
Consolidated, lightweight, functional metal parts	Metal	DMLS
Bezels (**Dashboard interface**)	Photopolymer	Material jetting
Headlight prototypes	Resin	SLA
High detail visual prototypes	photo-activated resin	Material jetting and SLA printing
Functional mounting brackets	PA12 nylon to titanium	SLS

(D. E. Cooper, Stanford, Kibble, & Gibbons, 2012; Holemans & Data, 2015; Lim, Le, Lu, & Wong, 2016; Mohammadizadeh, Fidan, Allen, & Imeri, 2018; Shunmugavel, Polishetty, & Littlefair, 2015)

Figure 8. Normal hip joint morphology and hip implant, (b)Internal plates and fracture: Geometry is necessarily complex

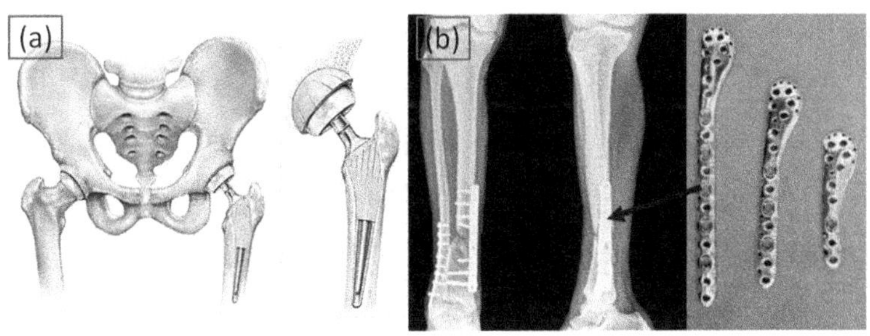

Figure 9. (a) 3D-printed circuit board, (b) UAV antenna part

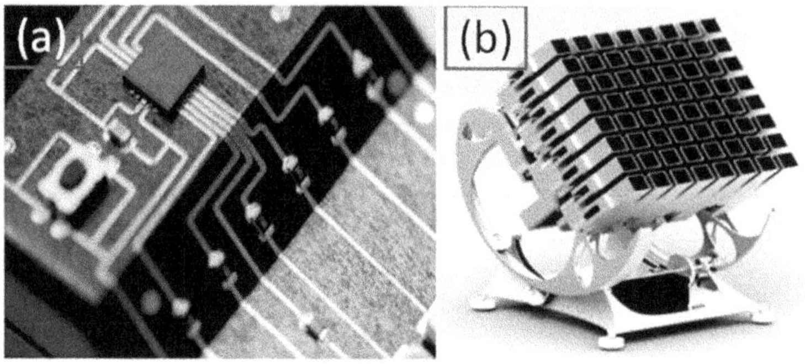

The additive manufacturing in electronic industry facilitates some advantages like: absence of etching by harmful chemical that use in traditional processes, Simplified assembly in a single build process, Reduced product size, external packaging for protruding parts is not required.

SUMMARY AND FUTURE WORK

3D printing technology can be used to create everything from airplane parts, automotive parts, medical implants to electronic equipment. The four commercial processes such as PBF, DMLM, EBM, BJP are well adopted by recent industries described in this chapter. All the advantages and disadvantages of 3D technologies are given for choosing specific structure and number of require product. In the future, food could be printed with customized nutritional content, optimized based on biometric and genomic data. Clearly, additive manufacturing will impact engineering jobs. Engineers working in biomedicine, food, auto and avionics in addition to civil engineering and industrial design will see increasing uses for 3D printing. Advances in chemical science will lead to more advanced plastics being manufactured by 3D printers. No matter what your area of engineering interest, additive manufacturing (both 3D and 4D) will likely play a role in its future.

FUNDING DECLARATION

This research received no specific grant from any funding agency in the public, commercial, or not-for-profit sectors.

REFERENCES

Additive Manufacturing & 3D Printing Processes I GE Additive. (n.d.). Retrieved February 25, 2020, from https://www.ge.com/additive/additive-manufacturing/information/additive-manufacturing-processes

Additive Manufacturing for the Aerospace Industry. (2019). Elsevier.

Additive manufacturing in aerospace is growing - 3Dnatives. (n.d.). Retrieved February 25, 2020, from https://www.3dnatives.com/en/additive-manufacturing-aerospace-growing-061220184/

Additive Manufacturing in the Aerospace Industry > ENGINEERING.com. (n.d.). Retrieved February 25, 2020, from https://www.engineering.com/AdvancedManufacturing/ArticleID/14218/Additive-Manufacturing-in-the-Aerospace-Industry.aspx

Additive Manufacturing Industry overview I GE Additive. (n.d.). Retrieved February 25, 2020, from https://www.ge.com/additive/industry-overview

Aerospace 3D printing Applications I 3D Hubs. (n.d.). Retrieved February 25, 2020, from https://www.3dhubs.com/knowledge-base/aerospace-3d-printing-applications/

Airbus Gets 1st Production LEAP-1A Jet Engines - GE Reports. (n.d.). Retrieved February 25, 2020, from https://www.ge.com/reports/airbus-gets-1st-production-jet-engines-with-3d-printed-parts-from-cfm/

Barazanchi, A., Li, K. C., Al-Amleh, B., Lyons, K., & Waddell, J. N. (2017). Additive Technology: Update on Current Materials and Applications in Dentistry. *Journal of Prosthodontics*, *26*(2), 156–163. doi:10.1111/jopr.12510 PMID:27662423

Bourell, D., Kruth, J. P., Leu, M., Levy, G., Rosen, D., Beese, A. M., & Clare, A. (2017). Materials for additive manufacturing. *CIRP Annals*, *66*(2), 659–681. doi:10.1016/j.cirp.2017.05.009

Chang, J., He, J., Mao, M., Zhou, W., Lei, Q., Li, X., Li, D., Chua, C.-K., & Zhao, X. (2018). Advanced Material Strategies for Next-Generation Additive Manufacturing. *Materials (Basel)*, *11*(1), 166. doi:10.3390/ma11010166 PMID:29361754

Chaudhuri, B., Mondal, B., Ray, S. K., & Sarkar, S. C. (2016). A novel biocompatible conducting polyvinyl alcohol (PVA)-polyvinylpyrrolidone (PVP)-hydroxyapatite (HAP) composite scaffolds for probable biological application. *Colloids and Surfaces. B, Biointerfaces*, *143*, 71–80. doi:10.1016/j.colsurfb.2016.03.027 PMID:26998868

Chen, X., Ware, H. O. T., Baker, E., Chu, W., Hu, J., & Sun, C. (2017). The Development of an All-polymer-based Piezoelectric Photocurable Resin for Additive Manufacturing. *Procedia CIRP*, *65*, 157–162.

Compton, O. C., & Nguyen, S. T. (2010). Graphene Oxide, Highly Reduced Graphene Oxide, and Graphene: Versatile Building Blocks for Carbon-Based Materials. *Small*, *6*(6), 711–723. doi:10.1002mll.200901934 PMID:20225186

Cooper, A., Kang, S., Kietzman, J., Prinz, F., Lombardi, J., & Weiss, L. (1999). Automated fabrication of complex molded parts using Mold Shape Deposition Manufacturing. *Materials & Design*, *20*(2-3), 83–89. doi:10.1016/S0261-3069(99)00013-8

Cooper, D. E., Stanford, M., Kibble, K. A., & Gibbons, G. J. (2012). Additive Manufacturing for product improvement at Red Bull Technology. *Materials & Design*, *41*, 226–230. doi:10.1016/j.matdes.2012.05.017

Delgado Camacho, D., Clayton, P., O'Brien, W. J., Seepersad, C., Juenger, M., Ferron, R., & Salamone, S. (2018). Applications of additive manufacturing in the construction industry – A forward-looking review. *Automation in Construction*, *89*, 110–119. doi:10.1016/j.autcon.2017.12.031

Direct Metal Laser Melting (DMLM) | GE Additive. (n.d.). Retrieved February 25, 2020, from https://www.ge.com/additive/additive-manufacturing/information/direct-metal-laser-melting-technology

Dutta, B., & Froes, F. H. (2015). The additive manufacturing (AM) of titanium alloys. In Titanium Powder Metallurgy (pp. 447–468). Elsevier.

Engineering and manufacturing - great.gov.uk international. (n.d.). Retrieved February 25, 2020, from https://www.great.gov.uk/international/content/about-uk/industries/engineering-and-manufacturing/

ExOne | Binder Jetting Technology. (n.d.). Retrieved February 25, 2020, from https://www.exone.com/en-US/case-studies/what-is-binder-jetting

Fathi, S., & Dickens, P. (2013). Challenges in drop-on-drop deposition of reactive molten nylon materials for additive manufacturing. *Journal of Materials Processing Technology*, *213*(1), 84–93. doi:10.1016/j.jmatprotec.2012.08.006

Flowers, P. F., Reyes, C., Ye, S., Kim, M. J., & Wiley, B. J. (2017). 3D printing electronic components and circuits with conductive thermoplastic filament. *Additive Manufacturing*, *18*, 156–163. doi:10.1016/j.addma.2017.10.002

Frazier, W. E. (2014). Metal Additive Manufacturing: A Review. *Journal of Materials Engineering and Performance*, *23*(6), 1917–1928. doi:10.100711665-014-0958-z

Gaytan, S. M., Cadena, M. A., Karim, H., Delfin, D., Lin, Y., Espalin, D., MacDonald, E., & Wicker, R. B. (2015). Fabrication of barium titanate by binder jetting additive manufacturing technology. *Ceramics International*, *41*(5), 6610–6619. doi:10.1016/j.ceramint.2015.01.108

GE Aviation celebrates 30,000th 3D printed fuel nozzle - 3D Printing Industry. (n.d.). Retrieved February 25, 2020, from https://3dprintingindustry.com/news/ge-aviation-celebrates-30000th-3d-printed-fuel-nozzle-141165/

Gonzalez, J. A., Mireles, J., Stafford, S. W., Perez, M. A., Terrazas, C. A., & Wicker, R. B. (2019). Characterization of Inconel 625 fabricated using powder-bed-based additive manufacturing technologies. *Journal of Materials Processing Technology*, *264*, 200–210. doi:10.1016/j.jmatprotec.2018.08.031

Gross, B. C., Erkal, J. L., Lockwood, S. Y., Chen, C., & Spence, D. M. (2014). Evaluation of 3D Printing and Its Potential Impact on Biotechnology and the Chemical Sciences. *Analytical Chemistry*, *86*(7), 3240–3253. doi:10.1021/ac403397r PMID:24432804

Gu, D. (2015). Laser Additive Manufacturing (AM): Classification, Processing Philosophy, and Metallurgical Mechanisms. In *Laser Additive Manufacturing of High-Performance Materials* (pp. 15–71). Springer Berlin Heidelberg. doi:10.1007/978-3-662-46089-4_2

Guo, N., & Leu, M. C. (2013). Additive manufacturing: Technology, applications and research needs. *Frontiers of Mechanical Engineering*, *8*(3), 215–243. doi:10.100711465-013-0248-8

Herzog, D., Seyda, V., Wycisk, E., & Emmelmann, C. (2016). Additive manufacturing of metals. *Acta Materialia*, *117*, 371–392. doi:10.1016/j.actamat.2016.07.019

Holemans, W., & Data, R. U. S. A. (2015). *(12) United States Patent. 2.*

Huang, S. H., Liu, P., Mokasdar, A., & Hou, L. (2013). Additive manufacturing and its societal impact: A literature review. *International Journal of Advanced Manufacturing Technology*, *67*(5-8), 1191–1203. doi:10.100700170-012-4558-5

Huang, Y., Leu, M. C., Mazumder, J., & Donmez, A. (2015). Additive Manufacturing: Current State, Future Potential, Gaps and Needs, and Recommendations. *Journal of Manufacturing Science and Engineering*, *137*(1), 014001. Advance online publication. doi:10.1115/1.4028725

In, T., III. How A Walnut-Sized Part Changed The Way GE Aviation Builds Jet Engines - GE Reports. (n.d.). Retrieved February 25, 2020, from https://www.ge.com/reports/transformation-3d-walnut-sized-part-changed-way-ge-aviation-builds-jet-engines/

Irwin, J., Reutzel, E. W., Michaleris, P., Keist, J., & Nassar, A. R. (2016). Predicting Microstructure From Thermal History During Additive Manufacturing for Ti-6Al-4V. *Journal of Manufacturing Science and Engineering*, *138*(11), 111007. Advance online publication. doi:10.1115/1.4033525

Jared, B. H., Aguilo, M. A., Beghini, L. L., Boyce, B. L., Clark, B. W., Cook, A., Kaehr, B. J., & Robbins, J. (2017). Additive manufacturing: Toward holistic design. *Scripta Materialia*, *135*, 141–147. doi:10.1016/j.scriptamat.2017.02.029

Jiménez, M., Romero, L., Domínguez, I. A., Espinosa, M. del M., & Domínguez, M. (2019). Additive Manufacturing Technologies: An Overview about 3D Printing Methods and Future Prospects. *Complexity*, *2019*, 1–30. doi:10.1155/2019/9656938

Johannes Karl Fink. (2018). *3D Industrial Printing with Polymers*. John Wiley & Sons.

Khairallah, S. A., Anderson, A. T., Rubenchik, A., & King, W. E. (2016). Laser powder-bed fusion additive manufacturing: Physics of complex melt flow and formation mechanisms of pores, spatter, and denudation zones. *Acta Materialia*, *108*, 36–45. doi:10.1016/j.actamat.2016.02.014

Khan, M., & Dickens, P. (2010). Selective Laser Melting (SLM) of pure gold. *Gold Bulletin*, *43*(2), 114–121. doi:10.1007/BF03214976

Klosterman, D., Chartoff, R., Osborne, N., & Graves, G. (n.d.). *Laminated Object Manufacturing, a New Process for the Direct Manufacture of Monolithic Ceramics and Continuous Fiber CMCs*. Academic Press.

Kok, Y., Tan, X. P., Wang, P., Nai, M. L. S., Loh, N. H., Liu, E., & Tor, S. B. (2018). Anisotropy and heterogeneity of microstructure and mechanical properties in metal additive manufacturing: A critical review. *Materials & Design*, *139*, 565–586. doi:10.1016/j.matdes.2017.11.021

Kruth, J. P. (1991). Material Incress Manufacturing by Rapid Prototyping Techniques. *CIRP Annals*, *40*(2), 603–614. doi:10.1016/S0007-8506(07)61136-6

LaMonica, M. (2013). Additive Manufacturing-Innovations, Advances, and Applications. MIT Technology Review, 116. doi:10.1201/b19360

Li, Y., & Gu, D. (2014). Parametric analysis of thermal behavior during selective laser melting additive manufacturing of aluminum alloy powder. *Materials & Design*, *63*, 856–867. doi:10.1016/j.matdes.2014.07.006

Li, Y., Hu, Y., Cong, W., Zhi, L., & Guo, Z. (2017). Additive manufacturing of alumina using laser engineered net shaping: Effects of deposition variables. *Ceramics International*, *43*(10), 7768–7775. doi:10.1016/j.ceramint.2017.03.085

Lim, C. W. J., Le, K. Q., Lu, Q., & Wong, C. H. (2016). An Overview of 3-D Printing in Manufacturing, Aerospace, and Automotive Industries. *IEEE Potentials*, *35*(4), 18–22. doi:10.1109/MPOT.2016.2540098

Lipton, J. I., Cutler, M., Nigl, F., Cohen, D., & Lipson, H. (2015). Additive manufacturing for the food industry. *Trends in Food Science & Technology*, *43*(1), 114–123. doi:10.1016/j.tifs.2015.02.004

Liu, R., Wang, Z., Sparks, T., Liou, F., & Newkirk, J. (2017). Aerospace applications of laser additive manufacturing. In *Laser Additive Manufacturing* (pp. 351–371). Elsevier. doi:10.1016/B978-0-08-100433-3.00013-0

Manfredi, D., Calignano, F., Krishnan, M., Canali, R., Paola, E., Biamino, S., … Fino, P. (2014). Additive Manufacturing of Al Alloys and Aluminium Matrix Composites (AMCs). In Light Metal Alloys Applications. InTech.

Mohamed, O. A., Masood, S. H., & Bhowmik, J. L. (2017). Experimental investigation of time-dependent mechanical properties of PC-ABS prototypes processed by FDM additive manufacturing process. *Materials Letters*, *193*, 58–62. doi:10.1016/j.matlet.2017.01.104

Mohammadizadeh, M., Fidan, I., Allen, M., & Imeri, A. (2018). Creep behavior analysis of additively manufactured fiber-reinforced components. *International Journal of Advanced Manufacturing Technology*, *99*(5-8), 1225–1234. doi:10.100700170-018-2539-z

Murr, L. E., Gaytan, S. M., Ceylan, A., Martinez, E., Martinez, J. L., Hernandez, D. H., Machado, B. I., Ramirez, D. A., Medina, F., & Collins, S. (2010). Characterization of titanium aluminide alloy components fabricated by additive manufacturing using electron beam melting. *Acta Materialia*, *58*(5), 1887–1894. doi:10.1016/j.actamat.2009.11.032

Murr, L. E., Martinez, E., Amato, K. N., Gaytan, S. M., Hernandez, J., Ramirez, D. A., Shindo, P. W., Medina, F., & Wicker, R. B. (2012). Fabrication of Metal and Alloy Components by Additive Manufacturing: Examples of 3D Materials Science. *Journal of Materials Research and Technology*, *1*(1), 42–54. doi:10.1016/S2238-7854(12)70009-1

NASA 3-D Prints First Full-Scale Copper Rocket Engine Part I NASA. (n.d.). Retrieved February 25, 2020, from https://www.nasa.gov/marshall/news/nasa-3-D-prints-first-full-scale-copper-rocket-engine-part.html

NASA develops first 3-D printed copper rocket engine part - The Economic Times. (n.d.). Retrieved February 25, 2020, from https://economictimes.indiatimes.com/news/science/nasa-develops-first-3-d-printed-copper-rocket-engine-part/articleshow/47013482.cms?from=mdr

Ota, H., Emaminejad, S., Gao, Y., Zhao, A., Wu, E., Challa, S., Chen, K., Fahad, H. M., Jha, A. K., Kiriya, D., Gao, W., Shiraki, H., Morioka, K., Ferguson, A. R., Healy, K. E., Davis, R. W., & Javey, A. (2016). Application of 3D Printing for Smart Objects with Embedded Electronic Sensors and Systems. *Advanced Materials Technologies*, *1*(1), 1600013. doi:10.1002/admt.201600013

3D . Printing in Aerospace & Aviation I GE Additive. (n.d.). Retrieved February 25, 2020, from https://www.ge.com/additive/additive-manufacturing/industries/aviation-aerospace

Qin, Y., Qi, Q., Scott, P. J., & Jiang, X. (2019). Status, comparison, and future of the representations of additive manufacturing data. *Computer Aided Design*, *111*, 44–64. doi:10.1016/j.cad.2019.02.004

Quan, Z., Larimore, Z., Wu, A., Yu, J., Qin, X., Mirotznik, M., Suhr, J., Byun, J.-H., Oh, Y., & Chou, T.-W. (2016). Microstructural design and additive manufacturing and characterization of 3D orthogonal short carbon fiber/acrylonitrile-butadiene-styrene preform and composite. *Composites Science and Technology*, *126*, 139–148. doi:10.1016/j.compscitech.2016.02.021

Rahman, Z., Barakh Ali, S. F., Ozkan, T., Charoo, N. A., Reddy, I. K., & Khan, M. A. (2018). Additive Manufacturing with 3D Printing: Progress from Bench to Bedside. *The AAPS Journal*, *20*(6), 101. doi:10.120812248-018-0225-6 PMID:30209693

Rengier, F., Mehndiratta, A., von Tengg-Kobligk, H., Zechmann, C. M., Unterhinninghofen, R., Kauczor, H.-U., & Giesel, F. L. (2010). 3D printing based on imaging data: Review of medical applications. *International Journal of Computer Assisted Radiology and Surgery*, *5*(4), 335–341. doi:10.100711548-010-0476-x PMID:20467825

Seifert, T., Sowade, E., Roscher, F., Wiemer, M., Gessner, T., & Baumann, R. R. (2015). Additive Manufacturing Technologies Compared: Morphology of Deposits of Silver Ink Using Inkjet and Aerosol Jet Printing. *Industrial & Engineering Chemistry Research*, *54*(2), 769–779. doi:10.1021/ie503636c

Shahzad, K., Deckers, J., Zhang, Z., Kruth, J.-P., & Vleugels, J. (2014). Additive manufacturing of zirconia parts by indirect selective laser sintering. *Journal of the European Ceramic Society*, *34*(1), 81–89. doi:10.1016/j.jeurceramsoc.2013.07.023

Shi, Y., Wang, Y., Chen, J., & Huang, S. (2008). Experimental investigation into the selective laser sintering of high-impact polystyrene. *Journal of Applied Polymer Science*, *108*(1), 535–540. doi:10.1002/app.27686

Shunmugavel, M., Polishetty, A., & Littlefair, G. (2015). Microstructure and Mechanical Properties of Wrought and Additive Manufactured Ti-6Al-4V Cylindrical Bars. *Procedia Technology*, *20*, 231–236. doi:10.1016/j.protcy.2015.07.037

Snyder, J., Rin Son, A., Hamid, Q., & Sun, W. (2016). Fabrication of Microfluidic Manifold by Precision Extrusion Deposition and Replica Molding for Cell-Laden Device. *Journal of Manufacturing Science and Engineering*, *138*(4), 041007. Advance online publication. doi:10.1115/1.4031551

Spears, T. G., & Gold, S. A. (2016). In-process sensing in selective laser melting (SLM) additive manufacturing. *Integrating Materials and Manufacturing Innovation*, *5*(1), 16–40. doi:10.118640192-016-0045-4

Spoerk, M., Arbeiter, F., Cajner, H., Sapkota, J., & Holzer, C. (2017). Parametric optimization of intra- and inter-layer strengths in parts produced by extrusion-based additive manufacturing of poly(lactic acid). *Journal of Applied Polymer Science*, *134*(41), 45401. doi:10.1002/app.45401

Studart, A. R. (2016). Additive manufacturing of biologically-inspired materials. *Chemical Society Reviews*, *45*(2), 359–376. doi:10.1039/C5CS00836K PMID:26750617

Students Aim for Space with 3D-Printed Rocket Engine I Space. (n.d.). Retrieved February 25, 2020, from https://www.space.com/29808-students-3d-printed-rocket-engine.html

Tofail, S. A. M., Koumoulos, E. P., Bandyopadhyay, A., Bose, S., O'Donoghue, L., & Charitidis, C. (2018). Additive manufacturing: Scientific and technological challenges, market uptake and opportunities. *Materials Today*, *21*(1), 22–37. doi:10.1016/j.mattod.2017.07.001

Travitzky, N., Bonet, A., Dermeik, B., Fey, T., Filbert-Demut, I., Schlier, L., Schlordt, T., & Greil, P. (2014). Additive Manufacturing of Ceramic-Based Materials. *Advanced Engineering Materials*, *16*(6), 729–754. doi:10.1002/adem.201400097

Tuomi, J., Paloheimo, K.-S., Vehviläinen, J., Björkstrand, R., Salmi, M., Huotilainen, E., Kontio, R., Rouse, S., Gibson, I., & Mäkitie, A. A. (2014). A Novel Classification and Online Platform for Planning and Documentation of Medical Applications of Additive Manufacturing. *Surgical Innovation*, *21*(6), 553–559. doi:10.1177/1553350614524838 PMID:24616012

Vayre, B., Vignat, F., & Villeneuve, F. (2012). Metallic additive manufacturing: State-of-the-art review and prospects. *Mechanics & Industry*, *13*(2), 89–96. doi:10.1051/meca/2012003

Velu, R., Calais, T., Jayakumar, A., & Raspall, F. (2019). A Comprehensive Review on Bio-Nanomaterials for Medical Implants and Feasibility Studies on Fabrication of Such Implants by Additive Manufacturing Technique. *Materials (Basel)*, *13*(1), 92. doi:10.3390/ma13010092 PMID:31878040

Walker, J., Andani, M. T., Haberland, C., & Elahinia, M. (2014). *Additive Manufacturing of Nitinol Shape Memory Alloys to Overcome Challenges in Conventional Nitinol Fabrication. Volume 2A: Advanced Manufacturing*. American Society of Mechanical Engineers.

Williams, C. B., Mistree, F., & Rosen, D. W. (2011). A Functional Classification Framework for the Conceptual Design of Additive Manufacturing Technologies. *Journal of Mechanical Design*, *133*(12), 121002. Advance online publication. doi:10.1115/1.4005231

Willis, K., Brockmeyer, E., Hudson, S., & Poupyrev, I. (2012). Printed optics. In *Proceedings of the 25th Annual ACM Symposium on User Interface Software and Technology - UIST '12*, 589. New York: ACM Press. 10.1145/2380116.2380190

Wilts, E. M., Pekkanen, A. M., White, B. T., Meenakshisundaram, V., Aduba, D. C., Williams, C. B., & Long, T. E. (2019). Vat photopolymerization of charged monomers: 3D printing with supramolecular interactions. *Polymer Chemistry*, *10*(12), 1442–1451. doi:10.1039/C8PY01792A

Wu, A. S., Brown, D. W., Kumar, M., Gallegos, G. F., & King, W. E. (2014). An Experimental Investigation into Additive Manufacturing-Induced Residual Stresses in 316L Stainless Steel. *Metallurgical and Materials Transactions. A, Physical Metallurgy and Materials Science*, *45*(13), 6260–6270. doi:10.100711661-014-2549-x

Wu, H., Liu, W., He, R., Wu, Z., Jiang, Q., Song, X., Chen, Y., Cheng, L., & Wu, S. (2017). Fabrication of dense zirconia-toughened alumina ceramics through a stereolithography-based additive manufacturing. *Ceramics International*, *43*(1), 968–972. doi:10.1016/j.ceramint.2016.10.027

Yadollahi, A., Shamsaei, N., Thompson, S. M., Elwany, A., & Bian, L. (2017). Effects of building orientation and heat treatment on fatigue behavior of selective laser melted 17-4 PH stainless steel. *International Journal of Fatigue*, *94*, 218–235. doi:10.1016/j.ijfatigue.2016.03.014

Zander, N. E. (2019). *Recycled Polymer Feedstocks for Material Extrusion Additive Manufacturing*. Academic Press.

Zenou, M., & Grainger, L. (2018). Additive manufacturing of metallic materials. In *Additive Manufacturing* (pp. 53–103). Elsevier. doi:10.1016/B978-0-12-812155-9.00003-7

Zhai, Y., Lados, D. A., & LaGoy, J. L. (2014). Additive Manufacturing: Making Imagination the Major Limitation. *JOM*, *66*(5), 808–816. doi:10.100711837-014-0886-2

KEY TERMS AND DEFINITIONS

Bio-Printer: This an additive manufacturing process using biological cell/tissue.

Graphene: This is the basis structural unit of carbon based nanomaterials. Each carbon atoms in graphene are sp^2 hybridise bonded with other carbon atom looks like a dense one atomic thick structure.

Maraging Steel: Maraging steels are carbon free FeNi-alloys with additions of Co, Mo, Ti, Al. The term maraging comes from the strengthening mechanism that transforming the alloy to martensite with subsequent age hardening.

Shape Memory Effect: This is a property of shape memory alloy (SMA) that can regain its shape when the deformed shape is subjected to the temperature, higher than the austenitic temperature.

Superalloy: These are the alloys that can increase their strength with increase in temperature. Generally working under the temperature >0.7 Tm without creep.

Superelasticity: This is also known as pseudoelasticity property. This is a property of shape memory alloys trained at austenitic temperature to regain its shape without any deformation.

Chapter 3
Additive Manufacturing of Nickel-Based Superalloys Used in the Aero-Engines:
SLM of Inconel

Raja A.
Indian Institute of Technology, Madras, India

Mythreyi O. V.
Indian Institute of Technology, Madras, India

Jayaganthan R.
Indian Institute of Technology, Madras, India

ABSTRACT

Aero-engines contains parts that are generally lightweight, subjected to high performance at high temperature. Ni-based superalloys are widely used in engine turbines. Manufacturing these parts by additive manufacturing (AM) methods provides researchers a lot of creative space for complex design to improve efficiency. Powder bed fusion (PBF) and direct energy deposition (DED) are the two most widely-used metal AM methods. Both methods are influenced by the source, parameters, design, and raw material. Selective laser melting is one of the laser-based PBF techniques to create small layer thickness and complex geometry with greater accuracy and properties. The layer-by-layer metal addition generates epitaxial growth and solidification in the built direction. There are different second phases in the Ni-based superalloys. This chapter details the micro-segregation of these particles and its influence on the microstructure, and mechanical properties are dependent on the process influencing parameters, the thermal kinetics during the process, and the post-processing treatments.

DOI: 10.4018/978-1-7998-4054-1.ch003

INTRODUCTION

Superalloys, as its name implies, are special class of metallic materials that exhibits stable microstructure, good mechanical strength, toughness, resistance to degradation in corrosive environment at a temperature more than 80% of the melting point (Pollock et al., 2006). It is because of the superalloys, which mostly encompassed in aero engine gas-turbines, humans able to fly in the skies. The aero-engine, operates according to the Newton's third law of motion. The fans at the front of the engine sucks the air in, the compressor squeezes the air and sprayed along with fuel in to the combustion chamber. The air-fuel mixture is then ignited by an electric spark, the burning mixture will expand and leave the exit nozzle, after passing through the turbine blades, with a thrust that propels the engine. The cycle would continue, as the bursting out gases push the turbine blades to rotate the shaft which in turn rotate the compressor and fan at front to bring more air into the system. The cold air temperature rises to 600 °C during compression and can reach 1500 °C during combustion and drops down to approximately 800 °C after expansion, around the turbines, and before leaving the exhaust nozzle. Superalloys are used to withstand such high temperature for long operating hours. The world's first successful axial compressor turbojet engine, the Junkers Jumo 004, with a pull starter, run first during the world war II in 1940s to power Messerschmitt Me 262 fighter and the Arado Ar 234 reconnaissance/bomber. The Jumo 004 engine was made up of steel stator, aluminium coated mild steel combustion chamber, hollow turbine blades of Cromadur alloy (12% chromium, 18% manganese, and 70% iron) and cast magnesium engine casing (Kay & Couper, 2004). The usage of superalloys in aero-engine increased from about 10% in 1950 to approximately 50% in 1985 (Akca & Gursel, 2015).

There are three kinds superalloys classified based on the major element present in the alloy. They are Iron (Fe)-based, Cobalt (Co)-based and, Nickel (Ni)-based superalloys. Fe-based superalloys are cheaper amid the three superalloys. The Fe-based superalloys, such as, Cr-Mo-V stainless steel and Maraging steels are used as shaft material due to its good high temperature fatigue strength and they are less prone to segregation. Co-based superalloys are not as strong as Ni-based superalloys but retain their properties at temperatures as high as 1100 °C, therefore, they are used in low stress and high temperature applications such as stationary vanes in the gas-turbine. Ni-based superalloys are most complex and interesting among all the three because it can withstand applications at higher fraction of the melting temperature of the material. Ni-based superalloys contain γ-phase with face cantered cubic structure which precipitates out γ'-phase (a primitive cubic structure (Al, Ti) with Ni at face centre to strengthen the γ-matrix. While Fe-based superalloys required at least 25% Ni to stabilize γ–phase (austenite) and Co-based superalloys depends on carbide precipitates for strengthening. The γ'-precipitates may have one of the chemical formula; Ni_3Al, Ni_3Ti or $Ni_3(Al, Ti)$. When γ' precipitates out of γ matrix, they exist in coherent cube-cube relationship. A small negative misfit (γ' has a smaller lattice parameter than γ) between γ and γ' phase generates rafts in the microstructure that prevents climb which improves the creep property. The misfit can be controlled by changing the Al/Ti ratio or by the addition of rhenium (Re). With the addition of niobium (Nb), the superalloy called IN 718, the low temperature yield strength can be increased further due to the formation of γ'' phase in the form of discs with orientation relationship of (001)γ''||{001}γ and [100]γ''||<100>γ. The composition of γ'' phase is Ni_3Nb, with a crystal structure of body-centred tetragonal lattice (Sun et al., 2015). The elements Cr and Al are added to prevent oxidation. The metal carbides (MC) precipitates at grain boundary which reduces the tendency of grain boundary sliding (Tytko et al., 2012) at high temperatures. Re and Ru are used to develop single crystal superalloys. However, all the precipitates that form in superalloys are not beneficial as precipitates such as Laves and σ phases which have topologically closed pack crystal structure (TCP) imparts poor strengthening effect. TCP has stacking sequence similar to closed pack structure but actually not a closed pack one. TCP structured phases are brittle, when it precipitates, it depletes other useful elements in the microstructure and are detrimental. However, addition of Re promotes strengthening effect of TCP phases.

Figure 1. Flow chart of widely used processes to manufacture superalloy components
Adapted from (Akca & Gursel, 2015)

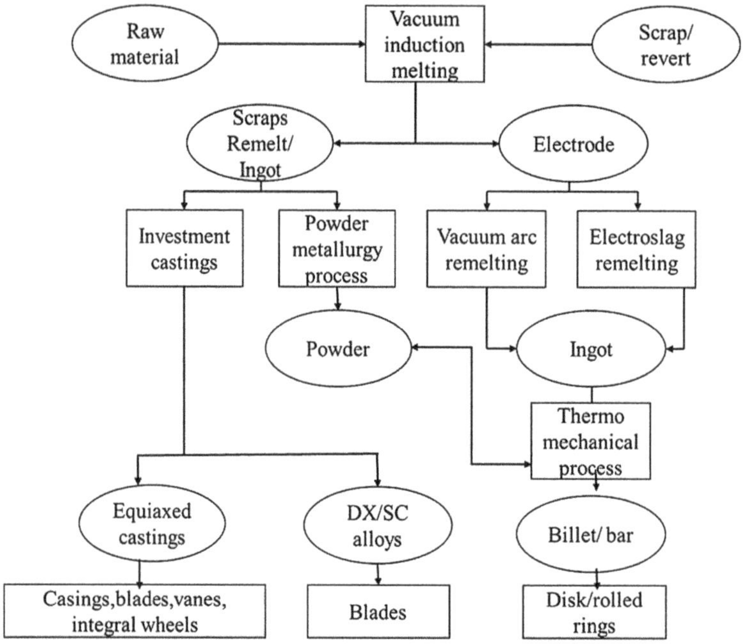

The conventional way of manufacturing a superalloy component is shown in the Figure 1. Initially ingots produced by vacuum induction melting (VIM). The ingots were re-melted, alloyed and fabricated in to desired shapes by investment casting, wrought processing and/or powder metallurgy, depending on the need and application of the component. These process are highly complex even to make a part with simple design. The aero-engine components require adequate cooling and continuous monitoring of parts for safe operation throughout its operating life. With the advent of additive manufacturing, any complex design with internal cooling passages and channels to fix measuring probes can be produced without any additional machining operations. It considerably reduces number of joints which prevents the part from galvanic corrosion susceptibility and reduces built-to-fly ratio.

AM OF Ni-BASED SUPERALLOYS

The important AM methods by which metalic parts can be manufactured are powder bed fusion (PBF) and direct energy deposition (DED). The schematics of PBF and DED methods are presented in Figure 2(a) and Figure 2(b) respectively. In PBF, the raw material, the powder in the container raise to an extent, according to the calculated layer thickness of the process. A recoater arm spread the raised powder from the container into the substrate on the processing bed. Once the powder spread on to the bed, the power source melt and solidify to form geometry according to the sliced 3D CAD model. The recoater then remove the remaining powder on the bed. The powder in the container will raise again for the deposition of next layer and the process will continue layer by layer till it get the complete geometry. The entire process will take place inside a designed chamber. The power source can be either laser or electron beam. If laser is the power source then the process may require argon or nitrogen as a shielding gas to control oxidation. The process is called selective laser melting (SLM). If the power source is electron beam then the entire process will take place in vacum chamber. The process is called electron beam melting (EBM). In DED, the raw material can be either powder or wire, which will be fed concentrically around the power source and the deposition will takes place layer by layer. The power source can either be laser or electron beam and the process can termed as laser metal deposition (LMD) and electron beam metal deposition (EMD).

Figure 2. Schematic of (a) PBF and (b) DED

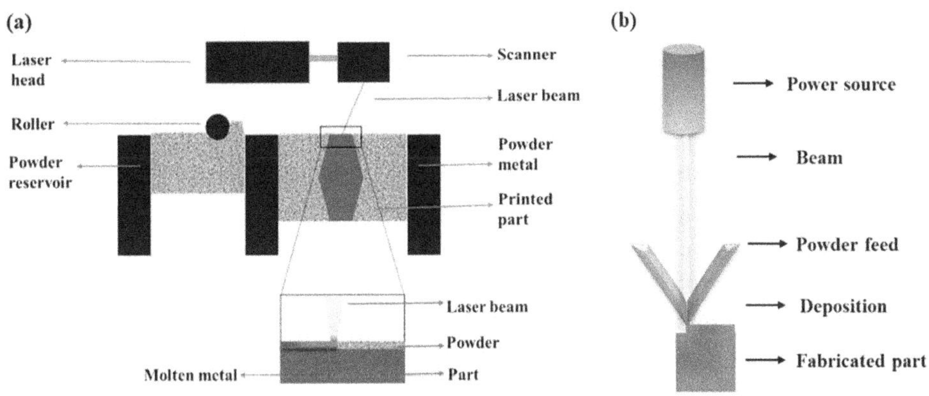

The advantages of AM are, in general, good for any product and material. So, what is so significant in manufcturing the Ni-based superalloys through additive manufacturing? Its metallurgy. As already mentioned in the introduction it contains different secondary phases in the microstrcuture to contribute to enhance certain properties of the material and it contains some other phases, like Laves and sigma phases, which are detrimental. It is highly possible to tailor the volume fraction, shape, size and location of these secondary phase particles in the microstructure by AM. Also, using AM functioanlly graded material (FGM) with different microstructure or different materials along its length/thickness can be fabricated. Carroll et al. (2016) fabricated FGM comprised of 304L stainless steel and IN 625 by powder based LMD method. While using LMD, the power used to fabricate IN 718 alloy was in the range of 1.0-1.8 kW (Abioye et al., 2013). In IN718, after LMD, it was reported that the samples contain oxides such as Al_2O_3, which was identified to be deposited from the wire and Cr_2O_3 films, which developed during the process (Zhang et al., 2013). Jia et al. (Jia & Gu, 2014b) fabricated IN718 alloy by SLM method using a laser power of 100 to 130 W and found the presence of Cr_2O_3 films inside the γ matrix. It was observed on altering the SLM process parameters the thickness of oxide film reduced. By modulating the laser profile from Gaussian to elliptical, during the process, in conduction mode laser scan it is possible to engineer specific microstructure (columnar or equiaxed) (Roehling et al., 2017). Above all, the non-equillibrium microstructure that developed out of high thermal gradient retains certain favourable microstructure such as martensitic α′phase in Ti6Al4V and austenitic phase in certain martensitic stainless steel (Herzog et al., 2016). Similarly, as built Ni-based superalloys can generate supersaturated γ-phase in the microstructure which improves the material's solid solution strength. Still the uniqueness in AM of Ni-based superalloys is that its performance enhancements at high temperatures. In an comparative study of IN625 alloy fabricated by SLM, EBM and binder jetting by Gonzalez et al. (2019), there were microstructural differences observed corresponding to the nature of the different processes and SLM outperformed other two technologies in most of the evaluated mechanical properties. Hence, it is worth discussing detailedly about the SLM of Ni-based superalloys.

SLM OF NICKEL BASED SUPERALLOYS

The schematics and functioning of SLM system was described in the previous session. There were several challenges pertaining to the geometrical accuracy of the parts, re-coater arm distortion of solidified layer, mixing of used and virgin powder, production rate, support structure generation for inclined parts, removal of the finished product, and surface finish of the part. These challenges are concerned to the design of the SLM system. Once a viable SLM system had been designed, the process generated challenges are important from the subject point of view. In order to ensure the flowability of the powder during SLM, the particles size need to be in the range of 15 to 63 μm, with a mean particle size (D_{50}) of 31 μm and for EBM it is 45 to 105 μm (Nguyen et al., 2017). If the particle size is larger, it will be difficult to melt completely as diameter of the laser beam is approximately 100 μm and if the particle size is too small then the particles flow away due to the inert gas flow (Herzog et al., 2016). Other than particle diameter the particle size distribution also plays a major role in the surface finish. Larger the range of particle sizes, higher is the surface roughness. Spierings et al. (2011, 2016) observed improvement in the mechanical properties on narrowing down the particle size distribution (15.26–55.54 μm, 19.84–41.13 μm, and 7.12–24.17 μm) in steels and Nickel alloys. Flowability of recycled or used powders are always found to be lesser than the virgin powder due its exposure to power and subjected to deformation. The shape of the powder particles also influence the flowability. Among the different shapes of power particles such as spherical, irregular, and angular, the spherical particles have higher flowability.

The SLM process parameters control the melt pool shape and size during SLM. If the melt pool is affected by the heat affected zone of previous layer, then it is more symmetric than the melt pool which is not affected by the heat affected zone of the previous layer (Criales et al., 2017a). Increasing the laser power or/and decreasing the scan velocity increases both melt pool depth and width. The difference is more significant in width rather than in depth. Increasing the hatch distance increase the distance between heat-affected zone of previous layer and the current layer which would result in reduced melt pool width. However, the changes in melt pool shape and size are non-linear. Another way to measure the melt pool shape is based on the volumetric energy density. The maximum temperature, heating rate, cooling rate, and spattering during the processing of every single layer depends on these process parameter and volumetric energy density. Özel et al. (2018) provided a temporal and spatial model to monitor over melting and spatters respectively, during the process. Thus the process parameters influences the thermal kinetics of the process. The heating rate during SLM of IN625 was predicted as 600 to 1000 °C/ms whereas the cooling rate was predicted as 150 °C/ms, while the process attained a maximum temperature of 1380 °C

(Criales et al., 2017b). These changes in melt pool shape, melt pool size, heating rate, cooling rate and maximum temperature as a consequence of process parameters controls the microstructure of the solidified material (Arısoy et al., 2019). Hence, it is significant to discuss the influence of these parameters on the SLMed Ni-based superalloys from the reported literatures.

INFLUENCE OF PROCESS PARAMETERS

Laser Power

On optimizing the process parameter for the SLM of IN625 alloys using single track deposit, Li, Guo, et al., (2017), continuous track was observed to form after 60 W and deep penetration was observed after 75 W laser power. The contact angle of melt track found to decrease with increasing laser power. Also, the width and depth of the melt pool increases with increasing laser power.

Scan Speed

On increasing the scan speed keeping all other process parameters, the density and dimensional accuracy decreased but the micro-hardness increased in laser sintered IN 625 alloy (Sateesh et al., 2014). Choi et al. (2017) also studied the influence of scan speed on the densification of SLMed microstructure of IN718 alloy at a constant laser power of 90 W. The scan speed, alone, varied from 100 mm/s to 1600 mm/s and the minimum porosity of 0.35% observed at the scanning speed of 800 mm/s. To obtain fully dense particle the scanning speed of 300 to 800 mm/s, which corresponding to the volumetric energy density of 60-150 J/mm^3, was suggested. Xia et al. (2017) studied the influence of scanning speed in IN718 alloy. On increasing the scanning speed from 200 mm/s to 500 mm/s, the peak temperature reduced from 2336 °C to 1738 °C while the maximum cooling rate increased from 1.21 x 10^7 °C/s to 5.12 x 10^7 °C/s, correspondingly. Thus, increasing scanning speed reduce the laser power interaction-time with surrounding powder materials. Metallurgical pores were observed on the surface of the layers deposited at higher scanning speed. As metallurgical pores are formed due to entrapment of gases from the environment or that evolving out of the metal vapour, lower scanning speed cause effective thermal convection that facilitate the escaping of entrapped gases. Acharya et al. (2017) reported that increasing the scan velocity can modify the oriented microstructure to misoriented microstructure. If the scanning direction remains same, a misorientation of approximately 27° was observed at a scanning speed of 1600 mm/s and at a laser power of 335 W. On increasing the scan velocity further leads to the formation of secondary dendrites in the scan direction. In general, the melt pool width and depth increases with decreasing laser power. However, the track height, the surface roughness and the contact angle were independent of scan speed (Li, Guo, et al., 2017).

Hatch Spacing

The density and geometrical accuracy of the SLMed parts are better at a hatch spacing of 0.3 when compared to that at 0.4, keeping layer thickness, laser power and beam diameter constant though it vary with the variation in scan speed. (Sateesh et al., 2014). Besides, hatch spacing, hatch length significantly influence the microstructure. Reducing the hatch length to ten times decrease the texture intensity twice.

Shorter hatch length developed compressive residual stress and longer hatch length developed residual stress gradient along the built direction in SLMed IN718 alloy (Nadammal et al., 2017).

Layer Thickness

Sufiiarov et al. (2017) reported that the yield strength (YS) and ultimate tensile (UTS) increases and the elongation to failure decreases with the decrease in layer thickness of the SLMed IN 718 alloy. Increasing the layer thickness promotes columnar grain growth which is favourable for manufacturing turbine blades of aero-engine (Popovich et al., 2018) while decreasing it leads to an equi-axed grain structure.

Energy Densities

In the work by Jia & Gu, (2014b), the effect of volumetric energy density on oxidation and relative density of SLMed IN718 was analysed. At low volumetric energy density of 70 J/mm^3, the relative density of the material was measured as 92.5%. On increasing the volumetric energy density to 110 J/mm^3 by reducing the scanning speed the relative density increased to 97.8% and on increasing it further to 130 J/mm^3 a nearly full dense material of 98.9% was achieved. The denser the material better is the high temperature oxidation resistance in the material. In another work by Jia & Gu, (2014a), at the line energy density (P/v) of 180 J/m the scan tracks of SLMed IN718 alloy appears discontinuous and open pores surrounding the large sized balls were observed. As the line density increased to 330 J/m, the surface was found to be free of pores and balls. The relative density increased from 73. 6% to 98. 4% from the former to the later. It was because the dynamic viscosity of the molten liquid dependent on the temperature of the liquid. Lesser the liquid temperature, higher would be the dynamic viscosity. Hence, at higher scan speed and relatively lower laser power, the high viscosity of the liquid is impeded by the scan speed resulting in improper melting and balling. Further, the hardness of the material that obtained at line energy density of 330 J/m was equivalent to the hardness value of heat-treated conventionally smelted IN 718 alloy. It is because, SLM create negative misfit between γ and γ¢ precipitates. A normalized-model based processing diagram was proposed by Thomas et al., (2016) which presents the isopleths of the normalized energy density E_0* to analyse any additive manufacturing process.

Scanning Strategy

Raghavan et al. (2017) employed a localized melt scan strategy during EBM of IN718 alloy which exhibited a consistent solidification microstructure. It partially eliminates the effect of geometry on the solidification microstructure. Carter et al. (2014) compared 5 mm x 5 mm island scanning strategy with "back-and-forth" scanning. The 5 x 5 island scan produced 1 mm x 1 mm diamond microstructured pattern where the over-etched boundaries contain elongated grains with cracks and equi-axed fine grains at the middle of the square on the XY plane (perpendicular to build direction). The cracks were due to ductility dip cracking (DDC) which will be explained later in the chapter under micro-cracks. On the XZ pane (parallel to build direction), bimodal grain distribution of coarse and elongated grains surrounded by fine and equi-axed grains were observed. On the other hand, the back and forth scan strategy produced more homogeneous microstructure with larger {001} columnar grains aligned in the build direction with some tilting due to the deviation in the heating direction from Z-axis of the sample. In the work by Lu et al. (2015), 3 mm x 3 mm island scanned IN718 alloy showed higher residual stress (210

MPa) than the 5 mm x 5 mm island scanned material (150 MPa) steeper thermal gradient in the former condition. However, on increasing the island area to 7 mm x 7 mm, the residual increased slightly to 160 MPa, because the thermal stresses resided in the material due to longer processing time. The 2 mm x 2 mm island scan material showed residual stress of only 100 MPa despites its higher thermal gradient, because the cracks that formed released such very high residual stresses. During SLM of IN718 using unidirectional scanning, the primary dendrites had grown at angle of 60° to the substrate while during the bidirectional scan, the dendrites grows at angle 45° to the substrate (Wei et al., 2015). A new method of fractal scan strategy was used by Smith et al., (2017) to reduce residual stress induced hot cracks in the unweldable Ni superalloy, IN734LC. Sun et al. (2018) studied the influence of three simple scan strategies during SLM of IN718 alloy. They are scanning along X- direction, scanning in XY direction with 90° rotation and with 67° rotational scan strategy. The X-scan strategy developed grain <101> grains elongated along the build direction, XY scan with 90° rotation generated cubic texture and that with 67° rotational strategy developed <001> fibre texture.

Beam Profile

In order to achieve columnar to equi-axed dendrites, increase in scan speed and laser power are generally preferred. In conduction mode laser heating, the laser beam profile influences the grain morphology. Gaussian beam profile generates columnar grains whereas elliptical beam profile generates equi-axed or mixed equi-axed and columnar grain structures. Roehling et al. (2017) achieved tailoring the grain morphology at moderate powers (150-450 W) by varying the laser beam profile with constant scan speed and laser power. The elliptical profile cause vortex of the molten pool due to Marangoni convection and recoil pressure effects. The vortex stirs the melt from the vortex depression to the comparatively cooler transition region at a high velocity. This high velocity molten melt fragments the dendrite tip. These solid fragments then act as an intrinsic nucleation sites for equi-axed grains.

MICROSTRUCTURE

As-Built Microstructure

The layer-by-layer deposition leads to high thermal gradient and repeated thermal cycles. The as-built SLMed IN718 alloy was usually characterized by elongated grains and equi-axed grains in parallel to build direction and perpendicular to it respectively. The Nb-rich Laves phase were present in the heavily segregated, interdendritic region (Nie et al., 2014). Popovich et al. (2017b) reported presence of Laves phase, rich in Nb and Mo, as well as MC rich in Nb and Ti in the interdendritic region. They are brittle and detrimental to mechanical properties. The Laves phase in the interdendritic region of columnar cells were in the form of long chain morphology whereas in equi-axed cellular dendritic Laves phase were distributed in the form of fine discrete particles. High cooling rate and low thermal gradient leads to equi-axed dendrites whereas low cooling rate and high thermal gradient produces columnar dendrites. The cooling rate and thermal gradient were controlled by the heat input and the velocity of growing dendritic tip (Kundin et al., 2015). Interestingly, in a work by Farber et al. (2018) on SLM of IN718 alloy, there were no Laves phase but γ¢ and γ¢¢ were observed in the interdendritic region in as-built condition. There were no reason mentioned for the absence of Laves phases in the work but the initial powder used in their work was recycled IN718 powder. Similar observation of uniformly distributed γ¢ particles in the γ-matrix and γ¢¢ particles in the interdendritic region was reported by Ni et al. (2017) but the initial powder was not recycled one. Among different AM methods, SLM solidifies microstructure at higher cooling rate due to the high velocity of the solidification front. The process of this kind leads to residual stress, micro-segregation and epitaxial growth. The columnar grain growth due to epitaxy was well discussed in the literatures, the influence of residual stress and micro-segregation that leads to hot cracking and solidification/liquefaction cracking respectively are discussed below in brevity.

Residual Stress

There are three types of residual arises in the material depends on its scale. Type I residual stress varies over the geometry of the part, type II occurs due to difference in phases and type III residual stress are in atomic scale due to dislocations. Among the three type of residual stresses, the material's mechanical properties were highly influenced by type I residual stress. In the process of layer by layer manufacturing, residual stresses arise due to thermal gradient mechanism and cool-down phase of top layer. In thermal gradient mechanism, the top layer expands on heating which would be restricted by the cooler bottom layer. This would induce compressive stress on the top and tensile stress on the bottom layer. In the cool-down phase of top layer, the top layer contract (shrinkage) on cooling which would be restricted by the cooler bottom layer and cause tensile stress on the top layer and compressive at the bottom. Hence, the upper part of the built was under compressive stress and the bottom part was under tensile stress. The substrate on which the material built influences the residual stress magnitude and direction. After the part removed from the substrate, the residual stress of the part relaxes, also both the top and bottom part of the built was subjected to tensile stress while the middle part was under compressive stress. Similar observation of tensile residual stresses at the edges of the surfaces and compressive stresses in the middle was observed in the SLMed Ti6Al4V and IN718 alloy (Ahmad et al., 2018). It was mentioned in the work of Barros et al. (2019) that tensile stresses at the top and bottom layer, while compressive at the middle layers was typical to Laser based PBF AM method. Even, similar distribution of residual stress in Laser based DED of Wasp alloy was reported by Moat et al. (2011). However, in the work by Mishurova et al. (2018), tensile residual stresses were observed, both before and after the removal of substrate, on the surfaces of prism built by SLM using IN718 alloy. In the work by Barros et al. (2019), anisotropy in residual stress was observed in IN718 alloy. The residual stress measured as 200-300 MPa in the horizontal orientation and 600-900 MPa in the vertical orientation, as thermal stresses are usually larger along the build direction than that along the scan direction. Further, after heat treatment (1065°C/1h (hour)/AC (air cooled) + double aging (DA)), the tensile residual stress on the surfaces become compressive. In IN718 alloy, Kromm et al., (2018) reported that redistribution of stress at every single layer lead to maximum residual stress on the surface (last deposited layer) which can reach the yield stress of the material. The part thickness and scan strategy can also influence the parts residual stress. Thicker the substrate lesser will be the residual stress for a specific part size. The YS of the alloying elements will also affect the residual stress. Higher the yield strength of the added material higher the residual stress. Hence, scan strategy, number of layers, layer thickness and the yield strength of alloying elements influences the magnitude of residual stress.

Micro-Cracking

Hot cracking, solidification/liquefaction cracking and DDC are the micro-cracks observed during the additive manufacturing of Ni-based superalloy. Hot cracking is the solid state cracking of material during layer-by-layer manufacturing when the tensile residual stress becomes higher than the ultimate tensile stress of the material. The hot cracking also depends on the thermal conductivity (k) and co-efficient of thermal expansion (α_{CTE}) of the material. Higher the k/α_{CTE} ratio, more stable is the material at specific temperature (Hunt et al., 2014). Thermal stresses and hot cracking can be reduced through control of platform pre-heat temperatures and scanning parameters (Harrison et al., 2015). The solidification/liquefaction cracking was attributed to micro-segregation. Such micro-cracks were observed in Ni-superalloys with high Al and Ti content. These alloys, such as IN738, CMSX-4, Rene 142 or CM247LC promote γ¢ precipitates at high temperatures are difficult to weld alloys. Micro-cracks due to gas porosity, aligned intergranular pores due to micro-shrinkage and liquefaction/solidification cracking was reported on selective EBM of Ni-based superalloy containing high Al and Ti content (Gault et al., 2018). Boron has high tendency to segregate at the grain boundaries. It lowers the liquid/solid interfacial energy, hence liquid films remain wetted to the dendrites till the last stage of solidification. Phosphor and Si have tendency to reduce the solidus temperature, thus increasing the solidification temperature range of Ni-based superalloys. Even a minor amount of Si (> 0.03 wt. %), during the directional solidification, promotes micro-segregation that leads to solidification cracks in SLMed IN738LC alloy (Engeli et al., 2016). Other than P and Si, Cloots et al. (2016) reported that Zr and B were also responsible for the reduction in solidus temperature that promotes solidification cracking. In Hastealloy X, increasing Si and C content increases the solidification temperature range that created hot tearing during SLM (Tomus et al., 2017). On the contrary, solidification cracks in Laser based DED of IN625 alloys were reported to reduce by increasing the Ti content up to 3-5% (Hu et al., 2017). Under equilibrium, IN 625 alloy melts at 1347°C and solidification begins at 1257° C, thus it have a solidification temperature range of 90°C. However, during additive manufacturing, the non-equilibrium conditions increases the solidification temperature range to 210°C which leads to solidification cracking. The element Ti has the ability reduce the solidification temperature range and form deposits with no cracks. Rickenbacher et al. (2013) employed hot isostatic pressing (HIP) treatment post SLM of IN738LC alloy to eliminate these hot cracking defects. However, Tillmann et al. (2017) reported that it would be impossible to obtain fully dense microstructure after HIP on SLMed IN718 alloy because of the hindrance due to entrapped inert gases. Hence, it was suggested that HIP on the alloy SLMed under vacuum (Tillmann et al., 2017) or optimization of alloying elements (Harrison et al., 2015) before SLM may yield nearly full dense microstructure. At temperatures of around 0.5 to 0.7 times the melting point of the material, some austenitic alloys experience a ductility drop. If these materials are processed in this temperature range, it leads to a solid state intergranular cracking called DDC (Ramirez et al., 2006).

Post Processing Microstructure

The as-built SLMed Ni-based superalloys generally contains elongated and equi-axed cellular structures with γ matrix and Laves phase in the cell boundaries. On aging the material in the temperature range of 600 – 900 °C γ¢ and γ¢¢ would precipitate out to aid in strengthening of the material. They can be distinguished by their shape. The γ¢ - $Ni_3(Al,Ti)$ were round shaped while γ¢¢ (Ni_3Nb) were disc shaped structure. As γ¢¢ (bcc) is a metastable structure it will transform into δ (Ni_3Nb) particles with orthorhombic crystal structure above 650°C (V. A. Popovich et al., 2017b). Deng et al. (2018) studied the influence of different post-heat treatments on the SLMed IN718 alloy. After direct aging (720°C/8h/FC (furnace cooled) at 50°C/h to 620°C + 620°C/8h/AC), the morphology of as-built microstructure remains same with γ dendrites and Laves at the interdendrites. The aging temperature was too low to dissolve Laves phase but some γ¢¢ precipitated out in the matrix. The melt pool tracks and cellular boundaries almost disappears after solution treatment (980°C/1h/WC (water cooled)) and aging. Also, decrease in size of the Laves phase due to partial dissolution and presence of needle-like δ particles were observed in the microstructure. On increasing the homogenization temperature (1080°C/1h/WC) and followed by aging leads to segregation of Nb from Laves and get homogeneously distributed in to the matrix. It leads to the precipitation of more γ¢¢ in the matrix. No δ particles was observed after homogenization and aging, as 1080°C was beyond the δ solvus temperature. On solution treating the SLMed material in between homogenization and aging leads to precipitation of δ-particles only at the grain boundaries with some γ¢¢ in the matrix. Thus the γ¢¢ increases with the increasing temperature and increasing holding time during the heat treatment. Presence of γ¢ and γ¢¢ particles were observed after heat treatment (Solution + aging) in EBMed IN718 alloy (Sun et al., 2018). On heat treating the as built IN625 alloy to 870°C, the δ phase rich in Nb and Mo were observed to precipitate at the interdendritic region after 0.5 hours. Further increasing the duration of heating increased precipitation of needle-like δ phases both inside the dendritic region and interdendritic region. Globular MC where also observed in the matrix (dendritic region) (Zhang et al., 2018). Precipitation of MC after heat treatment formed zigzag grain boundaries in SLMed IN625 alloy (Li et al., 2015). The zigzag grain boundaries are important for the alloy to improve its ductility, toughness and performance. Tucho et al. (2017) and, Tucho & Hansen (2019) studied the effect of solution treatment independently without double aging treatment on SLMed IN718 alloy. Solution treatment at 1100°C and 1250°C performed for 1h and 7h at each temperatures. The Laves phase got dissolved with the increase in temperature and time and at 1250 °C for 7h, Laves phases were completely dissolved. At heat treatment above 1150 °C, random grain growth accompanied by dislocation annihilation and twinning occurred in SLMed IN 625 alloy (Li, White, et al., 2017). Divya et al. (2016) reported 5 nm sized γ¢ precipitates dispersed in the γ matrix in as-built condition in SLMed CM247LC alloy. After heat treatment at 1230 °C/2h/AC, 500 nm sized γ¢ precipitates in the γ matrix and much larger γ¢ precipitates observed in the grain boundaries. Keller et al. (2017) predicted that the micro-segregation in SLMed IN 625 alloy depends on the thermodynamic driving force for nucleation of different phases. The thermodynamic driving forces for various secondary phases from γ matrix, at 870°C and 1150°C, were calculated in their work (Table. 1), ignoring kinetic obstacles and interfacial energy between the matrix and the segregating phase. Hence, at these temperatures, MC, BCC, δ, μ, and σ phases may get highly segregated and diffuse during long heat treatments. Thus the local thermodynamic conditions, alloy composition, temperature and holding time controls the volume fraction of these secondary phases.

Besides heat treatment, HIP at 1180 °C/3h at 150 MPa pressure and furnace cooling on SLMed IN718 alloy caused significant pore closure and grain growth (Popovich et al., 2017b). No Laves phases but NbC aligned along the interdendritic region was observed. The carbides randomly distributed in the as-built microstructures were observed to distribute along the grain boundaries after HIP. However, the columnar structure and equi-axed cellular structure in the as-built microstructure was preserved after HIP. Hence, HIP results in crack closure, microstructural homogenization and dissolution of metastable phases in the microstructure. Heat treatment (1065 °C/1 h/ AC + 760 °C/10 h/FC at 55 °C/h to 650 °C/8 h/AC) after HIP, contains microstructure of same grains size, as presence of NbC at boundaries

Table 1. Thermodynamic driving force for secondary phase formation at the stress relief (870°C) and homogenization temperature (1150°C) of Ni-based superalloy. Equilibrium phases are underlined.

Stress Relief Temperature 870°C		**Homogenization Temperature 1150°C**	
Phase	**$-\Delta G_{nuc}$ (kJ/mol)**	**Phase**	**$-\Delta G_{nuc}$ (kJ/mol)**
MC	20.5	MC	17.0
M_2C	15.6	M_2C	13.1
μ	8.0	M_6C	5.8
M_6C	7.9	BCC	3.3
BCC	6.3	μ	3.0
Σ	5.2	Laves	2.9
Laves	4.1	liquid	1.8
Δ	3.5	Σ	1.2
γ″	3.5	γ″	1.2
$M_{23}C_6$	3.4	Δ	1.1

Adapted from (Keller et al., 2017)

restricted the grain growth. Further, evenly distributed TiC, increased carbide density and δ – particles were observed in the microstructure.

MECHANICAL PROPERTIES

Tensile Properties

The as-built SLMed Ni-based superalloys containing anisotropic microstructure tend to express anisotropic mechanical behaviour. Ni et al., (2017) reported that the longitudinal samples showed lesser UTS of 1101 MPa and higher elongation of 24.5% than the transverse samples which exhibited UTS of 1167 MPa and elongation of 21.5%. On heat treating, the room temperature tensile test of SLMed IN625 alloy showed general trend of decreasing strength and increasing percentage elongation in the order of heat-treatment temperature; as-built, stress relief annealing, recrystallization annealing, solution treatment and HIP (Kreitcberg et al., 2017a). The as-built SLMed IN718 material exhibited lowest UTS (~1100 MPa) and highest percentage elongation (36%) while the direct aged materials exhibited highest UTS (1500 MPa) and lowest percentage elongation (15%) (Deng et al., 2018). The samples in other heat treated conditions such as solution treated + aged (SA), homogenized + aged (HA), and homogenized + solution treated + aged (HSA) samples exhibited UTS of approximately 1200 MPa and percentage elongation in the range of 20 – 25%. The high residual stress and dislocation density work hardened the as-built material. In aged materials, in addition to dislocations and residual stress, the presence of γ¢¢ must increase the strength while also increasing the local stresses at Laves phase and dislocation that reduces the ductility of the material. The residual stresses was reduced to 80% after SA and completely removed after HA and HSA treatments but increased with the fraction of γ¢¢ which was attributed to the difference in strength in these conditions. In HA/HSA conditions the cracks can cut through needle-like δ particles which created larger voids than the voids that initiated around Laves phases in AS and DA conditions. This fact attributed the difference in ductility in these conditions. In the work by Sui et al. (2019), the DA samples showed lesser UTS (1269 MPa) and elongation (15%) while HA sample heat treated at 1050°C/15min/WC + aged, exhibited higher UTS (1370 MPa) and elongation (22%). The HA sample which was treated at 1050°C/45min/WC + aged exhibited a UTS of 1341 MPa and 19% elongation. The reason for increased strength in HA samples was precipitation of γ¢¢ and γ¢ particles while the reason for increased ductility was due to the change in the shape of Laves phase from long chain or striped shaped structure to small globular precipitates which prevented easy crack initiation.

At high temperature (650°C) the SLMed IN718 alloy was tensile tested at both vacuum and air and found that the damage process was not influenced by oxides. Also the material exhibited anisotropic tensile properties and inter-dendritic phases was observed as crack initiation sites (Hilaire et al., 2019). The effect of heat treatment, HIP and HIP + heat treatment of SLMed IN718 alloy on the tensile properties were reported by Popovich et al. (2018). The HIP process improved the percentage elongation at the expense of strength while heat-treatment either before or after HIP increases the strength with the reduction in percentage elongation. Shot peening after heat treatment or HIP reduces percentage elongation to a greater extent, almost decreased by 5% (Farber et al., 2018). Chlebus et al. (2015) stated that solution treatment followed by double aging is necessary for SLMed IN718 alloy as it remarkably improve the mechanical properties of the material.

Kreitcberg et al. (2017a, 2017b) studied the tensile properties of SLMed IN625 alloy at room temperature and elevated temperature. The important difference between the tensile properties at room temperature and elevated temperature was that the change in fracture mode from transgranular to intergranular. At 760°C, the strength characteristics of SLMed IN625 alloy was equivalent to that of the wrought alloys but it showed lesser percentage elongation due to the presence of Laves phase in as-built material and MC in post heat treated materials at the grain boundaries. The percentage elongation of wrought material at room temperature is equivalent to that of elevated temperature elongation. On the other hand, in SLMed material, both in as-built condition and post heat treated conditions, the percentage elongation at the elevated temperature was lesser than that of the elongation at room temperature. The SLM material exhibit inferior elongation at elevated temperature, in fact, it is advantageous in creep aspect of the material. A fishbone diagram drawn with the factors contributing towards the mechanical properties of a laser additively manufactured sample is depicted in Figure. 3 and the comparison chart for tensile properties using different process parameters, heat treatment, scan strategy and orientations are presented in the Table. 2.

Fatigue and Fracture

The fatigue test on SLMed IN718 alloy performed on notched bend specimen in three different orientation. One with the loading plane perpendicular to the built-layers, another with loading plane in the plane of the built-layers and the third one with loading plane parallel to the built- layers. The material with loading plane perpendicular to build-layers exhibited highest surface roughness and least fatigue life (Konečná et al., 2016). The IN718 alloy which were solution treated, aged and HIPed after SLM and EBM, showed transgranular fatigue crack growth. The presence of second phase particles such as δ, TiN, and NbC did not influence the transgranular fracture (Gokuldoss et al., 2017). The effect of post-processing treatment

Figure 3. Fishbone diagram for mechanical properties in SLM

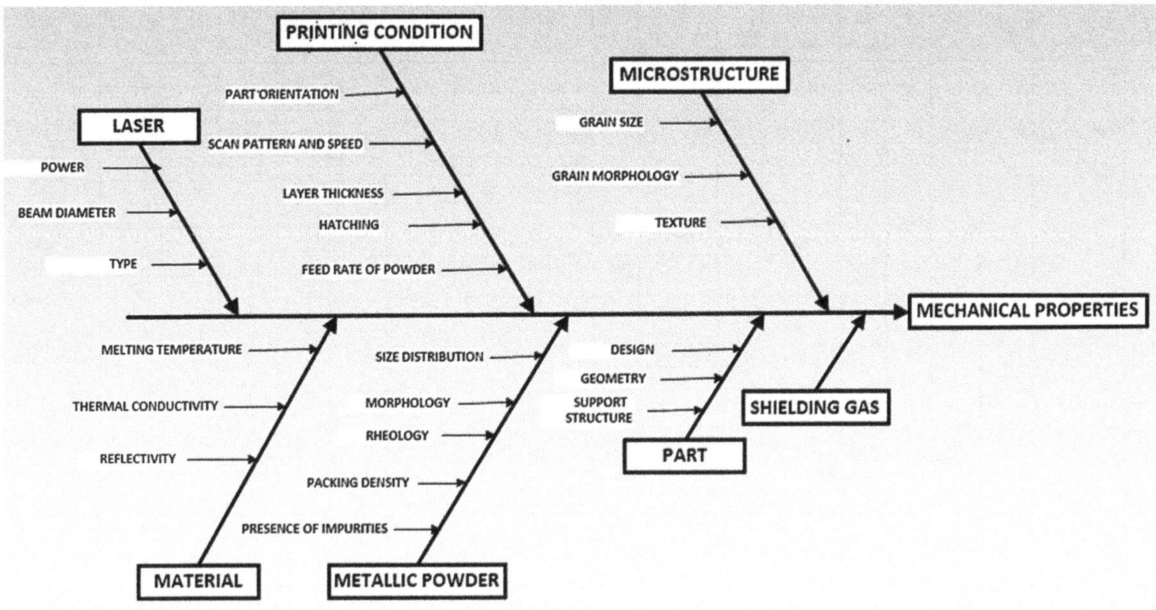

on Low cycle fatigue of SLMed IN718 was analysed by Aydinöz et al., (2016). The samples were tested after different post treatment, namely, solution annealing (1000°C/1h/AC), HIPed (1150°C/1000bar/4h/FC), solution annealed+aged, HIPed+aged, coated by arc-PVD+HIPed, arc–PVD+HIPed+aged. The aing was double aging; 720°C/8h/FC at 50°C/h to 621°C+621°C/8h/AC. Solutioned annealed sample at a temperature of 1000 °C for 1 hour and air cooled, showed lowest hysteresis width because of retained cellular structure and incomplete dissolution of precipitates at the grain boundaries restricts dislocation motion. HIPing and arc-PVD did not improve the fatigue properties. Aged samples exhibited least fatigue life due to the local stress concentrations around δ and γ¢¢ precipitates. Yoo et al. (2018) compared low cycle fatigue behaviour of SLMed IN718 after stress relieve annealing and different post-processing methods (homogenization and aging combinations). The samples were pre-strained to 1% then cyclically loaded for 1000 cycles at R = 0.05. The stress relieved samples restrict fatigue deformation by cellular boundaries that contain Ti and Nb carbides while the post processed samples restrict fatigue deformation by strain localization at grain boundaries and due to γ¢ and γ¢¢ precipitates.

The long fatigue crack growth (FCG) rate of SLMed IN718 alloys was comparable to the wrought alloys but the threshold stress intensity range ΔK_{th} was smaller (3 MPaÖm) than that of the conventional alloys (8-12 MPaÖm). The reason being the less B (boron) content of 32 ppm, finer grain structure with transgranular fracture and tensile residual stress that aided closure-free crack growth Konečná et al. (2016). There were two types of fracture surfaces observed during fatigue crack growth experiment of SLMed IN718 alloy after solution treatment and after HIP + solution treatment (Balachandramurthi et al., 2019). The solution treated SLM samples contains fine grain whereas the HIP + solution treated samples contain coarse grains. The fracture surfaces of solution treated samples exhibited striations. This is because, the crack-tip encounter more than one grain and the crack grows by simultaneous slip activity from two planes known as duplex slip. On the other hand the fracture surfaces of HIP+ solution treated samples exhibited planar facets. This was because, as the grain size was much larger than the crack-tip

Table 2. Tensile properties of SLMed Ni-based superalloys under different process parameters

S.No.	Alloy	Powder dia (μm)	Laser Power (W)	Laser Beam dia (μm)	Scan Speed (mm/s)	Hatch Spacing (μm)	Layer Thickness (μm)	Scan Strategy	Preheat	Heat Treatment	Test Temperature (°C)	Residual Stress (MPa)	Loading Axis/ Built Direction	YS (MPs)	UTS (MPa)	Elongation (%)	Ref
1	Nimonic-263	30	200	250	100	265	30	NA	250	As built	RT		0	818	1085	24	(Vilaro et al., 2012)
													90	653	860	70	
										600°C/8h AC			0	834	1136	29	
													90	697	910	52	
										1150°C/8h WQ+800°C/8h AC			0	843	1268	29	
													90	709	981	53	
2	IN738LC		50-100		50-500		20-60	Alternate x and y at 90°		As built	RT		0	1184	933	8.4	(Rickenbacher et al., 2013)
													90	1162	786	11.2	
											850		0	716	610	8	
													90	688	503	14.2	
3	IN 718	30	180		600	150	30	2x2 Island		As built	RT	100		804	1076	17	(Lu et al., 2015)
								3x3 Island				210		800	1075	21	
								5x5 Island				160		770	1064	22	
								7x7 Island				150		772	1064	25	
4	IN 718	25	400 (max)		1200				80	As Built	RT			849	1126	23	(Zhang et al., 2015)
										AB+SA(980°C1h/ AC+720°C 8h/ FC+620°C 8h/AC)				1084	1371	10	
										(AB+SA+1080°C 1.5h/AC) = HAS				1046	1371	12	
5	IN 718	30	100	180	85.7	160	50	Zig-zag, 90° rotated	2 pass	As built	RT		0	572	904	19	(Chlebus et al., 2015)
													45	643	991	13	
													45*45	590	954	20	
													90	723	1117	16	
										AB + 1100°C 1h/WC			0	1074	1320	19	
													45	1159	1377	8	
													45*45	1152	1371	15	
													90	1241	1457	14	
6	Hastealloy X	30	180		333	90	20	Meander Scan - 67° rotation		As built	RT			720	890	23	(Harrison et al., 2015)
											760			380	480	47	
	Modified Hastealloy X										RT			720	880	25	
											760			400	500	15	
7	IN718	30	175		620	120	30		200	As built (AB)	RT			580	845	19	(Aydinöz et al., 2016)
										AB+Solutionized (S) - (1000°C/1h/AC)				535	870	35	
										AB+S+aging (A) 720°C/8h/ FC at 50°C/h to 621°C+621°C/8h/AC				1240	1400	10	
										AB+HIP (H) 1150°C/1000bar/4h/ FC				430	875	40	
										AB+H+A				1100	1315	15	
										AB+ arc-PVD (P)+H				420	815	32	
										AB+P+H+A				1185	1300	15	
8	IN718	40	250	80	700	120	50	Horizontal scan 45° rotation		As built (AB)	RT			668	1011	22	(Popovich et al., 2017a)
			950	100	320	500	100							531	866	21	
9	IN718	40	250	80	700	120	50	Horizontal scan 45° rotation		AB + HT(850°C, 2h/AC)	RT			875	1153	17	(Popovich et al., 2017b)
			950	100	320	500	100							668	884	7	
		40	250	80	700	120	50	Horizontal scan 45° rotation		AB+HIP (1180°C, 3h, 150 MPa, FC)				645	1025	38	
			950	100	320	500	100							481	788	34	
		40	250	80	700	120	50	Horizontal scan 45° rotation		AB+HIP+HT				1145	1376	19	
			950	100	320	500	100							1065	1272	15	
10	IN 718	30	175		610	140	30			As built			0	646	1049	27	(Sufiiarov et al., 2017)
													90	609	949	32	
							50						0	807	1051	22	
													90	675	957	28	
11	IN 718	33	400		900	120	30	67° rotational scan		As built	RT		0	711	1110	25	(Ni et al., 2017)
													90	858	1167	22	

plasticity, the crack grows by single shear in the direction of primary slip. Thus the FCG doesn't seem to get influenced by the secondary phase particles but the grain size.

LEFM based fracture toughness test was carried out in SLMed IN 625 alloy with crack front along vertical (XZ) plane and horizontal (XY) plane using CT specimen by Hack et al. (2017). The average J_Q value for crack front in vertical plane was 408.6 kJ/m^2 (not valid) and for the crack front in horizontal plane it was 410.1 kJ/m^2 (valid J_{IC}). The J values of SLMed materials are higher than the rolled (247.3 kJ/m^3) and centrifugally cast materials (367.8 kJ/m^2).

Creep

In conventionally manufactured and heat treated Ni-based superalloys it was found that Cr content increased after creep (Kontis et al., 2019). The creep of the SLMed materials were reported superior to the conventional cast and wrought IN718 alloy. Pröbstle et al. (2016) tested creep of SLMed IN718 alloy after direct aging, solution heat treatment at 1000°C and 930°C and aging treatment. The aged materials exhibited superior creep properties while the samples solution treated at 1000 °C exhibited creep behaviour similar to conventional alloys. Rickenbacher et al. (2013) reported that the creep properties of SLMed IN738LC alloys were inferior to the cast material. Cho et al. (2018) developed a 3D woven, Ni-20Cr-3Ti-2Al (wt.%), Ni superalloy manufactured by SLM, which exhibited higher creep strain rate when compared to the bulk solid material at 850 °C and at a stress of 5 MPa but with an equivalent creep exponent.

Corrosion

The corrosion properties also influenced by the orientation of the built sample. The I_{corr} value decreased from 0.015 μA cm^2 to 0.006 μA cm^2 as the built orientation changes from 0° to 45° due to the changes in grain boundary area which contains second phase particles (Du et al., 2019). Li et al. (2018) stated that, after heat treatment the large non-precipitated grain boundaries coupled with δ particles to form a galvanic cell of large cathode and small anode. This galvanic cell accelerate the corrosion. The fatigue pre-cracked SLMed IN 625 CT specimen was tested in a 3.5% NaCl solution at constant displacement mode by (Hack et al., 2017). The reduction in load was observed as the indication for crack growth. The specimens loaded at 77 MPa Öm exhibited no reduction in load for 600-700 hours.

CONCLUSION

The chapter discussed briefly about the advancements in additive manufacturing of Ni-based superalloys which are used in aero-engines. Among the different AM methods, SLM was the one which had been well studied in the recent times. The influence of pre-alloyed powder, process parameters, and scanning strategy on the microstructure and mechanical properties in as-built condition, HIPed condition and after different heat treatments were studied in-depth. Heat treatment is necessary in SLMed Ni-based superalloys to tailor the microstructure for specific applications. The presence of Laves and δ phase and their orientation with respect to the built direction influences the fracture, fatigue and corrosion properties.

FUNDING INFORMATION

This research received no specific grant from any funding agency in the public, commercial, or not-for-profit sectors.

REFERENCES

Abioye, T. E., Folkes, J., & Clare, A. T. (2013). A parametric study of Inconel 625 wire laser deposition. *Journal of Materials Processing Technology*, *213*(12), 2145–2151. doi:10.1016/j.jmatprotec.2013.06.007

Acharya, R., Sharon, J. A., & Staroselsky, A. (2017). Prediction of microstructure in laser powder bed fusion process. *Acta Materialia*, *124*, 360–371. doi:10.1016/j.actamat.2016.11.018

Ahmad, B., van der Veen, S. O., Fitzpatrick, M. E., & Guo, H. (2018). Residual stress evaluation in selective-laser-melting additively manufactured titanium (Ti-6Al-4V) and inconel 718 using the contour method and numerical simulation. *Additive Manufacturing*, *22*, 571–582. doi:10.1016/j.addma.2018.06.002

Akca, E., & Gursel, A. (2015). A Review on Superalloys and IN718 Nickel-Based INCONEL Superalloy. *Periodicals of Engineering and Natural Sciences*, *3*(1), 15–27. doi:10.21533/pen.v3i1.43

Arısoy, Y. M., Criales, L. E., & Özel, T. (2019). Modeling and simulation of thermal field and solidification in laser powder bed fusion of nickel alloy IN625. *Optics & Laser Technology*, *109*, 278–292. doi:10.1016/j.optlastec.2018.08.016

Aydinöz, M. E., Brenne, F., Schaper, M., Schaak, C., Tillmann, W., Nellesen, J., & Niendorf, T. (2016). On the microstructural and mechanical properties of post-treated additively manufactured Inconel 718 superalloy under quasi-static and cyclic loading. *Materials Science and Engineering A*, *669*, 246–258. doi:10.1016/j.msea.2016.05.089

Balachandramurthi, A. R., Moverare, J., Dixit, N., Deng, D., & Pederson, R. (2019). Microstructural influence on fatigue crack propagation during high cycle fatigue testing of additively manufactured Alloy 718. *Materials Characterization*, *149*, 82–94. doi:10.1016/j.matchar.2019.01.018

Barros, R., Silva, F. J. G., Gouveia, R. M., Saboori, A., Marchese, G., Biamino, S., Salmi, A., & Atzeni, E. (2019). *Stress Analysis Before and After Heat Treatment*. Academic Press.

Carroll, B. E., Otis, R. A., Paul, J., Suh, J., Dillon, R. P., Shapiro, A. A., Hofmann, D. C., Liu, Z., & Beese, A. M. (2016). Functionally graded material of 304L stainless steel and inconel 625 fabricated by directed energy deposition : Characterization and thermodynamic modeling. *Acta Materialia*, *108*, 46–54. doi:10.1016/j.actamat.2016.02.019

Carter, L. N., Martin, C., Withers, P. J., & Attallah, M. M. (2014). The influence of the laser scan strategy on grain structure and cracking behaviour in SLM powder-bed fabricated nickel superalloy. *Journal of Alloys and Compounds*, *615*, 338–347. doi:10.1016/j.jallcom.2014.06.172

Catchpole-Smith, S., Aboulkhair, N., Parry, L., Tuck, C., Ashcroft, I. A., & Clare, A. (2017). Fractal scan strategies for selective laser melting of 'unweldable' nickel superalloys. *Additive Manufacturing*, *15*, 113–122. doi:10.1016/j.addma.2017.02.002

Chlebus, E., Gruber, K., Kuźnicka, B., Kurzac, J., & Kurzynowski, T. (2015). Effect of heat treatment on the microstructure and mechanical properties of Inconel 718 processed by selective laser melting. *Materials Science and Engineering A*, *639*, 647–655. doi:10.1016/j.msea.2015.05.035

Cho, H., Erdeniz, D., Sharp, K. W., & Dunand, D. C. (2018). Experimental and modeling study of compressive creep in 3D-woven Ni-based superalloys. *Acta Materialia*, *155*, 236–244. doi:10.1016/j.actamat.2018.05.067

Choi, J.-P., Shin, G.-H., Yang, S., Yang, D.-Y., Lee, J.-S., Brochu, M., & Yu, J.-H. (2017). Densification and microstructural investigation of Inconel 718 parts fabricated by selective laser melting. *Powder Technology*, *310*, 60–66. doi:10.1016/j.powtec.2017.01.030

Cloots, M., Uggowitzer, P. J., & Wegener, K. (2016). Investigations on the microstructure and crack formation of IN738LC samples processed by selective laser melting using Gaussian and doughnut profiles. *Materials & Design*, *89*, 770–784. doi:10.1016/j.matdes.2015.10.027

Criales, L. E., Arısoy, Y. M., Lane, B., Moylan, S., Donmez, A., & Özel, T. (2017a). Predictive modeling and optimization of multi-track processing for laser powder bed fusion of nickel alloy 625. *Additive Manufacturing*, *13*, 14–36. doi:10.1016/j.addma.2016.11.004

Criales, L. E., Arısoy, Y. M., Lane, B., Moylan, S., Donmez, A., & Özel, T. (2017b). Laser powder bed fusion of nickel alloy 625: Experimental investigations of effects of process parameters on melt pool size and shape with spatter analysis. *International Journal of Machine Tools & Manufacture*, *121*, 22–36. doi:10.1016/j.ijmachtools.2017.03.004

Deng, D., Peng, R. L., Brodin, H., & Moverare, J. (2018). Microstructure and mechanical properties of Inconel 718 produced by selective laser melting: Sample orientation dependence and effects of post heat treatments. *Materials Science and Engineering A*, *713*, 294–306. doi:10.1016/j.msea.2017.12.043

Divya, V. D., Muñoz-Moreno, R., Messé, O. M. D. M., Barnard, J. S., Baker, S., Illston, T., & Stone, H. J. (2016). Microstructure of selective laser melted CM247LC nickel-based superalloy and its evolution through heat treatment. *Materials Characterization*, *114*, 62–74. doi:10.1016/j.matchar.2016.02.004

Du, D., Dong, A., Shu, D., Zhu, G., Sun, B., Li, X., & Lavernia, E. (2019). Influence of build orientation on microstructure, mechanical and corrosion behavior of Inconel 718 processed by selective laser melting. *Materials Science and Engineering A*, *760*, 469–480. doi:10.1016/j.msea.2019.05.013

Engeli, R., Etter, T., Hövel, S., & Wegener, K. (2016). Processability of different IN738LC powder batches by selective laser melting. *Journal of Materials Processing Technology*, *229*, 484–491. doi:10.1016/j.jmatprotec.2015.09.046

Farber, B., Small, K. A., Allen, C., Causton, R. J., Nichols, A., Simbolick, J., & Taheri, M. L. (2018). Correlation of mechanical properties to microstructure in Inconel 718 fabricated by Direct Metal Laser Sintering. *Materials Science and Engineering A*, *712*, 539–547. doi:10.1016/j.msea.2017.11.125

Gault, B., Raabe, D., Chauvet, E., Kontis, P., Eric, A. J., Vayre, B., Tassin, C., & Blandin, J. (2018). Hot cracking mechanism affecting a non-weldable Ni-based superalloy produced by selective electron Beam Melting. *Acta Materialia*, *142*, 82–94. doi:10.1016/j.actamat.2017.09.047

Gokuldoss, P. K., Kolla, S., Eckert, J., Gokuldoss, P. K., Kolla, S., & Eckert, J. (2017). Additive Manufacturing Processes: Selective Laser Melting, Electron Beam Melting and Binder Jetting—Selection Guidelines. *Materials (Basel)*, *10*(6), 672. doi:10.3390/ma10060672 PMID:28773031

Gonzalez, J. A., Mireles, J., Stafford, S. W., Perez, M. A., Terrazas, C. A., & Wicker, R. B. (2019). Characterization of Inconel 625 fabricated using powder-bed-based additive manufacturing technologies. *Journal of Materials Processing Technology*, *264*, 200–210. doi:10.1016/j.jmatprotec.2018.08.031

Hack, H., Link, R., Knudsen, E., Baker, B., & Olig, S. (2017). Mechanical properties of additive manufactured nickel alloy 625. *Additive Manufacturing*, *14*, 105–115. doi:10.1016/j.addma.2017.02.004

Harrison, N. J., Todd, I., & Mumtaz, K. (2015). Reduction of micro-cracking in nickel superalloys processed by Selective Laser Melting: A fundamental alloy design approach. *Acta Materialia*, *94*, 59–68. doi:10.1016/j.actamat.2015.04.035

Herzog, D., Seyda, V., Wycisk, E., & Emmelmann, C. (2016). Additive manufacturing of metals. *Acta Materialia*, *117*, 371–392. doi:10.1016/j.actamat.2016.07.019

Hilaire, A., Andrieu, E., & Wu, X. (2019). High-temperature mechanical properties of alloy 718 produced by laser powder bed fusion with different processing parameters. *Additive Manufacturing*, *26*, 147–160. doi:10.1016/j.addma.2019.01.012

Hu, Y. L., Lin, X., Yu, X. B., Xu, J. J., Lei, M., & Huang, W. D. (2017). Effect of Ti addition on cracking and microhardness of Inconel 625 during the laser solid forming processing. *Journal of Alloys and Compounds*, *711*, 267–277. doi:10.1016/j.jallcom.2017.03.355

Hunt, J., Derguti, F., & Todd, I. (2014). *Ironmaking & Steelmaking Processes, Products and Applications Selection of steels suitable for additive layer manufacturing Selection of steels suitable for additive layer manufacturing*. doi:10.1179/0301923314Z.000000000269

Jia, Q., & Gu, D. (2014a). Selective laser melting additive manufacturing of Inconel 718 superalloy parts: Densification, microstructure and properties. *Journal of Alloys and Compounds*, *585*, 713–721. doi:10.1016/j.jallcom.2013.09.171

Jia, Q., & Gu, D. (2014b). Selective laser melting additive manufactured Inconel 718 superalloy parts: High-temperature oxidation property and its mechanisms. *Optics & Laser Technology*, *62*, 161–171. doi:10.1016/j.optlastec.2014.03.008

Kay, A. L., & Couper, P. (2004). *Junkers aircraft and engines, 1913-1945*. Putnam Aeronautical Books.

Keller, T., Lindwall, G., Ghosh, S., Ma, L., Lane, B. M., Zhang, F., Kattner, U. R., Lass, E. A., Heigel, J. C., Idell, Y., Williams, M. E., Allen, A. J., Guyer, J. E., & Levine, L. E. (2017). Application of finite element, phase-field, and CALPHAD-based methods to additive manufacturing of Ni-based superalloys. *Acta Materialia*, *139*, 244–253. doi:10.1016/j.actamat.2017.05.003 PMID:29230094

Konečná, R., Kunz, L., Nicoletto, G., & Bača, A. (2016). Long fatigue crack growth in Inconel 718 produced by selective laser melting. *International Journal of Fatigue*, *92*, 499–506. doi:10.1016/j.ijfatigue.2016.03.012

Konečná, R., Nicoletto, G., Kunz, L., & Bača, A. (2016). Microstructure and directional fatigue behavior of Inconel 718 produced by selective laser melting. *Procedia Structural Integrity*, *2*, 2381–2388. doi:10.1016/j.prostr.2016.06.298

Kontis, P., Kostka, A., Raabe, D., & Gault, B. (2019). Influence of composition and precipitation evolution on damage at grain boundaries in a crept polycrystalline Ni-based superalloy. *Acta Materialia*, *166*, 158–167. doi:10.1016/j.actamat.2018.12.039

Kreitcberg, A., Brailovski, V., & Turenne, S. (2017a). Effect of heat treatment and hot isostatic pressing on the microstructure and mechanical properties of Inconel 625 alloy processed by laser powder bed fusion. *Materials Science and Engineering A*, *689*, 1–10. doi:10.1016/j.msea.2017.02.038

Kreitcberg, A., Brailovski, V., & Turenne, S. (2017b). Elevated temperature mechanical behavior of IN625 alloy processed by laser powder-bed fusion. *Materials Science and Engineering A*, *700*, 540–553. doi:10.1016/j.msea.2017.06.045

Kromm, A., Cabeza, S., Mishurova, T., Nadammal, N., Thiede, T., & Bruno, G. (2018)... *Residual Stresses in Selective Laser Melted Samples of a Nickel Based Superalloy.*, *6*, 259–264. doi:10.21741/9781945291890-41

Kundin, J., Mushongera, L., & Emmerich, H. (2015). Phase-field modeling of microstructure formation during rapid solidification in Inconel 718 superalloy. *Acta Materialia*, *95*, 343–356. doi:10.1016/j.actamat.2015.05.052

Li, C., Guo, Y. B., & Zhao, J. B. (2017). Interfacial phenomena and characteristics between the deposited material and substrate in selective laser melting Inconel 625. *Journal of Materials Processing Technology*, *243*, 269–281. doi:10.1016/j.jmatprotec.2016.12.033

Li, C., White, R., Fang, X. Y., Weaver, M., & Guo, Y. B. (2017). Microstructure evolution characteristics of Inconel 625 alloy from selective laser melting to heat treatment. *Materials Science and Engineering A*, *705*, 20–31. doi:10.1016/j.msea.2017.08.058

Li, H., Feng, S., Li, J., & Gong, J. (2018). Effect of heat treatment on the δ phase distribution and corrosion resistance of selective laser melting manufactured Inconel 718 superalloy. *Materials and Corrosion*, *69*(10), 1350–1354. doi:10.1002/maco.201810159

Li, S., Wei, Q., Shi, Y., Zhu, Z., & Zhang, D. (2015). Microstructure Characteristics of Inconel 625 Superalloy Manufactured by Selective Laser Melting. *Journal of Materials Science and Technology*, *31*(9), 946–952. doi:10.1016/j.jmst.2014.09.020

Lu, Y., Wu, S., Gan, Y., Huang, T., Yang, C., Junjie, L., & Lin, J. (2015). Study on the microstructure, mechanical property and residual stress of SLM Inconel-718 alloy manufactured by differing island scanning strategy. *Optics & Laser Technology*, *75*, 197–206. doi:10.1016/j.optlastec.2015.07.009

Mishurova, T., Cabeza, S., Thiede, T., Nadammal, N., Kromm, A., Klaus, M., Genzel, C., Haberland, C., & Bruno, G. (2018). The Influence of the Support Structure on Residual Stress and Distortion in SLM Inconel 718 Parts. *Metallurgical and Materials Transactions. A, Physical Metallurgy and Materials Science*, *49*(7), 3038–3046. Advance online publication. doi:10.100711661-018-4653-9

Moat, R. J., Pinkerton, A. J., Li, L., Withers, P. J., & Preuss, M. (2011). Residual stresses in laser direct metal deposited Waspaloy. *Materials Science and Engineering A*, *528*(6), 2288–2298. doi:10.1016/j.msea.2010.12.010

Nadammal, N., Cabeza, S., Mishurova, T., Thiede, T., Kromm, A., Seyfert, C., Farahbod, L., Haberland, C., Schneider, J. A., Portella, P. D., & Bruno, G. (2017). Effect of hatch length on the development of microstructure, texture and residual stresses in selective laser melted superalloy Inconel 718. *Materials & Design*, *134*, 139–150. doi:10.1016/j.matdes.2017.08.049

Nguyen, Q. B., Nai, M. L. S., Zhu, Z., Sun, C. N., Wei, J., & Zhou, W. (2017). Characteristics of Inconel Powders for Powder-Bed Additive Manufacturing. *Engineering*, *3*(5), 695–700. doi:10.1016/J.ENG.2017.05.012

Ni, M., Chen, C., Wang, X., Wang, P., Li, R., Zhang, X., & Zhou, K. (2017). Anisotropic tensile behavior of in situ precipitation strengthened Inconel 718 fabricated by additive manufacturing. *Materials Science and Engineering A*, *701*, 344–351. doi:10.1016/j.msea.2017.06.098

Nie, P., Ojo, O. A., & Li, Z. (2014). Numerical modeling of microstructure evolution during laser additive manufacturing of a nickel-based superalloy. *Acta Materialia*, *77*, 85–95. doi:10.1016/j.actamat.2014.05.039

Özel, T., Shaurya, A., Altay, A., & Yang, L. (2018). Process monitoring of meltpool and spatter for temporal-spatial modeling of laser powder bed fusion process. *Procedia CIRP*, *74*, 102–106. doi:10.1016/j.procir.2018.08.049

Pollock, T. M., Rene, N., Rene, N., & Rene, N. (2006). Nickel-Based Superalloys for Advanced Turbine Engines : Chemistry, Microstructure, and Properties. *Journal of Propulsion and Power*, *22*(2), 361–374. doi:10.2514/1.18239

Popovich, A. A., Sufiiarov, Vs., Borisov, E., Polozov, I., & Masaylo, D. (2018). Design and manufacturing of tailored microstructure with selecive laser melting. *Materials Physics and Mechanics*, *38*, 1–10. doi:10.18720/MPM.3812018_1

Popovich, V. A., Borisov, E. V., Popovich, A. A., Sufiiarov, V. S., Masaylo, D. V., & Alzina, L. (2017a). Functionally graded Inconel 718 processed by additive manufacturing: Crystallographic texture, anisotropy of microstructure and mechanical properties. *Materials & Design*, *114*, 441–449. doi:10.1016/j.matdes.2016.10.075

Popovich, V. A., Borisov, E. V., Popovich, A. A., Sufiiarov, V. S., Masaylo, D. V., & Alzina, L. (2017b). Impact of heat treatment on mechanical behaviour of Inconel 718 processed with tailored microstructure by selective laser melting. *Materials & Design*, *131*, 12–22. doi:10.1016/j.matdes.2017.05.065

Pröbstle, M., Neumeier, S., Hopfenmüller, J., Freund, L. P., Niendorf, T., Schwarze, D., & Göken, M. (2016). Superior creep strength of a nickel-based superalloy produced by selective laser melting. *Materials Science and Engineering A*, *674*, 299–307. doi:10.1016/j.msea.2016.07.061

Raghavan, N., Simunovic, S., Dehoff, R., Plotkowski, A., Turner, J., Kirka, M., & Babu, S. (2017). Localized melt-scan strategy for site speci fi c control of grain size and primary dendrite arm spacing in electron beam additive. *Acta Materialia*, *140*, 375–387. doi:10.1016/j.actamat.2017.08.038

Ramirez, A. J., Sowards, J. W., & Lippold, J. C. (2006). Improving the ductility-dip cracking resistance of Ni-base alloys. *Journal of Materials Processing Technology*, *179*(1–3), 212–218. doi:10.1016/j.jmatprotec.2006.03.095

Rickenbacher, L., Etter, T., Hövel, S., & Wegener, K. (2013). High temperature material properties of IN738LC processed by selective laser melting (SLM) technology. *Rapid Prototyping Journal*, *19*(4), 282–290. doi:10.1108/13552541311323281

Roehling, T. T., Wu, S. S. Q., Khairallah, S. A., Roehling, J. D., Soezeri, S. S., Crumb, M. F., & Matthews, M. J. (2017). Modulating laser intensity pro fi le ellipticity for microstructural control during metal additive manufacturing. *Acta Materialia*, *128*, 197–206. doi:10.1016/j.actamat.2017.02.025

Sateesh, N. H., Kumar, G. C. M., & Prasad, K., C.K., S., & Vinod, A. R. (2014). Microstructure and Mechanical Characterization of Laser Sintered Inconel-625 Superalloy. *Procedia Materials Science*, *5*, 772–779. doi:10.1016/j.mspro.2014.07.327

Spierings, A. B., Herres, N., & Levy, G. (2011). Influence of the particle size distribution on surface quality and mechanical properties in AM steel parts. *Rapid Prototyping Journal*, *17*(3), 195–202. doi:10.1108/13552541111124770

Spierings, A. B., Voegtlin, M., Bauer, T., & Wegener, K. (2016). Powder flowability characterisation methodology for powder-bed-based metal additive manufacturing. *Progress in Additive Manufacturing*, *1*(1-2), 9–20. doi:10.100740964-015-0001-4

Sufiiarov, V. S., Popovich, A. A., Borisov, E. V., Polozov, I. A., Masaylo, D. V., & Orlov, A. V. (2017). The Effect of Layer Thickness at Selective Laser Melting. *Procedia Engineering*, *174*, 126–134. doi:10.1016/j.proeng.2017.01.179

Sui, S., Tan, H., Chen, J., Zhong, C., Li, Z., Fan, W., Gasser, A., & Huang, W. (2019). The influence of Laves phases on the room temperature tensile properties of Inconel 718 fabricated by powder feeding laser additive manufacturing. *Acta Materialia*, *164*, 413–427. doi:10.1016/j.actamat.2018.10.032

Sun, C., Kirk, M., Li, M., Hattar, K., Wang, Y., Anderoglu, O., Valdez, J., Uberuaga, B. P., Dickerson, R., & Maloy, S. A. (2015). Microstructure, chemistry and mechanical properties of Ni-based superalloy Rene N4 under irradiation at room temperature. *Acta Materialia*, *95*, 357–365. doi:10.1016/j.actamat.2015.04.061

Sun, S., Hagihara, K., & Nakano, T. (2018). Effect of scanning strategy on texture formation in Ni-25 at. % Mo alloys. *Materials & Design*, *140*, 307–316. doi:10.1016/j.matdes.2017.11.060

Sun, S.-H., Koizumi, Y., Saito, T., Yamanaka, K., Li, Y.-P., Cui, Y., & Chiba, A. (2018). Electron beam additive manufacturing of Inconel 718 alloy rods: Impact of build direction on microstructure and high-temperature tensile properties. *Additive Manufacturing*, *23*(August), 457–470. doi:10.1016/j.addma.2018.08.017

Thomas, M., Baxter, G. J., & Todd, I. (2016). Normalised model-based processing diagrams for additive layer manufacture of engineering alloys. *Acta Materialia*, *108*, 26–35. doi:10.1016/j.actamat.2016.02.025

Tillmann, W., Schaak, C., Nellesen, J., Schaper, M., Aydinöz, M. E., & Hoyer, K.-P. (2017). Hot isostatic pressing of IN718 components manufactured by selective laser melting. *Additive Manufacturing*, *13*, 93–102. doi:10.1016/j.addma.2016.11.006

Tomus, D., Rometsch, P. A., Heilmaier, M., & Wu, X. (2017). Effect of minor alloying elements on crack-formation characteristics of Hastelloy-X manufactured by selective laser melting. *Additive Manufacturing*, *16*, 65–72. doi:10.1016/j.addma.2017.05.006

Tucho, W. M., Cuvillier, P., Sjolyst-Kverneland, A., & Hansen, V. (2017). Microstructure and hardness studies of Inconel 718 manufactured by selective laser melting before and after solution heat treatment. *Materials Science and Engineering A*, *689*, 220–232. doi:10.1016/j.msea.2017.02.062

Tucho, W. M., & Hansen, V. (2019). Characterization of SLM-fabricated Inconel 718 after solid solution and precipitation hardening heat treatments. *Journal of Materials Science*, *54*(1), 823–839. doi:10.100710853-018-2851-x

Tytko, D., Choi, P., Klo, J., Kostka, A., Inden, G., & Raabe, D. (2012). Microstructural evolution of a Ni-based superalloy (617B) at 700 ° C studied by electron microscopy and atom probe tomography. *Acta Materialia*, *60*(4), 1731–1740. doi:10.1016/j.actamat.2011.11.020

Vilaro, T., Colin, C., Bartout, J. D., Nazé, L., & Sennour, M. (2012). Microstructural and mechanical approaches of the selective laser melting process applied to a nickel-base superalloy. *Materials Science and Engineering A*, *534*, 446–451. doi:10.1016/j.msea.2011.11.092

Wei, H. L., Mazumder, J., & DebRoy, T. (2015). Evolution of solidification texture during additive manufacturing OPEN. *Scientific Reports*, *5*(1), 1–7. doi:10.1038rep16446 PMID:26553246

Xia, M., Gu, D., Yu, G., Dai, D., Chen, H., & Shi, Q. (2017). Porosity evolution and its thermodynamic mechanism of randomly packed powder-bed during selective laser melting of Inconel 718 alloy. *International Journal of Machine Tools & Manufacture*, *116*, 96–106. doi:10.1016/j.ijmachtools.2017.01.005

Yoo, Y. S. J., Book, T. A., Sangid, M. D., & Kacher, J. (2018). Identifying strain localization and dislocation processes in fatigued Inconel 718 manufactured from selective laser melting. *Materials Science and Engineering A*, *724*, 444–451. doi:10.1016/j.msea.2018.03.127

Zhang, D., Niu, W., Cao, X., & Liu, Z. (2015). Effect of standard heat treatment on the microstructure and mechanical properties of selective laser melting manufactured Inconel 718 superalloy. *Materials Science and Engineering A*, *644*, 32–40. doi:10.1016/j.msea.2015.06.021

Zhang, F., Levine, L. E., Allen, A. J., Stoudt, M. R., Lindwall, G., Lass, E. A., Williams, M. E., Idell, Y., & Campbell, C. E. (2018). Effect of heat treatment on the microstructural evolution of a nickel-based superalloy additive-manufactured by laser powder bed fusion. *Acta Materialia*, *152*, 200–214. doi:10.1016/j.actamat.2018.03.017 PMID:31080354

Zhang, Y. N., Cao, X., Wanjara, P., & Medraj, M. (2013). Oxide films in laser additive manufactured Inconel 718. *Acta Materialia*, *61*(17), 6562–6576. doi:10.1016/j.actamat.2013.07.039

Chapter 4
Insights on Laser Additive Manufacturing of Invar 36

Mostafa Yakout
https://orcid.org/0000-0002-0887-0217
McMaster University, Canada

M. A. Elbestawi
https://orcid.org/0000-0003-0982-6127
McMaster University, Canada

ABSTRACT

Recently, additive manufacturing (AM) became a promising technology to manufacture complex structures with acceptable mechanical properties. The laser powder-bed fusion (L-PBF) process is one of the most common AM processes that has been used for producing a wide variety of metals and composites. Invar 36 is an austenite iron-nickel alloy that has a very low coefficient of thermal expansion; therefore, it is a good candidate for the L-PBF process. This chapter covers the state-of-the-art for producing Invar 36 using the L-PBF process. The chapter aims at describing research insights of using metal AM techniques in producing Invar 36 components. Like most of nickel-based alloys, Invar 36 is weldable but hard-to-machine. However, there are some challenges while processing these alloys by laser. This chapter also covers the challenges of using the L-PBF process for producing nickel-based alloys. In addition, it reports the L-PBF conditions that could be used to produce fully dense Invar 36 components with mechanical properties comparable to the wrought Invar 36.

INTRODUCTION

Since its discovery in 1920, Invar 36 is being used in the aerospace and electronic devices industries for its dimensional stability (Yakout, Cadamuro, Elbestawi, & Veldhuis, 2017). Although Invar 36 is a weldable material, it faces the following challenges during laser processing or welding (Corbacho, Suárez, & Molleda, 1998; Yakout, Elbestawi, & Veldhuis, 2018d):

DOI: 10.4018/978-1-7998-4054-1.ch004

1. Internal cracking due to its large heat-affected zone (HAZ).
2. Columnar grain formation at high temperatures.
3. Presence of inclusions, impurities, and interdendritic precipitates.
4. Migration of a grain boundary due to heat transferred from later layers.

The properties of Invar 36 are sensitive to laser processing conditions in laser-based manufacturing processes. Most importantly, the coefficient of thermal expansion (CTE) of additive parts could also be different than that of the wrought Invar 36 (G. Li, Gao, Chen, Zhang, & Zeng, 2014; Tepylo, Huang, & Patnaik, 2019). Accordingly, studying the process-structure-property relationships in laser-based manufacturing processes has been of utmost importance (DebRoy et al., 2018; Fereiduni, Yakout, & Elbestawi, 2019; Gallmeyer et al., 2020; Herzog, Seyda, Wycisk, & Emmelmann, 2016; Smith et al., 2016; Yakout, Elbestawi, Wang, & Muizelaar, 2019a; Yakout, Elbestawi, & Veldhuis, 2018b, 2019b, 2020a; Yan et al., 2018). This chapter focused on the Laser powder-bed fusion (L-PBF) of Invar 36. L-PBF is a common additive manufacturing (AM) process that produces metal parts using a focused laser source (DebRoy et al., 2018; Herzog et al., 2016; Yakout, Elbestawi, & Veldhuis, 2018c). L-PBF of Invar 36 has been presented in the open literature with some challenges (Harrison, Todd, & Mumtaz, 2017; Khanna, Mistry, Rahman Rashid, & Gupta, 2019; Qiu, Adkins, & Attallah, 2016; Wei et al., 2020; Yakout et al., 2018b, 2018d, 2019b; Yakout, Elbestawi, Veldhuis, & Nangle-Smith, 2020b). The L-PBF process is associated with internal cracking, spatter generation, pore formation, vaporization of some of the alloying elements, and residual stress formation (Bai, Yang, Wang, & Zhang, 2017; Collur, Paul, & Debroy, 1987; Fereiduni et al., 2019; He, DebRoy, & Fuerschbach, 2003; Herzog et al., 2016; Khan & Debroy, 1984; Yakout & Elbestawi, 2017, 2019; Yakout, Elbestawi, & Veldhuis, 2018a).

This book chapter focuses on the challenges and manufacturing flaws during the L-PBF of Invar 36. The laser processing conditions that provide stable melting, crack-free microstructures, and minimum residual stresses are discussed. The chapter starts with an introduction to Invar 36 and nickel-based alloys with regards to their properties and applications. Then, it explains the laser processing conditions for Invar 36, including their influences on part quality. It describes the properties of Invar 36 parts, including mechanical, thermal, and surface properties. In addition, applications, promises, and future work roadmaps of Invar 36 produced using additive manufacture are explored at the end of this chapter.

BACKGROUND

Introduction to Invar 36

Since its discovery in 1920, Invar 36 is a nickel-iron alloy that has been used in aerospace and automotive applications because it does not expand at high temperatures (Guillaume, 1904; Thakar & Trivedi, 2017). Invar 36 is hard-to-machine but weldable; therefore, producing complex structures of Invar 36 on a large scale using subtractive manufacturing methods is very challenging (Khanna, Gandhi, Nakum, & Srivastava, 2018; Nakamura, 1976; Thakar & Trivedi, 2017). Applications of Invar 36 include, but not limited to (Nakamura, 1976; Ogawa & Koseki, 1986; Shiga, 1996):

1. Orbiting satellites and lasers components.
2. Ring laser gyroscopes.

3. Precision instruments and electronic devices.
4. Microwave devices and waveguide tubes.
5. Structural components.
6. Mechanical and timing devices.

Compared to Inconel 718 that has a CTE of 13 µm/m.°C, Invar 36 has a very low CTE of approximately 1.2 µm/m.°C at temperatures below the Curie temperature of 279 °C. Invar 36 loses its permanent magnetic field, leaves only the effect of the lattice thermal contribution, and becomes nonmagnetic above the Curie temperature. This is likely due to the anharmonic lattice vibrations above this temperature (Delgadillo, Gollisch, & Feder, 1994; Endoh, 1979; Nakamura, 1976). Figure 1 shows that the CTE drops to its lowest value at 36% of nickel. Any other nickel-iron alloy below or above 36% of nickel shows high CTE.

Figure 1. Influence of nickel percentage on CTE
(Yakout et al., 2017)

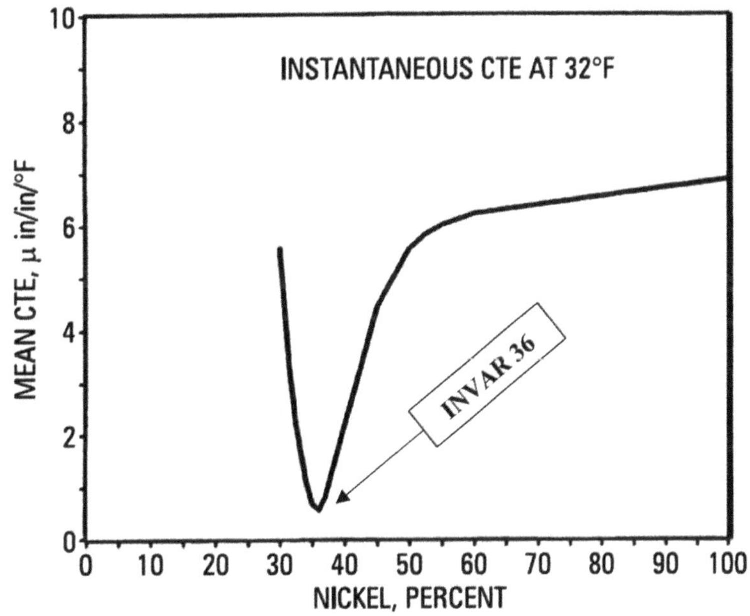

One of the main challenges in Invar 36 is maintaining its CTE after manufacturing. Any compositional inhomogeneities in the Invar 36 alloy will affect its mechanical and magnetic properties (Nakamura, 1976; Ogawa & Koseki, 1986; Shiga, 1996).

Additive Manufacturing of Invar 36

Metal AM techniques can be used for producing complex components and structures of Invar 36 (Harrison et al., 2017; Khanna et al., 2019; Qiu et al., 2016; Wei et al., 2020; Yakout et al., 2018b, 2018d, 2019b; Yakout et al., 2020b). The L-PBF process utilizes a laser source to fully melt predefined regions

Figure 2. (a) Melting chamber of a typical laser powder-bed fusion machine (Yakout et al., 2018d) and (b) 2D sketch of the melting process

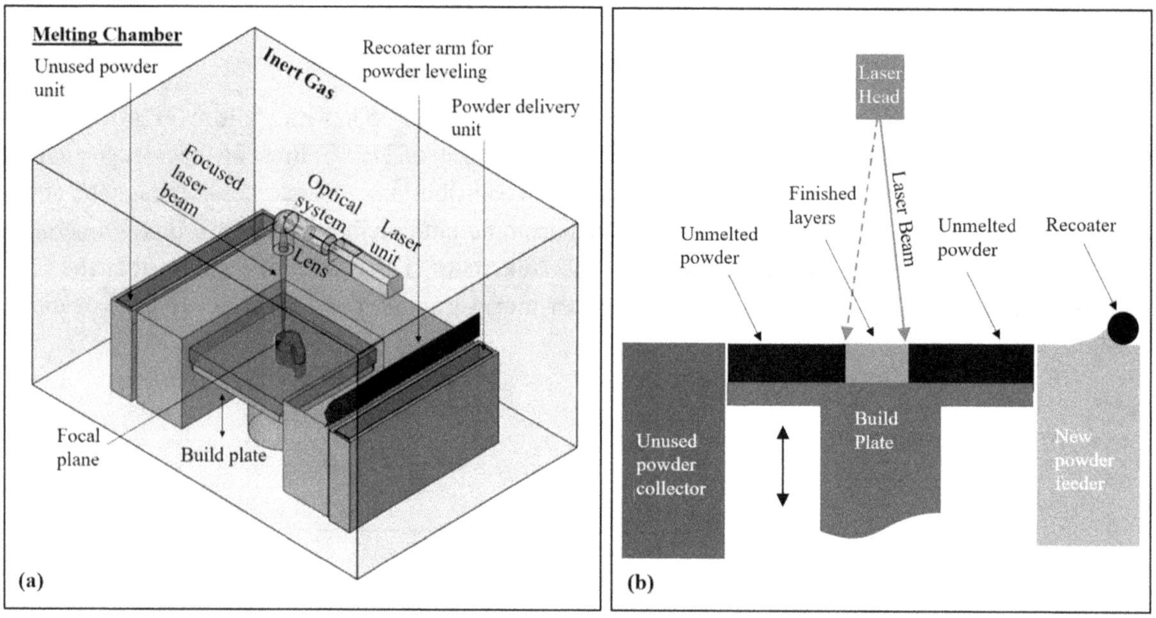

of powder layers to produce 3D functional products (ASTM International, 2015; H. Li, Chen, Tan, Song, & Feng, 2020; Yakout et al., 2018c). The melting chamber of a typical L-PBF machine consists of laser unit, optical lenses, powder delivery unit, powder recycling unit, recoater or leveling arm, and build platform (substrate), as shown in Figure 2a. The recoater builds a layer of metal powder; then the laser beam fully melts regions of each layer, as shown in Figure 2b. The melting process is fully shielded with an inert gas to avoid metal oxidation. Argon and nitrogen gases are commonly used in laser AM (Yakout et al., 2018d). Laser process parameters differ from one material to another material. These process parameters should be optimized for each new material before the actual production to avoid any manufacturing flaws during the L-PBF process. This area of research has been extensively discussed in the open literature (DebRoy et al., 2018; Fereiduni et al., 2019; Gallmeyer et al., 2020; Yakout & Elbestawi, 2017, 2019; Yakout et al., 2020b). The laser process parameters for L-PBF of Invar 36 will be discussed later in this chapter.

MATERIALS AND METHODS

The quality of the L-PBF parts depends on the following three main factors (Bartlett & Li, 2019; Bourell et al., 2017; Yang, Jamshidinia, Boulware, & Kelly, 2018):

1. Feedstock characteristics.
2. Laser processing conditions.
3. Post-processing procedures.

Feedstock Characteristics

In the L-PBF process, metal powders are used as the feedstock material in which their properties should be tested according to selected standards (ASTM International, 2014). These properties include:

1. Powder morphology.
2. Powder size distribution (PSD).
3. Chemical composition.
4. Flow characteristics.
5. Apparent density.

Morphology and Size Distribution of Feedstock

The PSD of feedstock can be measured using the sieving process, laser diffraction, light scattering, image analysis, or any other equivalent method (ASTM International, 2014, 2017). Measuring the PSD of the new virgin powder as received from the powder supplier is a common practice in metal AM. The PSD of the recycled powder after sieving should also be measured. Laser diffraction and light scattering are the most common techniques for measuring PSD in metal AM. A metal powder sample is spread in water within the path of a light source, then a photodetector collects the scattered light and converts it into electrical signals. These signals are then translated into a size distribution (ASTM International, 2017). The powder morphology can be tested using scanning electron microscopy (SEM), optical microscopy (OM), or any other equivalent imaging device.

A study presented the L-PBF of spherical particles, 15-45 μm in size, of Invar 36 powder. The PSD of both new and recycled powders of Invar 36 is shown in Figure 3a, and the SEM morphology of these powders is shown in Figure 3b (Yakout et al., 2018d). The PSD analysis did not show significant differences between both powders; however, the morphology analysis showed identical spherical particles in the virgin powder and agglomerated particles in the recycled powder. Another study showed the L-PBF of near-spherical particles, 25-50 μm in size, of Invar 36 powder, as shown in Figure 4a (Qiu et al., 2016). The cross-section of the powder shows some inhomogeneities in the powder, such as Fe-rich regions and micro pores, as shown in Figure 4b. It is recommended that the PSD and morphology of Invar 36 powder should be tested every time before use to avoid powder agglomeration (Cordova, Campos, & Tinga, 2019; Muñiz-Lerma, Nommeots-Nomm, Waters, & Brochu, 2018). Surface roughness and thermal properties of L-PBF parts are influenced by the PSD and powder morphology (Qiu et al., 2016; Yakout et al., 2018d).

Powder Chemistry

The following techniques could be used to identify the powder composition (ASTM International, 2017):

1. Energy-dispersive X-ray spectroscopy (EDS).
2. Inert gas fusion via determining the oxygen and hydrogen concentration.
3. X-ray fluorescence spectroscopy.
4. Atomic emission plasma spectrometry.
5. Combustion analysis via determining the carbon content.

Figure 3. (a) PSD, and (b) surface morphology of Invar 36 feedstock
(Yakout et al., 2018d)

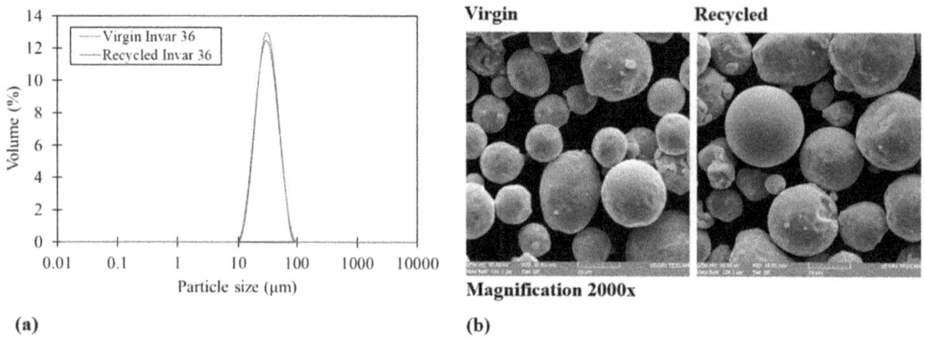

The chemical composition of Invar 36 powder was found to be within that of the wrought material (Yakout et al., 2018d). The wrought Invar 36 contains 35.5-36.5% Ni, 63.5-64.5% Fe, <0.5% Mn, <0.25% Si. Another study showed that the as-received Invar 36 powder has dark regions of 95% of Fe depleted in Ni, as shown in Figure 4b (Qiu et al., 2016). Any small changes in the concentration of nickel in Invar 36 would affect the material properties; therefore, the chemical composition of the powder should be tested before production (Nakamura, 1976; Ogawa & Koseki, 1986; Shiga, 1996). The powder flow is measured using either carney funnel or hall flowmeter funnel methods. A study showed the flowability and apparent density of Invar 36 powder for laser metal AM. Invar 36 powder is not free-flowing, which means that the particles stick together and form aggregates. The apparent density of the powder was relatively high compared to other AM powders because the powder is not free-flowing (Strauss & Stucky, 2016).

Figure 4. (a) Surface morphology of Invar 36 powder, and (b) its cross-section
(Qiu et al., 2016)

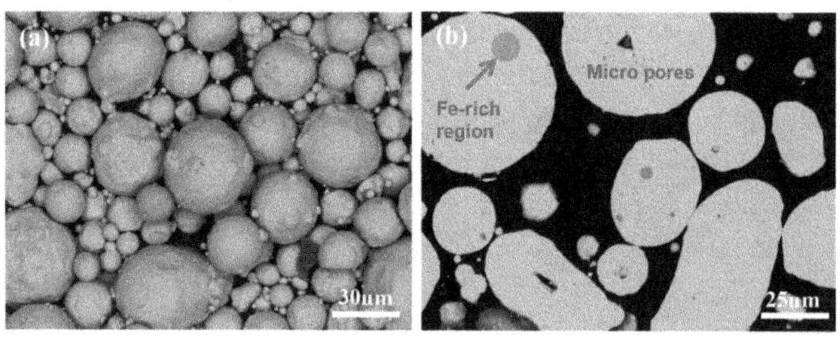

Laser Processing Conditions

The L-PBF processing conditions consist of four main categories: (i) laser process parameters, (ii) scanning strategies, and (iii) the use of supports, which will be discussed below in detail.

Laser Process Parameters

The machine-dependent parameters in L-PBF include laser type, laser wavelength, λ in (nm), beam spot size, ϕ in (mm), and continuous or pulsed laser operation (Xiong et al., 2019). The primary adjustable parameters include laser power, layer thickness, average scanning speed, stripe width, stripe overlap, and hatch spacing, as shown in Figure 5. These primary process parameters determine the laser energy density, E_v, using Equation (1).

$$E_v = \frac{P}{vht} = \frac{P}{\left(d/\theta\right)ht} \tag{1}$$

Where E_v is volumetric laser energy density in (J/mm^3), P is laser power in (W), v is laser scanning speed in (mm/s), d is point distance in (mm), θ is exposure time in (sec), h is hatch spacing in (mm), and t is layer thickness in (mm).

Figure 5. Primary process parameters of powder-bed fusion
(Yakout et al., 2019b)

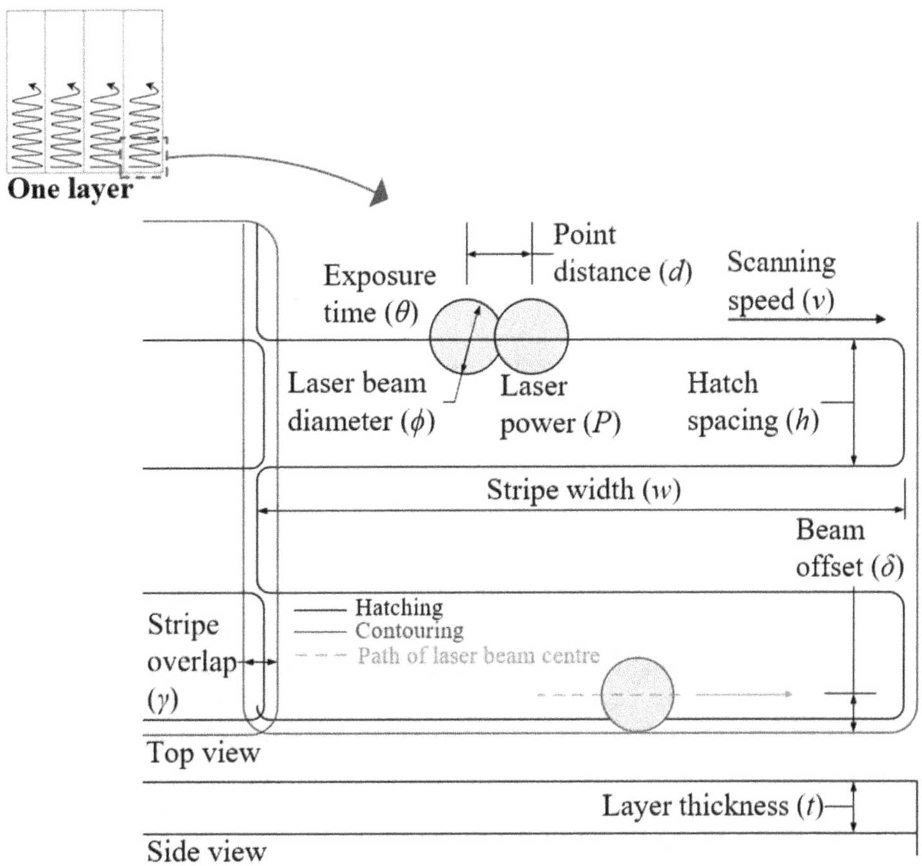

The secondary adjustable process parameters include beam offset, δ in (mm), and scanning angle between subsequent layers shown in Figure 6a. The primary and secondary parameters are those parameters used for the fusion of the core of parts produced. There are upskin parameters, contouring parameters, and downskin parameters, as shown in Figures 6a and 6b. The contouring parameters identify the laser process conditions that are used for finishing the outer boundary of every single layer. The upskin and downskin parameters identify the conditions for the fusion of the top and bottom surfaces of the part (Krishnan et al., 2014).

Figure 6. (a) Scanning pattern and (b) build regions
(Krishnan et al., 2014)

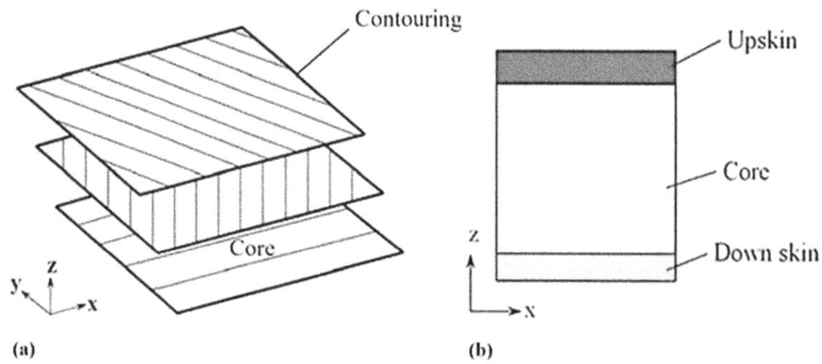

The laser energy density has been used as an important parameter for testing the material interaction with laser energy (Casalino, Mortello, & Campanelli, 2015; Mahmoudi et al., 2018; Pal, Lojen, Kokol, & Drstvensek, 2018; L.-z. Wang, Wang, & Hong, 2018; Y. Wang, Kamath, Voisin, & Li, 2018; Yakout et al., 2019b). It shows how much energy will be used in material volume to perform the melting process, but it does not include scanning pattern and material properties (Hitzler et al., 2017; Prashanth, Scudino, Maity, Das, & Eckert, 2017; Scipioni Bertoli, Wolfer, Matthews, Delplanque, & Schoenung, 2017). Table 1 presents the laser parameters reported for the L-PBF process of Invar 36. Several ranges of volumetric laser energy densities have been studied and reported in the open literature for the L-PBF of Invar 36 (Asgari, Salarian, Ma, Olubamiji, & Vlasea, 2018; Harrison et al., 2017; Khanna et al., 2019; Obidigbo, Tatman, & Gockel, 2019; Qiu et al., 2016; Wei et al., 2020; Yakout et al., 2018d, 2019b). The upskin, contouring, and downskin parameters were deactivated in most of these studies to eliminate the effect of post processing procedures. Although these studies showed a wide range of laser energy densities for Invar 36, one study showed two thresholds of laser energy densities to control the stability of the melting process (Yakout, 2019). The lowest limit defines the laser energy density required for transforming the parts produced from brittle to ductile, and the highest limit defines the laser energy density required for vaporizing some alloying elements and forming soot during the melting process. These thresholds are 52.1 J/mm^3 and 86.8 J/mm^3 for Invar 36. Brittle parts are produced at laser energy densities lower than the brittle-ductile transition energy density due to internal pores and cracks. Parts produced at or higher than the vaporization energy density have a lower amount of nickel. Both internal cracking and vaporization affect the functionality of Invar 36 parts.

Table 1. Laser processing conditions for powder bed fusion of Invar 36

Reference	Machine	P (W)	v (mm/s)	t (mm)	h (mm)	E_v (J/mm³)
Qiu et al. (2016)	Concept Laser M2	400	1800-4300	0.03	0.30	10-25
Strauss and Stucky (2016)	Phenix PXM	300	2500	0.03	0.09	44
Yakout et al. (2017)	EOS M290	150-300	700-2200	0.04	0.08-0.32	23-93
Harrison et al. (2017)	Renishaw AM125	180-200	333-1000	0.02	0.09	100-333
Yakout et al. (2018d)	EOS M280	200-300	600-1000	0.04	0.08-0.12	42-156
Asgari et al. (2018)	Renishaw AM400	250-400	N/A	0.03	N/A	119-190
Khanna et al. (2019)	Concept Laser M1	115-135	550-750	0.03-0.07	0.075-0.135	16-109
Obidigbo et al. (2019)	Custom Machine	200-300	1600-2600	N/A	0.08	N/A
Wei et al. (2020)	Renishaw AM250	200	161-1364	0.05	0.125	23-198

Other influencing parameters in the L-PBF process include inert gas and build plate preheating. Inert gases (e.g., argon or nitrogen) are used during the melting process to avoid oxidation. They also allow the removal of spatter particles and welding fumes (Ladewig, Schlick, Fisser, Schulze, & Glatzel, 2016). Some researchers have used argon gas shielding (Harrison et al., 2017; H. Li et al., 2020; Obidigbo et al., 2019; Qiu et al., 2016) and other researchers have used nitrogen gas shielding (Asgari et al., 2018; Yakout et al., 2017; Yakout et al., 2018d, 2019b) during the L-PBF of Invar 36. The influence of inert gas on spatter formation, oxidation, and process stability in L-PBF of Invar 36 has not been explored yet. On the other hand, build plate preheating affect residual stress formation (Zumofen, Beck, Kirchheim, & Dennig, 2018). Build plate preheating is commonly used in high carbon steels (Saewe, Gayer, Vogelpoth, & Schleifenbaum, 2019; Zumofen et al., 2018), tool steels (Mertens et al., 2016), and high thermal expansion alloys such as aluminum, copper, and titanium alloys (Saewe et al., 2019).

Scanning Strategies

Figure 7 shows the common scanning strategies in the L-PBF process. Meander and stripe strategies are commonly used for small areas, chessboard scanning is very common in large areas, and spiral scanning is preferable in irregular and complex cross-sections (Yakout et al., 2018a). The meander scanning is used when the laser path follows hatching vectors separated by the hatch distance, and the chessboard strategy (island scanning) is available for large cross-sections to avoid the formation of high residual stresses (Koutny et al., 2018). Figure 8 shows the hull and core and pre-sintering scanning strategies. The hull and core strategy is used for producing parts that have different process parameters within one layer; while the pre-sintering strategy can be used to scan each layer twice when more energy is needed for the adhesion between subsequent layers (Koutny et al., 2018). The island scanning strategy (Qiu et al., 2016; Wei et al., 2020), the meander scanning strategy (Asgari et al., 2018; Harrison et al., 2017; Strauss & Stucky, 2016), and the stripe scanning strategy (Obidigbo et al., 2019; Yakout et al., 2017; Yakout et al., 2018d) have been used in L-PBF of Invar 36.

Figure 7. (a) Meander, (b) stripe, (c) chessboard, and (b) spiral scanning strategies
(Yakout et al., 2018a)

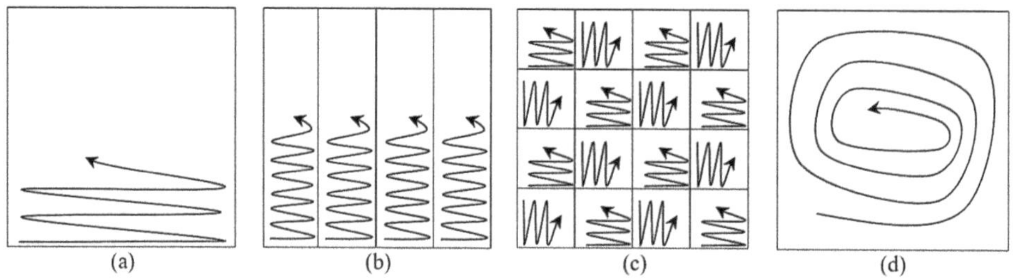

Build Supports

Several studies have explored the effect of supports and part orientation on the quality of parts produced (Calignano, 2018; Järvinen et al., 2014; Mfusi, Tshabalala, Popoola, & Mathe, 2018; Zhang et al., 2018). No data is available in the literature for examining supports in the L-PBF of Invar 36 because most of these preliminary studies included test coupons and samples that do not need supports. One study used supports for overhang cantilevers to quantify deflections of Invar 36. The influence of part supports on part deflection and mechanical properties of Invar 36 has not been explored yet (Yakout et al., 2019b).

Post-Processing Procedures

The L-PBF process is accompanied with several post-processing procedures, including air blasting (Brezinová et al., 2016), removing parts from the build plate (Yakout, 2019), surface finishing (Kaynak & Kitay, 2019), and heat treatment (Voznesenskaya, Kochuev, Chkalov, Kireev, & Morozov, 2020). Most researchers have used air blasting to clean Invar 36 parts and remove powder particles from the surface. They use electrical discharge machining to remove the parts off the substrate (Asgari et al., 2018; Har-

Figure 8. (a) Hull and core and (b) pre-sintering scanning strategies
(Koutny et al., 2018)

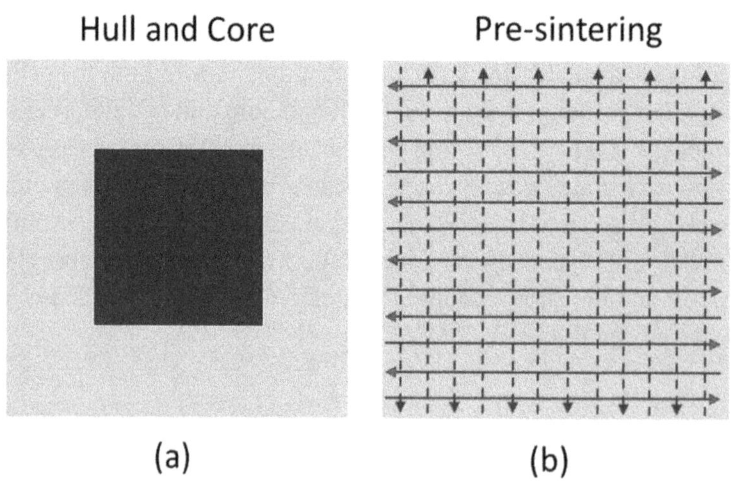

rison et al., 2017; H. Li et al., 2020; Obidigbo et al., 2019; Qiu et al., 2016; Yakout et al., 2017; Yakout et al., 2018d, 2019b). The effect of surface finishing and heat treatment procedures on the quality of Invar 36 parts has not yet been explored.

QUALITY OF INVAR 36 PARTS

The quality of L-PBF parts depends on the processing and post-processing conditions (Han et al., 2017; C. Li, Liu, Fang, & Guo, 2018; Schmutzler, Stiehl, & Zaeh, 2019). This section discusses the quality of Invar 36 parts and consists of four main subsections: (i) mechanical properties, (ii) thermal properties, (iii) material microstructure and metallurgy, and (iv) residual stresses.

Mechanical Properties

Tensile Mechanical Properties

Maintaining the mechanical properties of the as-built Invar 36 parts is challenging in the L-PBF process. Optimizing the process parameters will eliminate the formation of voids, inclusions, and internal cracks. A pioneer work on Invar 36 showed that varying the laser process parameters affects the tensile behavior of parts produced, as shown in Figure 9a (Yakout et al., 2019b). Some parts showed brittle fracture, while other parts showed ductile fracture based on the laser processing conditions. As shown in Figure 9b, producing parts at energy densities lower than the brittle-ductile transition energy density of Invar 36 (E_T = 52.1 J/mm^3) manifested brittle fracture due to voids formation, particularly lack-of-fusion porosity. Parts produced at or higher than the vaporization energy density (E_C = 86.8 J/mm^3) showed lower elongation and toughness. This is likely due to either the vaporization of some alloying elements and residual stress formation. It was recommended to use processing conditions that provide laser energy densities between E_T and E_C to avoid brittle fracture. Another study showed the effect of two post-processing procedures, heat treatment (HT) and hot isostatic pressing (HIP), on the mechanical properties of Invar 36 (Qiu et al., 2016). The tensile stress-strain results of Invar 36 parts produced both horizontally and vertically at two scanning speeds of 3200 mm/s and 1000 mm/s are shown in Figures 10a and 10b, respectively. It was found that parts produced at high scanning speeds manifest brittle fracture due to voids formation, and parts produced horizontally show high tensile properties compared to those produced vertically at high scanning speeds. Both HIP and HT post processing procedures did not improve the mechanical properties of parts produced. Samples produced horizontally at low scanning speeds showed similar mechanical properties to those produced horizontally at high scanning speeds. However, samples produced vertically at low scanning speeds showed extremely higher tensile properties than those produced vertically at high scanning speeds due to the absence of porosity at low scanning speeds. This means that the scanning speed affects the tensile properties in the vertical direction but does not affect it in the horizontal direction.

Figure 9. (a) Tensile test results of Invar 36 parts produced at various process parameters, and (b) the influence of laser energy density on Invar 36 toughness
(Yakout et al., 2019b)

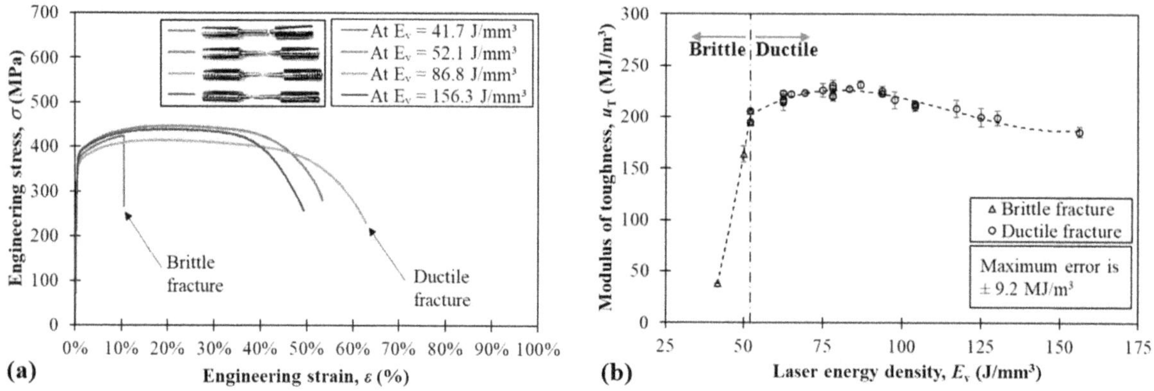

Figure 10. Stress-strain relationship for Invar 36 parts produced at a scanning speed of (a) 3200 mm/s, and (b) 1000 mm/s
(Qiu et al., 2016)

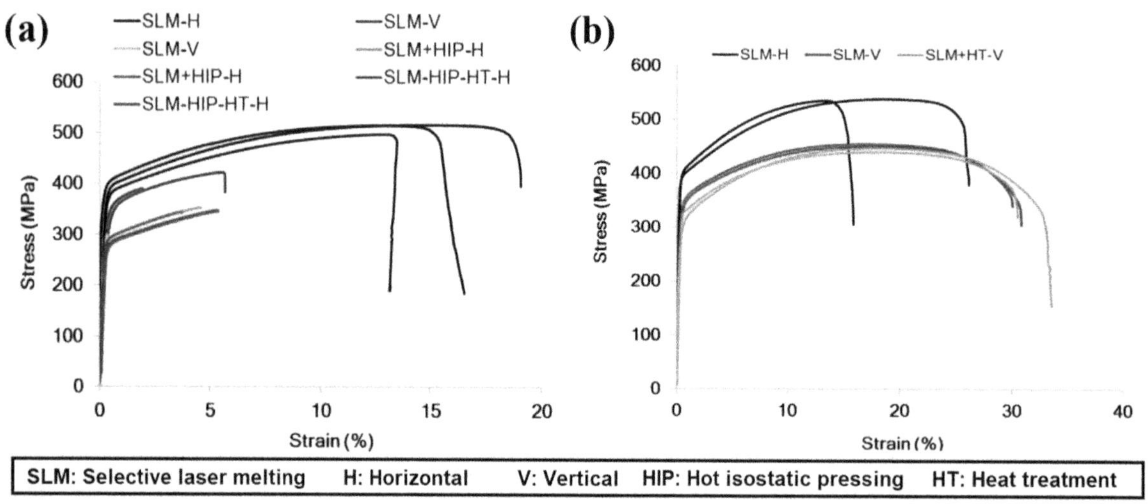

Vickers Hardness

On the other hand, the Vickers hardness of Invar 36 parts is affected by the laser process parameters in the L-PBF process (Wei et al., 2020). Increasing the exposure time or decreasing the point distance leads to a rise in the Vickers hardness of Invar 36, as shown in Figure 11a. Parts produced at low laser energy densities of 23.5-55.5 J/mm^3 showed low Vickers hardness of 0.5-1.25 GPa due to voids formation, as shown in Figure 11b. Raising the laser energy density from 55.5 to 99.2 J/mm^3 led to an increase in the Vickers hardness from 1.25 GPa to 1.86 GPa. The hardness of parts produced at 55.5-99.2 J/mm^3 is similar to that of the wrought Invar 36, which is 1.46 GPa. The Vickers hardness starts to decrease

Figure 11. Vickers hardness of Invar 36 as a function of (a) exposure time and point distance (PD), and (b) laser energy density
(Wei et al., 2020)

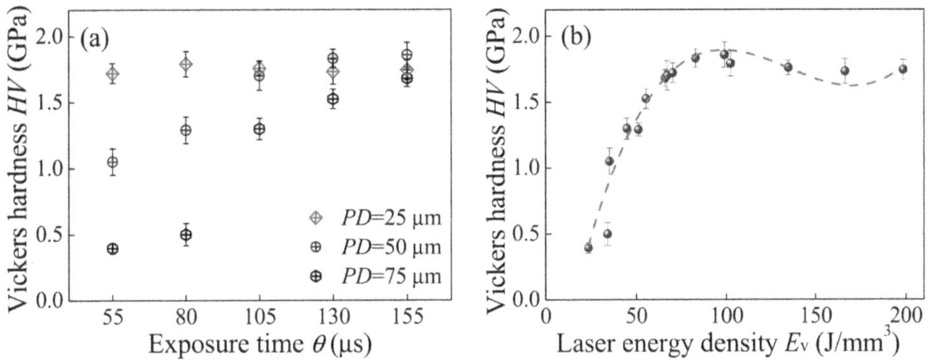

when using laser energy densities above 99.2 J/mm³, most likely because of the vaporization of some alloying elements and residual stress formation.

Thermal Properties

Invar 36 is commonly known for its unique thermal expansion property, so its thermal properties affect the application. A study showed that the linear thermal expansion properties of Invar 36 are isotropic, which means that thermal expansion of Invar 36 does not change with the orientation of the parts during the L-PBF process (Qiu et al., 2016). It was also found that the HIP process leads to a slight reduction in the CTE. This is attributed to that the HIP process leads to coarsening the α precipitates (Qiu et al., 2016). It was found in the literature that the HIP process causes an increase in the thermal conductivity (Chen et al., 2020), but the influence of the HIP on the CTE of AM parts is not explored yet. Another study showed that the CTE of L-PBF parts is lower than that of the wrought Invar 36 (Harrison et al., 2017). This is likely attributed to that the L-PBF parts contain internal pores that cause a reduction in the CTE (Al-Dabbagh, Al-Faluji, & Hashim, 2010). The L-PBF process produces Invar 36 parts with a CTE lower than that of the wrought Invar 36, as shown in Figure 12a (Yakout et al., 2018d). The laser energy density strongly affects the linear thermal expansion, as shown in Figure 12b. Parts produced at or lower than the vaporization energy density of Invar 36 (E_C = 86.8 J/mm³) show lower CTE. This is likely due to internal pores and cracks. However, the CTE of parts reduces above the vaporization energy density due to the vaporization of nickel. Any small changes in the nickel content of the alloy will affect its thermal expansion properties (Yakout et al., 2018d).

Material Microstructure and Metallurgy

Metallurgical properties of the L-PBF Invar 36 parts are reported in the literature (Asgari et al., 2018; Harrison et al., 2017; H. Li et al., 2020; Qiu et al., 2016; Yakout et al., 2018d). A study showed that the epitaxial growth of the grains in the build direction forms columnar structures, as shown in Figure 13. Two factors affect the epitaxial growth of the grains: (i) angle of grain growth with respect to the scan direction, and (ii) solute segregation (Harrison et al., 2017). Figure 14 shows the surface microstructures

Figure 12. (a) Curves of thermal expansion at several laser energy densities, and (b) dependency of thermal expansion of Invar 36 on laser energy density
(Yakout et al., 2018d)

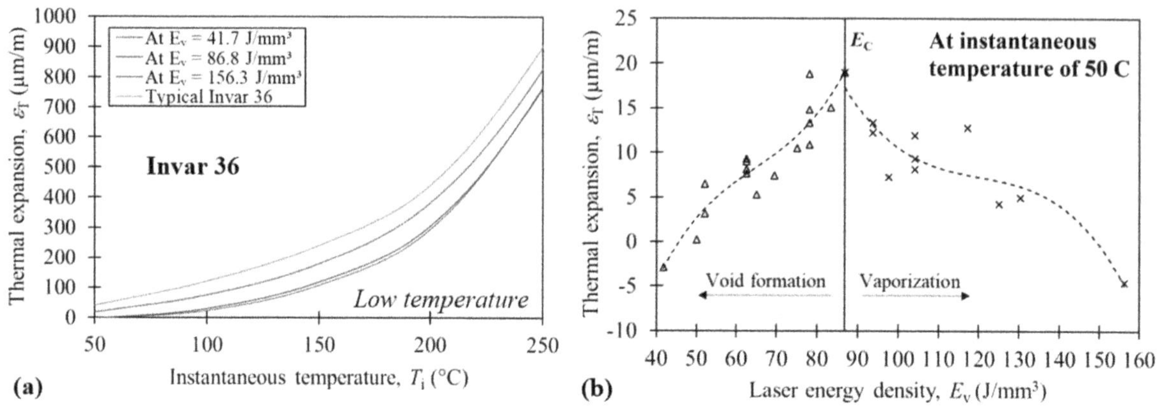

Figure 13. Micrograph of an etched Invar 36 sample showing columnar grains
(Harrison et al., 2017)

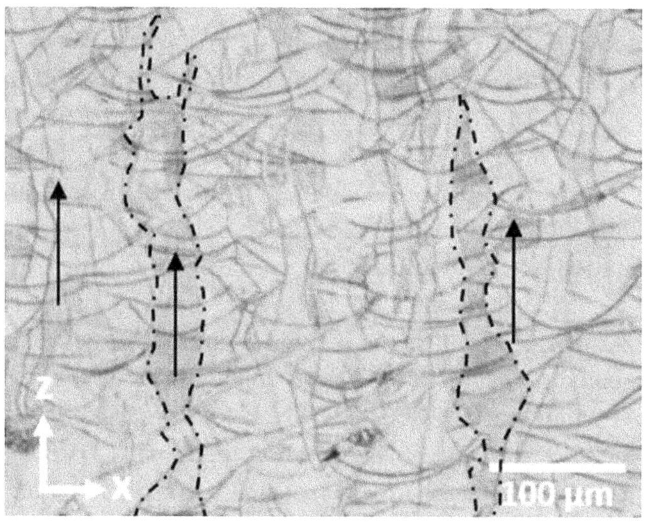

of parts built at various laser energy densities (Yakout et al., 2018d). It was found that low energy densities lead to the balling phenomenon, discontinuous beads, denudation zones, open pores, and micro cracks, as shown in Figures 14a, 14b, 14e, and 14f. However, high laser energy densities cause spatter formation, heat cracks, and keyhole pores, as shown in Figures 14d and 14h. It is recommended to use process parameters that provide a laser energy density of E_C = 86.8 J/mm^3 to maintain stable melting and minimize surface and internal defects, as shown in Figures 14c and 14g (Yakout et al., 2018d).

Residual Stress Formation

Thermal residual stresses are induced during the L-PBF process leading to part deflection (Han et al., 2017; Mercelis & Kruth, 2006; Van Hooreweder & Kruth, 2017). Two mechanisms affect residual stresses in the L-PBF process: (i) temperature gradient mechanism, and (ii) phase transformation mechanism. The first mechanism is based on the expansion and shrinkage in heating and cooling cycles. Phase transformation may also occur in one layer when melting new layers (Fereiduni et al., 2019; Kruth, Deckers, Yasa, & Wauthlé, 2012; Mercelis & Kruth, 2006; Yakout et al., 2020b). The amount of tensile residual stresses rapidly increases by increasing the amount of laser energy. In comparison with stainless steel and titanium alloys, Invar 36 showed the lowest amount of residual stress due to its lowest CTE (Yakout et al., 2020b). Thermal expansion and thermal diffusivity affect residual stress formation in the L-PBF process. Thermal expansion is a material property that shows how the material will expand when heat energy is applied. Thermal diffusivity represents the speed of heat conduction in the material. This phenomenon is very common in welding and forming processes (Birnbaum, Vikelic, & Lawrence Yao, 2010).

FUTURE RESEARCH DIRECTIONS

Nickel-based alloys are highly ductile materials with high melting temperatures, high corrosion resistance, and high magnetic properties at room temperature. Most nickel-based alloys are hard-to-machine but weldable. Therefore, there are lots of opportunities for the use of metal additive manufacturing techniques in processing nickel-based alloys to produce functional products and complex parts. The following research points are not covered in the literature yet for Invar 36:

1. Predicting residual stresses and part distortions in Invar 36 using finite element analysis.
2. Studying melt pool dynamics in L-PBF of Invar 36 using numerical and experimental methods.
3. Monitoring spatters and keyhole formation online during the L-PBF of Invar 36, as well as studying its influence on the metallurgy and microstructures of parts produced.
4. Highlighting more applications for complex components of Invar 36.
5. Mitigation of residual stresses in Invar 36 parts.
6. Eliminating voids formation and alloying elements vaporization during the L-PBF of Invar 36 by either process optimization or post processing

These research gaps need to be explored to produce functional components and products using Invar 36. The L-PBF of Invar 36 is found to be associated with voids formation, cracking, residual stress formation, vaporization of alloying elements, and soot formation. The following methods are recommended to avoid these manufacturing flaws:

1. Optimizing the laser process parameters to avoid reaching the vaporization laser energy density of Invar 36.
2. Avoiding building the parts on only one side of the build platform to ensure homogeneous distribution of the heat among the whole build platform.
3. Monitoring of voids and spatter formation during the L-PBF of Invar 36.
4. Heat treatment of Invar 36 parts after the L-PBF process to relief the internal residual stresses.

Figure 14. Microstructures of (a-d) top and (e-h) side surfaces of as-built Invar 36 parts (Yakout et al., 2018d)

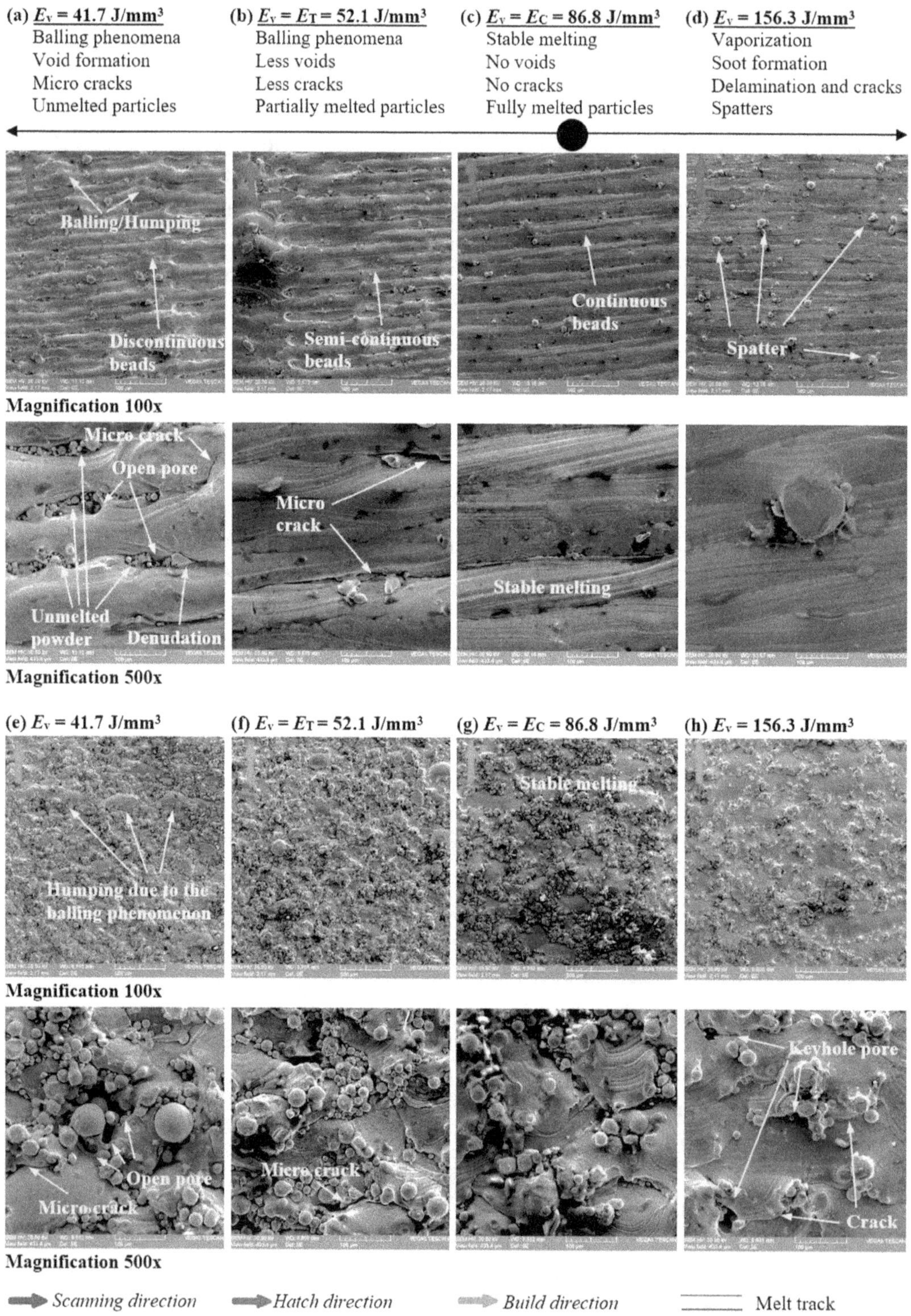

CONCLUSION

This book chapter summarized the process-structure-property relationships in laser powder-bed fusion of Invar 36. We presented the characteristics of Invar 36 powders (i.e., PSD, morphology, and composition). The processing conditions for the L-PBF of Invar 36 were also explained, showing the parameters that affect the mechanical and thermal properties of Invar 36 parts. Thresholds of laser energy density for the L-PBF of Invar 36 was also discussed. The thermal properties, mechanical properties, and metallurgy of Invar 36 parts were also explained. A brittle fracture was observed below a laser energy density of 52.1 J/mm^3 due to pores and discontinuous melt beads. A reduction in the CTE was observed above the 86.8 J/mm^3 due to vaporization of alloying elements. The following optimum process conditions are recommended for the laser powder-bed fusion of Invar 36:

1. Nitrogen gas shielding.
2. Stripe scanning strategy.
3. Laser process parameters that provide an energy density of 86.8 J/mm^3.
4. No preheating and less supports.
5. Invar 36 powder partciles 15-45 µm in size.

It is concluded that Invar 36 is a good candidate for additive manufacturing due to its unique thermal expansion property and its high weldability. Exploring various applications for Invar 36 components produced via laser additive manufacturing is recommended for future work.

ACKNOWLEDGMENT

This research received no specific grant from any funding agency in the public, commercial, or not-for-profit sectors.

REFERENCES

Al-Dabbagh, J. B., Al-Faluji, I. K., & Hashim, Y. B. (2010). *Negative Thermal Expansion in Ferromagnetic Fe-Ni Invar Alloys.* Paper presented at the Proceedings of ICSSST2010: 3rd International Conference on Solid State Science & Technology, Kuching, Sarawak, Malaysia.

Asgari, H., Salarian, M., Ma, H., Olubamiji, A., & Vlasea, M. (2018). On thermal expansion behavior of invar alloy fabricated by modulated laser powder bed fusion. *Materials & Design, 160*, 895–905. doi:10.1016/j.matdes.2018.10.025

ASTM International. (2014). *F3049 - 14 Standard guide for characterizing properties of metal powders used for additive manufacturing processes.* Author.

ASTM International. (2015). *ISO/ASTM 52900 Standard Terminology for Additive Manufacturing - General Principles - Terminology.* ASTM International.

ASTM International. (2017). *B822-17 Standard Test Method for Particle Size Distribution of Metal Powders and Related Compounds by Light Scattering*. Author.

Bai, Y., Yang, Y., Wang, D., & Zhang, M. (2017). Influence mechanism of parameters process and mechanical properties evolution mechanism of maraging steel 300 by selective laser melting. *Materials Science and Engineering A*, *703*, 116–123. doi:10.1016/j.msea.2017.06.033

Bartlett, J. L., & Li, X. (2019). An overview of residual stresses in metal powder bed fusion. *Additive Manufacturing*, *27*, 131–149. doi:10.1016/j.addma.2019.02.020

Birnbaum, A. J., Vikelic, S., & Lawrence Yao, Y. (2010). Advances in laser-induced plastic deformation processes. In J. Lawrence, J. Pou, D. K. Y. Low, & E. Toyserkani (Eds.), Advances in Laser Materials Processing (pp. 535-572). Woodhead Publishing.

Bourell, D., Kruth, J. P., Leu, M., Levy, G., Rosen, D., Beese, A. M., & Clare, A. (2017). Materials for additive manufacturing. *CIRP Annals*, *66*(2), 659–681. doi:10.1016/j.cirp.2017.05.009

Brezinová, J., Hudák, R., Guzanová, A., Draganovská, D., Ižaríková, G., & Koncz, J. (2016). Direct Metal Laser Sintering of Ti6Al4V for Biomedical Applications: Microstructure, Corrosion Properties, and Mechanical Treatment of Implants. *Metals*, *6*(7), 171. doi:10.3390/met6070171

Calignano, F. (2018). Investigation of the accuracy and roughness in the laser powder bed fusion process. *Virtual and Physical Prototyping*, *13*(2), 97–104. doi:10.1080/17452759.2018.1426368

Casalino, G., Mortello, M., & Campanelli, S. L. (2015). Ytterbium fiber laser welding of Ti6Al4V alloy. *Journal of Manufacturing Processes*, *20*, 250–256. doi:10.1016/j.jmapro.2015.07.003

Chen, J., Li, K., Wang, Y., Xing, L., Yu, C., Liu, H., Ma, J., Liu, W., & Shen, Z. (2020). The effect of hot isostatic pressing on thermal conductivity of additively manufactured pure tungsten. *International Journal of Refractory Metals & Hard Materials*, *87*, 105135. doi:10.1016/j.ijrmhm.2019.105135

Collur, M. M., Paul, A., & Debroy, T. (1987). Mechanism of alloying element vaporization during laser welding. *Metallurgical Transactions. B, Process Metallurgy*, *18*(4), 733–740. doi:10.1007/BF02672891

Corbacho, J. L., Suárez, J. C., & Molleda, F. (1998). Welding of invar Fe-36Ni alloy for tooling of composite materials. *Welding International*, *12*(12), 966–971. doi:10.1080/09507119809448543

Cordova, L., Campos, M., & Tinga, T. (2019). Revealing the Effects of Powder Reuse for Selective Laser Melting by Powder Characterization. *JOM*, *71*(3), 1062–1072. doi:10.100711837-018-3305-2

DebRoy, T., Wei, H. L., Zuback, J. S., Mukherjee, T., Elmer, J. W., Milewski, J. O., Beese, A. M., Wilson-Heid, A., De, A., & Zhang, W. (2018). Additive manufacturing of metallic components – Process, structure and properties. *Progress in Materials Science*, *92*, 112–224. doi:10.1016/j.pmatsci.2017.10.001

Delgadillo, I., Gollisch, H., & Feder, R. (1994). Lattice-vibration and spin-fluctuation effects on photoemission from ferromagnetic Ni. *Physical Review. B*, *50*(21), 15808–15813. doi:10.1103/PhysRevB.50.15808 PMID:9975948

Endoh, Y. (1979). Lattice dynamics in ferromagnetic invar alloys. *Journal of Magnetism and Magnetic Materials*, *10*(2), 177–182. doi:10.1016/0304-8853(79)90173-2

Fereiduni, E., Yakout, M., & Elbestawi, M. (2019). Laser-Based Additive Manufacturing of Lightweight Metal Matrix Composites. In B. AlMangour (Ed.), *Additive Manufacturing of Emerging Materials* (pp. 55–109). Springer International Publishing. doi:10.1007/978-3-319-91713-9_3

Gallmeyer, T. G., Moorthy, S., Kappes, B. B., Mills, M. J., Amin-Ahmadi, B., & Stebner, A. P. (2020). Knowledge of process-structure-property relationships to engineer better heat treatments for laser powder bed fusion additive manufactured Inconel 718. *Additive Manufacturing*, *31*, 100977. doi:10.1016/j.addma.2019.100977

Guillaume, C. E. (1904). Invar and Its Applications. *Nature, 71*(1832), 134-139.

Han, J., Yang, J., Yu, H., Yin, J., Gao, M., Wang, Z., & Zeng, X. (2017). Microstructure and mechanical property of selective laser melted Ti6Al4V dependence on laser energy density. *Rapid Prototyping Journal*, *23*(2), 217–226. doi:10.1108/RPJ-12-2015-0193

Harrison, N. J., Todd, I., & Mumtaz, K. (2017). Thermal expansion coefficients in Invar processed by selective laser melting. *Journal of Materials Science*, *52*(17), 10517–10525. doi:10.100710853-017-1169-4 PMID:32025047

He, X., DebRoy, T., & Fuerschbach, P. W. (2003). Alloying element vaporization during laser spot welding of stainless steel. *Journal of Physics. D, Applied Physics*, *36*(23), 3079–3088. doi:10.1088/0022-3727/36/23/033

Herzog, D., Seyda, V., Wycisk, E., & Emmelmann, C. (2016). Additive manufacturing of metals. *Acta Materialia*, *117*, 371–392. doi:10.1016/j.actamat.2016.07.019

Hitzler, L., Hirsch, J., Heine, B., Merkel, M., Hall, W., & Öchsner, A. (2017). On the Anisotropic Mechanical Properties of Selective Laser-Melted Stainless Steel. *Materials (Basel)*, *10*(10), 1136. doi:10.3390/ma10101136 PMID:28954426

Järvinen, J.-P., Matilainen, V., Li, X., Piili, H., Salminen, A., Mäkelä, I., & Nyrhilä, O. (2014). Characterization of Effect of Support Structures in Laser Additive Manufacturing of Stainless Steel. *Physics Procedia*, *56*, 72–81. doi:10.1016/j.phpro.2014.08.099

Kaynak, Y., & Kitay, O. (2019). The effect of post-processing operations on surface characteristics of 316L stainless steel produced by selective laser melting. *Additive Manufacturing*, *26*, 84–93. doi:10.1016/j.addma.2018.12.021

Khan, P. A. A., & Debroy, T. (1984). Alloying element vaporization and weld pool temperature during laser welding of AlSl 202 stainless steel. *Metallurgical Transactions. B, Process Metallurgy*, *15*(4), 641–644. doi:10.1007/BF02657284

Khanna, N., Gandhi, A., Nakum, B., & Srivastava, A. (2018). Optimization and analysis of surface roughness for INVAR-36 in end milling operations. *Materials Today: Proceedings*, *5*(2, Part 1), 5281–5288. doi:10.1016/j.matpr.2017.12.111

Khanna, N., Mistry, S., Rahman Rashid, R. A., & Gupta, M. K. (2019). Investigations on density and surface roughness characteristics during selective laser sintering of Invar-36 alloy. *Materials Research Express*, *6*(8), 086541. doi:10.1088/2053-1591/ab18bd

Koutny, D., Palousek, D., Pantelejev, L., Hoeller, C., Pichler, R., Tesicky, L., & Kaiser, J. (2018). Influence of Scanning Strategies on Processing of Aluminum Alloy EN AW 2618 Using Selective Laser Melting. *Materials (Basel, Switzerland)*, *11*(2), 298. doi:10.3390/ma11020298 PMID:29443912

Krishnan, M., Atzeni, E., Canali, R., Calignano, F., Manfredi, D., Ambrosio Elisa, P., & Iuliano, L. (2014). On the effect of process parameters on properties of AlSi10Mg parts produced by DMLS. *Rapid Prototyping Journal*, *20*(6), 449–458. doi:10.1108/RPJ-03-2013-0028

Kruth, J.-P., Deckers, J., Yasa, E., & Wauthlé, R. (2012). Assessing and comparing influencing factors of residual stresses in selective laser melting using a novel analysis method. *Proceedings of the Institution of Mechanical Engineers. Part B, Journal of Engineering Manufacture*, *226*(6), 980–991. doi:10.1177/0954405412437085

Ladewig, A., Schlick, G., Fisser, M., Schulze, V., & Glatzel, U. (2016). Influence of the shielding gas flow on the removal of process by-products in the selective laser melting process. *Additive Manufacturing*, *10*, 1–9. doi:10.1016/j.addma.2016.01.004

Li, C., Liu, Z. Y., Fang, X. Y., & Guo, Y. B. (2018). Residual Stress in Metal Additive Manufacturing. *Procedia CIRP*, *71*, 348–353. doi:10.1016/j.procir.2018.05.039

Li, G., Gao, M., Chen, C., Zhang, C., & Zeng, X. Y. (2014). Characterisation comparison of laser and laser–arc hybrid welding of Invar 36 alloy. *Science and Technology of Welding and Joining*, *19*(1), 30–37. doi:10.1179/1362171813Y.0000000159

Li, H., Chen, B., Tan, C., Song, X., & Feng, J. (2020). Microstructure evolution and mechanical properties of laser metal deposition of Invar 36 alloy. *Optics & Laser Technology*, *125*, 106037. doi:10.1016/j.optlastec.2019.106037

Mahmoudi, M., Tapia, G., Franco, B., Ma, J., Arroyave, R., Karaman, I., & Elwany, A. (2018). On the printability and transformation behavior of nickel-titanium shape memory alloys fabricated using laser powder-bed fusion additive manufacturing. *Journal of Manufacturing Processes*, *35*, 672–680. doi:10.1016/j.jmapro.2018.08.037

Mercelis, P., & Kruth, J. P. (2006). Residual stresses in selective laser sintering and selective laser melting. *Rapid Prototyping Journal*, *12*(5), 254–265. doi:10.1108/13552540610707013

Mertens, R., Vrancken, B., Holmstock, N., Kinds, Y., Kruth, J. P., & Van Humbeeck, J. (2016). Influence of Powder Bed Preheating on Microstructure and Mechanical Properties of H13 Tool Steel SLM Parts. *Physics Procedia*, *83*, 882–890. doi:10.1016/j.phpro.2016.08.092

Mfusi, B. J., Tshabalala, L. C., Popoola, A. P. I., & Mathe, N. R. (2018). The effect of selective laser melting build orientation on the mechanical properties of AlSi10Mg parts. *IOP Conference Series. Materials Science and Engineering*, *430*, 012028. doi:10.1088/1757-899X/430/1/012028

Muñiz-Lerma, A. J., Nommeots-Nomm, A., Waters, E. K., & Brochu, M. (2018). A Comprehensive Approach to Powder Feedstock Characterization for Powder Bed Fusion Additive Manufacturing: A Case Study on AlSi7Mg. *Materials (Basel)*, *11*(12), 2386. doi:10.3390/ma11122386 PMID:30486411

Nakamura, Y. (1976). The invar problem. *IEEE Transactions on Magnetics, 12*(4), 278–291. doi:10.1109/TMAG.1976.1059049

Obidigbo, C., Tatman, E.-P., & Gockel, J. (2019). Processing parameter and transient effects on melt pool geometry in additive manufacturing of Invar 36. *International Journal of Advanced Manufacturing Technology, 104*(5), 3139–3146. doi:10.100700170-019-04229-5

Ogawa, T., & Koseki, T. (1986). Weldability of Invar (Fe-36%Ni) and its Large-Diameter Pipe. *Welding Journal, 65*(28), 35–44.

Pal, S., Lojen, G., Kokol, V., & Drstvensek, I. (2018). Evolution of metallurgical properties of Ti-6Al-4V alloy fabricated in different energy densities in the Selective Laser Melting technique. *Journal of Manufacturing Processes, 35*, 538–546. doi:10.1016/j.jmapro.2018.09.012

Prashanth, K. G., Scudino, S., Maity, T., Das, J., & Eckert, J. (2017). Is the energy density a reliable parameter for materials synthesis by selective laser melting? *Materials Research Letters, 5*(6), 386–390. doi:10.1080/21663831.2017.1299808

Qiu, C., Adkins, N. J. E., & Attallah, M. M. (2016). Selective laser melting of Invar 36: Microstructure and properties. *Acta Materialia, 103*, 382–395. doi:10.1016/j.actamat.2015.10.020

Saewe, J., Gayer, C., Vogelpoth, A., & Schleifenbaum, J. H. (2019). Feasability Investigation for Laser Powder Bed Fusion of High-Speed Steel AISI M50 with Base Preheating System. *BHM. Berg- und Hüttenmännische Monatshefte, 164*(3), 101–107. doi:10.100700501-019-0828-y

Schmutzler, C., Stiehl, T. H., & Zaeh, M. F. (2019). Empirical process model for shrinkage-induced warpage in 3D printing. *Rapid Prototyping Journal, 25*(4), 721–727. doi:10.1108/RPJ-04-2018-0098

Scipioni Bertoli, U., Wolfer, A. J., Matthews, M. J., Delplanque, J.-P. R., & Schoenung, J. M. (2017). On the limitations of Volumetric Energy Density as a design parameter for Selective Laser Melting. *Materials & Design, 113*, 331–340. doi:10.1016/j.matdes.2016.10.037

Shiga, M. (1996). Invar alloys. *Current Opinion in Solid State and Materials Science, 1*(3), 340–348. doi:10.1016/S1359-0286(96)80023-4

Smith, J., Xiong, W., Yan, W., Lin, S., Cheng, P., Kafka, O. L., Wagner, G. J., Cao, J., & Liu, W. K. (2016). Linking process, structure, property, and performance for metal-based additive manufacturing: Computational approaches with experimental support. *Computational Mechanics, 57*(4), 583–610. doi:10.100700466-015-1240-4

Strauss, J. T., & Stucky, M. J. (2016). Laser Additive Manufacturing Processing of a Mixture of Iron and Nickel Powders. *Proceedings of the 27th Annual International Solid Freeform Fabrication Symposium – An Additive Manufacturing Conference.*

Tepylo, N., Huang, X., & Patnaik, P. C. (2019). Laser-Based Additive Manufacturing Technologies for Aerospace Applications. *Advanced Engineering Materials, 21*(11), 1900617. doi:10.1002/adem.201900617

Thakar, R. A., & Trivedi, S. V. (2017). An Overview of Dimensional Stability of Invar 36 Material for Space Based Optical Mounting Applications. *International Conference on Ideas, Impact and Innovation in Mechanical Engineering (ICIIIME 2017), 5*(6), 146-151.

Van Hooreweder, B., & Kruth, J.-P. (2017). Advanced fatigue analysis of metal lattice structures produced by Selective Laser Melting. *CIRP Annals*, *66*(1), 221–224. doi:10.1016/j.cirp.2017.04.130

Voznesenskaya, A. A., Kochuev, D. A., Chkalov, R. V., Kireev, A. V., & Morozov, V. V. (2020). Research of post-processing approaches for parts obtained by the method of selective laser melting. *Journal of Physics: Conference Series*, *1439*, 012028. doi:10.1088/1742-6596/1439/1/012028

Wang, L., Wang, S., & Hong, X. (2018). Pulsed SLM-manufactured AlSi10Mg alloy: Mechanical properties and microstructural effects of designed laser energy densities. *Journal of Manufacturing Processes*, *35*, 492–499. doi:10.1016/j.jmapro.2018.09.007

Wang, Y., Kamath, C., Voisin, T., & Li, Z. (2018). A processing diagram for high-density Ti-6Al-4V by selective laser melting. *Rapid Prototyping Journal*, *24*(9), 1469–1478. doi:10.1108/RPJ-11-2017-0228

Wei, K., Yang, Q., Ling, B., Yang, X., Xie, H., Qu, Z., & Fang, D. (2020). Mechanical properties of Invar 36 alloy additively manufactured by selective laser melting. *Materials Science and Engineering A*, *772*, 138799. doi:10.1016/j.msea.2019.138799

Xiong, W., Hao, L., Li, Y., Tang, D., Cui, Q., Feng, Z., & Yan, C. (2019). Effect of selective laser melting parameters on morphology, microstructure, densification and mechanical properties of supersaturated silver alloy. *Materials & Design*, *170*, 107697. doi:10.1016/j.matdes.2019.107697

Yakout, M. (2019). *Process-Structure-Property Relationships in Selective Laser Melting of Aerospace Alloys*. McMaster University.

Yakout, M., Cadamuro, A., Elbestawi, M. A., & Veldhuis, S. C. (2017). The selection of process parameters in additive manufacturing for aerospace alloys. *International Journal of Advanced Manufacturing Technology*, *92*(5), 2081–2098. doi:10.100700170-017-0280-7

Yakout, M., Elbestawi, M., Wang, L., & Muizelaar, R. (2019a). *Selective laser melting of soft magnetic alloys for automotive applications.* Paper presented at the Joint Special Interest Group meeting between euspen and ASPE; Advancing Precision in Additive Manufacturing.

Yakout, M., & Elbestawi, M. A. (2017). *Additive Manufacturing of Composite Materials: An Overview*. Paper presented at the 6th International Conference on Virtual Machining Process Technology (VMPT), Montréal, Canada.

Yakout, M., & Elbestawi, M. A. (2019). *Selective laser melting of graphene-reinforced aluminum matrix composites for electrical batteries*. Paper presented at the HI-AM Annual Conference 2019, Vancouver, Canada.

Yakout, M., Elbestawi, M. A., & Veldhuis, S. C. (2018a). On the characterization of stainless steel 316L parts produced by selective laser melting. *International Journal of Advanced Manufacturing Technology*, *95*(5), 1953–1974. doi:10.100700170-017-1303-0

Yakout, M., Elbestawi, M. A., & Veldhuis, S. C. (2018b). *Process-Structure-Property relationship for selective laser melting of aerospace alloys*. Paper presented at the 7th International Conference on Virtual Machining Process Technology (VMPT), Hamilton, Canada.

Yakout, M., Elbestawi, M. A., & Veldhuis, S. C. (2018c). A Review of Metal Additive Manufacturing Technologies. *Diffusion and Defect Data, Solid State Data. Part B, Solid State Phenomena*, *278*, 1–14. doi:10.4028/www.scientific.net/SSP.278.1

Yakout, M., Elbestawi, M. A., & Veldhuis, S. C. (2018d). A study of thermal expansion coefficients and microstructure during selective laser melting of Invar 36 and stainless steel 316L. *Additive Manufacturing*, *24*, 405–418. doi:10.1016/j.addma.2018.09.035

Yakout, M., Elbestawi, M. A., & Veldhuis, S. C. (2019b). Density and mechanical properties in selective laser melting of Invar 36 and stainless steel 316L. *Journal of Materials Processing Technology*, *266*, 397–420. doi:10.1016/j.jmatprotec.2018.11.006

Yakout, M., Elbestawi, M. A., & Veldhuis, S. C. (2020a). A study of the relationship between thermal expansion and residual stresses in selective laser melting of Ti-6Al-4V. *Journal of Manufacturing Processes*, *52*, 181–192. doi:10.1016/j.jmapro.2020.01.039

Yakout, M., Elbestawi, M. A., Veldhuis, S. C., & Nangle-Smith, S. (2020b). Influence of thermal properties on residual stresses in SLM of aerospace alloys. *Rapid Prototyping Journal*, *26*(1), 213–222. doi:10.1108/RPJ-03-2019-0065

Yan, W., Lin, S., Kafka, O. L., Yu, C., Liu, Z., Lian, Y., Wolff, S., Cao, J., Wagner, G. J., & Liu, W. K. (2018). Modeling process-structure-property relationships for additive manufacturing. *Frontiers of Mechanical Engineering*, *13*(4), 482–492. doi:10.100711465-018-0505-y

Yang, Y. P., Jamshidinia, M., Boulware, P., & Kelly, S. M. (2018). Prediction of microstructure, residual stress, and deformation in laser powder bed fusion process. *Computational Mechanics*, *61*(5), 599–615. doi:10.100700466-017-1528-7

Zhang, K., Fu, G., Zhang, P., Ma, Z., Mao, Z., & Zhang, D. Z. (2018). Study on the Geometric Design of Supports for Overhanging Structures Fabricated by Selective Laser Melting. *Materials (Basel, Switzerland)*, *12*(1), 27. doi:10.3390/ma12010027 PMID:30577654

Zumofen, L., Beck, C., Kirchheim, A., & Dennig, H.-J. (2018). Quality Related Effects of the Preheating Temperature on Laser Melted High Carbon Content Steels. *Proceedings of Additive Manufacturing in Products and Applications - AMPA2017*.

Chapter 5
Metal-Arc Welding Technologies for Additive Manufacturing of Metals and Composites

Fredrick M. Mwema
https://orcid.org/0000-0001-6116-5587
University of Johannesburg, APK Campus, South Africa

Esther T. Akinlabi
University of Johannesburg, South Africa

ABSTRACT

Additive manufacturing (AM) technology has been extensively embraced due to its capability to produce components at lower cost while achieving complex detail. There has been considerable emphasis on the development of low-cost AM technologies and investigation of production of various materials (metals, polymers, etc.) through AM processes. The most developed techniques for AM of products include stereolithography (SLA), fused deposition modelling (FDM), laser technologies, wire-arc welding techniques, and so forth. In this chapter, a review of the wire-arc welding-based technologies for AM is provided in two-fold perspective: (1) the advancement of the arc welding process as an additive manufacturing technology and (2) the progress in the production of metal/alloys and composites through these technologies. The chapter will provide important insights into the application of arc welding technology in additive manufacturing of metals and composites for advanced applications in the era of Industry 4.0.

DOI: 10.4018/978-1-7998-4054-1.ch005

INTRODUCTION

The additive manufacturing (AM) technology (also known as rapid prototyping, additive process and 3D printing), which involves production of 3D parts via layering of materials, has advanced over time. As such, several techniques are in existence for AM processing of metals, composites, and polymers. The choice of the manufacturing technology depends on the functionality, cost and dimensional accuracy of the 3D components. In most existing market problems require manufacturing solutions which can produce 3D parts tight dimensional and shape accuracy (Klobčar, Lindič, & Bušić, 2018). Additive manufacturing of plastics and polymers, through fused deposition modelling (FDM), (such as PLA, ABS, graphene-doped PLA, etc.) is the most developed technology. The plastic 3D printers have become so affordable such that most households in Western world and Asia own at least one. The FDM technology is user-friendly, cheap and simple to learn. However, the technology is challenged by dimensional accuracy, high roughness and lack of finer details, although continuous research is ongoing on improvement for FDM for production of 3D parts for advanced applications such as in biomedical, etc. (Katti, Sharma, & Katti, 2017).

There are several methods in existence for metal and composite additive manufacturing and these methods are classified based on the source of heat and they include, beam (laser and electron beam) and arc (tungsten inert gas (TIG), metal arc inert gas (MIG), and Plasma arc welding (PAW)) (Klobčar et al., 2018). A chart detailing these classifications is shown in Figure 1, although metal AM process can further be classified as reported by Ding et al. (Ding, Pan, Cuiuri, & Li, 2015). A lot of work is currently being advanced in laser-based AM of parts through selective laser sintering, laser metal deposition and laser cladding (Graf, Marko, Petrat, Gumenyuk, & Rethmeier, 2018; Mahamood, Akinlabi, Shukla, & Pityana, 2012). However, laser-based technologies are expensive and require extensive knowledge and experience as compared to arc-based technologies. The metal-arc welding additive technology proves to

Figure 1. Classifications of additive manufacturing methods for metal fabrication (adapted from Klobčar et al., 2018).

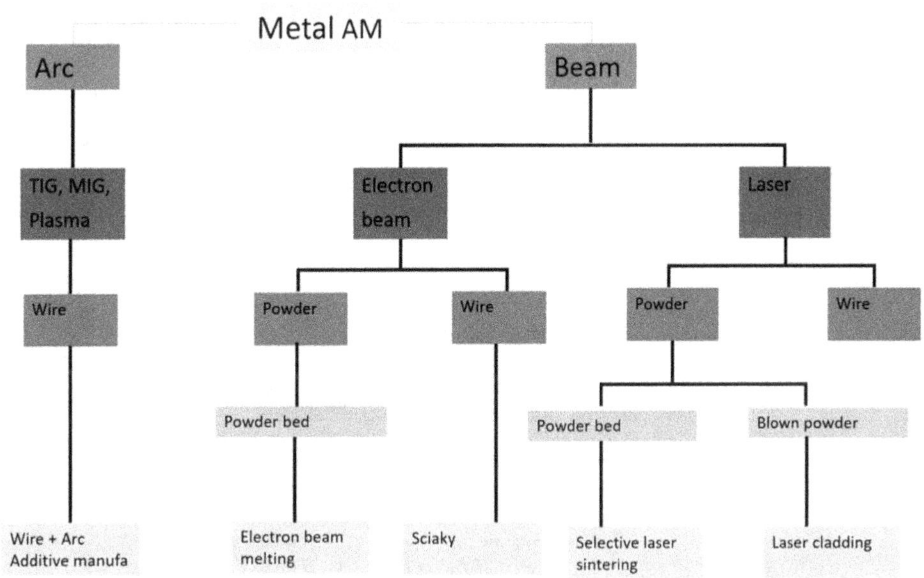

be easily adopted, especially in the developing world due to the simplicity and availability of the systems. Additionally, the method allows for fabrication of larger parts as compared to laser and electron beam AM processes. In this article therefore, the focus is on the metal-arc welding-based additive manufacturing technology for fabrication of metallic and composite parts, with emphasis on the various techniques and progress in research on the technology. The method is also evaluated for its future and place in Industry 4.0 for fabrication of high-end products.

METAL-ARC WELDING-BASED ADDITIVE MANUFACTURING METHODS

Techniques

Arc-welding based AM techniques have the capability to fabricate larger components as compared to other AM methods. It is also possible to produce components at lower manufacturing cost and the techniques are associated with low lead time and higher productivity. Wire-arc additive manufacturing (WAAM) techniques are some of the most popular systems for metal fabrication (Pan et al., 2018). The technology is probably among the oldest method for AM of metals but due to limitations associated with the process during the initial stages after its patent in 1920, it has been generally neglected for long time. The reason for its neglect could be due to very high residual stress induced on materials during arc welding, poor dimensional and shape accuracy and very high roughness associated with the process, high density of defects such as porosity, cracks and voids as compared to other AM methods, lack of advanced knowledge on CAD-to-product transfer related to WAAM and lack of monitoring and control systems during deposition (Ding et al., 2015). Basically, WAAM systems consists of the computer system, robot and its controls, power source and workpiece holding structure (Pan et al., 2018). The process involves continuous melting of a wire (as feedstock) through an electric arc (source of heat) and the molten wire material is extruded into bead, which are then layered on a substrate using a robot arm. The layering is repeated until the required thickness of the part is achieved. The major advantage with WAAM method is that it has large build envelop which is only limited by the movement distance of the robot arm and therefore can create larger parts than powder-based AM techniques. In cost terms, the WAAM technology is based on arc welding, which is a well-established process and its hardware usually consist of readily available welding facilities, hence making the WAAM method less expensive. Since it is based on arc welding technology, WAAM proves to be the most effective AM process for repairing cracks and other defects on functional components, such as moulds, turbine blades, shafts, and so forth.

Figure 2. Wire-arc additive manufacturing (WAAM) techniques

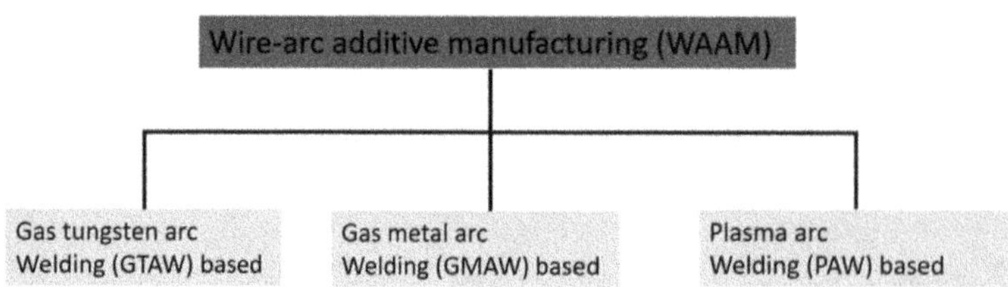

There are three major techniques for WAAM, and the classification is based on the source of power (arc) as shown in Figure 2. Here, we briefly describe the technology of each of the three techniques based on the published literature.

Gas Metal Arc Welding (GMAW) Based Additive Manufacturing

The process is also known as metal inert gas (MIG)/metal active gas (MAG) and it is the most popular weld-based AM technology. Figure 3 shows a schematic illustration of GMAW process as a welding process of metals. As shown, the process involves creation of an arc between the tip of the electrode and workpiece. The process occurs within an inert environment created by the shielding gas (usually inert gases, e.g. argon). The arc melts the electrode (feedstock/consumable wire) to form a pool of weld on the workpiece being welded. The electrode is continuously fed perpendicular to the workpiece (Lu et al., 2017; Rodrigues, Duarte, Miranda, Santos, & Oliveira, 2019). As an AM process, the continuous formation of the weld spool is enhanced by the motion of the nozzle across the surface of the substrate via the robotic manipulation and other computer controls. The welding torch (containing the nozzle) is held by a robot arm such that the nozzle can be moved horizontally across the surface of the substrate as shown in Figure 4. The process has been used for production of both simple and complex shaped parts (Figure 5). As shown in both Figures 4 and 5, the layering of the feedstock wire on the substrate is clearly observable. There are various variants of the GMAW based additive manufacturing process and have been described in literature (Klobčar et al., 2018; Pan et al., 2018).

Figure 3. A schematic diagram of Gas metal arc welding (GMAW) process
(Adapted from Ding et al., 2015).

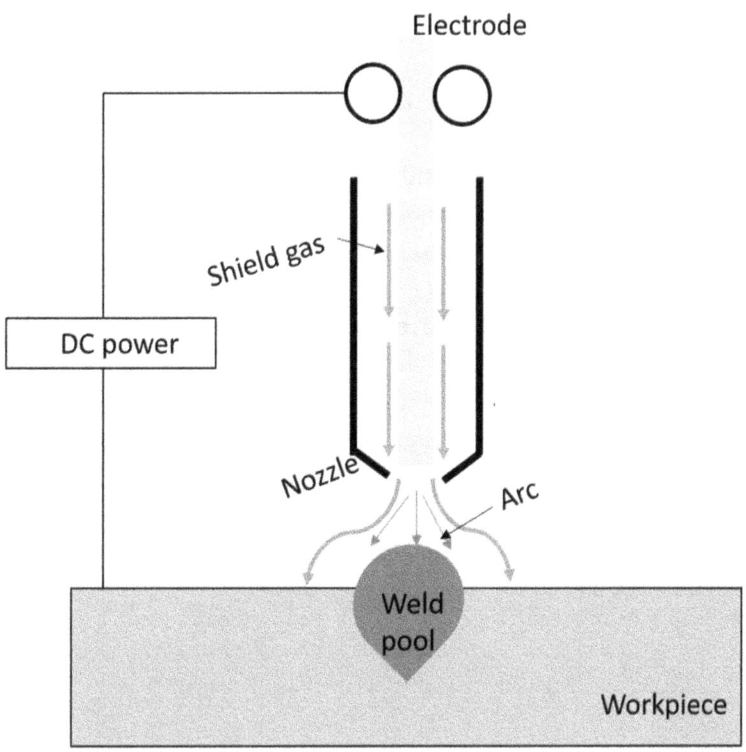

Figure 4. A schematic diagram for additive manufacturing process during GMAW technique (Adapted from Rodrigues et al., 2019).

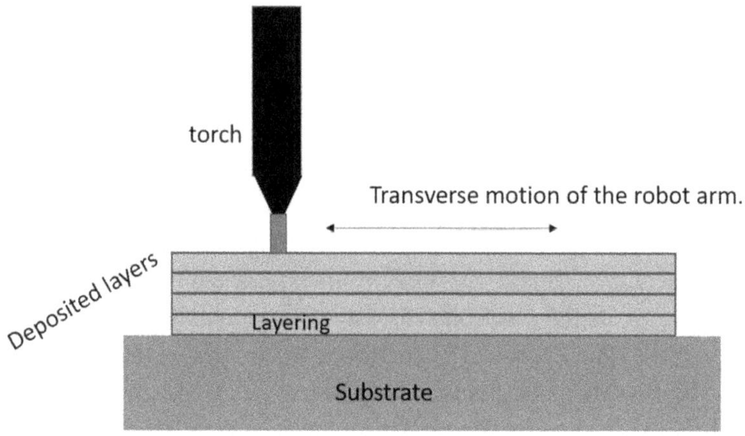

Figure 5. (a) Simple and (b) complex metallic parts produced via GMAW based AM process (Accesed under open access from Klobčar et al., 2018; Ma, Zhao, Li, Yang, & Xiao, 2019).

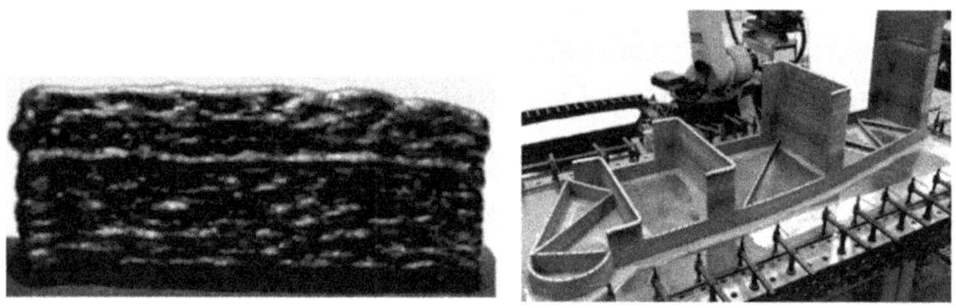

Gas Tungsten Arc Welding (GTAW) Based Additive Manufacturing

The process utilizes two wires; namely, a tungsten electrode and a feedstock, which are fed separately. The tungsten wire is non-consumable and contributes to energy dissipation to the consumable feedstock wire. Just like the GMAW, the electrode is housed in a contact tube inside a nozzle as illustrated by a schematic diagram in Figure 6. A shielding gas is passed through the nozzle to shield the electrode and feedstock wire from oxygen contamination during welding process. The electrode is usually perpendicular to the substrate whereas the feedstock wire can be fed on any orientation (it can be back, front or side feeding) based on the desired results and experience. In fact, in GTAW based AM, feedstock orientation greatly influences the material transfer during deposition and the quality of the layering and an optimization study has been presented (Geng, Li, Xiong, Lin, & Zhang, 2017). The system usually a gas lens which controls the flow of the shielding gas such that there are reduced chances of oxidation of the arc. The process is also influenced by other factors such as gas flowrate, arc length (distance between the workpiece and the nozzle), and so forth (Geng et al., 2017).

Figure 6. A schematic diagram showing basics features of gas tungsten arc welding (GTAW) process (Adapted from Ding et al., 2015).

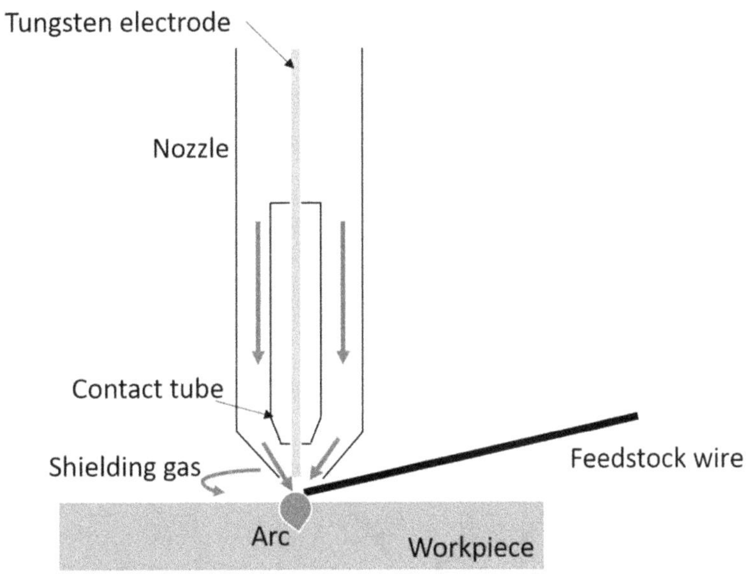

Plasma Arc Welding (PAW) Based Additive Manufacturing

Plasma arc welding (PAW) is very similar to the GTAW, since the electrode is non-consumable. However, PAW is considered a higher density process than the GTAW since the electrode is placed inside the torch body such that the plasma arc is separated from the shielding gas. The equipment is designed such that the plasma gas (argon or helium gas) is injected through the copper nozzle orifice separate from the shielding gas (argon with addition hydrogen gases) that ensures that the arc is constricted and intensified. Usually, there are three operating modes of PAW, namely, micro-plasma (operating at low welding currents ranging between 0.1 and 15 A), medium current (15-200 A) and keyhole plasma (using over 100 A). The heat affected zone (HAZ) of this process is very narrow as compared to the other welding processes, which allows for easy control of the geometry of the weld (Rao & Prasad, 2012). The micro-plasma PAW presents attractive characteristics as an additive manufacturing process since it is able to deposit thin parts of as low as 2 mm thicknesses (Rodrigues et al., 2019).

Its Place in Industry 4.0

In the modern industry, equipment, processes or systems must adopt the concept of a 'smart factory', which envisions automation, flexibility, sustainability and integration of information technology. Industry 4.0 emphasizes on the conversion of digital information to physical component/product. There are nine most important technologies that are significantly motivating the growth and advancement of Industry 4.0 across the world, namely, simulation, big data and analytics, robotics, augmented reality, the cloud, industrial internet of things (IIOTs), additive manufacturing, cybersecurity and horizontal and vertical integration (Rojko, 2017).

Figure 7. Plasma arc welding process
(Adapted from Ding et al., 2015)

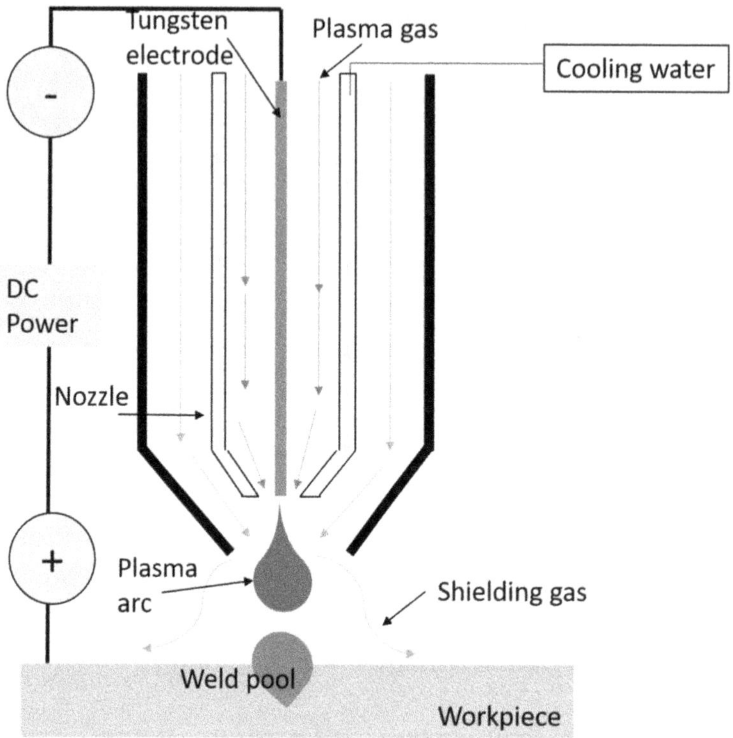

The WAAM based AM systems described above have adopted automation and digital technologies. The robot arm(s) perform most of the human-to-machine (H2M) operations. The design of the parts for AM in these systems is undertaken on computer aided tools and then transferred to the actual production on the WAAM equipment. Additionally, the application of digital and robotic technology in WAAM systems has enhanced control and management of AM parameters such as heating, which significantly enhances sustainability of these manufacturing processes. Just like most Industry 4.0 systems, remote control, incorporation of IIOTs and the cloud would enhance the efficiency and capabilities of WAAM processes.

In terms of reconfigurable system (flexible), some of WAAM based AM systems can easily configure their hardware and software for customized mass production of AM parts according to various market demands and manufacturing requirements (Koren et al., 1999). There are several WAAM systems in the market which consists of manipulation systems and CAD/CAM software. Usually, one system can be able to perform two or more of the WAAM processes since the arc welding source and robot arm manipulator can easily be combined. The flexibility can also be offered by integrating different sources of arc power such that a wide resource of materials can be fabricated since the type of arc for WAAM process depends on the material to be deposited. Additionally, incorporation of subtractive manufacturing tools onto the WAAM systems in future can significantly enhance their reconfiguration and hence alignment to the Industry 4.0 concept (Królikowski & Krawczyk, 2019).

MANUFACTURING OF METALS/ALLOYS AND COMPOSITES

As illustrated in the fundamentals of the WAAM processes, arc-welding based additive manufacturing is influenced by several factors some of which generally include, arc (power source) properties, electrode properties, feedstock wire, deposition speed, and so forth (Geng et al., 2017; Ryan, Sabin, Watts, & Whiting, 2018; Venturini, Montevecchi, Scippa, & Campatelli, 2016). These parameters require careful selection for manufacturing of defect-free additive manufacturing (AM) products. Obviously, optimization studies are critical for this to be achieved and continuous improvement of WAAM systems is very necessary; an example of such improvement is that of using compulsively constricted WAAM to reduce the heat affected zone (HAZ) on the deposited parts (Guo, Jia, Zhou, Liu, & Wu, 2020; Liu, Jia, Guo, Gao, & Wu, 2019). There are various studies reporting on WAAM manufacturing of metals and composite materials/parts; these studies clearly illustrate the complex interrelationships among the process parameters and properties of the manufactured part. Here, we illustrate some of the metals and composites, which have been successfully fabricated through WAAM technology while paying attention to the key insights on the quality enhancement of the products. Challenges influencing WAAM as AM processing of materials and possible strategies for improvement of the product quality were detailed in a recent invited review (Cunningham, Flynn, Shokrani, Dhokia, & Newman, 2018).

In composite fabrication, WAAM generally utilizes more than one feedstock wire or arc torches. For instance, a titanium alloy comprising of commercial pure Ti and titanium grade 5 alloy (Ti6Al4V) was fabricated using WAAM system, which is able to deposit the two feedstock wire of the two materials alternatively (Davis et al., 2019). This study investigated the influence of wife feed speed, deposition speed and interlayer thickness on the microstructural, layer fusion and mechanical properties of the Ti-based composite. In another study (Deng, Li, Wu, Yang, & Chen, 2019), TiB_2-reinforced Al-7Si-Cu-Mg composites were manufactured via casting and WAAM methods and their behaviour compared. It was shown that WAAM enhanced solid solubility of the reinforcements since there was better dispersion as compared to the casting process.

In metals and alloy fabrication, usually a single filament wire containing the required material chemistry and properties is used. A study (J. Gu et al., 2019) successfully fabricated aluminium alloy (A2319) using WAAM method with an objective of investigating the porosity characteristics during WAAM deposition. In another study, Cu-9Al-4Fe-4Ni-1Mn alloy was manufactured through wire-arc additive manufacturing and the properties of the parts were comparable to the cast alloy of similar material (Dharmendra, Hadadzadeh, Amirkhiz, Janaki Ram, & Mohammadi, 2019). WAAM has also successfully fabricated Ti6Al4V alloys for advanced applications (Wang, Williams, Colegrove, & Antonysamy, 2013). Inter-layer fusion and strength is the main limitation of WAAM process compared to casting and other AM methods and it causes porosity and weakness of the fabricated parts. Studies such as Gu et al. (J. Gu, Ding, Williams, Gu, Bai, et al., 2016; Gu, Ding, Williams, Gu, Ma, et al., 2016) have proposed cold rolling and post-heat treatment as the procedures for improving properties of WAAM additively fabricated Al-6.3-Cu, 5087 and 2319 aluminium alloys. It has been reported that post-heat treatment is more effective in enhancing the inter-layer fusion and properties of WAAM alloys as compared to cold working processes. However, cold rolling process was shown to eliminate inter-layer distortion and residual stresses in Al alloy (6082-T6) and Ti6Al4V manufactured through WAAM method (J.R. Hönnige et al., 2018; Jan Roman Hönnige, Williams, Roy, Colegrove, & Ganguly, 2017). Numerical simulations depicting the layer behaviour of various alloys additively manufactured through WAAM were reported (Abbaszadeh et al., 2019). The residual stresses resulting from WAAM were shown to enhance crack growth and failure in additively manufactured Ti6Al4V alloys (Zhang, Wang, Paddea, & Zhang, 2016). The quality of the feedstock wire has been shown to influence the microstructure, composition and hardness properties of WAAM additively manufactured Al alloys (J. L. Gu et al., 2014).

CONCLUSION

In this article, an overview of wire and arc welding based additive technologies has been provided. The fundamental concepts of the three arc welding technologies on which the technology is based have been highlighted. These welding technologies are metal arc, tungsten arc and plasma arc welding methods. The latter two utilizes a non-consumable wire whereas the former uses the electrode as the wire feedstock to create the weld pool. Various modifications have been introduced into these methods to create their capabilities as additive manufacturing processes. These modifications include incorporation of computer interface, robot manipulation system, and so forth. There are already various systems for AM processes based on the arc welding technologies and The Welding Institute (TWI) is among the leaders in enhancing this technology and currently they ae developing reconfigurable and more sustainable systems.

The WAAM AM has been used successfully in production of various alloys and composites and their performance evaluated. There are two major limitations associated with these processes in fabrication of AM parts: (i) weak inter-layer strength necessitated by insufficient fusion between adjacent layers and (ii) residual stresses created by the high heating of arc. Cold working through rolling and post-heat treatment have been proposed as the possible solutions to relieving the stresses as well as enhancing inter-layer fusion and hence improving the mechanical properties of the WAAM manufactured parts. Based on this overview, the authors note the following as possible areas of future focus:

- The WAAM systems are not fully aligned to the concept of Industry 4.0. There are extensive efforts to incorporate related technologies for maximum efficiency and effectiveness of the process. Incorporation of simulations in the process can enhance understanding on the distribution of the properties of the fabricated components. There are few simulations utilizing finite element tools to understand the inter-layer response to stress and strains as highlighted in the article.
- Although several alloys have been fabricated through WAAM technology, studies on composite fabrication are very few. The process seems promising in producing thick coatings for surface engineering purposes and therefore this need to be explored.
- WAAM has the superiority of fabricating large parts which can be directly used in aircraft, space, automotive, etc. For such applications however, advanced analyses such as utilization of synchrotron technology are required. Of course, neutron studies detailing residual stresses have been reported, however, most of the studies have reported on the average residual stress values. Stress and strain maps from experimental measurements can indeed provide comprehensive understanding of the WAAM processing.

FUNDING

There is no funding provided towards this work.

ACKNOWLEDGMENT

We acknowledge Global Excellence Scholarship (GES) 4.0 of the University of Johannesburg for the postdoctoral funding.

REFERENCES

Abbaszadeh, M., Hönnige, J. R., Martina, F., Neto, L., Kashaev, N., Colegrove, P., & Klusemann, B. (2019). Numerical Investigation of the Effect of Rolling on the Localized Stress and Strain Induction for Wire + Arc Additive Manufactured Structures. *Journal of Materials Engineering and Performance*, *28*(8), 4931–4942. doi:10.100711665-019-04249-y

Cunningham, C. R., Flynn, J. M., Shokrani, A., Dhokia, V., & Newman, S. T. (2018). Invited review article: Strategies and processes for high quality wire arc additive manufacturing. *Additive Manufacturing*, *22*(June), 672–686. doi:10.1016/j.addma.2018.06.020

Davis, A. E., Breheny, C. I., Fellowes, J., Nwankpa, U., Martina, F., Ding, J., & Prangnell, P. B. (2019). Mechanical performance and microstructural characterisation of titanium alloy-alloy composites built by wire-arc additive manufacture. *Materials Science and Engineering A*, *765*(July), 138289. doi:10.1016/j.msea.2019.138289

Deng, Y., Li, X., Wu, L., Yang, Q., & Chen, Y. (2019). Microstructure and Performance of WAAM TiB2-Reinforced Al–Si-Based Composites. In Physics and Engineering of Metallic Materials (Vol. 217, pp. 321–328). doi:10.1007/978-981-13-5944-6_31

Dharmendra, C., Hadadzadeh, A., Amirkhiz, B. S., Janaki Ram, G. D., & Mohammadi, M. (2019). Microstructural evolution and mechanical behavior of nickel aluminum bronze Cu-9Al-4Fe-4Ni-1Mn fabricated through wire-arc additive manufacturing. *Additive Manufacturing*, *30*(March), 100872. doi:10.1016/j.addma.2019.100872

Ding, D., Pan, Z., Cuiuri, D., & Li, H. (2015). Wire-feed additive manufacturing of metal components: Technologies, developments and future interests. *International Journal of Advanced Manufacturing Technology*, *81*(1–4), 465–481. doi:10.100700170-015-7077-3

Geng, H., Li, J., Xiong, J., Lin, X., & Zhang, F. (2017). Optimization of wire feed for GTAW based additive manufacturing. *Journal of Materials Processing Technology*, *243*, 40–47. doi:10.1016/j.jmatprotec.2016.11.027

Graf, B., Marko, A., Petrat, T., Gumenyuk, A., & Rethmeier, M. (2018). 3D laser metal deposition: Process steps for additive manufacturing. *Welding in the World*, *62*(4), 877–883. doi:10.100740194-018-0590-x

Gu, J., Ding, J., Williams, S. W., Gu, H., Bai, J., Zhai, Y., & Ma, P. (2016). The strengthening effect of inter-layer cold working and post-deposition heat treatment on the additively manufactured Al–6.3Cu alloy. *Materials Science and Engineering A*, *651*, 18–26. doi:10.1016/j.msea.2015.10.101

Gu, J., Ding, J., Williams, S. W., Gu, H., Ma, P., & Zhai, Y. (2016). The effect of inter-layer cold working and post-deposition heat treatment on porosity in additively manufactured aluminum alloys. *Journal of Materials Processing Technology*, *230*, 26–34. doi:10.1016/j.jmatprotec.2015.11.006

Gu, J., Gao, M., Yang, S., Bai, J., Ding, J., & Fang, X. (2019). Pore formation and evolution in wire + arc additively manufactured 2319 Al alloy. *Additive Manufacturing*, *30*, 100900. doi:10.1016/j.addma.2019.100900

Gu, J. L., Ding, J. L., Cong, B. Q., Bai, J., Gu, H. M., Williams, S. W., & Zhai, Y. C. (2014). The Influence of Wire Properties on the Quality and Performance of Wire+Arc Additive Manufactured Aluminium Parts. *Advanced Materials Research*, *1081*, 210–214. doi:10.4028/www.scientific.net/amr.1081.210

Guo, M., Jia, C., Zhou, J., Liu, W., & Wu, C. (2020). Investigating the generation process of molten droplets and arc plasma in the confined space during compulsively constricted WAAM. *Journal of Materials Processing Technology, 275*(February), 116355. doi:10.1016/j.jmatprotec.2019.116355

Hönnige, J. R., Williams, S., Roy, M. J., Colegrove, P., & Ganguly, S. (2017). Residual Stress Characterization and Control in the Additive Manufacture of Large Scale Metal Structures. In Residual Stresses 2016 (Vol. 2, pp. 455–460). doi:10.21741/9781945291173-77

Hönnige, J. R., Colegrove, P. A., Ganguly, S., Eimer, E., Kabra, S., & Williams, S. (2018). Control of residual stress and distortion in aluminium wire + arc additive manufacture with rolling. *Additive Manufacturing*, *22*, 775–783. doi:10.1016/j.addma.2018.06.015

Katti, D. R., Sharma, A., & Katti, K. S. (2017). Predictive Methodologies for Design of Bone Tissue Engineering Scaffolds. In *Materials for Bone Disorders* (pp. 453–492). Elsevier. doi:10.1016/B978-0-12-802792-9.00010-0

Klobčar, D., Lindič, M., & Bušić, M. (2018). Wire arc additive manufacturing of mild steel. *Materials and Geoenvironment*, *65*(4), 179–186. doi:10.2478/rmzmag-2018-0015

Koren, Y., Heisel, U., Jovane, F., Moriwaki, T., Pritschow, G., Ulsoy, G., & Van Brussel, H. (1999). Reconfigurable Manufacturing Systems. *CIRP Annals*, *48*(2), 527–540. doi:10.1016/S0007-8506(07)63232-6

Królikowski, M. A., & Krawczyk, M. B. (2019). Does Metal Additive Manufacturing in Industry 4.0 Reinforce the Role of Substractive Machining? In Lecture Notes in Mechanical Engineering (Vol. 1, pp. 150–164). doi:10.1007/978-3-030-18715-6_13

Liu, W., Jia, C., Guo, M., Gao, J., & Wu, C. (2019). Compulsively constricted WAAM with arc plasma and droplets ejected from a narrow space. *Additive Manufacturing*, *27*(March), 109–117. doi:10.1016/j.addma.2019.03.003

Lu, X., Zhou, Y. F., Xing, X. L., Shao, L. Y., Yang, Q. X., & Gao, S. Y. (2017). Open-source wire and arc additive manufacturing system: Formability, microstructures, and mechanical properties. *International Journal of Advanced Manufacturing Technology*, *93*(5–8), 2145–2154. doi:10.100700170-017-0636-z

Ma, G., Zhao, G., Li, Z., Yang, M., & Xiao, W. (2019). Optimization strategies for robotic additive and subtractive manufacturing of large and high thin-walled aluminum structures. *International Journal of Advanced Manufacturing Technology*, *101*(5–8), 1275–1292. doi:10.100700170-018-3009-3

Mahamood, R., Akinlabi, E., Shukla, M., & Pityana, S. (2012). Functionally Graded Material: An Overview. In *Proceedings of the World Congress on Engineering* (Vol. 3, p. 2065). Retrieved from http://researchspace.csir.co.za/dspace/handle/10204/6548

Pan, Z., Ding, D., Wu, B., Cuiuri, D., Li, H., & Norrish, J. (2018). Arc Welding Processes for Additive Manufacturing: A Review. In Transactions on Intelligent Welding Manufacturing, Transactions on Intelligent Welding Manufacturing (pp. 3–24). doi:10.1007/978-981-10-5355-9_1

Rao, C. S., & Prasad, K. S. (2012). Advances in Plasma Arc Welding : A Review. *Journal of Mechanical Engineering and Technology*, *4*(1), 35–60.

Rodrigues, T. A., Duarte, V., Miranda, R. M., Santos, T. G., & Oliveira, J. P. (2019). Current Status and Perspectives on Wire and Arc Additive Manufacturing (WAAM). *Materials (Basel)*, *12*(7), 1121. doi:10.3390/ma12071121 PMID:30987382

Rojko, A. (2017). Industry 4.0 Concept: Background and Overview. *International Journal of Interactive Mobile Technologies*, *11*(5), 77. doi:10.3991/ijim.v11i5.7072

Ryan, E. M., Sabin, T. J., Watts, J. F., & Whiting, M. J. (2018). The influence of build parameters and wire batch on porosity of wire and arc additive manufactured aluminium alloy 2319. *Journal of Materials Processing Technology*, *262*(April), 577–584. doi:10.1016/j.jmatprotec.2018.07.030

Venturini, G., Montevecchi, F., Scippa, A., & Campatelli, G. (2016). Optimization of WAAM Deposition Patterns for T-crossing Features. *Procedia CIRP*, *55*, 95–100. doi:10.1016/j.procir.2016.08.043

Wang, F., Williams, S., Colegrove, P., & Antonysamy, A. A. (2013). Microstructure and Mechanical Properties of Wire and Arc Additive Manufactured Ti-6Al-4V. *Metallurgical and Materials Transactions. A, Physical Metallurgy and Materials Science*, *44*(2), 968–977. doi:10.100711661-012-1444-6

Zhang, J., Wang, X., Paddea, S., & Zhang, X. (2016). Fatigue crack propagation behaviour in wire+arc additive manufactured Ti-6Al-4V: Effects of microstructure and residual stress. *Materials & Design*, *90*, 551–561. doi:10.1016/j.matdes.2015.10.141

Chapter 6
Wire + Arc Additive Manufacturing of Metals:
State of the Art and Challenges

Krishna Kishore Mugada
Indian Institute of Technology, Delhi, India

Aravindan Sivanandam
Indian Institute of Technology, Delhi, India

Ravi Kumar D.
Indian Institute of Technology, Delhi, India

ABSTRACT

Wire + Arc additive manufacturing (WAAM) processes have become popular because of their proven capabilities to produce large metallic components with high deposition rates (promoted by arc-based processes) compared to conventional additive manufacturing processes such as powder bed fusion, binder jetting, direct energy deposition, etc. The applications of WAAM processes were constantly increasing in the manufacturing sector, which necessitates an understanding of the process capability to various metals. This chapter outlines the significant outcomes of the WAAM process for most of the engineering metals in terms of microstructure and mechanical properties. Discussion on various defects associated with the processed components is also presented. Potential application of WAAM for different metals such as aluminum and its alloys, titanium, and steels was discussed. The research indicates that the components manufactured by the WAAM process have significant microstructural changes and improved mechanical properties.

DOI: 10.4018/978-1-7998-4054-1.ch006

Figure 1. Conventional MIG Welding setup for WAAM process.

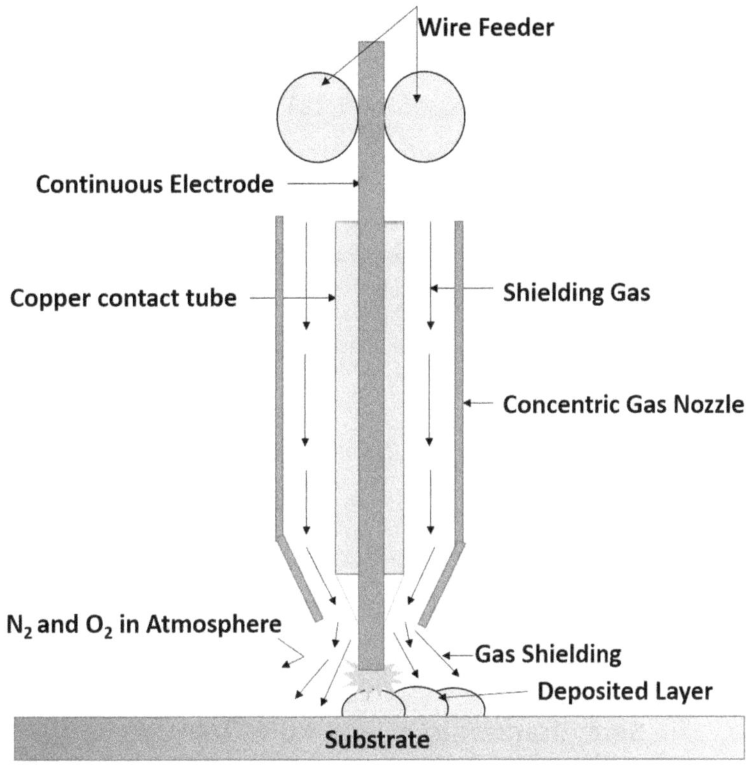

INTRODUCTION

Additive manufacturing (AM) a promising technology in reducing the waste, cost and time involved in the manufacturing of the components. A general AM process system has three parts (a) moving system, (b) energy source and (c) feedstock (Cunningham et al., 2018).

Wire + Arc additive manufacturing (WAAM) processes utilizes the electric arc (energy source) and wire feed stock (for melting) to deposit layer by layer and produce/create the components (Williams et al., 2016). WAAM process is also popularly known as shaped metal deposition (SMD) processes. WAAM is quite useful for producing large scale metallic components or parts in a cost-effective way compared to the conventional additive manufacturing techniques such powder bed fusion (PBF), material Jetting, direct energy deposition (DED), material extrusion, binder jetting and sheet lamination, etc., (Derekar, 2018)

The heat sources used in the WAAM processes should have high heat input. Therefore, conventional welding heat sources such as Metal inert gas welding (MIG), Cold metal transfer welding (CMT), Tungsten inert gas welding (TIG) and Plasma arc welding (PAW) are widely used. The conventional MIG welding setup used for the WAAM process is shown in Figure 1 and commonly referred to as Gas metal arc welding (GMAW). The arc is generated between the electrode and the workpiece. The wire is consumable, and no separate feeding is required. The MIG torch itself feeds the wire into the weld pool. Different modes used in MIG welding are globular transfer mode, short circuiting mode, spray mode and pulsed spray mode as shown in Figure 2. Cold metal transfer welding (CMT) is the modified MIG welding process and has better weld pool control, and high deposition rate at comparatively low heat

Figure 2. Different modes of transfer of molten metal in MIG welding process.

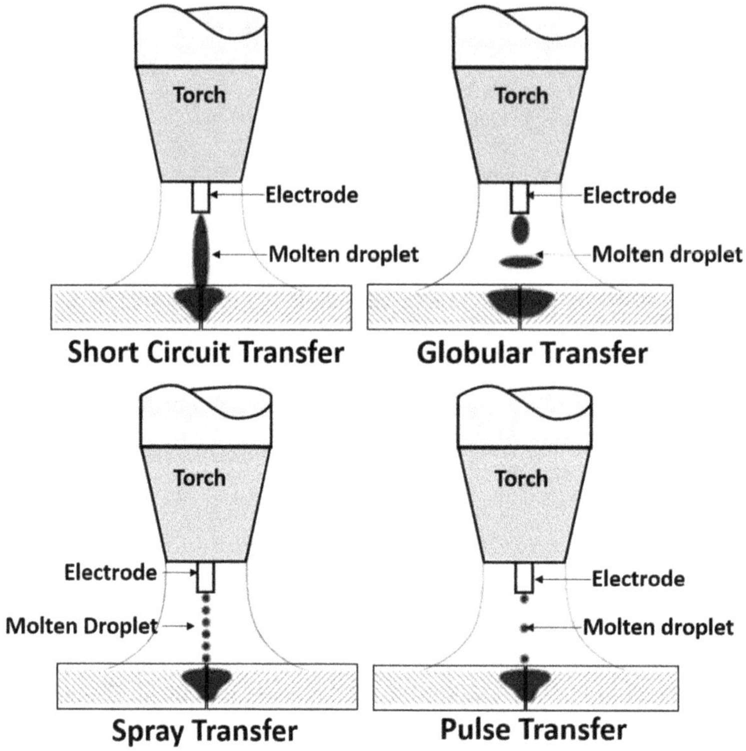

input (J. L. Z. Li et al., 2019). The conventional CMT welding setup with interfacial rolling attachment (for cold work between each deposited layers) used in the WAAM process is depicted in Figure 3.

Tungsten inert gas welding (TIG)- a process in which the electric arc is produced in between the non-consumable (tungsten) electrode and workpiece. TIG is generally termed as Gas Tungsten arc welding (GTAW). The conventional TIG welding setup used for the WAAM process is shown in Figure 4. Plasma arc welding (PAW) – similar to the GTAW process but it generates an electric arc which is three times higher than TIG process and the heating zone was quite narrow compared to TIG. PAW is widely used for deposition of smaller components (thin components). When the current in the PAW was less than 30A it is called as micro-plasma arc welding (MPAW) (Rodrigues et al., 2019). The conventional PAW welding setup used for the WAAM process is shown in Figure 5. Unlike in the MIG/CMT process, the wire is externally feed using the feed stock devices for TIG and PAW processes (for consistent deposition). The deposition layers produced in the WAAM processes for example a height of 1-2mm has a roughness of 500μm with single track deposit (Bekker et al., 2016). WAAM process is termed as a near net shaped processes because of the poor surface finish and therefore secondary operations like machining is essential for WAAM components.

Typical classification of AM process and various WAAM processes are shown in Figure 6. WAAM is suitable for (a) low to medium level complexity of component and (b) medium to large scale of production. Heat input calculations in WAAM process can be done using the traditional formulae which are linked with the current, voltage and traverse speed, which is similar to the conventional fusion welding process (Williams et al., 2016).

Figure 3. CMT WAAM deposition setup with interface rolling

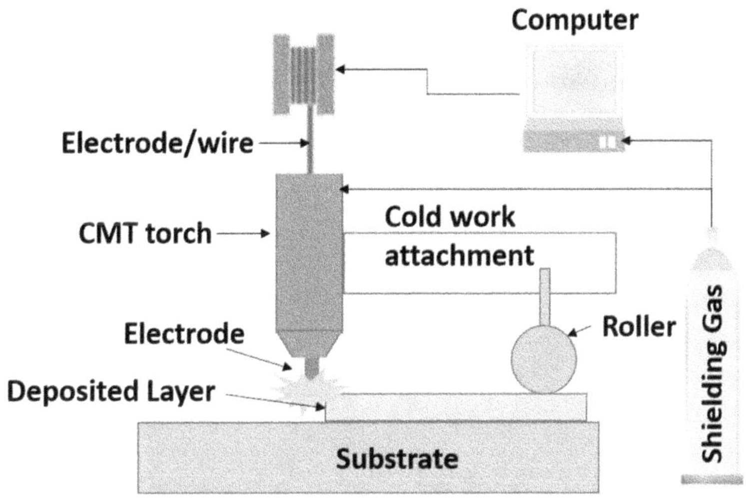

Figure 4. Conventional TIG welding setup for WAAM process.

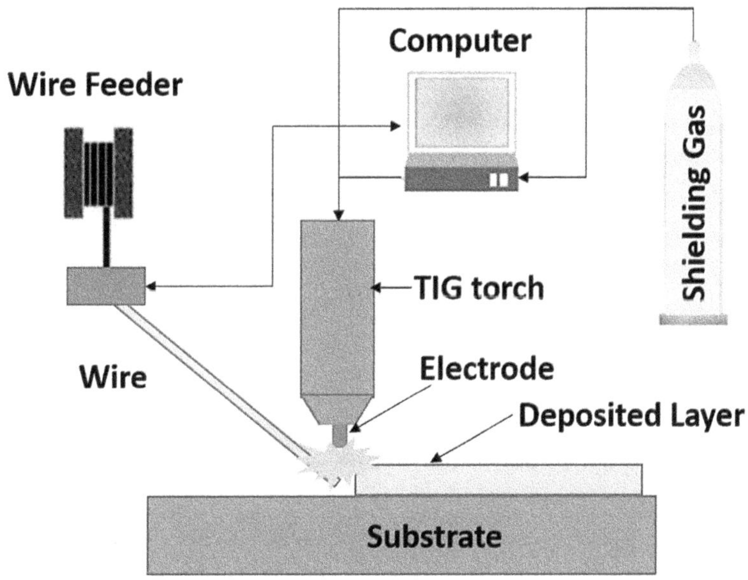

$$\text{Heat input} = (\text{Voltage x current}) / \text{Travel Speed} = \eta \frac{V \times I}{v_f} \tag{1}$$

where I =current, V= voltage, v_f = travel speed, and η =efficiency of the welding process.

The simple step by step procedure or cyclic loop of the general WAAM processes is shown in the Figure 7. The procedure involves the start of arc, wire feed and finally deposition of the molten liquid on to the substrate, the entire process will be done under the shielding gas protection. The process was repeated for continuous deposition until the required dimensions of the component was produced.

Figure 5. Conventional Plasma arc welding setup for WAAM process.

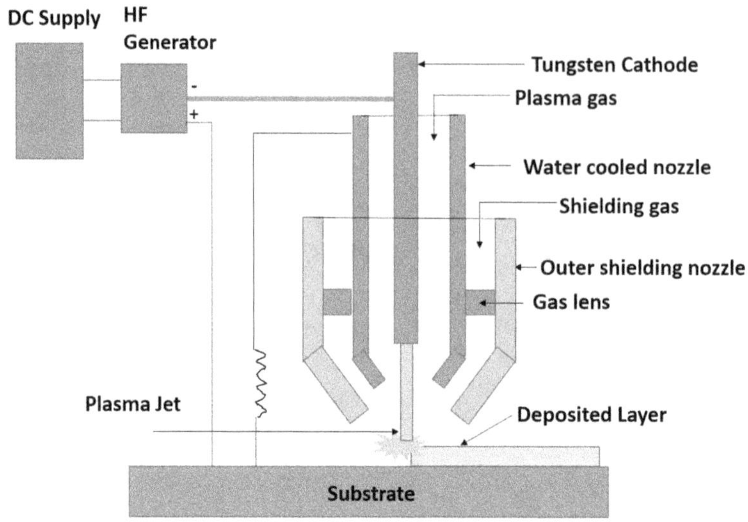

Figure 6. Typical Wire Arc Additive Manufacturing (WAAM) processes classification.

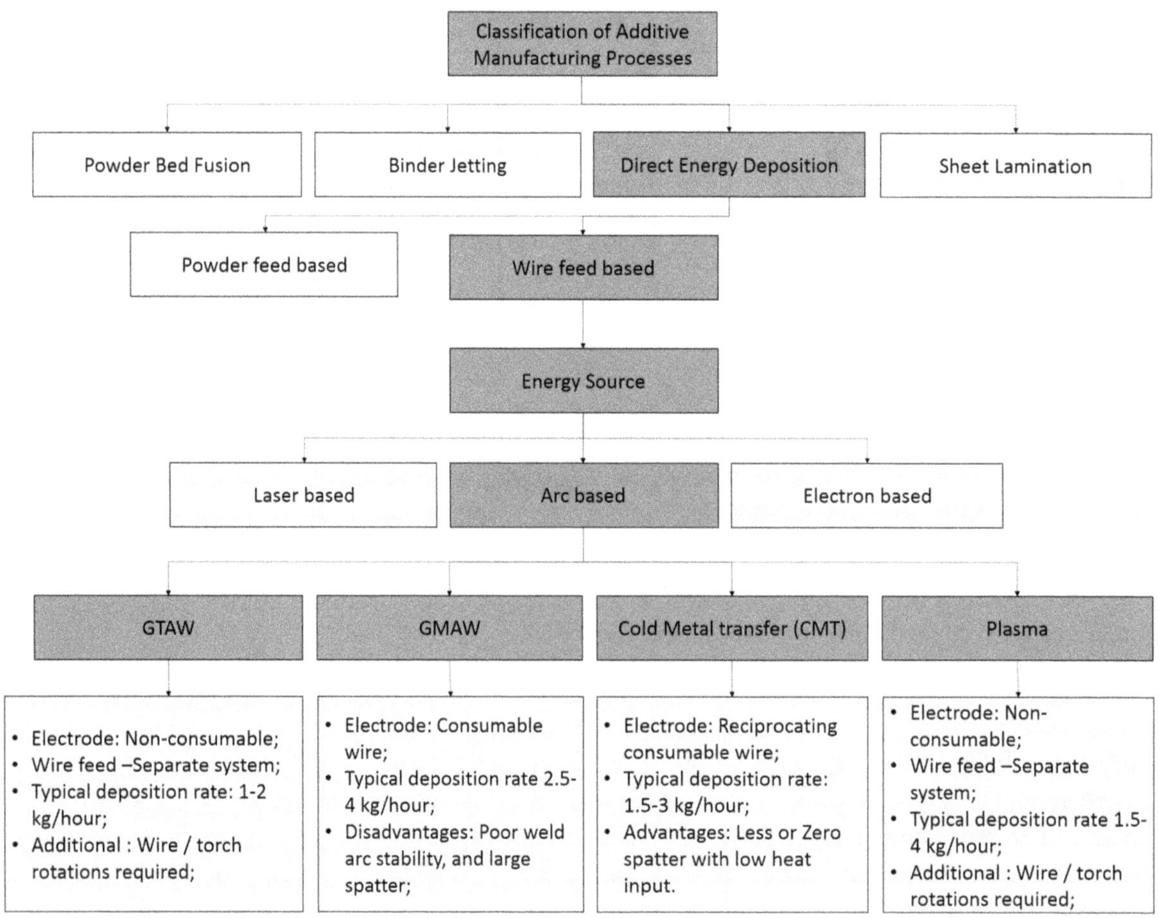

Figure 7. The cyclic loop of the WAAM process.

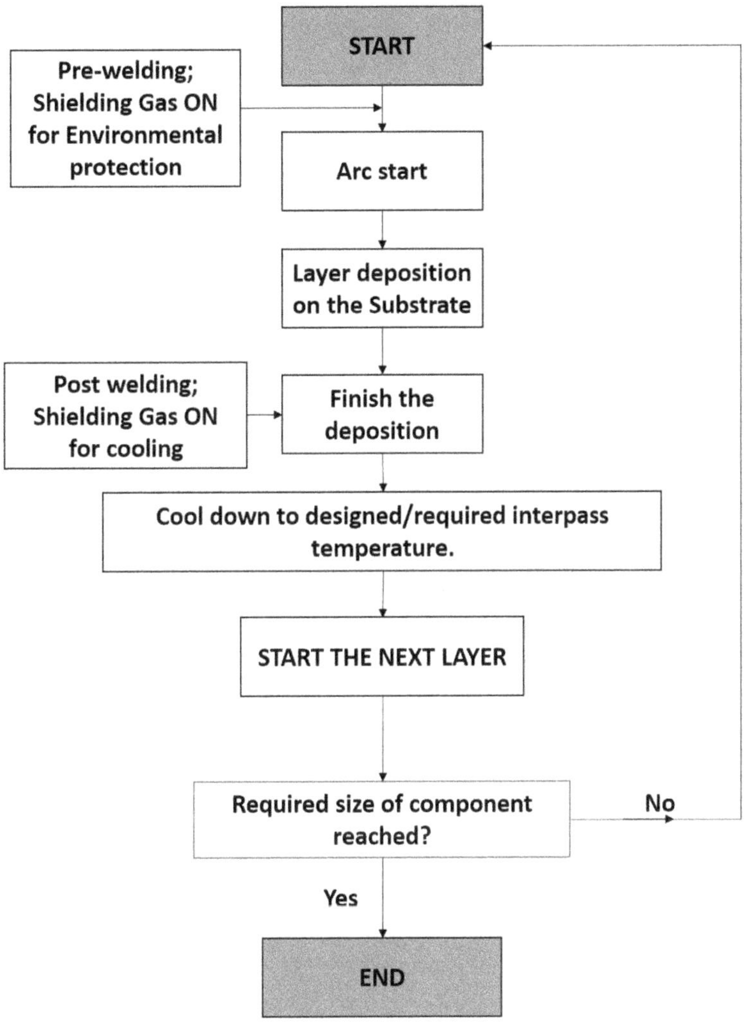

WAAM FOR VARIOUS METALS AND ALLOYS

Aluminum Based Alloys

Gu, Ding, Williams, Gu, Bai, et al., (2016) deposited Al 2219 alloy using WAAM process with Cold metal transfer as heat source. Further investigated the strengthening mechanisms of as-deposited components with (a) rolling in between inter-layer and (b) post deposit heat treatment (T6). The 2219 alloy was strengthened by precipitation mechanism due to which the post welded deposition WAAM sample exhibited superior strength. The typical microstructure of the as-deposited and post deposited heat treated (T6) samples obtained by WAAM process were shown in Figure 8.

Ayarkwa et al., (2017) worked on the alternating current TIG process to study the effects on current cycle time during the WAAM of thin walls with aluminum alloy. The electrode positive time cycle effects the heat input (shown in equation (1) with efficiency η =0.6) significantly and the microstruc-

Figure 8. Schematic sketch of microstructural changes of 2219 alloys by WAAM process (a) as-deposited layer (b) post deposited heat treated (T6) condition
Adapted from (Gu, Ding, Williams, Gu, Bai, et al., 2016)

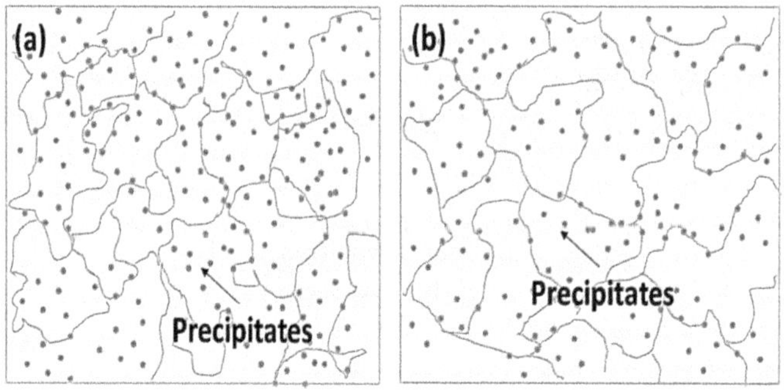

tural changes were observed. However, the electrode positive time cycles effects were less on the final strength and hardness of the deposited layers.

Gu, Ding, Williams, Gu, Ma, et al., (2016) worked on the WAAM of aluminum alloy (Al 5087 and Al 2319 alloys) and studied the effect of interlayer cold work and post deposition heat treatments on porosity levels. In post deposition heat-treated specimens, the hydrogen diffusion and Ostwald ripening dominates the growth of the pores. Inter rolling has successfully flattened the pores at higher loads (45kN) as depicted in Figure 9.

Figure 9. Schematic example of microstructure changes in 2319 alloy (a) as-deposited layer (b) enlarged view of a pore in as-deposited layer (c) flattened pore after inter-layer rolling at 45kN
Adapted from (Gu et al., 2018).

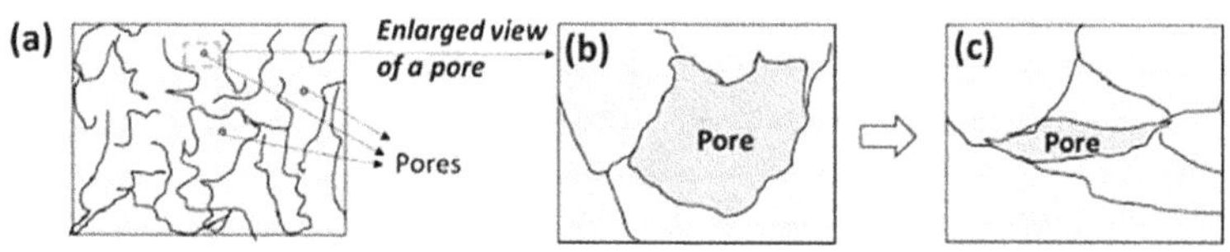

Geng et al., (2017) worked on the GTAW based WAAM process of 5A06 aluminum alloy and studied the wire feed optimization to improve the deposition accuracy. The parameters (angle between wire feed and deposited layer, distance between electrode tip and wire) associated with the deposition accuracy were shown in Figure 10. Increasing the wire feed angle, there was a shift in the start position, A smooth layer deposition was observed when both the layer size and layer surface were better which was achieved by the use of low wire feed angle and height of wire rotation axes. Qi, Qi, et al., (2018) studied the double wire feed deposition (WAAM process) of Al 4043 and Al 5087 alloys in GTAW process.

The WAAM of mixed composition wire deposition for Al-3.1Mg-2Si has resulted the microstructure which contains the α-Al, Mg_2Si and Al_9Si phases. The strength of the WAAM manufactured component was around 176 MPa (UTS). Qi, Cong, Qi, Sun, et al., (2018) studied the double wire arc manufacturing of Al-Cu-Mg alloys with Al-Cu wire of 2319 alloy and Al-Mg wire of 5087 alloy. The microstructural evolution of as-deposited Al-Cu-Mg alloys consists of coarse columnar and fine equiaxed grains at inter

Figure 10. Wire feed unit parameters
Adapted from (Fixter et al., 2017).

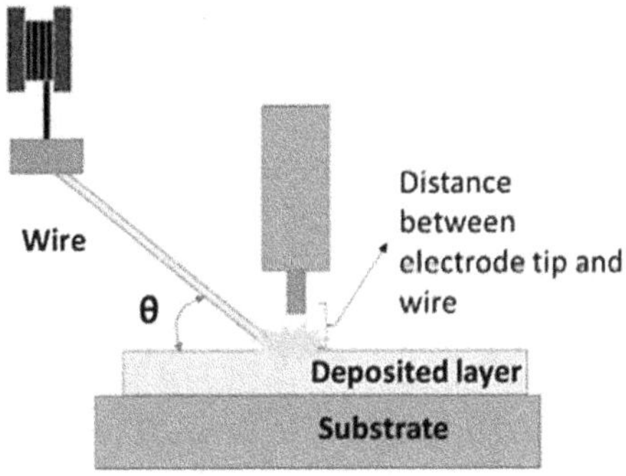

and inner layer regions respectively. A general representation of inter and inner layers were shown in Figure 11. The variations in the microstructure were due the solidification characteristics of Al-Cu-Mg alloys. The strength of the deposits was observed to be 280 MPa with typical brittle nature compared to the base alloys. Qi, Cong, Qi, Zhao, et al., (2018) studied the properties of 2024 deposits with WAAM process using GTAW as heat source. The post-deposition heat treatment dispersed the second phase precipitates in the inner/inter layers. Sun et al., (2018) investigated the as-deposited Al 2319 alloy fabricated with WAAM process (Heat source as TIG welding) and cold worked the samples with laser shock peening. The results show that the laser shock peening has improved the mechanical properties compared to the as-deposited build structure.

Figure 11. A general schematic illustration of as-deposited layers in WAAM process (a) macrostructure (b) microstructure with inter and inner layers
Adapted from (Qi, Cong, Qi, Sun, et al., 2018).

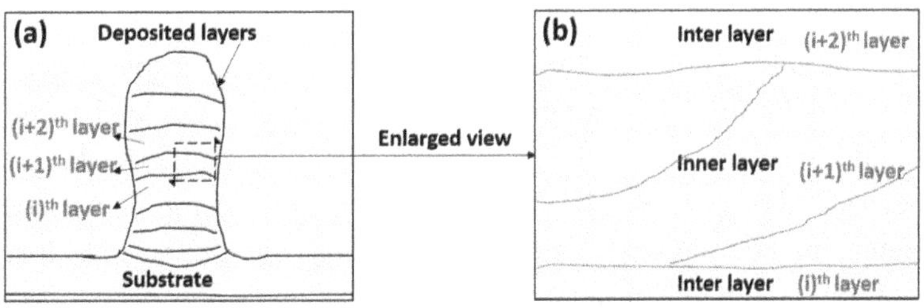

Horgar et al., (2018) investigated the WAAM process of Al 5183 alloy using conventional GMAW technique. The multilayers deposition initiates some cracking and porosity in between the deposited layers. By controlling the torch parameters (as facilitated by the cold metal transfer welding) and varying the travel speed from 12mm/s to 22 mm/s, and wire feed rate from 10 m/min to 15 m/min, the porosity and cracking in the layers were eliminated. Zhou et al., (2019) studied the effect of travel speed on wire arc deposited 2219 alloy in terms of microstructural evolution. As the travel speed or velocity of the arc increases the surface finish of the layers were good due to refined grain size and increased solidification rate. The equiaxed and coarse columnar grains were noticed at the inner and interface of deposited Al 2219 layers.

Zhong et al., (2019) investigated with WAAM of 2050 Al-Li alloy with heat source as variable polarity GTAW (VP-GTAW) and analyzed the microstructural evolution. The deposited samples were post heat treated with T6 condition and studied the microstructural effects. Due to the variation in cooling rate and thermal gradient near the surroundings of molten pool distinct zones in microstructure were observed. The inner layers have fine equiaxed grains (non-dendrite) and the inter layers have the coarse column grains. In the post heat treatment condition (T6), the secondary phases Al_2Cu and Al_3Li phases were dispersed along the grain boundary.

Qi et al., (2019) simultaneously fed the wires of Al 2319 alloy and Al 5087 alloy to produce the deposits of Al 2024 aluminum alloy with the double wire WAAM process. The dendrites in the as-deposited condition were disappeared after the post heat treatment. In the as deposited 2024 alloy, the precipitates were distributed in the entire region of deposit. After heat treatment the inter layer has less second phase particles compared to the other regions due to diffusion and discontinuous distribution characteristics.

Gu et al., (2020) manufactured the Al-Cu4.3-Mg1.5 alloys in WAAM using CMT process. The results indicate that the as deposited microstructure has the columnar dendrites and the post heat treatment affected the microstructure by resulting with the equiaxed grains and fine dispersion of the second phase precipitates. The heat treatment has enormously improved the strength and hardness by more than 50% compared to the as deposited sample.

Miao et al., (2020) deposited the 4043-wire using the WAAM process (TIG welding as source), and laser arc welding to study the microstructural evolution. The samples produced with laser arc welding facilitated superior properties. However, in the WAAM samples, segregation of second phase precipitates was finer in the deposited layers of 4043 alloy.

As a summary, the typical mechanical properties for various aluminum alloys manufactured by different WAAM processes (various heat sources) reported in the available literature are presented in Table 1.

Titanium Based Alloys

Dutta & Froes, (2017) broadly discussed the additive manufacturing capabilities for titanium alloys with emphasis on Ti-6Al-4V alloy. The strength comparison for Ti-6Al-4V alloy with various additive manufacturing techniques were discussed and shown that WAAM process as one of the best methods in AM for improved mechanical properties. The typical research and development associated with WAAM of Titanium components include the study of the arc characteristics such as arc force, molten pool geometry, deposition stability and deposition rates etc.,

Baufeld et al., (2011) studied the Ti-6Al-4V metal deposits manufactured by the WAAM process (TIG energy source) and analyzed the mechanical properties associated. The dense components of α/β phases were observed in post deposition microstructure of the Ti layers. Further, the heat treatment at 843^0C has a significant change in the microstructural properties which increases the strain at failure. Gou et al., (2019) investigated the WAAM deposits of Ti-6Al-4V alloy using the CMT source and analyzed the microstructural evolution for as-deposited condition and heat-treated condition. The acicular α martensite with lamellar structure was observed in the microstructure of as deposited layers.

Ma et al., (2016) worked on the post heat treatment studies of deposited γ-Ti Al alloy using the WAAM process (GTAW) and analyzed the microstructural changes during the heat treatment. The mi-

Table 1. Mechanical properties of various aluminum alloys produced with WAAM processes.

Filler Wire	Wire Dia. (mm)	Substrate	Heat Source	Condition	YS (MPa)	UTS (MPa)	EL (%)	Ref.
Al 2319	1.2	Al 2219-T87	*CMT-PA	As deposited	125	260	16	(Gu, Ding, Williams, Gu, Bai, et al., 2016)
				Heat treated (T6)	300	450	13	
				Rolling at a load of 15 kN	140	270	14.5	
				Rolling at a load of 30 kN	185	285	11	
				Rolling at a load of 45 kN	245	315	9	
				Rolling at a load of 45 kN + T6 heat treated	310	460	16	
Al 5556	1.2	Al 6082-T6	TIG AC	20% electrode positive cycle timing (EPCT)	145	235±10	35	(Ayarkwa et al., 2017)
				50% EPCT	150	250±10	35	
Al 5087	1.2	Al 5083-H32	CMT	As deposited	142	291	22	(Gu et al., 2018)
				Rolling at a load of 15 kN	169	320	35	
				Rolling at a load of 30 kN	149	311	39	
				Rolling at a load of 45 kN	200	344	47	
Al 2024	1.2	-	CMT	As deposited	175	290	12	(Fixter et al., 2017)
				Rolled 45 kN	315	375	8	
				T4	335	465	15	
				T6	415	505	8	
Al-3.1Mg-2Si (Al 4043 + 5087)	1.2	2A12	GTAW	As deposited	76.6	176	11.4	(Qi, Qi, et al., 2018)
Al-Cu-Mg (Al 2319 + Al 5087)	1.2	2A12	GTAW	As deposited	187	280	6	(Qi, Cong, Qi, Sun, et al., 2018)
Al 2024 (Al 2319 + Al 5087)	1.2	2A12	GTAW	As deposited	325	470	20	(Qi, Cong, Qi, Zhao, et al., 2018)
				Post heat treatment	330	497	16	
Al 5183	1.2	Al 6082	GMAW	As deposited	145	293	-	(Horgar et al., 2018)
Al 2219	1.2	Al 6061	TIG	As deposited + 350 mm/min torch speed.	175	275	9	(Zhou et al., 2019)
Al 2050	1.2	2A12	#VP-GTAW	Heat treated	260	400	5	(Zhong et al., 2019)
Al 2024 (Al 2319 + Al 5.87)	1.2	2A12	VP-GTAW	Base material	300±10	450±10	13	(Qi et al., 2019)
				As deposited	160±10	300±10	7	
				T4	300±10	450±10	12	
				T6	300±10	450±10	13	
Al-Cu4.3%-Mg1.5%	1.2	Al 2024	CMT	As deposited	175	285	12	(Gu et al., 2020)
				Heat treated	400	475	8	
Al 4043	1.2	Al 6061	TIG	As deposited	69	151	16	(Miao et al., 2020)

*CMT-PA (CMT-Phased Array), #VP-GTAW (Variable polarity GTAW) crostructural observation shows a full lamellar structure near the substrate during heat treatment at 1200^{0}C temp for 24 hours. The homogenous microstructure was noticed in the post heat treatment as compared to the as-deposited layers as depicted in Figure 12.

Figure 12. Schematic sketch of microstructural changes of the deposited layers (a) as deposited (b) post heat treatment
Adapted from (Gou et al., 2019).

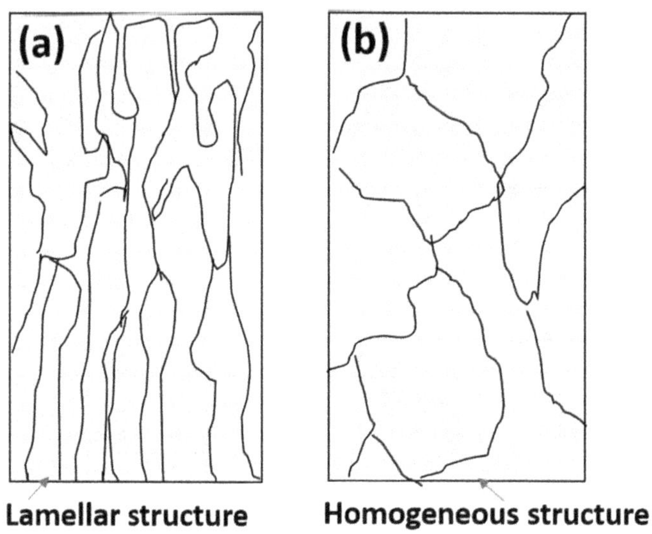

The post heat treatment primarily controls the near substrate zone microstructure by the reaction (discontinuous). The α_2 phase decomposition and formation of equiaxed γ grains were favored by the post deposition heat treatment process.

Steels

Shen et al., (2015) produced iron rich Fe-Al using WAAM process with GTAW heat source and studied the mechanical properties, and microstructural evolution of the deposited layer. The WAAM process has consistent composition as of the original Fe_3Al intermetallics. The WAAM has segregated and distributed the intermetallics in the entire deposited layer uniformly resulting in enhancement of the yield strength.

Abe & Sasahara, (2016) worked on the deposition of stainless steel and nickel-based alloys in WAAM process and investigated the mechanical properties and microstructural changes in the deposits. Comparable properties of deposited layers were indicated with base materials. In the YS308L (18%Cr-8%Ni steel) weld deposit region the grain size of δ -ferrite was more compared to the base metal. In the Ni6082 weld deposit region the austenite grains were observed. The dissimilar deposition of 308L steel and Ni based alloys are shown in Figure 13.

Chen et al., (2018) worked on the WAAM process and deposited the 316L stainless steel (austenitic) and studied the heat treatment effects on the microstructural, mechanical and corrosion properties. Heat treatment causes the σ phase formation that increases the sensitivity which becomes weak zones for

Figure 13. Cross-section of built layers (dissimilar 308L steel and Ni based alloy)
Adapted from (Ma et al., 2016).

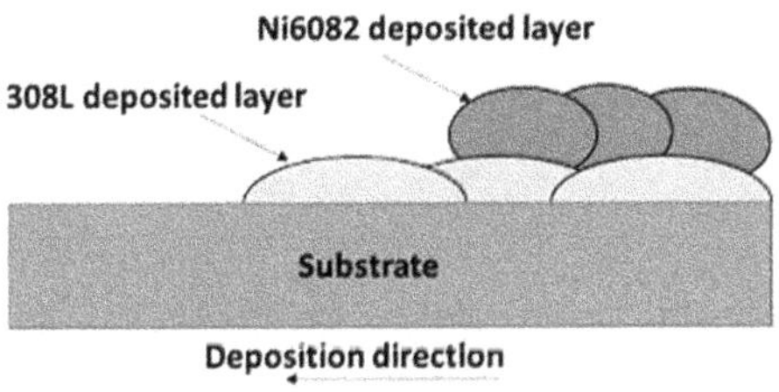

deformation and corrosion as well. The micro cracks generate at the σ interfaces and propagates during the deformation as shown in Figure 14.

Chen et al., (2017) worked on WAAM of austenitic 316L stainless steel using GMAW as energy source and studied the microstructural evolution and mechanical properties. The as-deposited microstructure of 316L contains the δ, γ, And σ phases (Iron-chromium phase diagram). At the interphases, the γ/δ combinations were observed. The cracks were initiated at the σ phases during the tensile testing and leading the deposits to failure as shown in Figure 14.

Figure 14. Crack behavior in the deposited 316 layers during deformation tests (σ, γ, and δ are phases)
Adapted from (Chen et al., 2017).

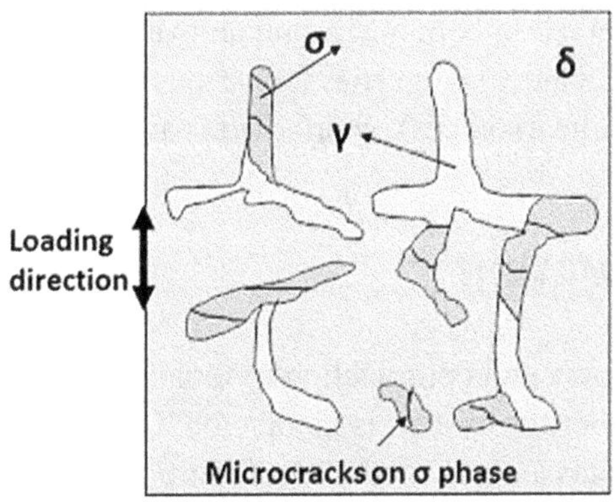

From fourth layer onwards, the δ and σ in the deposited layers were turned in to the fine vermicular morphology due to the thermal effects induced during WAAM deposition.

Y. Li et al., (2019) worked on the H08Mn2Si steel deposits using the GMAW based WAAM process and studied the molten pool stability for various wire feed rates and traverse speeds. The bead width enlarges with increase in the torch inclination angle, and the position of the GMAW torch (forward or backward positions) has significant effect on the arc force on the weld pool.

Rodriguez et al., (2018) worked on the WAAM and deposited stainless steel by two energy sources CMT and Top TIG and compared the two processes with mechanical properties and microstructural changes. Both the processes are capable to deposit straight walls with good accuracy and reasonable flatness. The deposition rates were higher for CMT (3.7kg/h) and comparatively less for Top TIG process (2 kg/h). Ge et al., (2019) investigated the deposited layers of 2Cr13 stainless steels fabricated by the WAAM process and studied the microstructural evolution along with the defect distribution by developing the numerical models.

Figure 15. Transient stress distribution in the deposited layers.

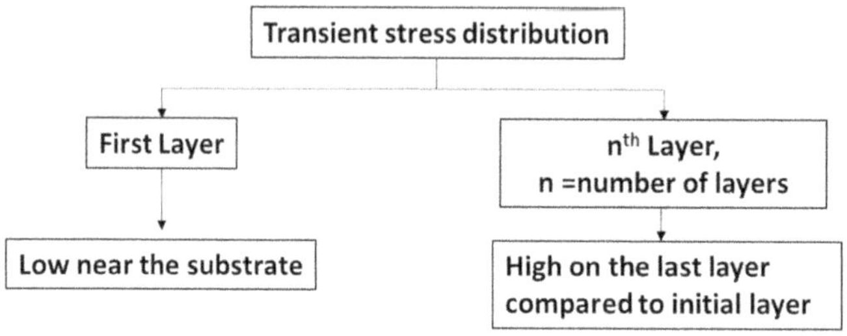

The stress distribution evolved regularly with the increment of the deposited layers and observed maximum for the last deposited layer, which was transformed into the residual stress upon solidification. A tree chart for transient stress distribution was shown in Figure 15. Rafieazad et al., (2019) studied the as-deposited microstructural changes and mechanical properties of the low carbon alloy steels deposited using WAAM process (GMAW). The predominant phases in the base metal were fine ferrite and laminar pearlite. After the deposition the equiaxed grains were observed with weak cubic texture. Comparable properties of deposits were observed in both horizontal and vertical conditions.

As a summary the typical mechanical properties for various steels and Ti alloys manufactured by different WAAM processes (various heat sources) reported in the available literature are presented in Table 2.

WAAM PROCESS AUTOMATION

Optimization of process parameters and optimization of tool path are important aspects in the WAAM processes for which automation of the system is key factor. The WAAM process Monitoring systems consists of the various components as explained in the Figure 16.

Limited online measurements and tool path monitoring make the process as a near net shape process because of poor surface finish on the final manufactured components which can be controlled by the full monitoring system as shown in the Figure 17.

The full automation monitoring system is sub-categorized into different layers as illustrated in Figure 17. Physical devices required for various signal measurements that control the final deposition quality will be captured in the sensor layer. The raw signals received by the sensors should be processed with data processing tools for complete automation of the WAAM processes.

Data extraction layer signifies the use of data processing algorithms to process and analyze the signal data and to activate the controls or control systems. Figure 17 shows the analysis (or) feature extraction layer with five factors (defects, molten pool characteristics etc.,) as illustrated. Further these will pass the

Table 2. Mechanical properties of various steels and Ti alloys produced with WAAM processes.

Filler Wire	WD (mm)	Substrate	ES	Condition	YS (MPa)	UTS (MPa)	EL (%)	Ref.
Fe3Al	0.9	DH36	GTAW	As deposited	847	944	3.27	(Shen et al., 2015)
316L	1.2	-	GMAW	As deposited	235	533	48	(Abe & Sasahara, 2016)
				1000ºC/1hr, WQ	255	549	41	
				1100ºC/1hr, WQ	323	498	56	
				1200ºC/1hr, WQ	215	474	57	
				1200ºC/4hr, WQ	204	494	70	
				Wrought 316L	222-265	505	56-63	
316L	1.2	-	GMAW	As deposited	235	533	48	(Chen et al., 2017)
				Cold worked	255-310	525-632	30	
				Solution treated	222-265	505-578	56-63	
316L	1.2		CMT +Top TIG	CMT-Vertical- Continuous Mode (CM)	336	574	42	(Rodriguez et al., 2018)
				Vertical- Pulsed Mode (PM)	331	536	45.6	
				Horizontal-CM	364	577	43.4	
				Horizontal-PM	374	588	45.1	
				Base material	346	651	47	
				Top TIG -Vertical	322	539	43.1	
				Top TIG - Horizontal	365.5	590.3	42.3	
ER70S-6	0.889	A36	GMAW	Horizontal	400	500	-	(Rafieazad et al., 2019)
				Vertical	385	500	-	
Ti6Al4V	1.2	Ti6Al4V	CMT	As deposited	-	1000	7.5	(Gou et al., 2019)
				900ºC/4h+FC	-	850	7	
				1200ºC/4h+FC	-	760	5.8	
Ti and Al	1.2		GTAW	As deposited +Y-direction.	485	550	-	(Ma et al., 2016)
				As deposited+ Z	450	500	-	
				1060ºC+Y	490	600	-	
				1060ºC+Z	490	590	-	
				1200ºC+Y	420	500	-	
				1200ºC+Z	410	500	-	

WQ- Water Quenching; FC-Furnace Cooling

decision-making stage by adjusting the process parameters for improved deposition quality. In special deposition criteria cases, the WAAM monitoring system should be altered based on the application.

ADVANTAGES AND CHALLENGES IN THE WAAM PROCESSES

The cost of powders used in conventional additive manufacturing process is much higher compared to the wire costs used in the WAAM processes, for example in case of wires 180-1,300 Rs/kg for steels and 9,000 to 22,000 Rs/kg for titanium etc., where for powders 5,000-8,000 Rs/kg in the case of steel powders and 24,000 to 68,000 Rs/kg in the case of titanium powders. Therefore, WAAM processes were cost efficient than powder-based AM techniques (Derekar, 2018). Deposition rates are higher for example

Figure 16. Monitoring parameters/devices used in the WAAM processes.

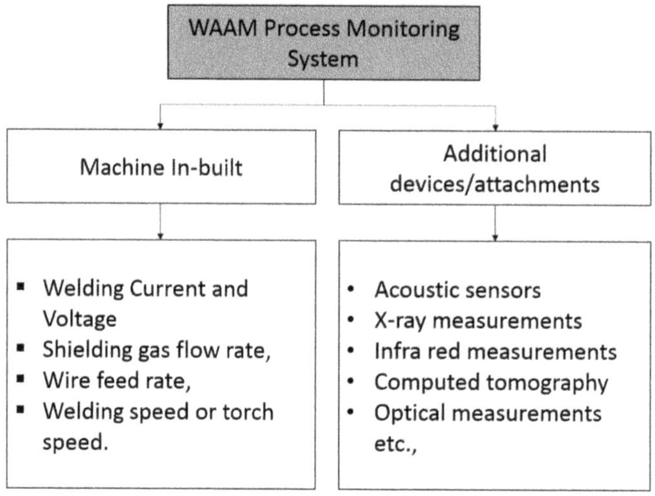

Figure 17. Schematic sketch of the complete monitoring system required for Wire Arc Additive Manufacturing processes
Adapted from (Xu et al., 2018).

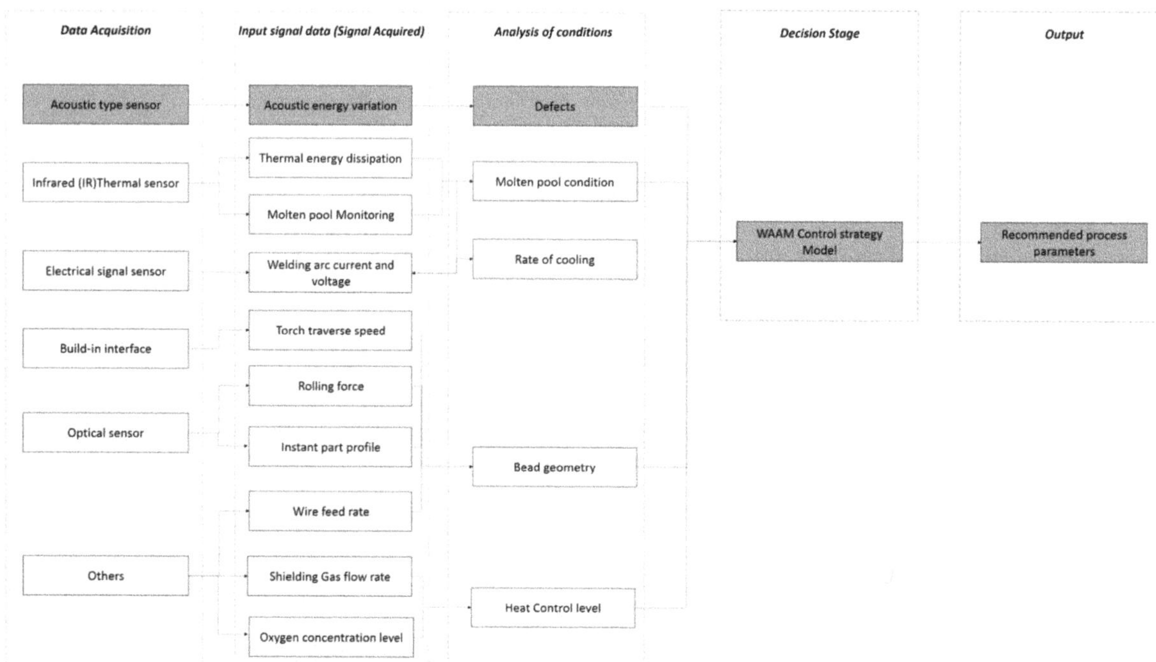

steels were deposited at a rate of up to 10 kg/h in WAAM as compared to 600 gram/h in conventional AM techniques. The typical challenges associated in the WAAM process are highlighted in the Figure 18.

Figure 18. Challenges in the WAAM processes
Adapted from (Williams et al., 2016).

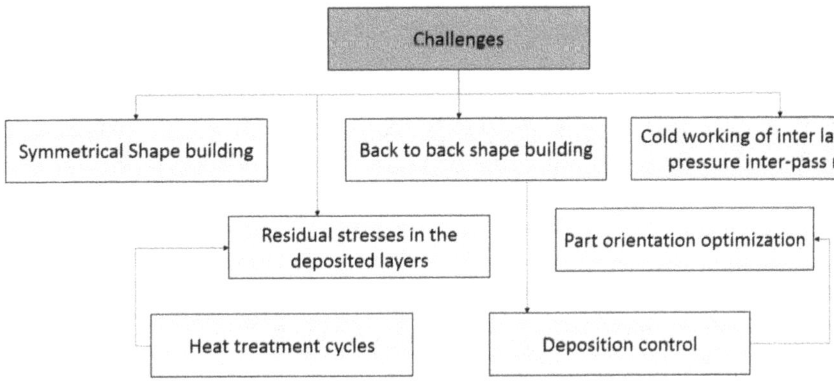

However, controlling the large deposition rates (large molten liquid) is quite challenging. In addition to that the layer by layer processes require stringent automatic control of the system where the thermal cycle effects cause non uniform properties in layers.

CONCLUSION

- WAAM processes are successful in the fabrication of complex product shapes. A lot of research and development is going on in the academia and industries in the areas of materials applicability, microstructural changes, thermal effects, defect (porosity) reduction, inner and interlayer cooling in the deposited layers etc.
- Larger deposition rates with cost effectiveness increases the usage of WAAM process in the industries for deposition of steels, titanium, nickel, magnesium, aluminum, and other alloys compared to the conventional AM processes.
- WAAM automation and modeling advancements improve the process penetration into the market for producing wide range of products.
- WAAM processes with subsequent processes such as cold working or post deposition heat treatments has shown significant improvement in the properties compared to the as-deposited layers (for example rolling minimizes the porosity and heat treatment refines the grain structure and promotes segregation of precipitates).
- Conventional heat sources such as GMAW, GTAW, Plasma arc welding processes have high heat inputs which are quite suitable as energy sources in WAAM processes. Because of controllable deposition (deposition stability) and defects minimization, CMT has become a popular heat source for WAAM processes.

REFERENCES

Abe, T., & Sasahara, H. (2016). Dissimilar metal deposition with a stainless steel and nickel-based alloy using wire and arc-based additive manufacturing. *Precision Engineering*, *45*, 387–395. doi:10.1016/j.precisioneng.2016.03.016

Ayarkwa, K. F., Williams, S. W., & Ding, J. (2017). Assessing the effect of TIG alternating current time cycle on aluminium wire + arc additive manufacture. *Additive Manufacturing*, *18*, 186–193. doi:10.1016/j.addma.2017.10.005

Baufeld, B., Brandl, E., & Van Der Biest, O. (2011). Wire based additive layer manufacturing: Comparison of microstructure and mechanical properties of Ti-6Al-4V components fabricated by laser-beam deposition and shaped metal deposition. *Journal of Materials Processing Technology*, *211*(6), 1146–1158. doi:10.1016/j.jmatprotec.2011.01.018

Bekker, A. C. M., Verlinden, J. C., & Galimberti, G. (2016). *Challenges in Assessing the Sustainability of Wire + Arc Additive Manufacturing for Large Structures*. Academic Press.

Chen, X., Li, J., Cheng, X., He, B., Wang, H., & Huang, Z. (2017). Microstructure and mechanical properties of the austenitic stainless steel 316L fabricated by gas metal arc additive manufacturing. *Materials Science and Engineering A*, *703*(January), 567–577. doi:10.1016/j.msea.2017.05.024

Chen, X., Li, J., Cheng, X., Wang, H., & Huang, Z. (2018). Effect of heat treatment on microstructure, mechanical and corrosion properties of austenitic stainless steel 316L using arc additive manufacturing. *Materials Science and Engineering A*, *715*(October), 307–314. doi:10.1016/j.msea.2017.10.002

Cunningham, C. R., Flynn, J. M., Shokrani, A., Dhokia, V., & Newman, S. T. (2018). Invited review article: Strategies and processes for high quality wire arc additive manufacturing. *Additive Manufacturing*, *22*(June), 672–686. doi:10.1016/j.addma.2018.06.020

Derekar, K. S. (2018). A review of wire arc additive manufacturing and advances in wire arc additive manufacturing of aluminium. *Materials Science and Technology (United Kingdom)*, *34*(8), 895–916. doi:10.1080/02670836.2018.1455012

Dutta, B., & Froes, F. H. S. (2017). The Additive Manufacturing (AM) of titanium alloys. *Metal Powder Report*, *72*(2), 96–106. doi:10.1016/j.mprp.2016.12.062

Fixter, J., Gu, J., Ding, J., Williams, S. W., & Prangnell, P. B. (2017). Preliminary investigation into the suitability of 2xxx alloys for Wire-Arc Additive Manufacturing. *Materials Science Forum*, *877*, 611–616. . doi:10.4028/www.scientific.net/MSF.877.611

Ge, J., Ma, T., Han, W., Yuan, T., Jin, T., Fu, H., Xiao, R., Lei, Y., & Lin, J. (2019). Thermal-induced microstructural evolution and defect distribution of wire-arc additive manufacturing 2Cr13 part: Numerical simulation and experimental characterization. *Applied Thermal Engineering*, *163*(100), 114335. doi:10.1016/j.applthermaleng.2019.114335

Geng, H., Li, J., Xiong, J., Lin, X., & Zhang, F. (2017). Optimization of wire feed for GTAW based additive manufacturing. *Journal of Materials Processing Technology*, *243*, 40–47. doi:10.1016/j.jmatprotec.2016.11.027

Gou, J., Shen, J., Hu, S., Tian, Y., & Liang, Y. (2019). Microstructure and mechanical properties of as-built and heat-treated Ti- 6Al-4V alloy prepared by cold metal transfer additive manufacturing. *Journal of Manufacturing Processes, 42*(October), 41–50. doi:10.1016/j.jmapro.2019.04.012

Gu, J., Ding, J., Williams, S. W., Gu, H., Bai, J., Zhai, Y., & Ma, P. (2016). The strengthening effect of inter-layer cold working and post-deposition heat treatment on the additively manufactured Al-6.3Cu alloy. *Materials Science and Engineering A, 651*, 18–26. doi:10.1016/j.msea.2015.10.101

Gu, J., Ding, J., Williams, S. W., Gu, H., Ma, P., & Zhai, Y. (2016). The effect of inter-layer cold working and post-deposition heat treatment on porosity in additively manufactured aluminum alloys. *Journal of Materials Processing Technology, 230*, 26–34. doi:10.1016/j.jmatprotec.2015.11.006

Gu, J., Gao, M., Yang, S., Bai, J., Zhai, Y., & Ding, J. (2020). Microstructure, defects, and mechanical properties of wire + arc additively manufactured Al[sbnd]Cu4.3-Mg1.5 alloy. *Materials & Design, 186*, 108357. doi:10.1016/j.matdes.2019.108357

Gu, J., Wang, X., Bai, J., Ding, J., Williams, S., Zhai, Y., & Liu, K. (2018). Deformation microstructures and strengthening mechanisms for the wire+arc additively manufactured Al-Mg4.5Mn alloy with inter-layer rolling. *Materials Science and Engineering A, 712*(November), 292–301. doi:10.1016/j.msea.2017.11.113

Horgar, A., Fostervoll, H., Nyhus, B., Ren, X., Eriksson, M., & Akselsen, O. M. (2018). Additive manufacturing using WAAM with AA5183 wire. *Journal of Materials Processing Technology, 259*(November), 68–74. doi:10.1016/j.jmatprotec.2018.04.014

Li, J. L. Z., Alkahari, M. R., Rosli, N. A. B., Hasan, R., Sudin, M. N., & Ramli, F. R. (2019). Review of wire arc additive manufacturing for 3d metal printing. *International Journal of Automotive Technology, 13*(3), 346–353. doi:10.20965/ijat.2019.p0346

Li, Y., Xiong, J., & Yin, Z. (2019). Molten pool stability of thin-wall parts in robotic GMA-based additive manufacturing with various position depositions. *Robotics and Computer Integrated Manufacturing, 56*(August), 1–11. doi:10.1016/j.rcim.2018.08.002

Ma, Y., Cuiuri, D., Li, H., Pan, Z., & Shen, C. (2016). The effect of postproduction heat treatment on γ-TiAl alloys produced by the GTAW-based additive manufacturing process. *Materials Science and Engineering A, 657*, 86–95. doi:10.1016/j.msea.2016.01.060

Miao, Q., Wu, D., Chai, D., Zhan, Y., Bi, G., Niu, F., & Ma, G. (2020). Comparative study of microstructure evaluation and mechanical properties of 4043 aluminum alloy fabricated by wire-based additive manufacturing. *Materials & Design, 186*, 108205. doi:10.1016/j.matdes.2019.108205

Qi, Z., Cong, B., Qi, B., Sun, H., Zhao, G., & Ding, J. (2018). Microstructure and mechanical properties of double-wire + arc additively manufactured Al-Cu-Mg alloys. *Journal of Materials Processing Tech., 255*(December), 347–353. doi:10.1016/j.jmatprotec.2017.12.019

Qi, Z., Cong, B., Qi, B., Zhao, G., & Ding, J. (2018). Properties of wire + arc additively manufactured 2024 aluminum alloy with different solution treatment temperature. *Materials Letters, 230*, 275–278. doi:10.1016/j.matlet.2018.07.144

Qi, Z., Qi, B., Cong, B., Sun, H., Zhao, G., & Ding, J. (2019). Microstructure and mechanical properties of wire + arc additively manufactured 2024 aluminum alloy components: As-deposited and post heat-treated. *Journal of Manufacturing Processes, 40*(May), 27–36. doi:10.1016/j.jmapro.2019.03.003

Qi, Z., Qi, B., Cong, B., & Zhang, R. (2018). Microstructure and mechanical properties of wire + arc additively manufactured Al-Mg-Si aluminum alloy. *Materials Letters*, *233*, 348–350. doi:10.1016/j.matlet.2018.09.048

Rafieazad, M., Ghaffari, M., Vahedi Nemani, A., & Nasiri, A. (2019). Microstructural evolution and mechanical properties of a low-carbon low-alloy steel produced by wire arc additive manufacturing. *International Journal of Advanced Manufacturing Technology*, *105*(5-6), 2121–2134. doi:10.100700170-019-04393-8

Rodrigues, T. A., Duarte, V., Miranda, R. M., Santos, T. G., & Oliveira, J. P. (2019). Current Status and Perspectives on Wire and Arc Additive Manufacturing (WAAM). *Materials (Basel)*, *12*(7), 1121. doi:10.3390/ma12071121 PMID:30987382

Rodriguez, N., Vázquez, L., Huarte, I., Arruti, E., Tabernero, I., & Alvarez, P. (2018). Wire and arc additive manufacturing: A comparison between CMT and TopTIG processes applied to stainless steel. *Welding in the World*, *62*(5), 1083–1096. doi:10.100740194-018-0606-6

Shen, C., Pan, Z., Ma, Y., Cuiuri, D., & Li, H. (2015). Fabrication of iron-rich Fe – Al intermetallics using the wire-arc additive manufacturing process. *Additive Manufacturing*, *7*, 20–26. doi:10.1016/j.addma.2015.06.001

Sun, R., Li, L., Zhu, Y., Guo, W., Peng, P., & Cong, B. (2018). *Microstructure, residual stress and tensile properties control of wire- arc additive manufactured 2319 aluminum alloy with laser shock peening.* doi:10.1016/j.jallcom.2018.02.353

Williams, S. W., Martina, F., Addison, A. C., Ding, J., Pardal, G., & Colegrove, P. (2016). *Wire þ Arc Additive Manufacturing.* doi:10.1179/1743284715Y.0000000073

Xu, F., Dhokia, V., Colegrove, P., McAndrew, A., Williams, S., Henstridge, A., & Newman, S. T. (2018). Realisation of a multi-sensor framework for process monitoring of the wire arc additive manufacturing in producing Ti-6Al-4V parts. *International Journal of Computer Integrated Manufacturing*, *31*(8), 785–798. doi:10.1080/0951192X.2018.1466395

Zhong, H., Qi, B., Cong, B., Qi, Z., & Sun, H. (2019). Microstructure and Mechanical Properties of Wire + Arc Additively Manufactured 2050 Al–Li Alloy Wall Deposits [English Edition]. *Chinese Journal of Mechanical Engineering*, *32*(1), 92. Advance online publication. doi:10.118610033-019-0405-z

Zhou, Y., Lin, X., Kang, N., Huang, W., Wang, J., & Wang, Z. (2019). Influence of travel speed on microstructure and mechanical properties of wire + arc additively manufactured 2219 aluminum alloy. *Journal of Materials Science and Technology*. Advance online publication. doi:10.1016/j.jmst.2019.06.016

ADDITIONAL READING

Gisario, A., Kazarian, M., Martina, F., & Mehrpouya, M. (2019). Metal additive manufacturing in the commercial aviation industry : A review. *Journal of Manufacturing Systems*, *53*(June), 124–149. doi:10.1016/j.jmsy.2019.08.005

Javadi, Y., Macleod, C. N., Pierce, S. G., Gachagan, A., Lines, D., Mineo, C., Ding, J., Williams, S., Vasilev, M., Mohseni, E., & Su, R. (2019). Ultrasonic phased array inspection of a Wire + Arc Additive Manufactured (WAAM) sample with intentionally embedded defects. *Additive Manufacturing*, *29*(July), 100806. doi:10.1016/j.addma.2019.100806

Karmuhilan, M., & Kumar, A. (2018). ScienceDirect Intelligent process model for bead geometry prediction in WAAM. *Materials Today: Proceedings*, *5*(11), 24005–24013. doi:10.1016/j.matpr.2018.10.193

Kelly, S. M., & Kampe, S. L. (2004). Microstructural Evolution in Laser-Deposited Multilayer Ti-6Al-4V Builds : Part II. *Thermal Modeling.*, *35*(June), 1869–1879. doi:10.100711661-004-0095-7

Lindgren, L., Lundbäck, A., Fisk, M., Pederson, R., & Andersson, J. (2016). Simulation of additive manufacturing using coupled constitutive and microstructure models. *Additive Manufacturing*, *12*, 144–158. doi:10.1016/j.addma.2016.05.005

Michopoulos, J. G., Iliopoulos, A. P., Steuben, J. C., Birnbaum, A. J., & Lambrakos, S. G. (2018). On the multiphysics modeling challenges for metal additive manufacturing processes. Additive Manufacturing, 22(February 2015), 784–799. doi:10.1016/j.addma.2018.06.019

Neikter, M., Åkerfeldt, P., Pederson, R., Antti, M., & Sandell, V. (2018). Materials Characterization Microstructural characterization and comparison of Ti-6Al-4V manufactured with di ff erent additive manufacturing processes. *Materials Characterization*, *143*(January), 68–75. doi:10.1016/j.matchar.2018.02.003

Ngo, T. D., Kashani, A., Imbalzano, G., Nguyen, K. T. Q., & Hui, D. (2018). Additive manufacturing (3D printing): A review of materials, methods, applications and challenges. Composites Part B, 143(December 2017), 172–196. doi:10.1016/j.compositesb.2018.02.012

Stavropoulos, P., Foteinopoulos, P., Papacharalampopoulos, A., & Bikas, H. (2018). Addressing the challenges for the industrial application of additive manufacturing : Towards a hybrid solution. *International Journal of Lightweight Materials and Manufacture*, *1*(3), 157–168. doi:10.1016/j.ijlmm.2018.07.002

Wang, J., Pan, Z., Wei, L., He, S., Cuiuri, D., & Li, H. (2019). Introduction of ternary alloying element in wire arc additive manufacturing of titanium aluminide intermetallic. *Additive Manufacturing*, *27*(February), 236–245. doi:10.1016/j.addma.2019.03.014

Wang, J. F., Sun, Q. J., Wang, H., Liu, J. P., & Feng, J. C. (2016). Effect of location on microstructure and mechanical properties of additive layer manufactured Inconel 625 using gas tungsten arc welding. *Materials Science and Engineering A*, *676*, 395–405. doi:10.1016/j.msea.2016.09.015

KEY TERMS AND DEFINITIONS

Additive Manufacturing (AM): AM is the process of manufacturing parts by melting and depositing the material in layer by layer fashion.

Binder Jetting: Process in which a liquid bonding agent is selectively deposited to join powder materials.

Direct Energy Deposition (DED): Process in which focused thermal energy is used to fuse materials by melting and are being deposited.

Heat Source: which melts the metal (wire), it can be of GTAW, GMAW, or CMT, etc.

Inner Layer: Is the deposited layer in between the two inter layers.

Inter Layer: Is the interface layer present in between the two deposited layers.

Powder Bed Fusion: Process in which thermal energy selectively fuses regions of a powder bed.

Sheet Lamination: Process in which sheets of material are bonded to form an object.

Wire Arc Additive Manufacturing (WAAM): WAAM processes utilizes the electric arc (energy source) and wire feed stock (for melting) to deposit the metal layer by layer.

Chapter 7
Additive Manufacturing of Multi-Material and Composite Parts

V. Senthilkumar
National Institute of Technology, Trichy, India

Velmurugan C.
Indian Institute of Information Technology, Tiruchirappalli, India

K. R. Balasubramanian
https://orcid.org/0000-0001-7179-5566
National Institute of Technology, Tiruchirappalli, India

M. Kumaran
National Institute of Technology, Tiruchirappalli, India

ABSTRACT

Additive manufacturing (AM) technology can be employed to produce multimaterial parts. In this approach, multiple types of materials are used for the fabrication of a single part. Custom-built functionally graded, heterogeneous, or porous structures and composite materials can be fabricated thorough this process. In this method, metals, plastics, and ceramics have been used with suitable AM methods to obtain multi-material products depending on functional requirements. The process of making composite materials by AM can either be performed during the material deposition process or by a hybrid process in which the combination of different materials can be performed before or after AM as a previous or subsequent stage of production of a component. Composite processes can be employed to produce functionally graded materials (FGM).

DOI: 10.4018/978-1-7998-4054-1.ch007

INTRODUCTION

Manufacturing industries have been faced many difficulties such as *manufacturing speed* and *cost* over the years. Therefore, industrialists were focused on alternative manufacturing techniques to increase the manufacturing speed and to control the cost. The restrictions in traditional manufacturing processes have been diminished by the development of different AM processes. The AM process is a rapid manufacturing technology that has the ability to synthesis a component by adding the required composition of material continuously still final part is finished (Boddeti, Ding, Kaijima, Maute, & Dunn, 2018). Commercially, many 3D fabrication techniques are available, comprising electron beam melting, fused deposition modeling (FDM), selective laser sintering and melting, laser engineered net shaping, polyjet 3-D printing, and projection micro-stereo lithography etc. (Spowart, Gupta, & Lehmhus, 2018; Chai et al., 2017; Liu et al., 2017, Stull et al., 2018).

In these methods, metals, plastics, and ceramics are used with a suitable AM process to obtain multi-material products depending on functional requirements (Bandyopadhyay, & Heer, 2018). The development of composite materials by AM can either be achieved during the material deposition process or by a hybrid process. Heterogeneous scaffolds and functionally graded materials (FGM) are fabricated through composite structures which can be obtained various properties in a single integrated component (Toursangsaraki, 2018). The parts can be built with the combination of different materials in the hybrid process of AM. Multi-material AM process is fascinating to the researchers due to its potentials such as customized design and structural applications of rapid manufacturing. It can be achieved the property benefits as similar to the hybridized products. Moreover, the multi-material AM process can be developed the products with fast and robust structures (Ngo, Kashani, Imbalzano, Nguyen, & Hui, 2018; Muguruza et al., 2017).

In conventional manufacturing, the combined materials structures are developed using the joining of mechanical fasteners such as screws or rivets depending on the applications. These kinds of additional setups are not required in a multi-material AM process which can be fabricated multi-material layers or multi-composite parts continuously (Chen, & Zheng, 2018). This technique mainly classified under two categories such as multi-layer method and multi-material structure. In the first technique, the components can be developed layer by layer, where the material of second layer deposits over the first layer mate-

Figure 1. Schematic comparison conventional manufacturing and additive manufacturing processes

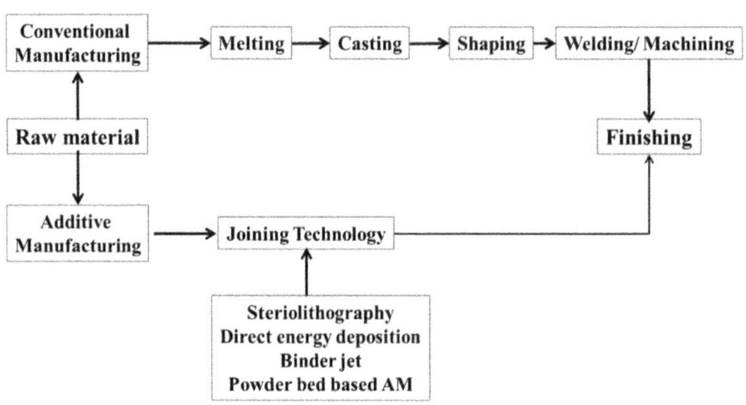

rial. The second method is called a multi-material structure, where the AM process builds the vertical structure simultaneously using different materials (Edgar, & Tint, 2015).

The multi-material structure is developed using number of steps in conventional manufacturing process. The initial sets of components are fabricated using primary processes where the secondary operations or post processes executed to make the multi-material component. In industry, the basic products are developed using casting, melting and forging process followed by some finishing operation of machining which bring it into desired final product. Thereafter, they allowed to joining processes such as different welding techniques to add the secondary component with that primary part. The difficulties such as time consumption and wastage of materials are observed from the traditional manufacturing processes. In order to avoid such difficulties, additive manufacturing processes are being implemented to prepare a multi-model structure in a single step. The raw powder elements are allowed to make the final multi-parts from the design stage (Bandyopadhyay, & Heer, 2018). A clear illustration of the comparison of conventional and additive manufacturing processes is shown in Figure 1.

Composite structures can be made with different materials in a single step which allows to synthesis the required product directly from the part design. The main reason behind this multi-material concept was initiated to improve the functional properties of a specific component. Ti6Al4V alloy has potential applications in automotive, chemical, aerospace and marine industries because of great specific strength, high stiffness and better corrosion resistance. Moreover, it is used in many implantations and biomedical applications (Attar et al., 2019). Titanium alloys have been faced a major failure in the formation of oxide layer which caused adhesive wear on the surface due to high reactivity with the oxygen. In this condition, it is essential to improve the wear resistance in the abrasion environment. Therefore, TiB_2 was introduced to prepare the multi-material of TiB_2/ Ti6Al4V which could be improved the hardness and wear resistance characteristics of titanium alloy (Wang et al., 2019).

This chapter aims to consolidate the report of present additive manufacturing technologies supported to develop the multi-material and composite structures. The beneficial characteristics of multi-material processed through different additive manufacturing processes are discussed. The effects process parameters on functional characteristics of AM developed multi-material structure are analysed. The factors followed to improve the functional characteristics of various multi-material combinations are discussed under different fabrication techniques with the schematic illustrations.

ADDITIVE MANUFACTURING

Recently, the technology of additive manufacturing is incessantly being restructured and customized in various fields such as automotive, aerospace, medical, and food processing industries. The additive manufacturing technology can be built the material with complex shapes using the layer-wise elegant concept. Moreover, the AM process can construct products using a wide variety of material compositions. In addition to the strength, surface finish and geometric tolerance are the major influencing factors while considering the results of AM process (Udroiu, Braga, and Nedelcu, 2019). These parameters can be modified for the same input at different AM processes. These changes are happened by the variation in AM methodology, type of materials and orientation of the structure (Gao et al., 2015). Different alterations of AM and its combination focused on creating multi-material or composite structures. Each AM processes carries unique benefits; therefore, the AM process could be selected based on the requirement. The presence of voids among the layers needs to be controlled by proper selection process and parameters. In addition to this void formation, staircase effect in developing part using AM technique is the greatest challenge. Another challenge in the AM is the survival of mechanical properties and microstructure in the printed part.

ADDITIVE MANUFACTURING FOR DEVELOPING MULTI-MATERIAL AND COMPOSITE

Additive manufacturing has high potential applications due to design and manufacturing flexibility. Moreover, AM techniques can be developed the components with intricate shapes without any additional processes. AM process has the potential to meet fine solidified microstructures as per the requirement due to the rapid cooling provisions. The AM process can be fabricated the products with multi-material and composite structures under various techniques as listed in Figure 2.

Figure 2. Classification of additive manufacturing process for metals

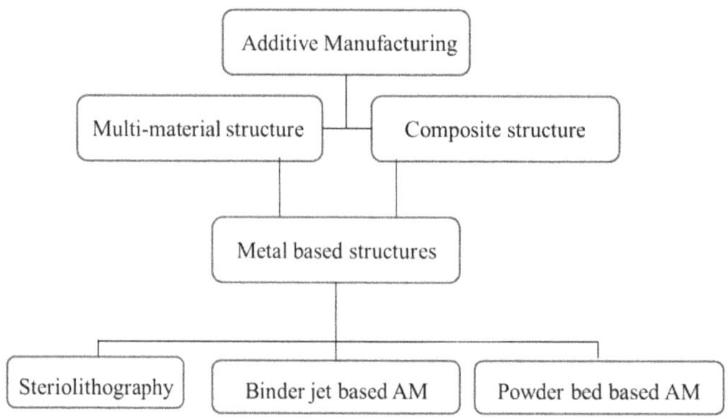

Stereolithography Methods

Stereolithography is generally defined as SLA 3D printing which is the most popular and widespread technique in the field of additive manufacturing. It prints the products using a high-powered laser to harden the liquid resin, which is confined in a reservoir to develop the desired 3D structure. Stereolithography based AM technique has distinct benefits includes high-resolution, no need of additional set-up, and economic printing. The primary technology of SLA is offered a good quality of products with accuracy, and access of different materials comprises biomedical materials (Mu et al., 2017).

A skinny layer can be printed with high accuracy using the SLA process. Moreover, the multi-material part has processed using the stereolithography process. The application of SLA process in the development of metal renovations has augmented in recent days. The development of metal coping using conventional casting process can be replaced by SLA technique to improve the efficiency. The cobalt-chromium coping has been prepared using lost wax technique. The same internal spacing and discrepancies in the composite have been rectified using SLA technique on cobalt-chromium metal copings. The results suggested that the metal copings developed with SLA process has obtained more efficient than the conventional fabrication techniques (Kim, Kim, Kim, & Moon, 2018).

Now-a-days, rapid prototyping and manufacturing processes are extensively used in all engineering fields to fabricate the products with complex shapes. The composite structure developed through SLA process can be coated with metal element to improve the mechanical properties. In such a way, the SLA developed components have been coated with nickel. The material coated on SLA sintered parts exhibited better strength, flexural strength and Young's modulus (Zhou, Li, Zeng, & Zhang, 2007). Thus, the SLA process is supported to improve the accuracy and quality of surface developed through rapid manufacturing processes. The parts fabricated using SLA process involved to secondary operations when they have improper surface. Therefore, the coating is applied on the SLA printed parts to improve the mechanical and functional properties.

Binder Jet Based Additive Manufacturing Techniques

Binder jetting (BJ) is one of the AM methods which manufacture products that could be enabled in different fields such as prototypes, bone implants and foundry moulds. The BJ based process is a subcategory of AM technique that can sinter the products using ceramics, metals and polymer composites (Dini, Ghaffari, Jafar, Hamidreza, & Marjan, 2019). In this technique, the powder material is spread layer by layer with the help of a rotating roller. Thereafter, the liquid binding agent from the head is passed on the powder bed to develop the structure. In some cases, the heater is involved in controlling moisture in the powder system or liquid binder agent. Afterward, a new layer is developed on the existing layer which is lowered from the initial position. This process is repeated to construct the required product and its schematic is shown in Figure 3 (Ziaee, & Crane, 2019). The mechanical properties of the binder jetted metallic parts can be improved with processing treatments.

Figure 3. Schematic view of binder jetting method
(Adapted from (Ziaee, & Crane, 2019))

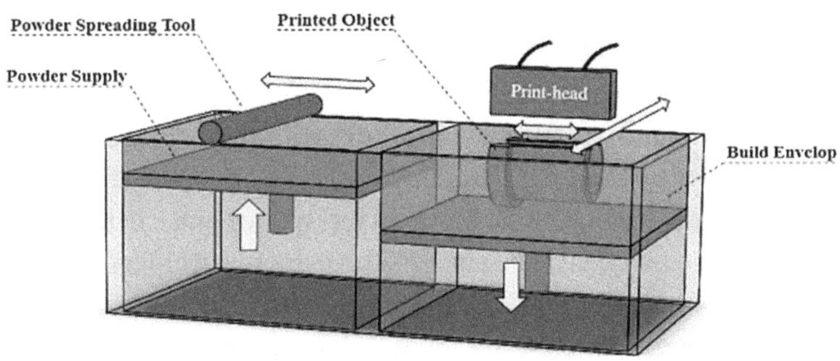

The parameters such as powder size, flowability and volume fraction are very important in binder jetting based AM technique. Hence, these factors are influenced by the porous structure in the sintered products. The control of porosity is essential to increase the mechanical behaviours of the jetted parts. The powder bed of binder jet process is constructed in the z-axis direction which could be considered as layer thickness. The range of layer thickness can be varied from 0.0889 to 0.2286 mm as per the applications. A very thin part can be printed in this BJ method by selecting the lower thickness of the layer which caused more time consumption (Butscher et al., 2012).

The development of amorphous structure has fascinated in many fields due to its excellent mechanical properties, permeability, corrosion and wear resistance. The binder jet based additive manufacturing is supported to develop the intricate shapes with amorphous structures using metal powders. The structural amorphous based metal composite has been fabricated using ferrous, chromium and molybdenum powder elements with the help of binder jet 3D printing process. The micro-porous structure was observed from the interiors of the powder particles which could be supported to develop the amorphous porous structural components (Ilogebe, Waters, Elliot, & Shackleford, 2019).

The binder jet based AM technique can be implemented to develop the multi-functional materials that exhibit unique properties. The nickel, manganese and gallium based multi-material has been sintered using binder jet 3D printing process using the powder elements of respective materials. The powder elements which possessed the mixture of nickel, manganese and gallium prepared for binder jet process. The binder jet equipment was used to develop the structure with layer by layer construction up to 15 layers. The sintering of multi-material components by binder jet process would enable better functional characteristics such as magnetic and shape memory effect even for intricate structures (Mostafaei et al., 2017).

Powder Bed Fusion Techniques

The development of functionally graded materials is executed based on the multi-material concept where the part can be constructed layer by layer. The laser or electron beam is used to melt or sinter the powder materials together which is called powder bed fusion technique. The major techniques classified under this concept are selective laser sintering and melting, shape deposition manufacturing (SDM), laser engineered net shaping (LENS) and electron beam melting (EBM). These techniques can be fabricated products using metals, polymers and ceramics (Sing, An, Yeong, & Wiria, 2016).

Selective laser sintering (SLS) is one of the AM techniques that practice laser energy to sinter powder elements composed to form solid components. A schematic view of SLS process is shown in Figure 4. This SLS process is mainly influenced the microstructure evolution of the sintered 3D parts. The parameters such as scanning speed and laser intensity are needed to optimize to obtain the great microstructure on the SLS processed products. The powder elements are collected in a chamfer and involved to the heating process less than the melting temperature of that material. The SLS process led to the formation of solid structure of multi-materials composition with shorter time compared to conventional sintering processes (Yuan, Zheng, Chua, Yan, & Zhou, 2018).

Selective Laser Melting (SLM) is one of the best AM techniques that can be fabricated the metal part in a single step process. SLM technique can be sintered the parts by layer formation process whereas the metals are fabricated in the sequence of deposition, solidification and stack on top surface. The quality of the final product sintered using SLM process can be influenced by some important factors as listed in Table 1 (Aboulkhair, Everitt, Ashcroft., & Tuck, 2014). The functional component has been developed using SLM process with the composition aluminium, silicon and magnesium. The SLM processed component has attained a maximum relative density of ~99% due to its better flowability. The mechanical properties such as hardness and yield strength were improved through this SLM process. The maximum density has been achieved by this SLM process by the selection of optimum SLM parameters such as hatching distance 0.13 mm, laser power of 350 watts and scanning speed of 1650 mm/s (Raus et al., 2017).

Figure 4. The schematic view of SLS process

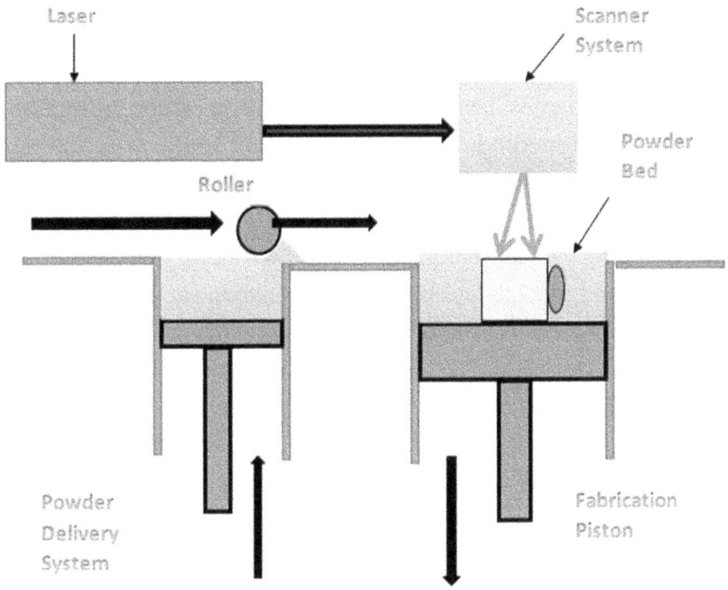

The same composition of material was allowed to direct metal laser sintering (DMLS) process which could be produced a structure with near net shape. The mixture of powder elements possessed aluminium, silicon and manganese. The intensity of laser beam has been modulated in DMLS process that caused the melting of powder elements on top new layer. The new layer has penetrated slightly on the previous layer which supported to produce a better link between layers. Moreover, the molten powder metals exhibited a small melt pools which led to the solidification rapidly. Therefore, a maximum density can be achieved from the part sintered using DMLS process. The results suggested that the corrosion resistance of DMLS sintered structure has been increased by involving them in to shot peening and polishing process. Therefore, DMLS fabricated components can be involved any kind of precise applications than the SLM fabricated structure (Cabrini et al., 2016).

The composite and multi-material components can be sintered using the selective laser melting (SLM) process by melting one material into others. The tungsten composite has been prepared to develop the

Table 1. Influencing parameters in SLM process

Process Parameter			
Laser Related	**Power Related**	**Scan Related**	**Temperature Related**
Laser power	Particle size	Scan speed	Powder bed temperature
Spot size	Particle shape & distribution	Scan spacing	Temperature uniformity
Pulse duration	Powder bed density	Scan pattern	Powder bed temperature
Pulse frequency	Layer thickness		
	Material properties		

(Adapted from (Aboulkhair, Everitt, Ashcroft., & Tuck, 2014))

copper and tin based composite structure with great density technique. A three dimensional composite of W-Cu10Sn was developed by melting the tungsten element using SLM. The copper and tin were involved in melting process and induced 3D structure which caused the formation of W-Cu10Sn composites with higher density (Zhou et al., 2020).

Additive manufacturing techniques have emerged as a revolution for prompt manufacturing of parts by an incremental layer-on-layer method. The development of multi-material part has fascinating the manufacturing sector compared to that of traditional manufacturing processes. A hybrid-structure has been developed with AlSi10Mg and Al-Cu-Ni-Fe-Mg cast alloy substrate using SLM process as displayed in Figure 5. The selection of parameters and building direction are major factors which could be decided the microstructural and mechanical behaviours between the hybrid materials. The AlSi10Mg was placed on the Al-Cu-Ni-Fe-Mg substrate by stripe scanning of turning the laser with 67° between the consecutive layers. The SLM printed AlSi10Mg part exhibited a fine dendritic structure in the building direction. The better mechanical properties such as hardness and tensile strength have been attained in SLM processed AlSi10Mg due to its fine microstructure. Hence, the mechanical properties of 3D printed SLM part is mainly influenced by the microstructure.

Figure 5. Schematic view of SLM printed

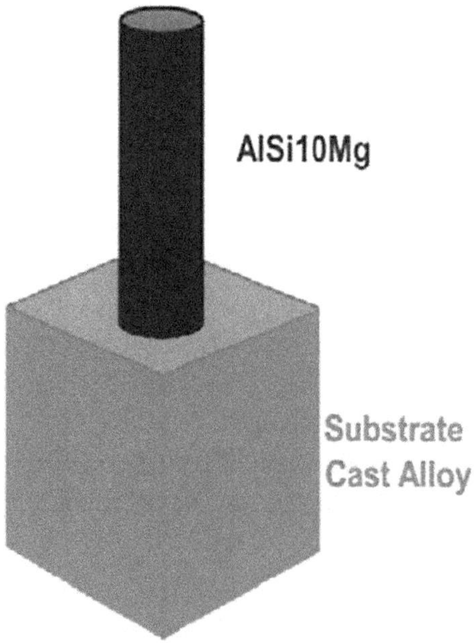

Laser engineered net shaping (LENS) is one of the powder bed-based techniques formed from the laser shield concept. The powder spreading concept can be eliminated in this technique. The powder elements are feed through an axial nozzle focussed on the part to be constructed. Moreover, the laser energy is focussed at the same point through the nozzle. The nozzle is fixed with a multi-axis movement table which enables as the part is being fabricated (Revathi et al., 2019). The schematic view of the LENS process is shown in Figure 6.

Figure 6. Schematic view of the LENS process

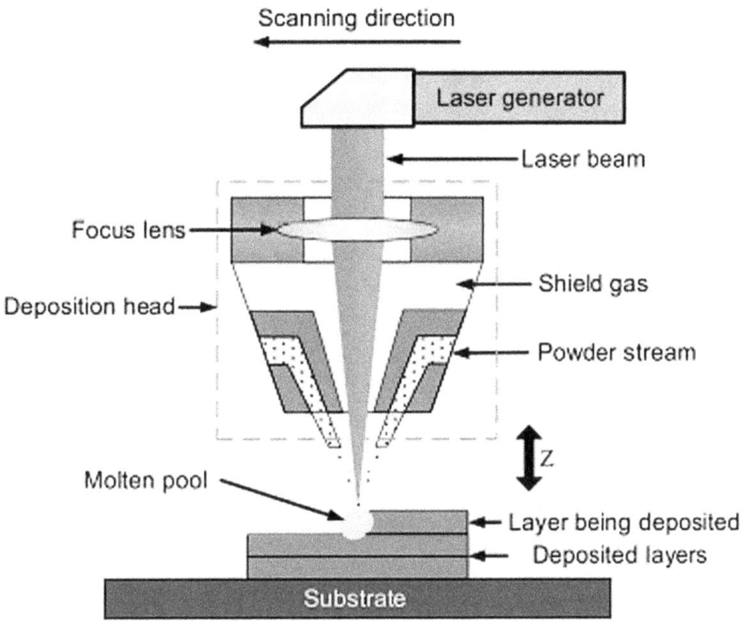

Laser engineered net shaping is a technique that enables the metal AM technique with the multi-material concept. An alloy of silver incorporated titanium has been developed using the LENS process with different content of silver (0.5 to 2%). The LENS sintered Ag-Ti alloy was achieved higher hardness and lower ductility than the pure titanium. The results of this study suggested that Ti-Ag alloy was developed as an antimicrobial solution for orthopedic fields. Moreover, it has enough potential to use in biomedical applications (Maharubin, Hu, Sooriyaarachchi, Cong, & Tan, 2019). The multi-material components developed using the LENS process can be exhibited better mechanical properties and they are feasible to use in biomedical instruments.

The shape deposition manufacturing (SDM) technique developed the heterogeneous parts layer by layer using 3D concept. Each layer in the structures is modified with some material removal and secondary processes. The SDM printed parts are allowed to material removal process to make them into required near net shape (Pham, & Dimov, 2012). A schematic view is shown in Figure 7 where the sequence of processes is listed under the hybrid SDM technique. The SDM can formulate a part with multi-material structures by assembling various metal layers. In SDM, component or stacks are constructed by the series of alternating layers of base and support elements (Vaezi, Chianrabutra, Mellor, & Yang, 2013). The main limitation was observed from the SDM process that the multi-material structure formed at a same layer is not feasible.

Electron Beam Melting (EBM) is used to sinter the products using powder bed fusion concept where the electron beam supplied the heat energy to melt the materials. Traditional melting processes have faced several issues such as *high melting point* and *oxygen affinity* which can be resolved by the EBM process (Galati, Snis, & Iuliano, 2019). The EBM based AM technique needs to do some pre-procedures to print the products successfully that consist of preheating of powder bed using high electron beam current. This pre-process would be supported to control the temperature gradient between surrounding and powder bed (Murr et al., 2012). Moreover, the composite structures used in tribo-corrosion resistive

Figure 7. Schematic view of shape deposition manufacturing technique
(Adapted from (Vaezi, Chianrabutra, Mellor, & Yang, 2013))

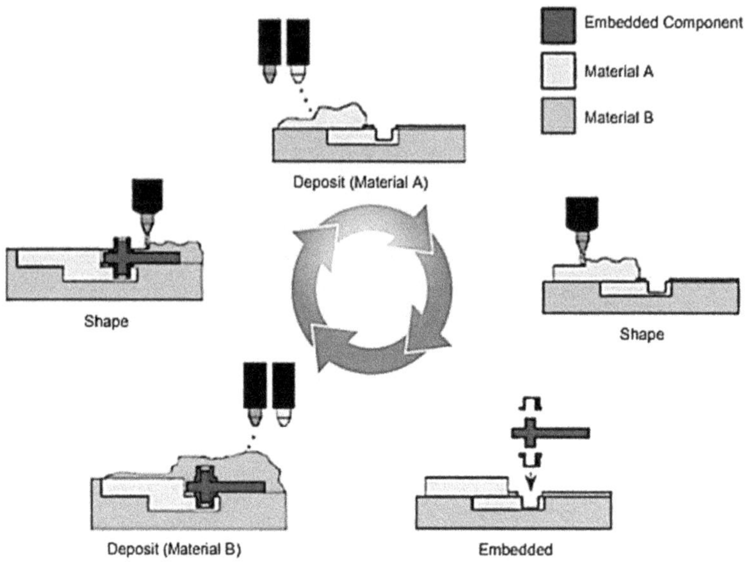

application can be developed using electron beam melting process. The main advantage of this process is the in-situ thermal treatment by the preheating process which could be removed the thermal stress as well as the tearing cracks (Sames, Unocic, Dehoff, Lolla, & Babu, 2014).

The metal matrix composite structure has been developed with the powders such as WC and NiBSi using EBM process. The powder had blended to obtain sufficient flowability to use in EBM process. The EBM developed structure revealed a high dense with crack-free condition where inter-diffusion had occurred perfectly among the powder elements. The EBM sintered composite exhibited a great flexural and compressive strength of 930 MPa and 1850 MPa respectively (Peng et al., 2016). The results suggested the EBM process to print the multi-material and composite part with desired properties.

The EBM technique can be applied to develop near-net-shape structure of nanocomposite to improve the mechanical and tribological behaviours. The TiC/Nb-TiAl nanocomposite structure has been developed by varying the TiC composition using EBM process. In this mechanism, the nano-TiC elements were used to diminish the charge accumulation due to its low electrical conductivity. The base form of $\alpha 2$ structure has been changed into γ phase by the dispersion of nano-composite of TiC element which led to the formation of rod structure. Thus, the mechanical properties of composite have been enhanced by the nano-TiC dispersion by the electron beam melting process (Kan, Chen, Peng, Liang, & Lin, 2020). Researchers have focused to merge two or more AM techniques to overcome the issues. Electron beam melting and selective laser melting jointly used to develop the functionally graded material. This concept was used to develop the TiAl and Ti based FGM with three different overlap scan distances such as 0.5, 1.0 and 2.0 mm. The width of interfacial transition zone is mainly influenced the mechanical properties of printed parts. In this case, the width of interfacial transition zone was increased with overlap scan distance. Therefore, the mechanical and microstructural properties have been strengthened (Zhou et al., 2020). In recent days, research is going to combine the mechanism of various AM techniques to resolve the failures.

FUNCTIONALLY GRADED MULTI-STRUCTURE

Initially, functionally graded materials were developed using traditional manufacturing processes such as centrifugal casting, powder metallurgy with conventional sintering process and vapour deposition, etc.,. But, these manufacturing methods have some interferences to fabricate the bulk structure except centrifugal casting which can be fabricated only the product with cylindrical structure. Moreover, the powder metallurgy route consists of multi-stages process such as powder preparation, milling process, consolidation of powder elements followed by sintering and cooling which consumes more time. Therefore, the traditional manufacturing processes consume more time, multi-tools and equipment for developing FGM with multi-material gradient structure. In recent days, additive manufacturing process has become a primary method in fabricating the FGM structure (Zhang, Jaiswal, Rai, & Nelaturi, 2018; Zhang et al., 2019; Yan, Chen, & Liou, 2019).

Functionally graded materials (FGM) has developed with inhomogeneous structure to obtain the benefits than the normal composite structures. The chemical composition of the inhomogeneous structure varies gradually from bottom layer to top surface. Therefore, the desired functional behaviours can be induced in the FGM as per the applications (Pei et al., 2017). The reason behind this implementation of multi-material concept in FGM is to tailor the superior properties of various materials in a single component. Recently developed concept of multi-material development in FGM using AM process is supported to improve the properties of component (Chen, & Zheng, 2018).

Figure 8. Schematic view of (a) discontinuous and (b) continuous FGM

The composition of material has varied continuously within the product which is developed as FGM. The schematic illustration of functionally graded material that possesses different chemical compositions is shown in Figure 8. The multi-material is sintered using number of layers as shown in Figure 8 (a). The FGM developed using continuous step is revealed in Figure 8 (b). The interface surface poised the mixture of composition at different ratio. However, the material can be reflected as a FGM even if the composition distributes inhomogeneous state.

The aluminium based functionally graded composite structure can be developed using friction stir additive manufacturing. The composite structure was developed with the aluminium plate where the holes with 1mm diameter were drilled to keep the titanium carbide. The local mechanical properties of FGM fabricated using friction stir based AM process have been mainly influenced by the volume fraction. In addition to these, the factors such as particle size, number of stir tool passes and addition of external element with matrix were influenced the properties of FGM (Sharma et al., 2017). Similarly, the functionally graded component has been developed using 316L stainless steel and H13 tool steel which have various lattice properties such as austenitic lattice and ferritic lattice respectively. The multi-material was developed using selective laser melting process with inline mechanism. The AM sintered multi-material structure has been involved to the analysis of equiaxed grain growth which resulted in the formation excellent interface layers and bonding strength. The solidification of SLM processed muti-structured component mainly influenced by the convection compared with diffusion process. Finally, the multi-material FGM specimen can be fabricated using selective laser melting process with tailored properties to meet the requirements (Hengsbach et al., 2018).

Moreover, the additive manufacturing offers many facilities rather than the traditional manufacturing processes, particularly in the geometries of shape and composition of materials. The recent developed concept in additive manufacturing is the control of composition based on location which could be developed the functionally graded materials. This could be analysed with point by point changes of the composition i.e. gradual variation which can be tailored the mechanical properties and microstructures in FGM (Wang et al., 2018; Zhang, Wei, Shi, & Xi, 2008). In an earlier investigation, the bulk component has been developed with the combination of three materials such as chromium, nickel and 316 stainless steel. The laser based direct energy deposition technique was involved to develop the gradient structure. Moreover, the results of FGM was validated using modelling software to investigate the effect of each elements presented in the system. The detrimental phases in FGM has influenced mechanical and metallurgical properties. Therefore, the process parameters were optimized in direct energy deposition technique for constructing the FGM. The printed FGM was subjected to heat treatment; the composition has been characterized before and after the heat treatment process. The detrimental effect was observed from the heat treated functionally graded materials in the form of σ phase. Therefore, a slight variation was observed in the composition of nickel due to the combined effects of discrepancy in evaporation and powder flowability. This could be suggested that the required gradient paths can be formed in FGM to achieve the expected properties such as strength, hardness and heat dissipation etc. (Eliseeva et al., 2019).

The functionally graded material structure can be developed using hybrid or multi-additive manufacturing process to enhance the properties as per the applications. The hybrid additive manufacturing processes have implemented by combining the selective laser melting and cold spraying techniques. This setup was implemented to develop the gradient structure of titanium and aluminium. The aluminium oxide was deposited on the selective laser melted titanium alloy (Ti6Al4V) through cold spraying concept. The thick and dense FGM structure was developed by this hybrid method without formation of any intermetallic structure. The Ti6Al4V produced via selective laser melting possessed acicular martensite structure and cold sprayed aluminium oxide exhibited a structure without any changes in grain size. These can be attributed to enhance the strength of the FGM with phase transformation and work hardening effect. This hybrid AM process has been avoided the formation of brittle intermetallic structure. The defect of pore was presented with large size in SLM printed structure than the cold sprayed part which led to high plastic deformation. Moreover, better adhesive and cohesive properties were observed from the fractured surface of the FGM (Yin et al., 2018).

Basically, functionally graded materials belong to the classification of advanced smart materials which possessed excellent properties to implement them in the thermal barrier coating applications. Therefore, FGM has fascinated both industrialists and research community by the presence of locally dependent gradient properties which directed to implement in high temperature applications (Li, Sun, Ao, Gu, & Xiao, 2007). As a new concept in wire arc additive manufacturing, FGM has been developed with iron and aluminium composition. Two wire feeders were installed with wire arc additive manufacturing setup to supply the iron and aluminium wires. The wire feeders were functioned with different speed controller to supply the material composition as per the formation of Fe-FeAl structure. Reasons behind this Fe-Al structure were observed as high fracture resistance and corrosion resistance properties. The results of this study revealed that the structure has been built with the variation of chemical composition by adjusting the wire feed ratio aluminium and iron. The mechanical properties of Fe-Al FGM have improved in the transverse direction with homogeneous condition which makes them as lower corrosion resistance (Tortorelli, & Natesan, 1998; Shen, Pan, Cuiuri, D., Roberts, & Li, 2016).

Titanium and its alloys have fascinated in different fields due to their tremendous properties such as excellent heat resistance, light weight, corrosion resistance and high strength to weight ratio. In order to fabricate a better FGM, titanium alloy can be combined with any structural or stainless steel which has excellent weldability. But, joining of these two alloys has some difficulties in conventional manufacturing processes (Cheepu, Ashfaq, & Muthupandi, 2017). In such cases, laser metal deposition can be used as additive manufacturing process which can be developed the FGM with the combination of stainless steel and titanium alloys. An investigation has been made on the combining of titanium and SS316 using laser metal deposition process. In this study, a thin wall has been constructed in the route of Ti-6Al-4V→V→Cr→Fe→SS316 to form as a FGM where the formation intermetallic prevented to enhance mechanical properties. The results showed that there was no formation of any brittle phases in the FGM developed via laser metal deposition. Therefore, it was suggested that to form a FGM with SS316and titanium alloy, laser metal deposition process can be followed to form the structure without any intermetallic phases. Moreover, the FGM was developed with titanium alloy and invar steel using direct energy deposition additive manufacturing. Unfortunately, the intermetallic such as FeTi, Fe_2Ti, Ni_3Ti, and $NiTi_2$ were found in the gradient region FGM. The formation of cracks in the FGM was observed by the presence of these intermetallics (Li et al., 2017). Therefore, the control of secondary phases in the AM developed component is very essential to prevent the failures.

FUTURE RESEARCH DIRECTION

The main issues in the implementation of additive manufacturing in industry are quality standards and management. The quality of additive manufactured products that can be easily modified by changing the process parameters which influenced the quality standards also. Therefore, customary design and quality standards are need to be investigated in additive manufacturing processes particularly for developing multi-material components. The earlier investigations on development of multi-material have used the powder bed fusion additive manufacturing techniques such as selective laser sintering, melting, laser engineered net shaping and direct metal laser sintering. The fabrication of multi-material is feasible in wire arc additive manufacturing process also. Therefore, researchers and industrialist can focus on the wire and arc based additive manufacturing process to develop the multi-material structure. The additive manufacturing based SLA process requires an additional setup to hold the structure which could be avoided in other process like SLS to make it as an economically. Many research works have focused on the development of functionally gradient materials using two dissimilar materials such as titanium alloy-stainless steel, titanium-invar and aluminium-titanium. With continue of these works, the researcher can focus on the development of functionally gradient materials using three or more materials to enhance the functional properties. Moreover, the formation of intermetallics created many issues in the functional properties of the FGM. Therefore, the control of secondary phases by proper control of process parameters or selection of additive manufacturing processes in the development of FGM can focus.

SUMMARY AND CONCLUSION

The development of products with multi-material and composite structures using different additive manufacturing techniques has been reviewed in this chapter. The unique feature which could be provided the benefits to the 3D printed component rather than the traditional manufacturing process was discussed. The products that can be manufactured using AM techniques for various fields such as medical devices, electronics, biomedical implementations, and robotics and prosperities obtained by the AM process have been discussed. The mechanical, metallurgical and functional properties enhanced through various AM processes were discussed with their results. The development of functionally graded materials using additive manufacturing processes was reviewed with their benefits. The benefits of functionally graded materials developed through hybrid AM process rather than the existing additive manufacturing systems were discussed. The influence of intermetallics on properties of functionally graded material has been deliberated. The selection of specific process from the various additive manufacturing processes is an important to improve the properties and feasibility of the part. The selection would be based on factors such as size of the component, number of materials involved in developing the part, complexity of the shape and application to be focussed. However, much probable of AM techniques in the development of new products are still to be explored.

This research received no specific grant from any funding agency in the public, commercial, or not-for-profit sectors.

REFERENCES

Aboulkhair, N. T., Everitt, N. M., Ashcroft, I., & Tuck, C. (2014). Reducing porosity in AlSi10Mg parts processed by selective laser melting. *Additive Manufacturing*, *1*, 77–86. doi:10.1016/j.addma.2014.08.001

Attar, H., Bermingham, M. J., Ehtemam-Haghighi, S., Dehghan-Manshadi, A., Kent, D., & Dargusch, M. S. (2019). Evaluation of the mechanical and wear properties of titanium produced by three different additive manufacturing methods for biomedical application. *Materials Science and Engineering A*, *760*, 339–345. doi:10.1016/j.msea.2019.06.024

Bandyopadhyay, A., & Heer, B. (2018). Additive manufacturing of multi-material structures. *Materials Science and Engineering R Reports*, *129*, 1–16. doi:10.1016/j.mser.2018.04.001

Boddeti, N., Ding, Z., Kaijima, S., Maute, K., & Dunn, M. L. (2018). Simultaneous digital design and additive manufacture of structures and materials. *Scientific Reports*, *8*(1), 15560. doi:10.103841598-018-33454-3 PMID:30348943

Butscher, A., Bohner, M., Roth, C., Ernstberger, A., Heuberger, R., Doebelin, N., Von Rohr, P. R., & Müller, R. (2012). Printability of calcium phosphate powders for three-dimensional printing of tissue engineering scaffolds. *Acta Biomaterialia*, *8*(1), 373–385. doi:10.1016/j.actbio.2011.08.027 PMID:21925623

Cabrini, M., Lorenzi, S., Pastore, T., Pellegrini, S., Manfredi, D., Fino, P., Biamino, S., & Badini, C. (2016). Evaluation of corrosion resistance of Al–10Si–Mg alloy obtained by means of Direct Metal Laser Sintering. *Journal of Materials Processing Technology*, *231*, 326–335. doi:10.1016/j.jmatprotec.2015.12.033

Chai, X., Chai, H., Wang, X., Yang, J., Li, J., Zhao, Y., Cai, W., Tao, T., & Xiang, X. (2017). Fused deposition modeling (FDM) 3D printed tablets for intragastric floating delivery of domperidone. *Scientific Reports*, *7*(1), 1–9. doi:10.103841598-017-03097-x PMID:28588251

Cheepu, M., Ashfaq, M., & Muthupandi, V. (2017). A new approach for using interlayer and analysis of the friction welding of titanium to stainless steel. *Transactions of the Indian Institute of Metals*, *70*(10), 2591–2600. doi:10.100712666-017-1114-x

Chen, D., & Zheng, X. (2018). Multi-material additive manufacturing of metamaterials with giant, tailorable negative Poisson's ratios. *Scientific Reports*, *8*(1), 9139. doi:10.103841598-018-26980-7 PMID:29904093

Dini, F., Ghaffari, S. A., Jafar, J., Hamidreza, R., & Marjan, S. (2019). A review of binder jet process parameters; powder, binder, printing and sintering condition. *Metal Powder Report*.

Edgar, J., & Tint, S. (2015). Additive manufacturing technologies: 3D printing, rapid prototyping, and direct digital manufacturing. *Johnson Matthey Technology Review*, *59*(3), 193–198. doi:10.1595/205651315X688406

Eliseeva, O. V., Kirk, T., Samimi, P., Malak, R., Arróyave, R., Elwany, A., & Karaman, I. (2019). Functionally Graded Materials through robotics-inspired path planning. *Materials & Design*, *182*, 107975. doi:10.1016/j.matdes.2019.107975

Galati, M., Snis, A., & Iuliano, L. (2019). Powder bed properties modelling and 3D thermo-mechanical simulation of the additive manufacturing Electron Beam Melting process. *Additive Manufacturing*, *30*, 100897. doi:10.1016/j.addma.2019.100897

Gao, W., Zhang, Y., Ramanujan, D., Ramani, K., Chen, Y., Williams, C. B., & Zavattieri, P. D. (2015). The status, challenges, and future of additive manufacturing in engineering. *Computer Aided Design*, *69*, 65–89. doi:10.1016/j.cad.2015.04.001

Hengsbach, F., Koppa, P., Holzweissig, M. J., Aydinöz, M. E., Taube, A., Hoyer, K. P., Starykov, O., Tonn, B., Niendorf, T., Tröster, T., & Schaper, M. (2018). Inline additively manufactured functionally graded multi-materials: Microstructural and mechanical characterization of 316L parts with H13 layers. *Progress in Additive Manufacturing*, *3*(4), 221–231. doi:10.100740964-018-0044-4

Ilogebe, A. B., Waters, C. K., Elliot, A. M., & Shackleford, C. (2019). Morphology of Binder-Jet Additive Manufactured Structural Amorphous Metal Matrix Composites. *International Journal of Engineering Science*, *8*, 15–24.

Kan, W., Chen, B., Peng, H., Liang, Y., & Lin, J. (2020). Fabrication of nano-TiC reinforced high Nb-TiAl nanocomposites by electron beam melting. *Materials Letters*, *259*, 126856. doi:10.1016/j.matlet.2019.126856

Kim, S. B., Kim, N. H., Kim, J. H., & Moon, H. S. (2018). Evaluation of the fit of metal copings fabricated using stereolithography. *The Journal of Prosthetic Dentistry*, *120*(5), 693–698. doi:10.1016/j.prosdent.2018.01.012 PMID:29807735

Li, J. Q., Sun, W. A., Ao, W. Q., Gu, K. M., & Xiao, P. (2007). Al2O3–FeCrAl composites and functionally graded materials fabricated by reactive hot pressing. *Composites. Part A, Applied Science and Manufacturing*, *38*(2), 615–620. doi:10.1016/j.compositesa.2006.02.012

Li, W., Karnati, S., Kriewall, C., Liou, F., Newkirk, J., Taminger, K. M. B., & Seufzer, W. J. (2017). Fabrication and characterization of a functionally graded material from Ti-6Al-4V to SS316 by laser metal deposition. *Additive Manufacturing*, *14*, 95–104. doi:10.1016/j.addma.2016.12.006

Liu, Y. J., Wang, H. L., Li, S. J., Wang, S. G., Wang, W. J., Hou, W. T., Hao, Y. L., Yang, R., & Zhang, L. C. (2017). Compressive and fatigue behavior of beta-type titanium porous structures fabricated by electron beam melting. *Acta Materialia*, *126*, 58–66. doi:10.1016/j.actamat.2016.12.052

Maharubin, S., Hu, Y., Sooriyaarachchi, D., Cong, W., & Tan, G. Z. (2019). Laser engineered net shaping of antimicrobial and biocompatible titanium-silver alloys. *Materials Science and Engineering C*, *105*, 110059. doi:10.1016/j.msec.2019.110059 PMID:31546421

Mostafaei, A., Kimes, K. A., Stevens, E. L., Toman, J., Krimer, Y. L., Ullakko, K., & Chmielus, M. (2017). Microstructural evolution and magnetic properties of binder jet additive manufactured Ni-Mn-Ga magnetic shape memory alloy foam. *Acta Materialia*, *131*, 482–490. doi:10.1016/j.actamat.2017.04.010

Mu, X., Bertron, T., Dunn, C., Qiao, H., Wu, J., Zhao, Z., Saldana, C., & Qi, H. J. (2017). Porous polymeric materials by 3D printing of photocurable resin. *Materials Horizons*, *4*(3), 442–449. doi:10.1039/C7MH00084G

Muguruza, A., Bo, J. B., Gómez, A., Minguella-Canela, J., Fernandes, J., Ramos, F., & Cirera, A. (2017). Development of a multi-material additive manufacturing process for electronic devices. *Procedia Manufacturing*, *13*, 746–753. doi:10.1016/j.promfg.2017.09.180

Murr, L. E., Martinez, E., Amato, K. N., Gaytan, S. M., Hernandez, J., Ramirez, D. A., Shindo, P.W., Medina, F. & Wicker, R. B. (2012). Fabrication of metal and alloy components by additive manufacturing: examples of 3D materials science. *Journal of Materials Research and Technology, 1*(1), 42-54.

Ngo, T. D., Kashani, A., Imbalzano, G., Nguyen, K. T., & Hui, D. (2018). Additive manufacturing (3D printing): A review of materials, methods, applications and challenges. *Composites. Part B, Engineering*, *143*, 172–196. doi:10.1016/j.compositesb.2018.02.012

Pei, E., Loh, G. H., Harrison, D., de Amorim Almeida, H., Verona, M. D. M., & Paz, R. (2017). A study of 4D printing and functionally graded additive manufacturing. *Assembly Automation*, *37*(2), 147–153. doi:10.1108/AA-01-2017-012

Peng, H., Liu, C., Guo, H., Yuan, Y., Gong, S., & Xu, H. (2016). Fabrication of WCp/NiBSi metal matrix composite by electron beam melting. *Materials Science and Engineering A*, *666*, 320–323. doi:10.1016/j.msea.2016.04.079

Pham, D., & Dimov, S. S. (2012). *Rapid manufacturing: the technologies and applications of rapid prototyping and rapid tooling*. Springer Science & Business Media.

Raus, A. A., Wahab, M. S., Ibrahim, M., Kamarudin, K., Ahmed, A., & Shamsudin, S. (2017, April). Mechanical and physical properties of AlSi10Mg processed through selective laser melting. In AIP Conference Proceedings (Vol. 1831, No. 1, p. 020027). AIP Publishing LLC. doi:10.1063/1.4981168

Revathi, A., Mitun, D., Balla, V. K., Dwaipayan, S., Devika, D., & Manivasagam, G. (2019). Surface properties and cytocompatibility of Ti-6Al-4V fabricated using Laser Engineered Net Shaping. *Materials Science and Engineering C*, *100*, 104–116. doi:10.1016/j.msec.2019.02.099 PMID:30948044

Sames, W. J., Unocic, K. A., Dehoff, R. R., Lolla, T., & Babu, S. S. (2014). Thermal effects on microstructural heterogeneity of Inconel 718 materials fabricated by electron beam melting. *Journal of Materials Research*, *29*(17), 1920–1930. doi:10.1557/jmr.2014.140

Sharma, A., Bandari, V., Ito, K., Kohama, K., & Ramji, M., & BV, H. S. (. (2017). A new process for design and manufacture of tailor-made functionally graded composites through friction stir additive manufacturing. *Journal of Manufacturing Processes*, *26*, 122–130. doi:10.1016/j.jmapro.2017.02.007

Shen, C., Pan, Z., Cuiuri, D., Roberts, J., & Li, H. (2016). Fabrication of Fe-FeAl functionally graded material using the wire-arc additive manufacturing process. *Metallurgical and Materials Transactions. B, Process Metallurgy and Materials Processing Science*, *47*(1), 763–772. doi:10.100711663-015-0509-5

Sing, S. L., An, J., Yeong, W. Y., & Wiria, F. E. (2016). Laser and electron-beam powder-bed additive manufacturing of metallic implants: A review on processes, materials and designs. *Journal of Orthopaedic Research*, *34*(3), 369–385. doi:10.1002/jor.23075 PMID:26488900

Spowart, J. E., Gupta, N., & Lehmhus, D. (2018). Additive Manufacturing of Composites and Complex Materials. *JOM*, *70*(3), 272–274. doi:10.100711837-018-2742-2

Stull, J. A., Hill, M. A., Lienert, T. J., Tokash, J., Bohn, K. R., & Hooks, D. E. (2018). Corrosion characteristics of laser-engineered net shaping additively-manufactured 316L stainless steel. *JOM*, *70*(11), 2677–2683. doi:10.100711837-018-3123-6

Tortorelli, P. F., & Natesan, K. (1998). Critical factors affecting the high-temperature corrosion performance of iron aluminides. *Materials Science and Engineering A*, *258*(1-2), 115–125. doi:10.1016/S0921-5093(98)00924-1

Toursangsaraki, M. (2018). *A review of multi-material and composite parts production by modified additive manufacturing methods*. arXiv preprint arXiv:1808.01861

Udroiu, R., Braga, I. C., & Nedelcu, A. (2019). Evaluating the Quality Surface Performance of Additive Manufacturing Systems: Methodology and a Material Jetting Case Study. *Materials (Basel)*, *12*(6), 995. doi:10.3390/ma12060995 PMID:30917576

Vaezi, M., Chianrabutra, S., Mellor, B., & Yang, S. (2013). Multiple material additive manufacturing–Part 1: A review. *Virtual and Physical Prototyping*, *8*(1), 19–50. doi:10.1080/17452759.2013.778175

Wang, J., Pan, Z., Ma, Y., Lu, Y., Shen, C., Cuiuri, D., & Li, H. (2018). Characterization of wire arc additively manufactured titanium aluminide functionally graded material: Microstructure, mechanical properties and oxidation behaviour. *Materials Science and Engineering A*, *734*, 110–119. doi:10.1016/j.msea.2018.07.097

Wang, R., Gu, D., Xi, L., Lin, K., Guo, M., & Zhang, H. (2019). Selective laser melted TiB2/Ti6Al4V graded materials and first-principle calculations. *Materials Letters, 254*, 33–36. doi:10.1016/j.matlet.2019.07.015

Yan, L., Chen, Y., & Liou, F. (2019). Additive Manufacturing of Functionally Graded Metallic Materials Using Laser Metal Deposition. *Additive Manufacturing, 100901.*

Yin, S., Yan, X., Chen, C., Jenkins, R., Liu, M., & Lupoi, R. (2018). Hybrid additive manufacturing of Al-Ti6Al4V functionally graded materials with selective laser melting and cold spraying. *Journal of Materials Processing Technology, 255*, 650–655. doi:10.1016/j.jmatprotec.2018.01.015

Yuan, S., Zheng, Y., Chua, C. K., Yan, Q., & Zhou, K. (2018). Electrical and thermal conductivities of MWCNT/polymer composites fabricated by selective laser sintering. *Composites. Part A, Applied Science and Manufacturing, 105*, 203–213. doi:10.1016/j.compositesa.2017.11.007

Zhang, B., Jaiswal, P., Rai, R., & Nelaturi, S. (2018). Additive manufacturing of functionally graded material objects: A review. *Journal of Computing and Information Science in Engineering, 18*(4), 041002. doi:10.1115/1.4039683

Zhang, C., Chen, F., Huang, Z., Jia, M., Chen, G., Ye, Y., ... Zhang, L. (2019). Additive manufacturing of functionally graded materials: A review. *Materials Science and Engineering A, 764*, 138209. doi:10.1016/j.msea.2019.138209

Zhang, Y., Wei, Z., Shi, L., & Xi, M. (2008). Characterization of laser powder deposited Ti–TiC composites and functional gradient materials. *Journal of Materials Processing Technology, 206*(1-3), 438–444. doi:10.1016/j.jmatprotec.2007.12.055

Zhou, J., Li, H., Yu, Y., Firouzian, K., Qian, Y., & Lin, F. (2020). Characterization of interfacial transition zone of functionally graded materials with graded composition from a single material in electron beam powder bed fusion. *Journal of Alloys and Compounds, 832*, 154774. doi:10.1016/j.jallcom.2020.154774

Zhou, Z., Li, D., Zeng, J., & Zhang, Z. (2007). Rapid fabrication of metal-coated composite stereolithography parts. *Proceedings of the Institution of Mechanical Engineers. Part B, Journal of Engineering Manufacture, 221*(9), 1431–1440. doi:10.1243/09544054JEM827

Zhou, Z., Tan, Z., He, D., Zhou, Z., Cui, L., Wang, Y., Shao, W., & Wang, G. (2020). Fabrication of three-dimensional connected W-Cu10Sn composites by selective laser melting. *Materials Letters, 264*, 127377. doi:10.1016/j.matlet.2020.127377

Ziaee, M., & Crane, N. B. (2019). *Binder jetting: A review of process, materials, and methods*. Additive Manufacturing.

ADDITIONAL READING

Bourell, D., Kruth, J. P., Leu, M., Levy, G., Rosen, D., Beese, A. M., & Clare, A. (2017). Materials for additive manufacturing. *CIRP Annals*, *66*(2), 659–681. doi:10.1016/j.cirp.2017.05.009

Francois, M. M., Sun, A., King, W. E., Henson, N. J., Tourret, D., Bronkhorst, C. A., ... & Gibbs, J. W. (2017). Modeling of additive manufacturing processes for metals: Challenges and opportunities. Current Opinion in Solid State and Materials Science, 21(LA-UR-16-24513; SAND-2017-6832J).

Herderick, E. (2011). Additive manufacturing of metals: A review. *Materials Science and Technology*, 1413.

Krivilyov, M. D., Mesarovic, S. D., & Sekulic, D. P. (2017). Phase-field model of interface migration and powder consolidation in additive manufacturing of metals. *Journal of Materials Science*, *52*(8), 4155–4163. doi:10.100710853-016-0311-z

Kruth, J. P., Dadbakhsh, S., Vrancken, B., Kempen, K., Vleugels, J., & Van Humbeeck, J. (2015). Additive manufacturing of metals via selective laser melting: process aspects and material developments.

Kumar, S., & Pityana, S. (2011). Laser-based additive manufacturing of metals. *Advanced Materials Research*, *227*, 92–95. doi:10.4028/www.scientific.net/AMR.227.92

Mahmoodkhani, Y., Ali, U., Shahabad, S. I., Kasinathan, A. R., Esmaeilizadeh, R., Keshavarzkermani, A., ... Toyserkani, E. (2019). On the measurement of effective powder layer thickness in laser powder-bed fusion additive manufacturing of metals. *Progress in Additive Manufacturing*, *4*(2), 109–116. doi:10.100740964-018-0064-0

Milewski, J. O. (2017). Additive manufacturing of metals. From Fundamental Technology to Rocket Nozzles, Medical Implants, and Custom Jewelry,, 134-157.

Yampolskiy, M., Schutzle, L., Vaidya, U., & Yasinsac, A. (2015, March). Security challenges of additive manufacturing with metals and alloys. In *International Conference on Critical Infrastructure Protection* (pp. 169-183). Springer, Cham. 10.1007/978-3-319-26567-4_11

Yang, L., Hsu, K., Baughman, B., Godfrey, D., Medina, F., Menon, M., & Wiener, S. (2017). Additive manufacturing of metals: the technology, materials, design and production (pp. 65-70). Springer.

KEY TERMS AND DEFINITIONS

3D Printing: Techniques to build model in three dimensional by adding the material in successive layer.

Binder Jetting: Technique in which liquid agent of binders are deposited for binding the powder particles together to form a required finished component.

Additive Manufacturing: Creation of complex 3D parts from digital data with the help of 3D printing techniques.

Advanced Manufacturing: Construction of existing and new products by applying innovative techniques in automation, sensing, and computation.

Composites: Materials possessed of two or more components with various physical and chemical properties to enhance the any specific property of finished product.

Directed Energy Deposition: Method of additive manufacturing processes that builds the component by coaxial feeding of metals in the form of wire or powder.

Rapid Prototyping: Quick process to fabricate model/prototype, presently it is used to fabricate final products and assembly.

Chapter 8
Direct Laser Fabrication of Compositionally Complex Materials:
Challenges and Prospects

Jithin Joseph
https://orcid.org/0000-0003-4878-1320
Deakin University, Australia

ABSTRACT

Additive manufacturing (AM) opens up the possibility of a direct build-up of components with sophisticated internal features or overhangs that are difficult to manufacture by a single conventional method. As a cost-efficient, tool-free, and digital approach to manufacturing components with complex geometries, AM of metals offers many critical benefits to various sectors such as aerospace, medical, automotive, and energy compared to conventional manufacturing processes. Direct laser fabrication (DLF) uses pre-alloyed powder mix or in-situ alloying of the elemental powders for metal additive manufacturing with excellent chemical homogeneity. It, therefore, shows great promise to enable the production of complex engineering components. This technique allows the highest build rates of the AM techniques with no restrictions on deposit size/shape and the fabrication of graded and hybrid materials by simultaneously feeding different filler materials. The advantages and disadvantages of DLF on the fabrication of compositionally complex metallic alloys are discussed in the chapter.

INTRODUCTION

Additive Manufacturing (AM) was defined by ASTM (American Society for Testing & Materials) as "the process of making parts by joining materials layer-by-layer from a 3D-CAD model (ASTM: F2792-12a)" (Sugavaneswaran & Arumaikkannu, 2014). The process of fabricating objects by the in-situ melting followed by the rapid solidification of a powder layer has created a new scientific frontier in physical metallurgy for the fabrication of compositionally complex alloy components with excellent

DOI: 10.4018/978-1-7998-4054-1.ch008

strength-ductility synergy (Joseph et al., 2015). This process opens up new technological opportunities for the manufacturers to directly build-up of parts with sophisticated internal features or overhangs that are difficult to manufacture by a single conventional method. The ASTM 52900:2015 classifies various AM processes into following categories (Bandyopadhyay & Heer, 2018; Tofail et al., 2018) such as Material/Binder jetting; Laser engineered net shaping (LENS)/ direct laser fabrication (DLF); Material extrusion; Vat photopolymerization; Selective laser melting (SLM) and; Sheet lamination.

As a cost-efficient, tool-free and digital approach to manufacturing components with complex geometries, AM of metals offer vital benefits to aerospace, medical, automotive, and energy sectors compared to traditional manufacturing processes (Bourell, 2016; Conner et al., 2014a). It includes the reduced fabrication time and cost, material waste and enables low-volume production of complicated part designs with minimal geometric constraints and a high degree of customization (Conner et al., 2014b). Parts can be compact and multi-functional with the elimination of joints and fewer assembled components, leads to the savings of production/assembly lead time, cost and environmental impact (Berdine, DiPaola, & Weinberg, 2019; Bogers, Hadar, & Bilberg, 2016; Fera, Macchiaroli, Fruggiero, & Lambiase, 2018). The researchers of Monash University (in association with Deakin University and CSIRO, Australia), fabricated world's first AM printed jet engine with a significantly reduced number of components and was on display at Paris Airshow-2015 (Figure 1 (Wu, 2015)). A well-known practical example of such compact design is the 3D-printed (SLM) fuel nozzle of the LEAP jet engine with a compact design (GE Aviation-2016, one component instead of 18) with sophisticated cooling pathways and support ligaments (Kellens et al., 2017). This light and cost-effective engine resulted in improved fuel efficiency and durability of the aircraft.

Figure 1. The world's first 3D printed prototype of jet engine that was on display at Paris Airshow-2015 (Adapted from https://www.monash.edu/mcam/news/articles/paris-le-bourget-airshow (Wu, 2015); Used with permission from MCAM, Monash University, Australia).

AM showed the potential to change the face of conventional manufacturing industry, combining advantages of large-scale manufacturing with mass customization production of unique, complex components (Eyers & Potter, 2017; Fera et al., 2018; Kellens et al., 2017; Manyika et al., 2013). AM is already playing a pivotal role in critical engineering sectors (aerospace/medical) in the fabrication and supply-chain management of complex customized components anticipated for a long service life (Bogers et al., 2016). Currently, many industries integrate metallic 3D printing into the conventional production lines to eliminate time-consuming, complex manufacturing processes, and to meet the desire for product customization at no extra cost (Zhu, Dhokia, Nassehi, & Newman, 2013). The success of the metal AM industry is reflected by the growth in the fabrication of components for both general and critical engineering applications with an annual impact of ~$200bn to the global economy by 2025 (Eyers & Potter, 2017).

DIRECT LASER FABRICATION

The important metal AM technologies with the local melting of a powder layer, and subsequent rapid solidification is SLM, DLF/LENS and selective electron beam melting (SEBM). For a metallic alloy to be fabricated using AM, it should satisfy two critical criteria to ensure quality depositions: availability of homogeneous spherical alloy powders (average diameter of 20-200 microns) and good weldability. Powder bed-based SLM technology is the most advanced AM technique for fabricating high-value components with the highest complexity where a high-energy laser beam scans and selectively melts the powder in a layered fashion until the part is completed (Gorsse, Hutchinson, Gouné, & Banerjee, 2017). However, SLM requires the gas/plasma atomized powders of the alloy to be printed. It is possible to alloy metal powders in-situ to form a complex alloy by supplying metal powders from different feeding sources into the focal point of a high-power laser beam. They melt/fuse with a previously deposited layer to produce engineering components with complex geometries and a high degree of accuracy. This process is known as directed energy deposition or DLF (Agarwala, Bourell, Beaman, Marcus, & Barlow, 1995; Conner et al., 2014a). The advantages and disadvantages of DLF on the fabrication of compositionally complex metallic alloys will be discussed in the following sections.

In DLF, a small quantity of powder is melted in-situ with the aid of a computer-controlled high-power laser; the moving laser beam follows a prescribed path, and the melt pool is added in a layer-by-layer fashion to obtain the final shape of the component (Khaing, Fuh, & Lu, 2001; Lewis & Schlienger, 2000; Simchi, Petzoldt, & Pohl, 2003). DLF uses pre-alloyed powder mix, or in-situ alloying of the metal powders for metal AM with excellent chemical homogeneity and therefore shows great promise to enable the production of complex engineering components from chemically complex alloys (Figure 2). This technique allows the highest build rates of the AM techniques with limited restrictions on deposit size/shape and the fabrication of graded and hybrid materials. It allowed the design, rapid alloy screening and the high-throughput investigation of complex alloys by varying the feed rates of component elements (Ewald et al., 2019; Haase, Tang, Wilms, Weisheit, & Hallstedt, 2017; Joseph et al., 2015; Mahamood, Akinlabi, Shukla, & Pityana, 2013).

Figure 2. Deposition trials of (a) thin-walled structures and (b) components with overhanging features using Optomec MR-7 DLF system
(Adapted from (Joseph, 2016)).

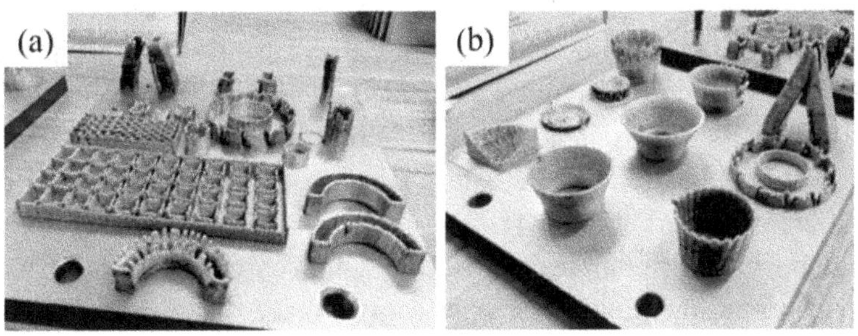

The flexibility of DLF allows for starts and stops, in-situ adjustments, increase or decrease of powder flow rates and the possibility of functionally-graded depositions during the process. The process offers excellent potential in the repairing or remanufacturing of engineering components leading to the extended service life of the component with significant savings in economic costs and resources (Lewis & Schlienger, 2000). This process allows for the cost-efficient repair of complex geometries to treat extensive in-service damages (due to wear) and to restore design shape with a minimum amount of post-processing when compared to conventional welding techniques. Another prospect of DLF is to build and repair tooling surfaces with complex curvatures (gear teeth, Figure 3) that involve 3D-scanning the damaged surface, creating a CAD profile that would match the shape of the scanned surface and finally the printing/repairing of the surface. The principal manufacturers of DLF platforms on industrial scale AM technologies in competition with 'conventional' approaches to manufacturing include Trumpf and Optomec.

Figure 3. Repair of damaged gear teeth using DLF
(Adapted from https://optomec.com/wp-content/uploads/2019/03/DEDWebinar-slides.pdf (Cobbs, 2019); Used with permission from Optomec Inc.)

PRINTING COMPOSITIONALLY COMPLEX ALLOYS

DLF has shown the capability to print components from pre-alloyed powders of Ni-base superalloys, alloy steels, Ti–alloys, functionally graded materials and metal matrix composites. DLF is principally attractive for the fabrication of compositionally complex alloys using elemental powders as feedstock to fabricate chemically homogenous alloys under controlled powder flow rates and optimized processing conditions. This powder-fed system enables the production of numerous alloy compositions, surface coatings and functionally gradient components in a single batch by adjusting the flow rates of metal powders from the different hoppers (Figure 4).

A new class of compositionally complex alloys, called high entropy alloys (HEAs) consists of a high concentration of four or more alloying elements. Fabrication of HEAs using traditional techniques required

Figure 4. Fabrication of numerous alloy compositions in a single batch using Optomec MR-7 DLF system by controlling the powder flow rates from the different hoppers
(Adapted from (Joseph, 2016)).

multiple remelting and regressive processing and was able to produce bulk homogeneous HEA samples with limited size and shapes. However, DLF overcomes the size/shape limits during the fabrication HEAs by conventional production routes (Joseph, 2016; Joseph et al., 2015). Apart from the homogeneous microstructure and enhanced mechanical properties of bulk/coating HEAs, DLF was proved to be a powerful tool to validate the thermodynamic calculations of compositionally complex alloys (CCA) by high-throughput production of bulk HEA samples (Moorehead et al., 2020). A case study for this approach is shown in Figure 5 were numerous $Al_xCoCrFeNi_{2.1}$ HEAs were fabricated by DLF to analyze the change in microstructure with the increase in Al-content. This sort of high-throughput experiments is essential for the development of potential HEAs from the vast multidimensional composition space. The high energy input in DLF certainly facilitated homogeneous chemical composition due to high dynamic mixing in the larger melt pool size compared to other 3-D printing techniques such as SLM and EBM.

Figure 5. High-throughput exploration of novel $Al_xCoCrFeNi_{2.1}$ high entropy alloys with varying compositions
(Adapted from (Joseph, 2016)).

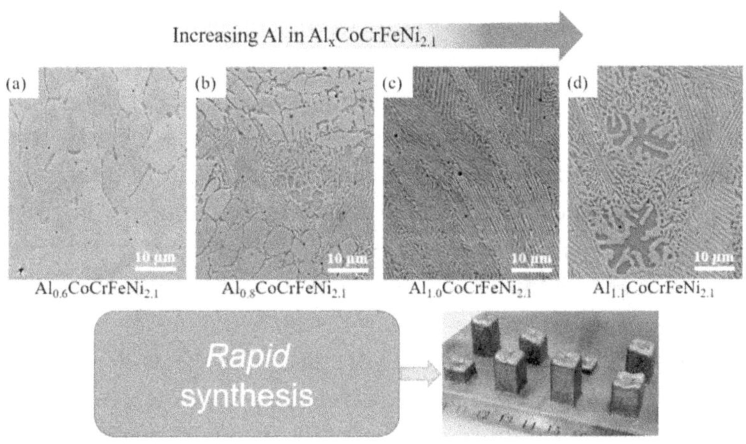

SOLIDIFICATION MICROSTRUCTURE

The solidification microstructure during DLF is significantly influenced by a diverse set of laser processing parameters (Shi, Gu, Xia, Cao, & Rong, 2016; Thijs, Verhaeghe, Craeghs, Humbeeck, & Kruth, 2010) coupled with highly complex heat/mass transport phenomena such as heat conduction from melt pool to the substrate, Marangoni convection in the melt pool, and radiation during deposition (Collins, Brice, Samimi, Ghamarian, & Fraser, 2016; Gockel, Beuth, & Taminger, 2014; Kobryn & Semiatin, 2003; Romano, Ladani, Razmi, & Sadowski, 2015; Shamsaei, Yadollahi, Bian, & Thompson, 2015; Thompson, Bian, Shamsaei, & Yadollahi, 2015). Important laser processing parameters such as laser power, laser beam spot diameter, powder feed rate, scanning strategy, scanning speed, and layer dimensions (width and thickness) affect the microstructure and part quality. The Marangoni convective heat transfer is due to the differences in surface tension as a result of the high thermal gradient in the melt pool and determines the melt pool flow dynamics in as-deposited DLF components. Other forces such as aerodynamic drag (outward drag forces above melt pool) and buoyancy (upward movements of molten metal due to density changes as a result of thermal gradients inside the melt pool) generated during the process further complicate the melt pool dynamics. The thermal conditions that exist locally around the laser-induced melt pool are directly impacted by the alloy chemistry and thermal properties of the as-deposited material (Collins et al., 2016; Gockel et al., 2014; Kobryn & Semiatin, 2003; Romano et al., 2015; Shamsaei et al., 2015; Thompson et al., 2015).

The additive manufacturing process involves a rapid cooling rate (10^3-10^6 Ks^{-1}) and a large thermal gradient (10^5-10^7 Km^{-1}) in the melt pool (Shi et al., 2016). The heat transfer kinetics in the melt pool during DLF is extremely complex due to the highly localized heat input on the substrate over a very short time (Shi et al., 2016; Thijs et al., 2010; Yin et al., 2012). It is exacerbated by the formation of a new heat-affected zone in the vicinity of the deposited melt pool as a result of the remelting of the previously deposited layer (Thompson et al., 2015). Dilution can be defined as the ratio of the depth of the melt pool below the previously deposited layer (d), and the sum of d and the height of the deposited material above the previously deposited layer (h). A high powder flow rate or low energy input corresponds to a high value of h, where the laser energy is not sufficient to enough for the remelting of the previously deposited layer. A high value of d may be due to a low powder feed rate or high energy input. In the case of laser fabricated alloys, the high energy density of the laser results in a directional heat transfer from the deposited molten metal to the already consolidated layers opposite to the building direction (Dinda, Dasgupta, & Mazumder, 2012). However, the cooling rate during this process is determined by the size of the melt pool; smaller melt pools result in rapid cooling rates and finer microstructures, and larger melting pools induce comparatively slower cooling rates (Thompson et al., 2015). The size and the morphology of the melt pool can be optimized by the proper selection of the laser processing parameters.

Like any other solidification processes, the microstructures of the as-deposited components by DLF may be described using thermal gradient, G (K/m), and solidification front velocity, R (m/s) (Kurz & Trivedi, 1994). A high scanning speed or a high metal powder feed rate or a low laser power will result in a low value of G and a high value of R, and vice versa. A high ratio of these parameters ($G/R > 1$) corresponds to a low nucleation rate and a low value for the ratio ($G/R < 1$) corresponds to a high nucleation rate in the as-deposited structure. Proper control of these parameters facilitates the development of fully columnar or fully equiaxed or a mixed microstructure in the as-deposited component (Figure 6). The preferential texture observed in the as-deposited components is due to the preferential growth parallel to the direction of maximum thermal gradient in the melt pool that is almost parallel to the built direction. A detailed understanding of the local physical/thermal properties of the melt pool and the interfacial properties of the molten metal/substrate may facilitate the design of preferential texture and microstructural features in the deposited components. The as-deposited samples by any AM techniques are characterized by metastable/unusual microstructure and high strength but relatively low ductility compared to the cast counterparts. However, post-thermo-mechanical treatments such as hot isostatic pressing may improve the mechanical properties without distorting the geometry of the as-deposited components.

Another important factor is the alloy chemistry. Additive manufactured alloys with single-phase structure (such as steels and Ni-alloys (Kunze, Etter, Grässlin, & Shklover, 2015)) with high heat extraction rate from the melt pool to the underlying layer exhibit columnar structure with strong <001> texture due to rapid heat transfer from the deposited layer and the remelted zone (Collins et al., 2016; Song, Dong, Coddet, Liao, & Coddet, 2014; Thijs et al., 2010) as a result of epitaxial growth (Figure 6(a)). Multi-phase alloys (such as Ti-6Al-4V (Kobryn & Semiatin, 2003) and $Al_{0.6}$CoCrFeNi, Figure 6(b)) with low heat extraction rate at the melt pool-bottom layer produce a randomly textured equiaxed grain structure with fine Widmanstätten plates (Shamsaei et al., 2015). Another multi-phase alloy $Al_{0.6}CoCrFeNi_{2.1}$ by DLF showed a coarse dendritic structure with strong <001> texture for the dendritic FCC phase, Figure 6(c)). The as-deposited structures of AlCoCrFeNi and $Al_{1.2}$CoCrFeNi HEAs by DLF showed a mixed columnar/equiaxed (Figure 6(d)) and equiaxed structures (Figure 6(e)), respectively. It was shown in the literature that for a laser-assisted manufacturing process, the thermal conductivity of the deposit and its variation with temperature influences the melt pool size and the thermal gradient (Wang & Felicelli, 2006). High thermal conductivity materials result in pronounced thermal gradients across the melt pool and have a high heat extraction rate from the melt pool towards the underlying solidified layer (Figure 6(f)) (Romano et al., 2015).

Figure 6. EBSD maps of numerous solidification microstructures in as-deposited alloy components (a-e) and the schematic representation of the deposition process (f) by DLF
(Adapted from (Joseph, 2016)).

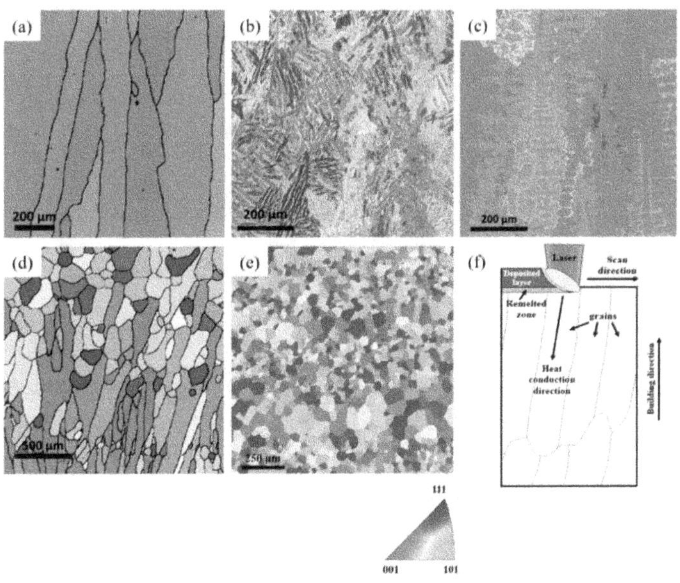

The laser-metal powder interaction determines the morphology of the melt pool, especially the attenuation of the laser beam by the metal powder. This interaction may be controlled by the process parameters such as laser power (P) and laser scan speed (v). Increasing P improves melt pool wetting due to the availability of more energy for powder consolidation. Reducing v increases the laser-metal powder interaction time and hence the high energy transferred to the powder particles. An increase in v reduces the interaction time and the energy transferred to metal powders. A reduction in the molten pool temperature hinder the wetting process due to the increase in surface tension and will result in a discontinuous deposition track. A decrease in P and increase in v will result in the transition from a continuous to an interrupted deposition track. Hence it is very important to adjust the P and v for the formation of a continuous track by promoting the wetting of the molten pool, where a combination of high v and low P is preferred.

The quality of the as-deposited components is primarily determined by the density (with a minimum density of microstructural defects) of the deposition. High laser power improves the part density, whereas increasing the layer deposition thickness likely decreases the final part density (Figure 7). Even though a minimum layer thickness if preferred, the minimum layer thickness should be at least four times the maximum powder size. AM component with high density can be obtained by using fine spherical powder particles with low oxygen content (less than 2 ppm is possible with modern DLF systems) within the operational window. A high density can also be possible by increasing laser energy density during the deposition process, achieved by decreasing the scan speed and hatch distances, resulting in a high volumetric mass density of the deposition with improved mechanical properties.

Figure 7. SEM macrograph of the cross-sections of as-deposited structures at (a) 700 W and (b) 800 W keeping all other laser processing parameters constant
(Adapted from (Joseph, 2016)).

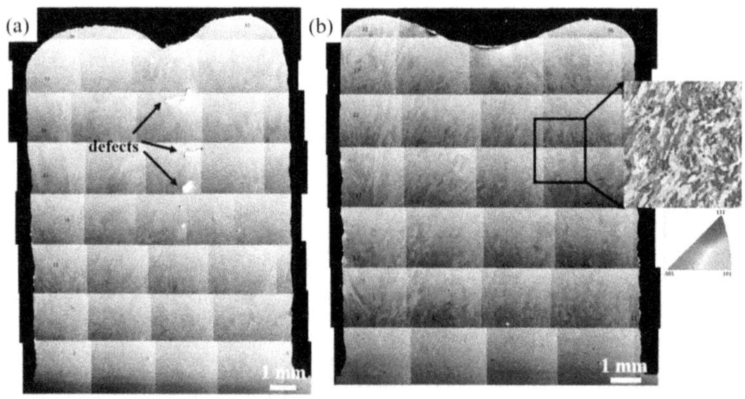

A major challenge associated with DLF is the utilization of the metal powder during the deposition process. Under normal processing conditions, only less than one-third of the metal powder fed to the laser focal point is deposited, and the rest is scattered. When the high-energy laser beam interacts with the metal powder or the underlying deposited layer, material vaporizes and forms a cavity. The high-power laser also ionizes the inert (argon) gas in the chamber and forms a plasma plume. The interaction of the metal vapour and the plasma plume formed during the deposition scatters metal particles away from the melt pool (Wolff et al., 2019). Particle scattering leads to the attenuation and scattering of the laser beam, which affects the quality of the DLF deposition. However, proper control of the carrier gas pressure facilitates the reduced scatter of metal powders and the increased penetration of the metal particles through the vapour-plasma plume. Also, the surface tension of the melt pool facilitates the entrainment of metal particles on the surface into it or ejects hot particles and influences microstructure and part quality during the deposition. However, the large temperature gradient in the melt pool during DLF results in significant variation of the surface tension, and Marangoni convection causes the movement of molten metal to regions from lower to higher surface tension. The high viscosity and wettability of the molten metal will lead to the effective spreading of the melt pool over the underlying solidified layer. Hence a combination of the physical/thermal characteristics of the molten metal during DLF affects the solidification rates, microstructure and part quality (Thompson et al., 2015).

A detailed understanding of the fundamental metallurgical and scientific principles of DLF processes is critical to ensure high-quality depositions and to avoid defects, spattering and microcracks in the deposited components. It is imperative to optimize the range of processing parameters for a particular material system and to detect the defects as early as possible in the AM process to minimize the risk of failure of the component during service. The detailed analysis of the physical/thermal processes during the interaction between the laser and metal powder in DLF (such as the melt pool dynamics, remelting of the previously deposited/substrate surface and defect formation) requires high-resolution real-time monitoring to optimize the processing parameters. The analysis of the microstructure (shown in Figure 7) and mechanical properties of the DLF depositions is a time-consuming and tedious way to fulfil this aspect. The in-situ high-speed, high-resolution monitoring of the deposition process by DLF will lead to the identification of processing conditions will facilitate high powder utilization and improved microstructural quality of the components (Leung et al., 2018). This feedback, coupled with fundamental theories on the physics of fluid/powder flow and heat transfer may be used to optimize DLF platform into a ground-breaking means for printing complex scalable components with flexible and functional materials.

METAL POWDERS

The part quality, microstructure and the mechanical properties of components by DLF are considerably influenced by the size, morphology, surface chemistry and chemical homogeneity of the metal powders used for the deposition process. The fine and uniform size distribution of the gas/plasma atomized spherical metal powders may result in uninterrupted powder flow during DLF and will ensure high-quality depositions. The defects and impurities in the metal powders will transmit to the deposited component and will lower the quality of the deposited components due to the generation of new defects (porosity and inclusions). The particle sizes and shapes of the metal powder should be similar to ensure the reproducibility of high-quality depositions between different DLF platforms or for the same DLF platform at different time frames. Absorptivity of metal powder is another parameter that influences the critical energy density for the quality of the depositions. Absorptivity (fraction of the radiation absorbed from the incident radiation) is influenced by both the composition of the metal powder and the wavelength of the laser beam. It is critical to match the laser wavelength with the powder characteristics for high-quality depositions (Kruth, Wang, Laoui, & Froyen, 2003).

Expensive pre-alloyed powders are typically used for part depositions to ensure microstructural homogeneity and part quality. Ti-based alloys are the most extensively used 3D–printed alloys for aerospace/medical applications as a result of excellent specific strength, biocompatibility and low density. Ni-based alloys follow it due to superior high-temperature properties essential for jet engine/gas turbine components. However, the use of elemental powders for deposition results in cost savings by avoiding the use of expensive pre-alloyed powders and introduces flexibility in chemical composition to the deposited part. This unique approach facilitates the rapid design of novel alloys by high-throughput computation and experimental screening of compositionally complex alloys (Figure 5).

CHALLENGES AND FUTURE PERSPECTIVES

DLF showed the ability to build components with complex geometry and flexible chemical composition by locally varying the flow rate of elemental powders for functionally graded materials. Even though DLF aids the high-throughput production and screening of numerous novel bulk alloy samples in a single batch, the high solidification rates during the process may result in non-equilibrium phases and pronounced crystallographic textures (Figure 6). Even though AM showed the ability to customize the parts, the manufacturer has to manage and control that process. Also, every stage of the design has to be monitored as per standards and approved for safety. A possible consideration for the development of industrial 3D printing is to enable the sharing of standardized digital design files over a standard platform for global access considering the intellectual property (IP) implications (Bogers et al., 2016).

Dimensional inaccuracy, microstructural defects and rough surfaces for the as-deposited parts may hinder the implementation of DLF for safety-critical high-value components such as medical implants and turbine blades with a strict tolerance level. The quality of the as-deposited part is determined by the laser/metal powder interaction, melt pool dynamics and the local thermal gradients, which is a consequence of the process and material parameters. A comprehensive understanding of these phenomena is mandatory to reduce defect formation and can be facilitated with in-situ, high-speed monitoring experiments during DLF. The need for post-processing of as-deposited parts such as hot isostatic pressing to obtain the required microstructural homogeneity (Joseph et al., 2018) and surface finishing operations for strict dimensional tolerances is to be considered.

DLF was proposed as a feasible technique for the repair/remanufacture of components with complex contours that are decommissioned or damaged during service. DLF made it possible to extend the lifetime and sustainability of such components at reduced cost and resources required for a new component. The components with deep or internal cracks/defects were milled down to the affected area followed by the use of DLF to deposit the affected area with the same material to regain the actual shape with excellent tolerance. However, defects at the boundaries between the original part and the deposited material may result in deteriorated functionality of the component and require rigorous process monitoring/optimization for high-quality repairs.

The tackling of processing and post-processing issues associated with the 3D printing process requires best practices developed by research standardization facilitated by the in-situ monitoring of the AM process. A universal standard for the processing-structure-property relationships in DLF depositions requires information from microstructural/application-specific analysis of multiple part depositions. It should also include consensus regarding the minimum requirements for qualifying 3D printing machines and processes, establish qualitative/quantitative evaluation methods and establish recommendations for standards practices (Berdine et al., 2019). A handful of already existing global standards for different materials and methods related additive manufacturing (specifically DLF) include:

1. ASTM52900; Standard terminology for AM – General Principles – Terminology
2. ASTM F2792; Standard terminology for AM Technologies
3. ASTM52915; Standard specification for AM File Format Version 1.2
4. F3049; Standard guide for characterizing properties of metal powders used for AM Processes
5. F2971; Standard practice for reporting data for test specimens prepared by AM
6. F3187 – 16; Standard Guide for Directed Energy Deposition of Metals
7. F3413 – 19; Guide for Additive Manufacturing — Design — Directed Energy Deposition
8. F3122; Standard guide for evaluating mechanical properties of metal materials via AM processes
9. ASTM52921; Standard terminology for AM- Coordinate systems and test methodologies
10. WK49229; New guide for anisotropy effects in mechanical properties of AM Parts
11. WK49272; New test methods for characterization of powder flow properties for AM Applications
12. WK47031; New guide for non-destructive testing of AM Metal parts used in Aerospace applications

It is critical to assess the safety aspects of DLF. The usage of fine-sized metal powders reduced the defects in the printed components than the coarser powder particle, however, lead to increased safety issues during powder processing and handling. The material safety data sheets for most of the metal powders recognizes the chances of health issues such as severe eye and allergic skin reactions due to improper handling. The most recommended procedure is to handle micron-sized metal powders in the protective atmosphere of a glove box with proper dust collection and personal protective equipment such as gloves, safety glasses and masks. The prolonged observation of the high-power laser during the melting process should also be avoided as it causes excessive strain of the eyes.

In DLF, depositions are made on a base plate with similar physical/chemical properties as that of the depositions. The base plate also serves as a mechanical support and as a heat sink dor the deposition during the DLF process. The part design strategy to include a part or whole of the base plate in the deposited component may reduce cost and energy consumption. Otherwise, the base plate has to be removed from the part by machining after the deposition process.

Currently, the AM techniques (such as DLF) improved its status as a tool for rapid prototyping of engineering components as a result of the design freedom and the ability to print intricate/complex parts. However, the as-deposited components by DLF are associated with relatively poor surface quality, difficulty in processing small details (compared to SLM) high machinery/operational costs. The AM process is significantly beneficial for small batches of complex/customized components, or in cases with functional advantages such as lightweight designs and part repair/remanufacturing. Also, the challenges such as the life cycle inventory of AM feedstock and OH&S (occupational health and safety) issues have to be tackled before implementing DLF for industrial applications (Kellens et al., 2017). If these challenges can be properly addressed, the potential advantages of AM include reduced downtime, improved flexibility/robustness, lower overall operation costs, and potential for sustainability improvements.

CONCLUSION

DLF platforms offer many unique capabilities in modern manufacturing, especially for small production runs of complex customized parts compared to multi-step conventional manufacturing processes. It may provide obvious benefits to the healthcare sector to save lives, for example, the development towards the creation of high-quality prosthetic devices/implants. This technique can be used to repair/remanufacture damaged components within an assembly such as tool/dies and turbine blades and thus avoiding the necessity of the fabrication of new components. However, 3D printing using DLF technique is incredibly complex, considering numerous highly localized deposition processes along with associated thermal and stress cycles. The metallurgy associated with this process can be correlated to a complex welding process with hundreds of welding passes, and the idea of 'just press print' to fabricate complex-shaped components is not a true depiction of the current scenario. The future perspectives of DLF as a tool for the fabrication of high-quality, high-end components include the development of efficient laser sources, improved powder utilization during the deposition process, the design of flexible/protective process chambers for DLF platforms and effective powder recycling strategies to reduce the environmental impacts of the process. The ecological perspective of the process should also be taken into account, with particular attention on standardized AM feedstock (especially metal powders) production, OH&S issues relating to AM.

ACKNOWLEDGMENT

The author would like to acknowledge the financial support of the Department of Innovation, Industry, Science & Research, Australia through the Australian-India Strategic Research Fund (AISRF53731—Advanced manufacturing of new high entropy alloys). The role of Prof. Daniel Fabijanic, Deakin University, as a mentor at various stages of the author's research, is gratefully acknowledged. Also, the author is grateful to the Advanced Characterisation/Additive Manufacturing Facility of Deakin University and Monash University, Australia, for technical support.

REFERENCES

Agarwala, M., Bourell, D., Beaman, J., Marcus, H., & Barlow, J. (1995). Direct selective laser sintering of metals. *Rapid Prototyping Journal*, *1*(1), 26–36. doi:10.1108/13552549510078113

Bandyopadhyay, A., & Heer, B. (2018). Additive manufacturing of multi-material structures. *Materials Science and Engineering R Reports*, *129*, 1–16. doi:10.1016/j.mser.2018.04.001

Berdine, G. G., DiPaola, M., & Weinberg, M. (2019). Economic and Regulatory Perspectives on Additive Manufacturing. In M. Dipaola & F. M. Wodajo (Eds.), 3D Printing in Orthopaedic Surgery (pp. 41-48). Elsevier. doi:10.1016/B978-0-323-58118-9.00004-X

Bogers, M., Hadar, R., & Bilberg, A. (2016). Additive manufacturing for consumer-centric business models: Implications for supply chains in consumer goods manufacturing. *Technological Forecasting and Social Change*, *102*, 225–239. doi:10.1016/j.techfore.2015.07.024

Bourell, D. L. (2016). Perspectives on Additive Manufacturing. *Annual Review of Materials Research*, *46*(1), 1–18. doi:10.1146/annurev-matsci-070115-031606

Cobbs, T., Lucas, B., & Crandall, J. L. (2019). *How 3D Metal Printing Saves Time and Lowers Costs: DED for Repair of Industrial Components from Optomec Inc.* https://optomec.com/wp-content/uploads/2019/03/DEDWebinar-slides.pdf

Collins, P., Brice, D., Samimi, P., Ghamarian, I., & Fraser, H. (2016). Microstructural control of additively manufactured metallic materials. *Annual Review of Materials Research*, *46*(1), 63–91. doi:10.1146/annurev-matsci-070115-031816

Conner, B. P., Manogharan, G. P., Martof, A. N., Rodomsky, L. M., Rodomsky, C. M., Jordan, D. C., & Limperos, J. W. (2014a). Making sense of 3-D printing: Creating a map of additive manufacturing products and services. *Additive Manufacturing*, *1–4*, 64–76. doi:10.1016/j.addma.2014.08.005

Dinda, G. P., Dasgupta, A. K., & Mazumder, J. (2012). Evolution of microstructure in laser deposited Al–11.28%Si alloy. *Surface and Coatings Technology*, *206*(8–9), 2152–2160. doi:10.1016/j.surfcoat.2011.09.051

Ewald, S., Kies, F., Hermsen, S., Voshage, M., Haase, C., & Schleifenbaum, J. H. (2019). Rapid Alloy Development of Extremely High-Alloyed Metals Using Powder Blends in Laser Powder Bed Fusion. *Materials (Basel)*, *12*(10), 1706. doi:10.3390/ma12101706 PMID:31130684

Eyers, D. R., & Potter, A. T. (2017). Industrial Additive Manufacturing: A manufacturing systems perspective. *Computers in Industry*, *92-93*, 208–218. doi:10.1016/j.compind.2017.08.002

Fera, M., Macchiaroli, R., Fruggiero, F., & Lambiase, A. (2018). A new perspective for production process analysis using additive manufacturing—complexity vs production volume. *International Journal of Advanced Manufacturing Technology*, *95*(1), 673–685. doi:10.100700170-017-1221-1

Gockel, J., Beuth, J., & Taminger, K. (2014). Integrated control of solidification microstructure and melt pool dimensions in electron beam wire feed additive manufacturing of Ti-6Al-4V. *Additive Manufacturing*, *1–4*, 119–126. doi:10.1016/j.addma.2014.09.004

Gorsse, S., Hutchinson, C., Gouné, M., & Banerjee, R. (2017). Additive manufacturing of metals: A brief review of the characteristic microstructures and properties of steels, Ti-6Al-4V and high-entropy alloys. *Science and Technology of Advanced Materials*, *18*(1), 584–610. doi:10.1080/14686996.2017.1361305 PMID:28970868

Haase, C., Tang, F., Wilms, M. B., Weisheit, A., & Hallstedt, B. (2017). Combining thermodynamic modeling and 3D printing of elemental powder blends for high-throughput investigation of high-entropy alloys – Towards rapid alloy screening and design. *Materials Science and Engineering A*, *688*, 180–189. doi:10.1016/j.msea.2017.01.099

Joseph, J. (2016). *Study of direct laser fabricated high entropy alloys. (PhD Thesis)*. Deakin University. http://dro.deakin.edu.au/view/DU:30088804

Joseph, J., Hodgson, P., Jarvis, T., Wu, X., Stanford, N., & Fabijanic, D. M. (2018). Effect of hot isostatic pressing on the microstructure and mechanical properties of additive manufactured AlxCoCrFeNi high entropy alloys. *Materials Science and Engineering A*, *733*, 59–70. doi:10.1016/j.msea.2018.07.036

Joseph, J., Jarvis, T., Wu, X., Stanford, N., Hodgson, P., & Fabijanic, D. M. (2015). Comparative study of the microstructures and mechanical properties of direct laser fabricated and arc-melted AlxCoCrFeNi high entropy alloys. *Materials Science and Engineering A*, *633*, 184–193. doi:10.1016/j.msea.2015.02.072

Kellens, K., Baumers, M., Gutowski, T. G., Flanagan, W., Lifset, R., & Duflou, J. R. (2017). Environmental dimensions of additive manufacturing: Mapping application domains and their environmental implications. *Journal of Industrial Ecology*, *21*(S1), S49–S68. doi:10.1111/jiec.12629

Khaing, M., Fuh, J., & Lu, L. (2001). Direct metal laser sintering for rapid tooling: Processing and characterization of EOS parts. *Journal of Materials Processing Technology*, *113*(1), 269–272. doi:10.1016/S0924-0136(01)00584-2

Kobryn, P. A., & Semiatin, S. L. (2003). Microstructure and texture evolution during solidification processing of Ti–6Al–4V. *Journal of Materials Processing Technology*, *135*(2–3), 330–339. doi:10.1016/S0924-0136(02)00865-8

Kruth, J.-P., Wang, X., Laoui, T., & Froyen, L. (2003). Lasers and materials in selective laser sintering. *Assembly Automation*, *23*(4), 357–371. doi:10.1108/01445150310698652

Kunze, K., Etter, T., Grässlin, J., & Shklover, V. (2015). Texture, anisotropy in microstructure and mechanical properties of IN738LC alloy processed by selective laser melting (SLM). *Materials Science and Engineering A*, *620*, 213–222. doi:10.1016/j.msea.2014.10.003

Kurz, W., & Trivedi, R. (1994). Rapid solidification processing and microstructure formation. *Materials Science and Engineering A*, *179-180*, 46–51. doi:10.1016/0921-5093(94)90162-7

Leung, C. L. A., Marussi, S., Atwood, R. C., Towrie, M., Withers, P. J., & Lee, P. D. (2018). In situ X-ray imaging of defect and molten pool dynamics in laser additive manufacturing. *Nature Communications*, *9*(1), 1355. doi:10.103841467-018-03734-7 PMID:29636443

Lewis, G. K., & Schlienger, E. (2000). Practical considerations and capabilities for laser assisted direct metal deposition. *Materials & Design*, *21*(4), 417–423. doi:10.1016/S0261-3069(99)00078-3

Mahamood, R. M., Akinlabi, E. T., Shukla, M., & Pityana, S. (2013). Scanning velocity influence on microstructure, microhardness and wear resistance performance of laser deposited Ti6Al4V/TiC composite. *Materials & Design*, *50*, 656–666. doi:10.1016/j.matdes.2013.03.049

Manyika, J., Chui, M., Bughin, J., Dobbs, R., Bisson, P., & Marrs, A. (2013). *Disruptive technologies: Advances that will transform life, business, and the global economy* (Vol. 180). McKinsey Global Institute. http://www.mckinsey.com/insights/business_technology/disruptive_technologies

Moorehead, M., Bertsch, K., Niezgoda, M., Parkin, C., Elbakhshwan, M., Sridharan, K., Zhang, C., Thoma, D., & Couet, A. (2020). High-throughput synthesis of Mo-Nb-Ta-W high-entropy alloys via additive manufacturing. *Materials & Design*, *187*, 108358. doi:10.1016/j.matdes.2019.108358

Romano, J., Ladani, L., Razmi, J., & Sadowski, M. (2015). Temperature distribution and melt geometry in laser and electron-beam melting processes – A comparison among common materials. *Additive Manufacturing*, *8*, 1–11. doi:10.1016/j.addma.2015.07.003

Shamsaei, N., Yadollahi, A., Bian, L., & Thompson, S. M. (2015). An overview of Direct Laser Deposition for additive manufacturing; Part II: Mechanical behavior, process parameter optimization and control. *Additive Manufacturing*, *8*, 12–35. doi:10.1016/j.addma.2015.07.002

Shi, Q., Gu, D., Xia, M., Cao, S., & Rong, T. (2016). Effects of laser processing parameters on thermal behavior and melting/solidification mechanism during selective laser melting of TiC/Inconel 718 composites. *Optics & Laser Technology*, *84*, 9–22. doi:10.1016/j.optlastec.2016.04.009

Simchi, A., Petzoldt, F., & Pohl, H. (2003). On the development of direct metal laser sintering for rapid tooling. *Journal of Materials Processing Technology*, *141*(3), 319–328. doi:10.1016/S0924-0136(03)00283-8

Song, B., Dong, S., Coddet, P., Liao, H., & Coddet, C. (2014). Fabrication of NiCr alloy parts by selective laser melting: Columnar microstructure and anisotropic mechanical behavior. *Materials & Design*, *53*, 1–7. doi:10.1016/j.matdes.2013.07.010

Sugavaneswaran, M., & Arumaikkannu, G. (2014). Modelling for randomly oriented multi material additive manufacturing component and its fabrication. *Materials & Design, 54*, 779-785. doi:10.1016/j.matdes.2013.08.102

Thijs, L., Verhaeghe, F., Craeghs, T., Humbeeck, J. V., & Kruth, J.-P. (2010). A study of the microstructural evolution during selective laser melting of Ti–6Al–4V. *Acta Materialia*, *58*(9), 3303–3312. doi:10.1016/j.actamat.2010.02.004

Thompson, S. M., Bian, L., Shamsaei, N., & Yadollahi, A. (2015). An overview of Direct Laser Deposition for additive manufacturing; Part I: Transport phenomena, modeling and diagnostics. *Additive Manufacturing*, *8*, 36–62. doi:10.1016/j.addma.2015.07.001

Tofail, S. A. M., Koumoulos, E. P., Bandyopadhyay, A., Bose, S., O'Donoghue, L., & Charitidis, C. (2018). Additive manufacturing: Scientific and technological challenges, market uptake and opportunities. *Materials Today*, *21*(1), 22–37. doi:10.1016/j.mattod.2017.07.001

Wang, L., & Felicelli, S. (2006). Analysis of thermal phenomena in LENS™ deposition. *Materials Science and Engineering A*, *435*, 625–631. doi:10.1016/j.msea.2006.07.087

Wolff, S. J., Wu, H., Parab, N., Zhao, C., Ehmann, K. F., Sun, T., & Cao, J. (2019). In-situ high-speed X-ray imaging of piezo-driven directed energy deposition additive manufacturing. *Scientific Reports*, *9*(1), 962. doi:10.103841598-018-36678-5 PMID:30700736

Wu, X. (2015). *The world 1st 3D printed jet engine was on display at this years Paris Airshow.* https://www.monash.edu/mcam/news/articles/paris-le-bourget-airshow

Yin, J., Zhu, H., Ke, L., Lei, W., Dai, C., & Zuo, D. (2012). Simulation of temperature distribution in single metallic powder layer for laser micro-sintering. *Computational Materials Science*, *53*(1), 333–339. doi:10.1016/j.commatsci.2011.09.012

Zhu, Z., Dhokia, V. G., Nassehi, A., & Newman, S. T. (2013). A review of hybrid manufacturing processes–state of the art and future perspectives. *International Journal of Computer Integrated Manufacturing*, *26*(7), 596–615. doi:10.1080/0951192X.2012.749530

ADDITIONAL READING

Aucott, L., Dong, H., Mirihanage, W., Atwood, R., Kidess, A., Gao, S., Wen, S., Marsden, J., Feng, S., Tong, M., Connolley, T., Drakopoulos, M., Kleijn, C. R., Richardson, I. M., Browne, D. J., Mathiesen, R. H., & Atkinson, H. V. (2018). Revealing internal flow behaviour in arc welding and additive manufacturing of metals. *Nature Communications*, *9*(1), 5414. doi:10.103841467-018-07900-9 PMID:30575762

DebRoy, T., Wei, H. L., Zuback, J. S., Mukherjee, T., Elmer, J. W., Milewski, J. O., Beese, A. M., Wilson-Heid, A., De, A., & Zhang, W. (2018). Additive manufacturing of metallic components – Process, structure and properties. *Progress in Materials Science*, *92*, 112–224. doi:10.1016/j.pmatsci.2017.10.001

Guo, N., & Leu, M. C. (2013). Additive manufacturing: Technology, applications and research needs. *Frontiers of Mechanical Engineering*, *8*(3), 215–243. doi:10.100711465-013-0248-8

Huang, S. H., Liu, P., Mokasdar, A., & Hou, L. (2013). Additive manufacturing and its societal impact: A literature review. *International Journal of Advanced Manufacturing Technology*, *67*(5), 1191–1203. doi:10.100700170-012-4558-5

Khairallah, S. A., Anderson, A. T., Rubenchik, A., & King, W. E. (2016). Laser powder-bed fusion additive manufacturing: Physics of complex melt flow and formation mechanisms of pores, spatter, and denudation zones. *Acta Materialia*, *108*, 36–45. doi:10.1016/j.actamat.2016.02.014

Khairallah, S. A., Martin, A. A., Lee, J. R. I., Guss, G., Calta, N. P., Hammons, J. A., Nielsen, M. H., Chaput, K., Schwalbach, E., Shah, M. N., Chapman, M. G., Willey, T. M., Rubenchik, A. M., Anderson, A. T., Wang, Y. M., Matthews, M. J., & King, W. E. (2020). Controlling interdependent meso-nanosecond dynamics and defect generation in metal 3D printing. *Science*, *368*(6491), 660–665. doi:10.1126cience.aay7830 PMID:32381724

Kunce, I., Polanski, M., Karczewski, K., Plocinski, T., & Kurzydlowski, K. J. (2015). Microstructural characterisation of high-entropy alloy AlCoCrFeNi fabricated by laser engineered net shaping. *Journal of Alloys and Compounds*, *648*, 751–758. doi:10.1016/j.jallcom.2015.05.144

Li, M., & Flores, K. M. (2020). Laser processing as a high-throughput method to investigate microstructure-processing-property relationships in multiprincipal element alloys. *Journal of Alloys and Compounds*, *825*, 154025. doi:10.1016/j.jallcom.2020.154025

Li, M., Gazquez, J., Borisevich, A., Mishra, R., & Flores, K. M. (2018). Evaluation of microstructure and mechanical property variations in AlxCoCrFeNi high entropy alloys produced by a high-throughput laser deposition method. *Intermetallics*, *95*, 110–118. doi:10.1016/j.intermet.2018.01.021

Martin, A. A., Calta, N. P., Khairallah, S. A., Wang, J., Depond, P. J., Fong, A. Y., Thampy, V., Guss, G. M., Kiss, A. M., Stone, K. H., Tassone, C. J., Nelson Weker, J., Toney, M. F., van Buuren, T., & Matthews, M. J. (2019). Dynamics of pore formation during laser powder bed fusion additive manufacturing. *Nature Communications*, *10*(1), 1987. doi:10.103841467-019-10009-2 PMID:31040270

Ocelík, V., Janssen, N., Smith, S. N., & De Hosson, J. T. M. (2016). Additive Manufacturing of High-Entropy Alloys by Laser Processing. *JOM*, *68*(7), 1810–1818. doi:10.100711837-016-1888-z

Sames, W. J., List, F. A., Pannala, S., Dehoff, R. R., & Babu, S. S. (2016). The metallurgy and processing science of metal additive manufacturing. *International Materials Reviews*, *61*(5), 315–360. doi:10.1080/09506608.2015.1116649

KEY TERMS AND DEFINITIONS

ASTM: American Society for Testing and Materials (ASTM) provides additive Manufacturing Technology standards define terminology, measure the performance of different production processes, ensure the quality of the end products, and specify procedures for the calibration of additive manufacturing machines.

Functionally-Graded Materials: Functionally-Graded Materials (FGMs) exhibit mutually exclusive properties for multiple functions and may be achieved by varying the chemical composition across the component or by surface modification. For example, the gears should be tough enough inside to withstand the fracture and should also be hard on the outside to resist wear.

HEA: High entropy alloys (HEA) are a new class of alloys introduced in 2004. These alloys have four or more alloying elements in equiatomic or near-equiatomic ratio and form solid solution structures.

High-Throughput: High throughput technology refers to the efficient processing/analysis of large numbers of samples and variables for further developments.

MCAM: The Monash Centre for Additive Manufacturing (MCAM) is a strategic research centre of Monash University, Australia. The centre takes fundamental research from material science, alloy design and processing, surface engineering, corrosion and hybrid materials.

Optomec: Optomec Inc. is a US-based firm who is a leading manufacturer of next-generation advanced DLF Processing for Industrial Metal Additive Manufacturing.

Trumpf: Trumpf Group is a German industrial machine manufacturing company and a leading manufacturer of laser systems and 3D printing machines for metal components.

Chapter 9
What Is Design for Additive Manufacturing (DfAM)?

Seung Hwan Joo
Inha University, South Korea

Sung Mo Lee
Korea Shipbuilding and Offshore Engineering, South Korea

Jin Ho Yoo
Korea-Additive Manufacturing User Group, South Korea

Hyeon Jin Son
Winforsys, South Korea

Seung Ho Lee
Metal 3D, South Korea

ABSTRACT

In order to use 3D printing technology as a sanction, it is necessary to optimize topology, component unification, and reduce weight need for advanced manufacturing design. In the case of metal 3D printing, it is necessary to manage deformation and defects in the process cause of using laser, and support generation and design optimization must be accompanied for efficiency. Currently, design progresses through simulation before actual production in AM field. This chapter explores design in additive manufacturing.

NECESSITY OF DESIGN FOR ADDITIVE MANUFACTURING (DfAM)

When 3D printing technology was first introduced, it was raved by many as an omnipotent tool similar to the philosopher's stone. However, many of the enthusiasts seem to adjust their original thoughts from 'the technology that can create anything' to 'technology of still far future' after using 3D printers on the market. I sometimes think the following. Is the problem lies in the 3D technology that is not fully developed? or in the people who are not familiar with methods of utilizing the new technology called

DOI: 10.4018/978-1-7998-4054-1.ch009

3D printing? To answer this question, it is necessary to understand the concept of Design for Additive Manufacturing (DfAM). DfAM is the key not only to the method of using 3D printing in manufacturing fields but also to generating new service markets via 3D printing.

DfAM is an advanced concept from DfM (Design for Manufacturing) which maximizes the time and cost productivity in product manufacturing through maximizing the best features of the manufacturing method. It is a design approach combined with additive manufacturing technology. It minimizes the manufacturing steps in product development and manufacturing process and maximizes functionality and performance of the product to mass-produce the upgraded products quickly with lower unit cost.

According to Ian, G (2014), the meaning of 3D printing technology in the fields of design, planning, and manufacturing. Early on, 3D printing has been expected to change the existing industry paradigms. Not only the process from the idea to product manufacturing would be surprisingly shortened, but also the perspective of designing and planning would be changed 180 degrees, allowing innovative design methods to be applied to manufacturing that is enabled only by 3D printing technology (p.399-435).

Figure 1. Comparison between the traditional manufacturing method and 3D printer manufacturing method

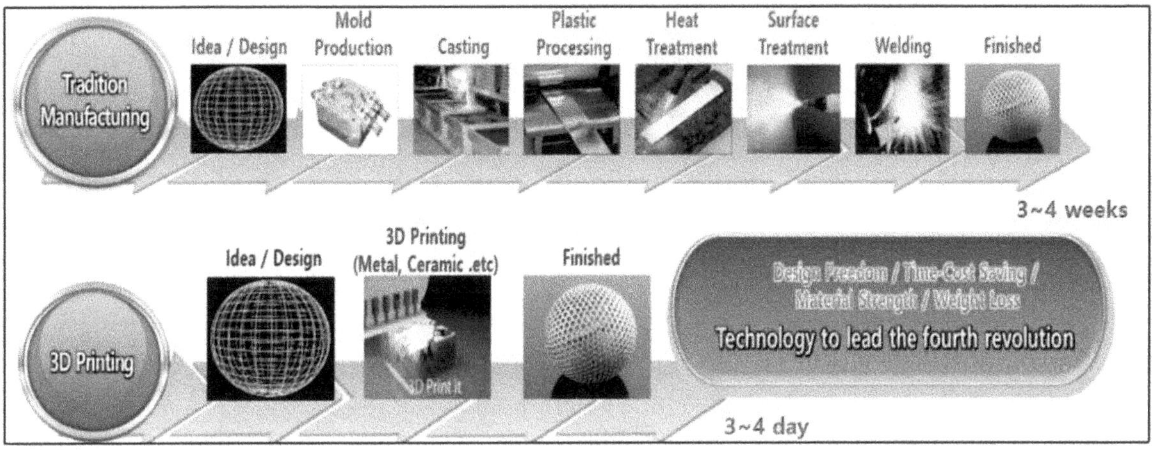

This also included the implementation of lightweight/high strength structure via optimized design, the possibility of one-process manufacturing without a complex assembly process to create products in complex forms, and the simultaneous application of composite material. This is DfAM, which is the design and engineering approach that can maximize the benefits of 3D printing technology. One of the most well-known limitations of 3D printers is its slow speed. With DfAM, productivity increases significantly, and the performance of product improves. People realized that it could increase productivity when it made it possible to drastically shorten the manufacturing process by 3-4 weeks.

Figure 2 exhibits the application case of 3D printing technology on the complexity of design shape, material, and dimension that DfAM can overcome.

DfAM is mainly used largely in four fields including

- Lightweight structure
- Parts unification of producing a number of parts into one part

Figure 2. Complexity of manufacturing surmountable by DfAM

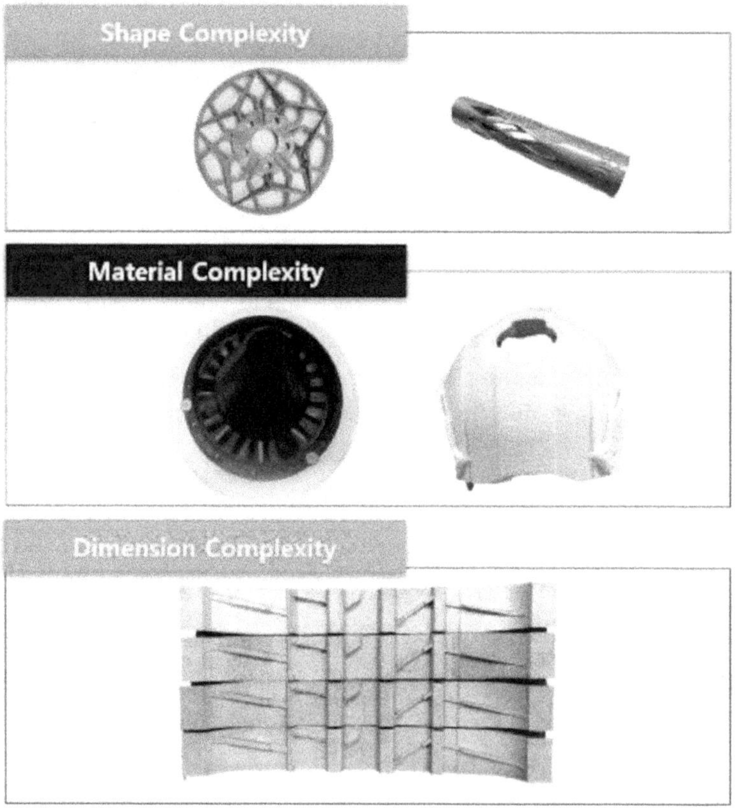

- Design that enables different properties from the material of one component
- Medical-specialized design.

There are various techniques in DfAM:

- "Multi Material" Design that realizes numerous properties and special functions by printing ingredients with different properties at the same time.

Figure 3. Definition of DfAM: a) Light weight b) Part unification c) Multi-property material d) Specialized part (Medical)

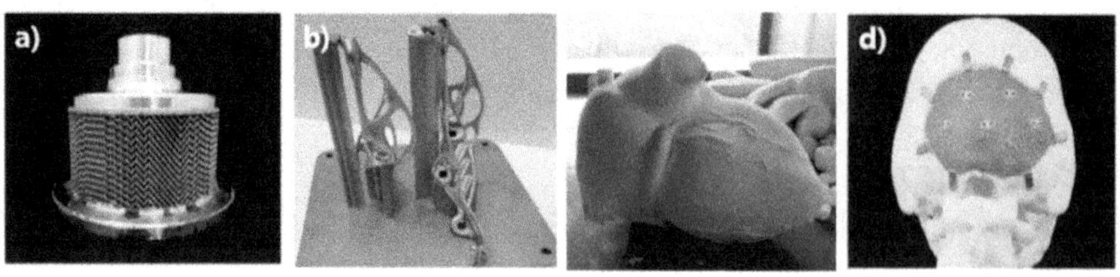

- "Phase Optimization" Design that calculates the optimal placement of materials with computer by considering design purposes and constraints.
- "Integrated Structure" Design that realizes efficient structure and improve the performance by reducing joint parts and merging them into one.
- "Grid Structure" that realizes the ultra-lightweight parts while maintaining rigidity through the grid structure composed of the repetitions of unit structure.

A producer can create profits through weight reduction, durability increase, and reduction of a number of parts.

According to Yoon O.S (2013), there are reasons why 3D printing technology is taking the center stage as a new manufacturing process in the transfer machine industry. DfAM technology allows integrated production without a separate assembly process of part modules of complex function and shape. Also, the energy efficiency can be improved with the designing and manufacturing of high-strength, low-vibration, and lightweight vehicle parts with complex internal structure. In particular, in the eco-friendly vehicle market including electric vehicles recently gaining attention, the market growth in 2020 is expected to double the growth in 2016.

DfAM APPLICATION CASE [INDUSTRIAL APPLICATION]

Automobile

According to Martinezmartin, A (2016), in fact, Korea and overseas well-renowned automobile companies making efforts to utilize 3D printing technology with plastic as well as metal and carbon composite materials to apply to automobile production. It is no longer a secret that numerous automobile companies and related parts companies overseas including world-renowned makers in Europe and large automobile companies in the U.S. and Japan are already actively utilizing 3D printing equipment in new vehicle development and mass-production.

FIT WEST of the U.S. uses DfAM technology to advance the function of F-1 automobile cylinder, successfully achieving 80% weight reduction.

Toyota has introduced a 3D printing lightweight automobile seat made of polymer and has attracted much attention.

Local Motors of U.S. introduced 3D printed electronic vehicles to the public and has received a lot of public attention (2014).

Also, by suggesting a cooperation platform for designers-developers-consumers to consider together with 3D printing production technology at the center, a new manufacturing industry business direction has been presented.

This shows that DfAM technology will play an essential engine role with 3D printing for all manufacturing industries including the future automobile industry proceeding in the direction of small-volume, multi-variety, lightweight, high-function, and tailored type.

Figure 4. Car wheel applied with DfAM Design Technology
(Source: KAMUG & Metal3D)

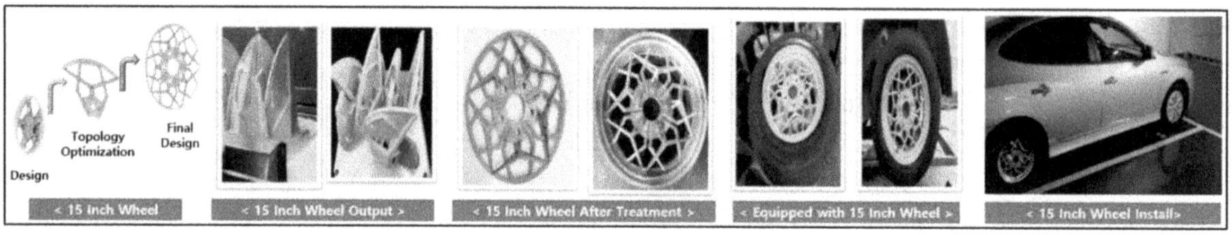

Aerospace Field

According to Bikas, H (2016), in aerospace industry is one of the promising high-tech industries that creates high returned value by application of cutting-edge technologies in several fields such as machinery, electronics, IT, and new materials (p.389-405).

There are ongoing search and development of utilization of 3D printing manufacturing technology for improving performance and decrease in the weight by the completed machine and Key part manufacturing companies such as Boeing Inc., GE Inc., etc. in U.S.

Boeing Inc. produced approximately 300 small aircraft parts, including the pipe that provides cold air to electronic equipment. UTC Inc. in U.S. also produced blades for aircraft engines utilizing 3D Printing.

According to Martinezmartin, A (2016), in the case of GE jet engine nozzle, 20 parts can be manufactured as one, which can increase the durability by 5 times and reduce production cost by 75%. GE Inc. acquired a German SLM Solution, and others for $1.4 billion to build the foundation for its gas turbine part production.

According to Wasserman, S (2015), GE attained Federal Aviation Administration (FAA) certificate for the engine temperature sensor housing in 2015 in the field of gas turbine engines and applied to commercial engines. By 2020, GE plans to increase the number of 3D printing produced parts in aviation engines could be up to 100,000.

NASA has successfully built and finished test flight a complex and delicate fuel injector of rocket with 3D printing technology. The number of parts reduced by 45% compared to conventional engines and decrease substantially both costs and about 1 year of manufacturing period to few months.

As a core technology for satellite projectile cost reduction, the rutherford engine is mostly manufactured using 3D printing technology due to its complex structure.

According to a 2016 German Aerospace Center (DLR) report, the spacecraft transport cost is more than 200,000 Euros per 1kg of cargo. Besides, the cost of shooting the spacecraft decreases with each reduction in spaceship, as the required amount of fuel decreases correspondingly. Accumulating in the excess weight, the parts must be shaved off as much as possible by aerospace engineers.

RUAG conglomerate needed an antenna bracket planned with an optimized design. The parts of aerospace need to satisfy all the conditions such as: weight optimization, frequency, inertia, stability, and stiffness.

RUAG ultimately combined Altair's intensive CAD and FED systems with guidelines for design and manufacturing using additive manufacturing from EOS to create a theoretically perfect antenna bracket form.

Due to this, the minimum stiffness requirements of the parts were exceeded by more than 30%, securing the stability due to the excess stiffness and a very uniform stress distribution, and the weight of components through lamination manufacturing was reduced by about 40% from 1.6Kg to 940g. The introduction of this innovative technology has improved the features of the parts and reduced costs.

Freeform Shapes

According to Lowa, A (2018), additive manufacturing has contributed significantly to industries with design freedom. Such geometric freedom enabling additive manufacturing provide esthetic, economic, and ergonomic benefits. The additive manufacturing function for producing complex form parts is applied to interior design, medicine, automobile, and aerospace industries (p.13-33).

Part Consolidation

According to Lowa, A (2018), the process of reducing the number of parts for assembly by combining several parts into one essential part is called part integration. With additive manufacturing, parts can be manufactured into one essential part (p.13-33). This process is not limited by the existing process restriction. The example of the integration of the part is shown in Figure 5. This is the Laminar Flow module, which was manufactured by wielding a number of aluminum parts, which required high cost and time, in addition to the problem of inaccurate form and measurement. By applying DfAM, the product which had been wielded with 5 parts was designed into 1 part, of which case is produced with a 3D printer. By integrating parts, the assembly process becomes easier and therefore is advantageous because the cost and time can be reduced significantly.

Figure 5. Laminar flow module
(Source: Winforsys)

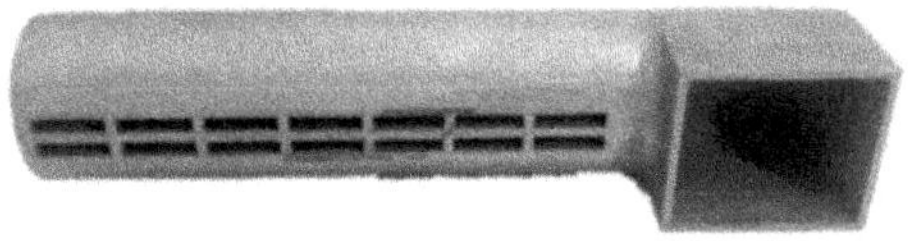

Internal Channels

Complex internal functions such as cooling channels, air ducts, and fluid channels that can improve the functionality and performance of components can be created via additive processing. The internal channel that is difficult to manufacture using existing manufacturing processes can be created with AM technology. An injection mold insert with a complex flood cooling channel was developed with Electron Beam Melting (EBM) process, which was shown to have a significantly higher cooling efficiency than conventional inserts used with the existing manufacturing process.

According to Kim, Y.S (2019), When parts are manufactured with additive processing technology, the materials are added by layer, through which parts can be built-in inside the printed parts. By using the Direct Metal Disposition (DMD) method, an injection mold frame of the conformal cooling method was manufactured. A copper cooling tube was embedded inside the substrate mold part to make a mold with conformal cooling channels. This cooling channel improved heat transfer and reduced the cooling time by 35% (Figure 6) (p.51-58).

Figure 6. Mold printed with a 3D printer
(Source: Metal 3D)

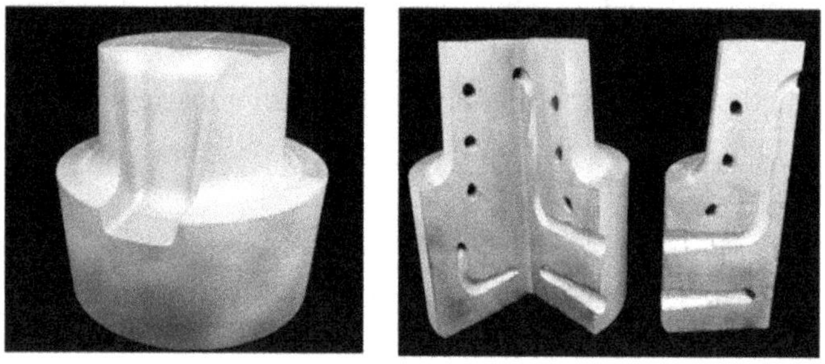

Topology Optimization

According to Brackett, D (2011), topology optimization is a method based on Finite Element Analysis (FEA) that optimizes the part shape to decrease weight while maintaining the strength. FEA software optimizes the density of each element by dividing the parts into elements. In all areas without stress, an optimized part shape is created by the software. This optimized shape is generally a complex shape that is difficult to be manufactured in an existing manufacturing process. Additive manufacturing can be used to create such a complex shape and, thus, topology optimization combined with additive manufacturing is used to produce strong lightweight components (p.348-362). Figure 7 shows the topology optimization process of a metal bracket.

Figure 7. Topology optimization process of metal bracket, all images courtesy of ANSYS
(Source: ANSYS)

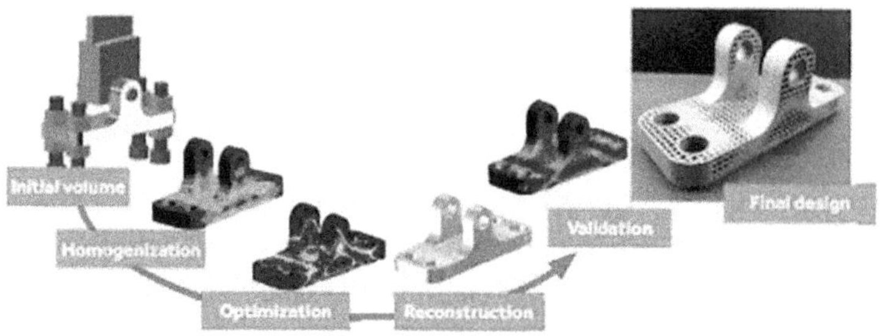

Figure 8. Dental prosthesis
(Source: Winforsys)

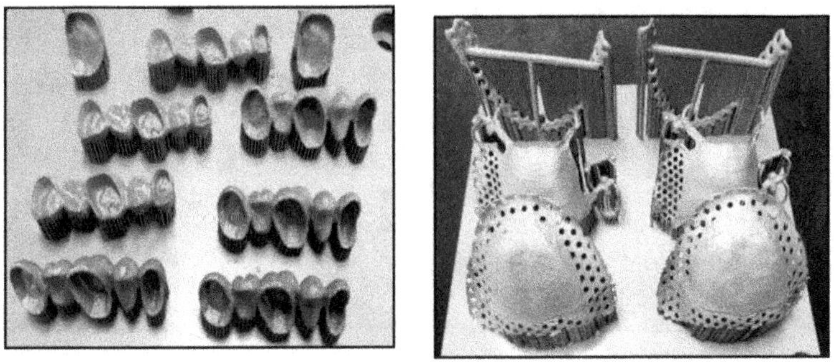

DfAM APPLICATION CASE [MEDICAL APPLICATION]

According to Ian, G.D (2015), medical industry is a field where the substantiation and commercialization are progressing at the fastest speed by utilizing the advantages of 3D printing technology, suitable for small-scale production of various kinds and customized production (p.107-145).

- Personalized implants such as hearing aids, dentures, prosthetic leg, and prosthetic arms, etc.
- Surgical guides and surgical models used directly in medical practice △ A variety of medical implants such as several bone implants and prostheses are manufactured with 3D printing technology and applied to patients.

According to Lowa, A (2018), currently, 3D printing medical devices have been devoted to verifying the safety and treatment effect of high returned value type medical devices such as implants. Based on successful clinical results, many hospitals are currently preparing to expand the market through the introduction of 3D printing medical services (p.13-33).

Dental Prosthesis

Renishaw Inc. in the UK is working with BioHorizons Inc. to manufacture and sell dental implants that have passed ISO 13485 quality control certification and biocompatibility testing. Groningen Inc. in the Netherlands has developed an implant that adds antibacterial function to existing dental implants.

Sternum

Anatomics Inc and Australia together with the Federal Institute of Science and Technology (CSIRO) produced a titanium skeleton that can be implanted in the body, and successfully transplanted part of the sternum and ribs of a 52-year-old osteosarcoma cancer patient.

Figure 9. Cranial implant
(Source: Metal3D)

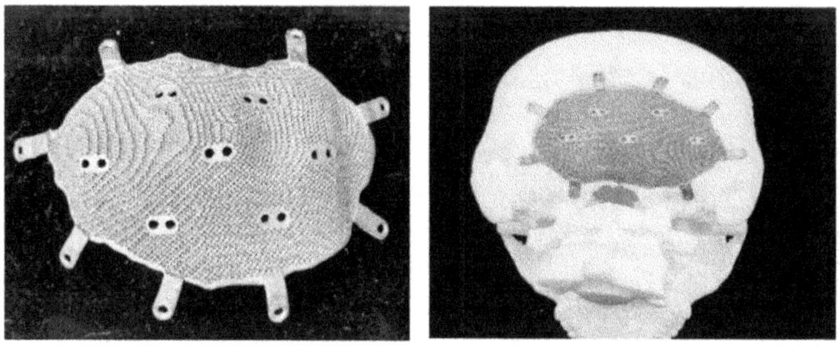

Skull Implant

Oxford Performance Materials Inc. in UK, manufactured a patient-specific skull implant made of medical high-performance polymer (PEEK) material, modeled it for the shape of a U.S. male patient's skull and received FDA approval.

The Chung Ang University Hospital in South Korea succeeded in transplantation using a titanium skull made of optimal thickness and strength by the Korea Institute of Industrial Technology, for a woman in her 60s whose skull was severely sunk the due to "subarachnoid hemorrhage".

Facial Skeleton

Belgium Materials Inc. has succeeded in restoring the face by manufacturing 3D printing technology for surgical reconstruction of a patient with a sunk face, and a surgical model and surgical guide necessary for the surgery.

Artificial Pelvis

The National Cancer Center of Korea has succeeded in pelvic bone reconstruction surgery by removing cancer from the pelvic bone and implanting a titanium alloy artificial bone made of bio-friendly 3D printing, which has excellent connection strength with surrounding normal tissue.

Surgery Simulation

Seoul Asan Hospital utilizes a 3D printing model made of the same size and structure as the patient's real heart for surgical simulation of patients with innate heart disease, thereby accurately establishing a surgical plan to improve accuracy and help patients and caretakers understand.

Production of Customized Implants

The combination of 3D printing technology across the medical field has been a great help in the production of customized human implants such as hearing aids, prosthetic arms and prosthetic legs. One of the most commonly used human implants through 3D printing technology is custom-made hearing aids. For building hearing aids that fits each different ear-shapes of patients, there is no technology more suited than 3D printing.

Hearing Aid

In the past, customized hearing aids were produced by cutting and trimming methods by skilled workers with at least 10 years of experience, taking a lot of time and money to produce. However, a custom-made hearing aid using 3D printing technology mimics the shape of the ear with silicone, then accurately recognizes the shape of the ear using a 3D scanner and print it with a 3D printer according to the model to produce a hearing aid that fits the patient's ear. Also, unlike conventional manual work, it can be mass-produced in a short period through equipment, making the product competitive in price, and storing the scanned data in the shape of the customer's ear so that it can be produced at any time.

A person related to a hearing aid said, "Since the introduction of 3D printing technology to produce sophisticated and delicate hearing aids, the defect rate has been significantly lowered and the customer's wearing feeling has also improved. "3D printer enabled fast mass production doing the works of 3-4 skilled workers."

Prosthetic Legs and Arms

3D printing technology is also used to tailor the cover of the prosthetic legs or prosthetic arms worn by patients who have lost their arms or legs. Bespoke Innovation, based in San Francisco, USA, incorporates 3D printing technology to create a cover that covers the patient's prosthetic legs and prosthetic arms. To make the prosthetic leg of the patient who has lost his leg, the shape of the leg is created for the prosthetic leg.

Existing prosthetic legs and prosthetic arms were pipe-like structures, or they were manufactured in a form that wraps the surface with a cover such as a sponge that does not fit the skin, so they tend not to have a good appearance. However, the prosthetic arms or prosthetic legs made using 3D printing technology not only have a tailored appearance that fits the patient's body shape, but also can express individual personality through cover of various materials and designs such as metal and leather.

DfAM SIMULATION TYPE

As DfAM simulation software, there are INSPIRE, ANSYS Suite, 3-Matic, Hyperworks, and SimSolid. The representative software among them are INSPIRE, ANSYS Suite, and 3-Matic.

INSPIRE

INSPIRE is a quick simulation solution that is strongest and easy to use in the industry that provides production design/topology optimization. Thus, the cost, development time, material consumption, and product weight can be reduced. By using INSPIRE by Altair company, which operates together with the existing CAD tool in creating the initial structure, it is possible to quickly, easily, and effectively structure interpretation results. At the initial stage of the design cycle, a concept design that satisfies structural functionality. Compared to existing methods, a significant amount of time can be saved to redesign to fulfill the requirements of design, verification, and structure. The connection with design space, load condition, and shape control can be modified or added with the user's what-if scenario, and it presents an excellent insight for the review of concept design results. INSPIRE is a useful software for functionality increase, reduction of design weight and delivery cost, and reduction of material cost because it effectively uses materials by allocating materials in structurally satisfying parts (Figure 10).

Figure 10. Example of implementation of topology optimization and lightweight by using INSPIRE, all images courtesy of Altair
(Source: Altair)

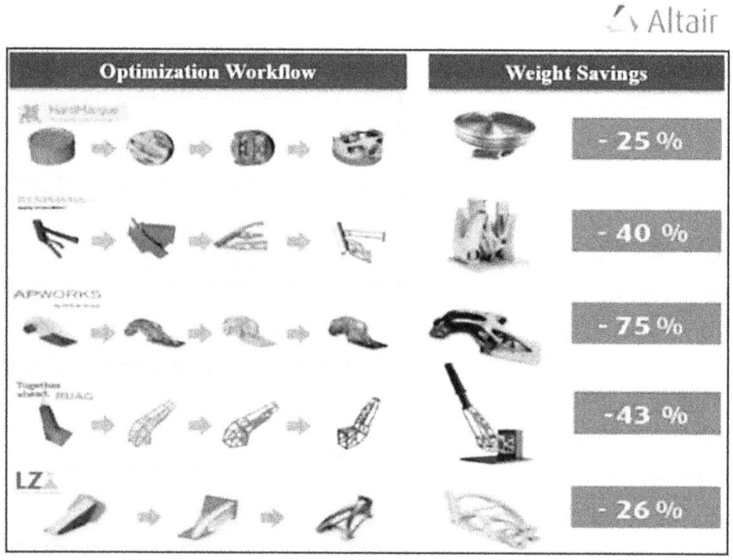

ANSYS Suite

ANSYS Suite simulates the process of validation after designing the shape that is over-designed as topology optimized shape physically or structurally. The AM process simulation workflow example using ANSYS additive suite.

The AM process simulation process using ANSYS additive suite is presented in Figure 11(a). The additive shape simulation using topology optimization conducts static structure analysis after giving the load condition on the over-designed shape. Based on this, it conducts optimization simulation of topology (Physical & Structural) by setting a permissible range of optimization elements (Mass and deformation, etc.) in a topology optimization environment. In addition, via another static structural analysis, a validation process is conducted to determine the additive shape. As such, the detailed simulation process using topology optimization is shown in Figure 7.

Figure 11. Topology optimization simulation process using ANSYS additive workbench, all images courtesy of ANSYS
(Source: ANSYS)

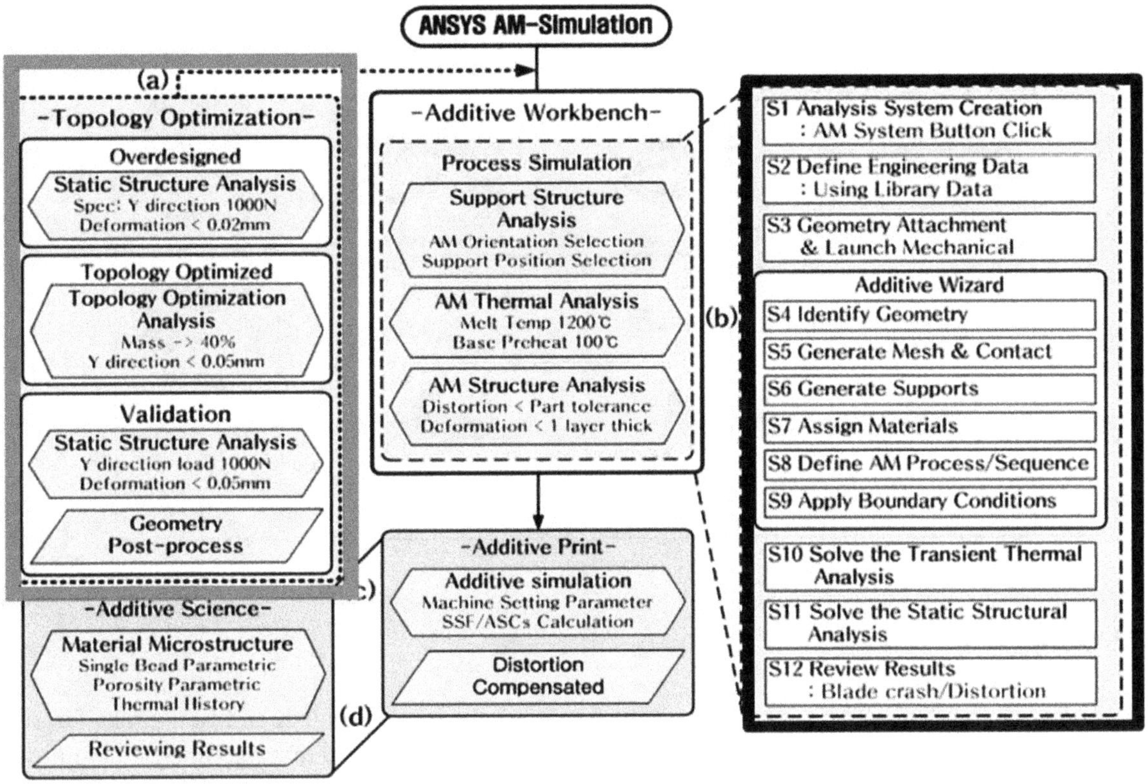

3-Matic

3-Matic software allows design modification and re-mesh, in addition to CAD data and scan data, and topology optimization data design for 3D printers. 3-Matic is an ideal software that enhances surface quality by cleaning up the topology data. With 3-Matic, you can create complex internal and external structures that are available for 3D printing with advantages such as increased strength, enhanced porous, and decreased design weight (Lightweight structure). In addition, the design can be easily improved with esthetic and functional texture as well as perforation and patterns. Because you can directly work on the STL data, FEA analysis can be immediately executed on the design or used for 3D printing. 3-Matic package is an effective software suitable for the post-treatment of the result of topology optimization. The suggested workflow can decrease the post-treatment time from several days to hours compared to other methods (Figure 12).

Figure 12. Lightweight bracket produced by european space agency using 3-Matic, all images courtesy of Materialise
(Source: Materialise)

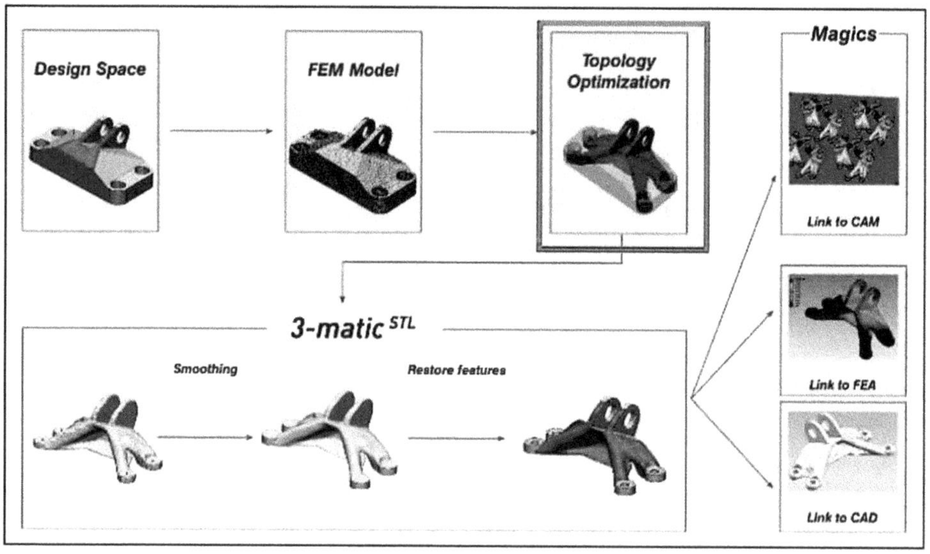

CONCLUSION

As mentioned earlier, DfAM provides the solution to overcome the problem and limitations faced in conventional design and manufacturing process as a design methodology that maximizes the intrinsic advantages of 3D printing. The author, believe that although DfAM is a design method that overcomes the disadvantages of 3D printing and highlights the advantages, it is not appropriate to apply DfAM to the production of all products and parts. If due to the problems such as complexity of structure, material, scale, and etc., manufacturing is not possible with the conventional methods or if it takes too much time or money, I believe that DfAM would be the key to overcome these limitations. Bringing 3D printer equipment in one process doesn't make a difference. The proper use of this technology requires a completely different approach to traditional design and production methods. This is not a problem for the design department or the production department. This means that a third organization that includes both departments or the new engineer is needed. DfAM believes it is important to train DfAM experts who understand both the design and 3D printing process and can produce optimal results because of how well it reflects the limitations and requirements of 3D printing production during design, analysis, optimization, and production.

What Is Additive Manufacturing (AM) Simulation? Why Additive Manufacturing Simulation Is Necessary?

Additive Manufacturing (AM) can be easily controlled to enable the basic design to be directly implemented as speculation features and dual-material. Thus, it can be significantly useful for advancement and customization. The ability to produce a customized product at once will significantly improve corporate margins and customer satisfaction while drastically reducing production costs and material waste.

The application of additive production technology at the development stage of a metal product has been known to have both significant strategic and financial effects. According to Kim, Y.S (2019), however, it has not been generalized for all companies and rather has been applied mainly by large companies such as an aviation corporation in need of high-cost and high-precision product design. This has been known to be due to the high price of printing equipment for metal additive manufacturing, significantly high cost of material powdering, and the high likelihood of technical failure at the initial development environment causing the establishment costs to be very high (p.51-58).

The problems that occur in developing metal AM parts are as follows. The existing additive method goes through pre-processing in order to prepare for the build by bringing the designed part in advance. During the pre-processing, a supporter is created, upon which the build stage proceeds by directly putting on the device. However, when building progresses as such, most customers must go through numerous trials and errors. As shown in Figure 8, customers experience the common yet vicious cycle of destruction/ supporter damage/ deformation after wire cutting due to the residual stress. The problem of trial and error is serious involving the acquisition of the desired product at the end after experiencing a variety of problems (Significant time and Costs incurred).

According to Materialise has said, over 75% of the total costs of additive manufacturing using metal 3D printing go into the printing process and a lot of costs are incurred by test printing and failed printing materials. In addition, in the case of 3D printing of complex and geometric structures, 15% on average is a failure. On average, trials of 3-5 times are required to complete metal printing work.

According to Oh, J.W (2017), the cause of defects occurring in the metal 3D printing AM process can be largely divided into structural factors and thermal factors. The deformation and residual stress that appear based on the combination of AM process conditions (Parameters) are elements that directly influence the quality of output. Deformation of outputs that appear during the AM process cannot be assembled with other parts after printing the product and can also damage the output or suspend the printing by causing a collision with recoat blade (In the case of Powder Bed Fusion (PBF) method). In addition, residual pressure can also cause product defects by causing a crack in the output or contraction of the output. Metal AM process repeats heating whereby a melt pool is made by melting the powder material with high-energy (Laser) and cooling whereby the melt pool is hardened and fused. Due to the repeated heating and cooling, thermal deformation is caused by the output (p.1-14).

In order to reduce the impact of such thermal deformation, it prevents deformation by effectively supporting the output from the base plate. As such, the support that holds the additive part in the right position during metal AM process manufacturing is critical. When support is created more than necessary, the cost (Additive material / Time increase) is increased, while insufficiency can lead to output failure by lowering the output quality.

As such, the repeated failures in trial and error for not being able to make the printing material at once in metal 3D printing can be caused by new equipment, lack of accumulated technology, unfamiliar manufacturing parameter adjustments, lack of experience, and diverse faulty factors.

3D printing is a term used to collectively and broadly refer to all process of 3D printing from the dissolution of supply material to the actual tool path necessary for making a component. In particular, metal 3D printing simulation is forecasting the printing process in advance and designing by considering the deformation and residual stress. In other words, the method reduces the trial and error at the development stage as well as time and cost by allowing direct printing from the initial stage.

According to Zhn, J.H (2015), for 3D printing technology to be used in the manufacturing field, advanced 3D printer, material, and software are essential. Most of the work of the main AM process is conducted via software including design to appropriate shape for 3D printing, generation of supporters that support and prevent the output from collapsing during the printing process, and calibrating or modifying by detecting crack or deformation of output in advance before printing. Especially, the importance is increasing because the materials for the software-based simulation in the AM process are being diversified, the output shape is becoming more complex and enlarged. The goal of Am process simulation is to predict macro deformation or stress of a part and thereby to prevent build-up failure. In order to achieve this goal, the determination of additive product direction, allocation of supporters, adjustment of size, and other overall AM process design improvement require the provision of optimized data (p. 175-192).

Figure 13. Problems of metal AM part manufacturing, all images courtesy of ANSYS
(Source:_ANSYS)

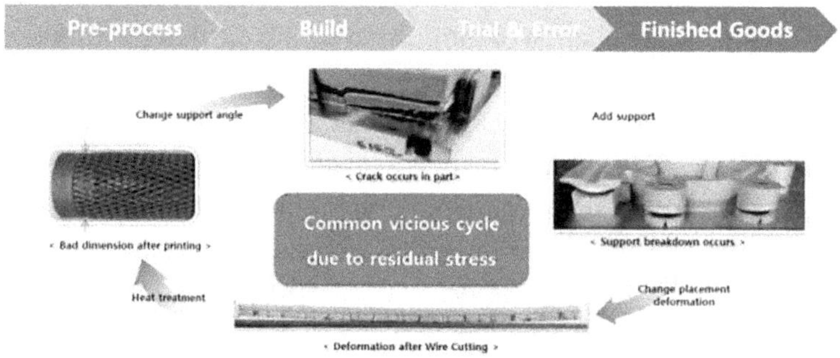

Types of Additive Manufacturing (AM) Process Simulation

According to Kim, Y.S (2019), as mentioned above, the AM process simulation can be conducted in a virtual place via software. As the traditional engineering and production process has been optimized for over 40 years with simulation technology, it is evolving gradually to solve new tasks in the additive manufacturing field. Recently, with the activation of AM technology, AM process simulation software is being developed and utilized in large numbers (p.51-58).

Major software products being utilized currently are organized in Table 1. According to the additive manufacturing technology trend, it could be confirmed that the PBF method is more advanced with technology development than the Directed Energy Deposition (DED) method in terms of equipment and materials. In addition, considering the demand of AM technology by the molding field, it is regrettable that the DED method-based simulation technology, which is advantageous for handling high-strength tool materials like in the field of mold, is at its initial stage.

A shown in Table. 1, there are many types of AM simulation, among which the most widely used are ANSYS Additive, Simufact Additive, and Flow-3D, which are discussed below.

Table 1. AM software for process simulation

NO.	Software Name (Company)	Supported AM Process	Software Characteristics	Remark
1	Additive Suite (Ansys Inc.)	PBF (DED)	All stages of AM process can be simulated	Additive Suite, Additive Print, Additive Science
2	Simufact Additive (MSC Software)	PBF (DMLS, EBM)	Modeling, Deformation compensation	Enable cloud
3	FLOW-3D (Flow Science, Inc.)	PBF, DED (BJ)	PBF and DED simulation possible	Analysis of material melting micro-structure
4	Netfabb (Autodek Inc.)	PBF, DED	Multi-scale simulation	Enable cloud
5	Amphyon (Altair Inc.)	PBF, (SLM, DMLS)	Buildup process, Shape compensation	Enable linkage of HyperWorks
6	GENOA 3DP (AlphaStar Corporation)	PBF	Simulation and analysis of polymers, metals and ceramics	ABAQUS FEM support
7	Magics (Materialise)	PBF	Insufficient reflection on thermal analysis	Simufact base technology
8	Siemens NX (Siemens, Inc.)	PBF (DED, BJ)	Pursuit of thermal deformation prevention	-
9	e-Xstream (e-Xstream engineering)	PBF (SLS, FFF, FDM)	Deformation compensation, AM special material database possession	Multi-material additive pursuit
10	SIMULIA (Dassault Systèmes)	PBF, DED (BJ)	Deformation compensation Multi-physics modeling	Enable cloud
11	COMSOL (COMSOL, Inc.)	PBF	For metal/plastic AM, Analysis of additive chemical process	Addition of electromagnetic analysis

(Source_Metal3D)

SLM(Selective Laser Melting), DMLS(Direct Metal Laser Sintering), EBM(Electron Beam Melting), PBF(Powder Bed Fusion), DED(Direct Energy Deposition), BJ(Binder Jet), SLS(Selective Laser Sintering), FFF(Fused Filament Fabrication), FDM(Fused Deposition Modeling), MJ(Material Jetting)

ANSYS Additive

In ANSYS Additive, there are two modules (ANSYS Workbench, ANSYS Additive Print) used to apply additive process simulation. Additive process simulation enables support creation and analysis, AM process parameter analysis, AM build-up process analysis, material characteristics and microtexture analysis.

First, ANSYS Workbench is a module that enables topology optimization as well as additive simulation. AM process simulation using Additive Workbench is composed of supporter analysis, thermal analysis, and structural analysis. The setting of such a simulation process is easily accessible from Workbench by using Additive Wizard.

The stage for AM simulation from Workbench is composed of 12 stages in total including the stage that uses Additive Wizard that sets the process parameter. This simulation stage enables simulation of Mesh on the additive shape, supporter, material property, process condition, equipment condition, temperature condition provision, and excess thermal analysis and structural analysis considering the build-up process, which allows analysis of the result. As such, the AM simulation process using Additive Wizard is shown in Figure 11(b).

Secondly, AM process simulation using Additive Print is conducted on the shape build-up process considering excess thermal interpretation and static structure interpretation. (The simulation results of Additive Workbench and Additive Print are similar). Moreover, the simulation included allows easy determination of Anisotropic Strain Coefficients (ASCs) and Strain Scaling Factor (SSF) about materials. The AM simulation process using Additive Print validates ASCs coefficient and optimal SSF from equipment and materials being used, supplements and detects, thereby differentiating itself from Additive Workbench by predicting and calibrating the distorted and final additive shape.

In Additive Print AM simulation, there are three types including Assumed Strain Simulation, Scan Pattern Simulation, and Thermal Strain Simulation, and the detailed simulation process. Assumed Strain Simulation assumes that equal directional deformation occurs in all positions of a part, enabling the quickest simulation. Scan Pattern Simulation uses an average deformation rate equal to an assumed uniform deformation rate. However, by atomizing a deformation rate for each component according to the local direction of scan vector inside components, lengthening simulation time. Thermal Strain Simulation predicts how thermal circulation influences deformation accumulation in each location within a component, thereby requiring the longest calculation time for simulation because thermal prediction on all scan vectors is required.

Calibration of SSF and ASCs coefficient using Additive Print is composed of the calculation process and verification process. For this process, the coefficient is determined by comparing and analyzing the measurement results via actual production using 3D printing equipment and simulation results.

Simufact Additive

Simufact Additive is a simulation software that helps the production of output with guaranteed quality from its first production by predicting part deformation and residual stress at additive manufacturing process (SLS, SLM, LBM, DMLS, EBM) of PBF method.

Figure 14 shows the general metal 3D printing process simulation. In the stage before the printing process simulation, it designs the print target shape by using the topology optimization method and conducts the 3D printing process simulation by using design files. Figure 14 shows additive manufacturing interpretation, heat treatment interpretation to remove residual stress, cutting of supporter from base plate and removal process analysis, and Hot Isostatic Press (HIP) process interpretation for high-densifying of output. It can conduct interpretation of strength and fatigue analysis by considering stress and deformation of output after 3D printing.

In order to obtain more accurate simulation result values considering material features of powder as well as characteristics of equipment including laser speed, size, and hatching pattern, Simufact Additive conducts the calibration process in advance.

The calibration process prints the specimen actually toward each direction (0°, 45°, 90°), cuts it, measures the change amount of specimen, reflects it to Simufact Additive simulation to calculate the innate strain. The innate strain includes the rate of change of plasticity/heat/creep/shape metamorphosis. Using the calculated innate strain, a more accurate 3D simulation is possible that includes the calibrated property values and process variables.

In addition, it is an advantage of the Simufact Additive to design deformation compensation. In the actual metal printing process, often, the output of shape different from the actual CAD file is created due to deformation. In order to solve this issue, methods of changing process conditions can be used including additive direction and support change. However, another method is to design deformation compensation. This method is to design after changing the initial shape by approving deformation in the opposite direction by predicting the deformation amount with simulation, based on which Simufact Additive can create changed CAD files in the form of STL. By using the generated STL file, simulation can be conducted again to conduct actual printing after confirming the deformation amount.

Figure 14. Metal 3D printing simulation process
(Source: Simufact Engineering, part of Hexagon's manufacturing intelligence division)

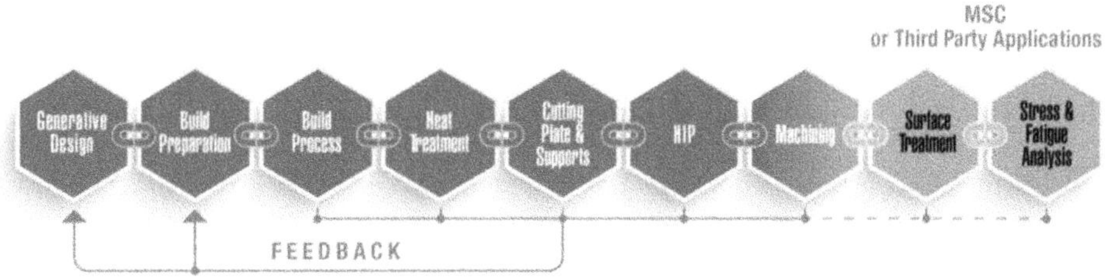

FLOW-3D

Metal 3D printing is referred to as additive manufacturing and is a process of manufacturing by adding metal powder or wire one layer each time in general. The attention on the metal additive manufacturing process has been continued for several years, and related research and development have been conducted actively. Metal 3D printing has the advantage of easy design and manufacturing of complex or special shapes, which is why widely used in diverse fields including shipbuilding, space, aviation, automobile, medical, and machinery. Metal 3D printing as such can be categorized into PBF process and DED process.

Process related to PBF and DED processes can be interpretatively implemented by using the weld and DEM module of FLOW-3D. FLOW-3D (Weld and DEM module included) enables the use of laser lighting conditions, heat flux, distribution of metal powder particle size, powder bed movement, evaporation pressure for base metal, shield gas effect, multiple laser reflection, and surface tension effect. In addition, FLOW-3D allows to easily and accurately determine at the design stage by an engineer based on the interpretation of metal 3D printing process because it allows analysis of metal powder distribution shape, volume fraction, powder melt movement, temperature occurring upon deposition, and thermal stress.

Powder Bed Fusion (PBF) Process

By using the Weld module of FLOW-3D, the laser lighting conditions on powder can be set to confirm the melting behavior by considering heat flux, laser spot size, laser movement and speed, shield gas, multi-reflection effect, reflectivity, evaporation pressure effect, and setting of surface tension.

Directed Energy Deposition (DED) Process

DED process interpretation is a method using the particle function of FLOW-3D, conducting melt and additive via laser heat flux as base metal particles fall. This method allows visual confirmation of the phenomenon of falling particles.

Laser Powder Bed Additive Manufacturing

A complex physical process is required for manufacturing Laser Powder Bed Fusion Additive (L-PBF). In particular, the absorbed laser beam energy forms a melting pool that occurs mainly due to surface tension inclination (or Marangoni shearing stress) via strong fluid flow by melting the particles. Heat transfer and fluid flow are influenced by the local distribution of powder particles within the powder bed, which may vary according to the location. Due to very temporary fluid flow, the melt pool surface (Free Surface) shape constantly evolves and influences the final surface quality.

CONCLUSION

We analyzed on the introduction of the metal 3D printing AM process, the current status of the AM process simulation software, and the AM process simulation. For 3D printing technology to be used at the manufacturing site, excellent 3D printers, materials, and software are essential. Most of the major AM processes are software, such as the design in a shape suitable for printing on a 3D printer, creation of a supporter to assist the prints from collapsing during the printing process, and pre-detection of cracks or deformations of the prints before correction and alteration are done in software. Moreover, in the AM process, software simulation is becoming more and more important as materials are diversified, and output shapes are more complicated and larger. The goal of the AM process simulation is to prevent build-up failures by predicting the macroscopic deformation and stress of the part. One might say that to achieve this goal, one must provide optimized data to improve the overall AM process design, including determining the orientation of the stack, adjusting the placement and size of the supporters. In general, the AM process simulation process consists of optimizing the layered design and optimizing the additive manufacturing parameters. The shape design simulation goes through a step of designing and verifying the over-designed shape as an optimized formation physically (Weight reduction) or structurally (Maintaining strength). The additive manufacturing process simulation will analyze AM build-up process parameters such as location and orientation of supporter generation and compact analysis, laser power required for equipment and materials, scan speed and temperature conditions. In addition, it analyzes the problems that may occur during the AM printing build process and analyzes the microstructure (Length, Depth, Width, Porosity) of the molten pool according to the AM equipment and materials. In the AM process simulation, structural deflection or thermal deformation easily occurs in the case where the shape of the additive manufacturing products is large and complex, so it is necessary to provide more optimized parameters for the build-up process and more precise production of supporters. Therefore, the AM process simulation must be performed when the shape of the stacked output is large and complex. Furthermore, the optimization of the AM process directly affects the quality and output cost of the output, hence the designer should carefully contemplate the decision.

ACKNOWLEDGMENT

This research received no specific grant from any funding agency in the public, commercial, or not-for-profit sectors.

REFERENCES

Agustí, J. I. (2017). Environmental design guidelines for digital fabrication. *Journal of Cleaner Production, 142*, 2780–2791. doi:10.1016/j.jclepro.2016.10.190

Anon Software. (2017). *Topology optimization guide.* Retrieved from Lecture Notes Online Web Site: http:// www.topology-opt.com/software-list/

ANSYS Software Company. (2019). *Ansys additive suite.* Retrieved from https:// www. ansys. com / products/struct-ures/ansys-additive-suite

Anton, W., Johan, P., & Johan, O. (2018). An optimization framework for additive manufacturing given topology optimization results. *Toolsand Methods for Competitive Engineering (TMCE) 2018 Proceedings,* 533-542.

Bikas, H., Stavropoulos, P., & Chryssolouris, G. (2016). Additive manufacturing methods and modelling approaches: A critical review. *International Journal of Advanced Manufacturing Technology*, *83*(1-4), 389–405. doi:10.100700170-015-7576-2

Brackett, D., Ashcroft, L., & Hague, R. J. M. (2011). Topology optimization for additive manufacturing. *Wolfson School of Mechanical and Manufacturing Engineering*, 348-362. Retrieved from https:// sffsymposium.engr.utexas.edu/Manuscripts/2011/2011-27-Brackett.pdf

Digital Engineering. (2018). *Altair Debuts inspire 2018 software.* Retrieved from https://www.digital-engineering247.com/article/altair-debuts-inspire-2018-software/

3. Dsim Company. (2017). *Additive manufacturing simulation.* Retrieved from Lecture Notes Online Web Site: http://3dsim.com

Geekie, L. & Jamie, J. (2018). Including simulation in the additive manufacturing work flow aa. *Industrlal Equipment*, 14-18. Retrieved from https://www.ansys.com/-/media/ ansys/ corporate/ resourcelibrary/ article/including-simulation-in-the-additive-manufacturing-workflow-aa-v12-i3.pdf

Ian, G. D., & Brent, S. (2015). Design for additive manufacturing. *Johnson Matthey Technology Review*, *59*(3), 107–145. doi:10.1007/978-1-4939-2113-3_17

Kim, D. C., Shin, S., Chung, W., Shin, C., & Lim, K. (2014). Application of 3D printing technology in seismic physical modeling. *J. Korean Soc. Miner. Energy Resour. Eng*, *56*(3), 260–269. doi:10.32390/ ksmer.2019.56.3.260

Kim, S. G. (2018). *Materialize to unveil metal 3D printing simulation module.* Retrieved from https:// www.zdnet.co.kr/view/?no=20181001104213

Kim, Y.S., Choi, S.W., & Yang, S.Y. (2019). Investigation to metal 3D printing additive manufacturing (AM) process simulation technology. *Journal of Drive and Control, 16*, 51–58. doi:10.7839/ksfc.2019.16.3.051

Lee, Y.S., & Zhang, W. (2019). *Laser Powder Bed Additive Manufacturing*. Retrieved from https://www.flow3d.co.kr/3d-%ED%94%84%EB%A6%B0%ED%8C%85-%ED%8C%8C%EC%9A%B0%EB%8D%94-%EB%B2%A0%EB%93%9C-%EC% 88% 98% EC%B9% 98% ED% 95%B4%EC%84%9D/

Lockwoodanthony, J. (2016). *Inspire, a true topological optimization tool that goes beyond its name!* Retrieved from https://blog.altair.co.kr/34946

Lowa, A. (2018). A design framework for additive manufacturing based on the integration of axiomatic design approach, inverse problem-solving and an additive manufacturing database. Iowa State University.

Martinezmartin, A. (2016). *Phase optimization + 3D printer = 48% mass reduction*. Retrieved from https://blog.altair.co.kr/34757

Materialise Software Company. (2020). *Materialise 3-Matic*. Retrieved from https:// www. Mater -ialise. com /en/ soft -ware/ magics/ modules

Ministry of Food and Drug Safety(MFDS). (2019). *GMP complaint guide by 3D printing customized medical device manufacturing process*. Academic Press.

MSC Software Company. (2018). *Simulation software for metal 3D printing*. Retrieved from https://www.mscsoftware.com/sites/default/files/simufact_additive_brochure_kr.pdf

Oh, J. W., Na, H., & Choi, H. (2017). Technology trend of the additive manufacturing (AM). *J. Korean Powder Metall, 24*(6), 1–14. doi:10.4150/KPMI.2017.24.6.494

3D . Systems Company. (2017). *3DXpert software for metal additive manufacturing*. Retrieved from Lecture Notes Online Web Site: https://www.3dsystems.com/software/3dxpert

Vayre, B., Vignat, F., & Villeneuve, F. (2012). Designing for additive manufacturing. *Procedia CIRP, 3*, 632–637. doi:10.1016/j.procir.2012.07.108

Wasserman, S. (2015, November 19). *3D Printing brings out the full potential of topology optimization.* Retrieved from https://www.engineering.com/ DesignSoftware/Design SoftwareArticles/ArticleID/11008/3D-Printing-Brings-Out-the-Full-Potential-of-TopologyOptimization.aspx?utm_source=solid Thinking&utm_campaign = c2b97 dfaab-solidThinking_Newsletter_Nov_2015_11_2015 &utm_medium =email&utm _term=0 _2eedfcce8a -c2b97dfaab-37115393

Yoon, O. S. (2013, September 11). *Issue quest publishes a report on business opportunities and response strategies in the spotlight 3D printing market*. Retrieved from http:// www. epnc. co. kr / news / articleView.html?idxno=10132

Zegard, T., & Paulino, G. H. (2016). Bridging topology optimization and additive manu-facturing. *Structural and Multidisciplinary Optimization, 53*(1), 175–192. doi:10.100700158-015-1274-4

Zhu, J.H., Zhang, W.H., & Xia, L. (2015). Topology optimization in aircraft and aerospace structures design. *Archives of Computational Methods in Engineering*, 1–28. doi:10.100711831-015-9151-2

Zwier, M. P., & Wits, W.W. (2016). Design for additive manufacturing: automated build orientation selection and optimization. *Procedia CIRP, 55*, 128-133.

ADDITIONAL READING

ANSYS Software Company. (2018). ANSYS releases additive print, suite for metal additive manufacturing. Retrieved from http://www.tenlinks.com/news/ansys-releases-additive-print-suite-for-metal-additive-mfg/

Çelebi, A(S)., & Appavurvther, Z.E. (2018). Analyzing the effect of voxel-based surface mesh application on residual stress with simufact additive software. *Düzce University Journal of Science & Technology, 6*, 930–940.

Cooka, P. S., & Murphyb, A. B. (2020). Simulation of melt pool behaviour during additive manufacturing: Underlying physics and progress. *Additive Manufacturing, 31*, 1–23. doi:10.1016/j.addma.2019.100909

Graf, M., Härtel, S., & Awiszus, B. (2019). Application of numerical simulation for lightweight design. *Jixie Gongcheng Xuebao, 8*(1), 176–191. doi:10.3390/met8121009

Marta, S. (2019). Simulation and validation of thermography inspection for components produced by additive manufacturing. *Applied Thermal Engineering, 159*, 113872. doi:10.1016/j.applthermaleng.2019.113872

Qian, X., Zhang, Y., Yin, Y., Zhou, J., Shen, X., & Ji, X. (2019). Optimization on the deformation of circular bosses of a compressor casing wax pattern in selective laser sintering process. *Procedia Manufacturing, 37*, 578–583. doi:10.1016/j.promfg.2019.12.091

Shane, E. (2018). Ansys Advantage Explores Simulation for Metal Additive Manufacturing. Retrieved from Lecture Notes Online Web Site: https://www.ansys.com/blog/aa-additive-intro

KEY TERMS AND DEFINITIONS

3D Printing: It is a three-dimensional object from a computer-aided design (CAD) model, usually by successively adding material layer by layer, which is why it is also called additive manufacturing.

Additive Manufacturing Simulations: The physics behind the manufacturing process can be accurately recreated in software platforms enabling end to end digitalization – predicting residual stresses, voids, cracks, and so on, factors which will be crucial in the service life of a part.

Design for Additive Manufacturing (DfAM): It is design for manufacturability as applied to additive manufacturing (AM). It is a general type of design methods or tools whereby functional performance and/or other key product life-cycle considerations such as manufacturability, reliability, and cost can be optimized subjected to the capabilities of additive manufacturing technologies.

Direct Metal Disposition (DMD): It is an extremely flexible technique with application in multiple areas from large scale component repair to medical implant manufacture.

Directed Energy Deposition (DED): It is an additive manufacturing process where metal wire or powder is combined with an energy source to deposit material onto a build tray or an existing part directly.

Electron Beam Melting (EBM): The technology manufactures parts by melting metal powder layer by layer with an electron beam in a high vacuum.

Lattice Structure: A lattice is an ordered array of points describing the arrangement of particles that form a crystal.

Microstructure: It is the very small scale structure of a material, defined as the structure of a prepared surface of material as revealed by an optical microscope.

Powder Bed Fusion (PBF): It is a subset of additive manufacturing (AM) whereby a heat source (eg, laser, thermal print head) is used to consolidate material in powder form to form three-dimensional (3D) objects.

Topology Optimization: It is a mathematical method that optimizes material layout within a given design space, for a given set of loads, boundary conditions and constraints with the goal of maximizing the performance of the system.

Chapter 10
Optimization and Simulation of Additive Manufacturing Processes:
Challenges and Opportunities – A Review

Deepak Kumar Sahini
https://orcid.org/0000-0001-5017-5181
Birla Institute of Technology, Mesra, India

Joyjeet Ghose
Birla Institute of Technology, Mesra, India

Sanjay Kumar Jha
Birla Institute of Technology, Mesra, India

Ajit Behera
National Institute of Technology, Rourkela, India

Animesh Mandal
Indian Institute of Technology, Bhubaneswar, India

ABSTRACT

Additive manufacturing (AM) has developed and gained popularity across the globe into a multi-billion-dollar industry that involves many materials and techniques. AM has created itself as a technology for the manufacturing of metallic parts with enhanced mechanical characteristics that are scientifically sound and commercially feasible. However, there are various challenges, from business point of view, like high machine and material costs. Considering the complexities involved, sustainable manufacturing, optimization tools, and simulation models are necessary in order to save time and costly trial and errors. Topology optimization and simulation of AM processes are commercially available and are receiving attention from scientists and industry. Thus, this chapter is designed to provide readers with a brief introduction to AM technologies with typical applications. The main objective of this chapter is to provide the current trends and innovations in the field of design for additive manufacturing (DFAM), topology optimization, and simulation technologies.

DOI: 10.4018/978-1-7998-4054-1.ch010

INTRODUCTION

Additive manufacturing (AM) has developed and gained popularity across the globe into a multi-billion-dollar industry that involves many materials and techniques (Kalita et al., 2019). Thus, AM also known as Rapid Prototyping (RP), 3D printing or freeform manufacturing, is defined as the "process of joining materials to produce parts from 3D model data, usually layer by layer, as opposed to subtractive manufacturing and formative manufacturing methodologies" as per International Organization for Standardization (ISO)/American Society for Testing and Materials (ASTM) 52900:2015 standards (Lee et al., 2017). RP normally refers to technology and tools that enable the manufacture of physical objects from the solid models produced in Computer Aided Design (CAD) using techniques of additive layer manufacturing without manufacturing process planning, tooling or any type of fixtures (Chang, 2015; Herzog et al., 2016; Ngo et al., 2018).

Generic Additive Manufacturing Process

Figure 1. General process of Additive Manufacturing from CAD to part
(Adapted from (Gibson et al., 2010)

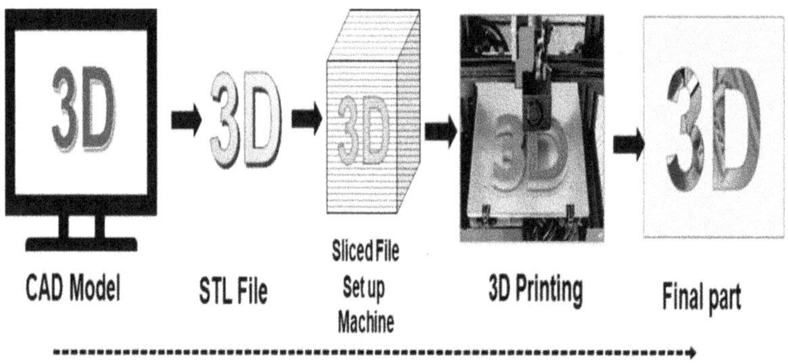

AM involves several steps, starting from a virtual CAD model to final physical component (Gibson et al., 2010). The general AM process, right from a CAD model to final manufacture of the part is illustrated in Figure 1.

Step 1: Concept design: In general AM process starts with a 3D CAD model created in a computer software. Most of the 3D systems are solid model systems with some surface modelling features (Gibson et al., 2010).

Step 2: Conversion to STL: The term STL is known as Stereolithography. It functions by eliminating all structural data, modelling history and approximates the model surfaces to a number of triangular facets. Thus, STL is basically a surface description of the model.

Step 3: Transfer to 3D printer and operation. The STL file once created, is sent directly to the 3D printer or the AM machine.

Step 4: Setup of AM machine: Most of the AM machineries have some specific built-in parameters, with respect to a particular machine or process. Complex cases have default settings, so as to accelerate the AM machine process setup and to minimize mistakes.

Step 5: Build: The AM process stages are semi-automated and requires significant manual control and decision-making. AM machines requires a follow-up in the arrangement of layer control, material deposition, and cross-section formation, until the build-up model is produced.

Step 6: Removal of parts and cleaning: The product of the AM-machine should be ready for use. The parts are separated/ removed from extra material surrounding the part.

Step 7: Post-processing: It refers to the stages of finishing, polishing, sandpapering, or application of coatings on the parts

Step 8: Application: AM parts are ready to use for different industrial applications.

Potential Applications of Additive Manufacturing (3D Printing)

The rapid prototyping has been widely used as a powerful tool for product creation in various industries with significant progress towards (Aboulkhair et al., 2019; Kosaraju et al., 2019)

Major areas of AM applications include:

- **Design:** In product design and development phase, a prototype is essential. RP parts help design revisions greatly to shorten the development time, reduces costs and increases the quality of the products (Larimian & Borkar, 2019).
- **Manufacturing:** RP models are used for making dies and moulds for sand casting and investment castings (Yasa & Ersoy, 2018).
- **Tool and mould making:** Tool holders of uniform pocket sizes, die casting molds, tooling in injection molds.
- **Automotive:** Components for motor aircraft, luxurious sports cars, vintage bikes, etc.
- **Electronics:** Radio Frequency Identification (RFID), embedded systems found in solid materials. Micro-electromechanical three-dimensional structures
- **Art:** In the fields of creating art works, and adds lifelike props to the screens.
- **Medical:** Recently, RP have brought a new dimension in the medicine industry. 3D printed medical devices, such as hearing aids, give patients access to personalized medical devices and implants. RP products has also gained popularity for dental applications (Yu et al., 2018).
- **Bioengineering:** Rapid prototyping offers new promises and better approach for fabricating tissue engineering applications (Singh & Ramakrishna, 2017; Yang et al., 2018).

Advantages and Limitations of AM

Advantages of AM

- Rapid prototyping produce parts with complex geometry and dimensional accuracy.
- Wide variety of materials (metals, polymers and ceramics) can be produced by AM (Liu & Shin, Y C, 2019).
- Best suited for high value spare parts.
- Elimination of design constraints.

- Increased build up speed
- Reduction in lead time is achieved due to increase in built-up speed.
- Flexibility in design
- AM has been widely used in various industries as a powerful tool for product development, and in making significant progress towards improvements.
- Direct transformation of design to part/component
- Customization of parts without any additional tooling requirements and manufacturing costs
- Parts of complex geometries can be produced. Example, honeycomb structures, ventilators, cooling channels, etc., related to complexity can be produced with no extra costs (Abdulhameed et al., 2019)
- Functional designs allow the development of complex features.
- Lightweight fabrication of parts of hollow or lattice structures.
- Capable of producing components directly to their near net shapes with no further processing (Colosimo et al., 2018).
- Potential to achieve zero wastage in production by increasing the usage of materials
- Substantial reduction the product development cycle and manufacturing lead times helps in quicker transfer to market (Ford & Despeisse, 2016).
- Less operating footprint to manufacture a wide range of parts
- Transformation from projection-based to on-demand manufacturing
- Excellent scalability (Frăţilă & Rotaru, 2017).
- Minimized design constraints
- Reduction in lead time is achieved due to increase in built-up speed.
- Well suited in manufacturing of cost-effective replacement and rework parts
- AM leads to green manufacturing by cleaning and minimizing waste (Guo & Leu, 2013).

Limitations of AM

- Rapid prototyping is not relevant for mass production.
- Parts made by AM show anisotropy.
- Limited material selection and limited component size
- High initial investment.
- Requires maintenance, removable support structures.

Classification of Additive Manufacturing Methods

The important factors for the enhancement of AM technology are rapid prototyping, capable of printing large structures, minimize defects and enhanced mechanical properties. In addition, a variety of materials are being used in 3D printing, including metals, polymers, ceramics and concrete (Aboulkhair et al., 2019; Martin et al., 2018; Ngo et al., 2018; C. Zhang et al., 2019). The most popular 3D printing machine that utilizes polymer as filaments is the Fused Deposition Modelling (FDM). Furthermore, the primary techniques of AM technologies are selective laser sintering (SLS), selective laser melting (SLM), inkjet printing, stereolithography, direct energy deposition (DED) and laminated object manufacturing (LOM) (Bhardwaj et al., 2019; Dass & Moridi, 2019; Manfredi et al., 2014; Promakhov et al., 2019; Saboori et al., 2017; Wang et al., 2017). As per ASTM Standards, AM processes are categorized into seven groups, listed in Table 1. The seven families of AM technology are demonstrated in Figure 2 and briefly described below:

Vat Photopolymerization

- **Process:** It is a process of taking a photosensitive resin and then curing that resin through selective exposure to light. The light can be by way of a laser or through a projector that can cure whole layer at time, initiating polymerization which leads to creation of part or component.
- **Strengths:** High accuracy as much as it can go up to 25μ, and smooth surface finish is achieved. More durable resins compared to material jetting.
- **Limitations:** Pool nature requires post processing curing and support removal.
- **Applications:** Fit and form testing, casting patterns. Vat photopolymerization technologies are great options for accurate parts with emerging capability for functional applications.

Powder Bed Fusion

- **Process:** Layers of powdered material are selectively melted by a focused energy source (laser, electron beam). Powder covering the combined component serves as a protective medium for the overhanging materials. The types used commonly are selective laser melting or electron beam melting, where powders of metal is used (Mehrpouya et al., 2019).
- **Strengths**: Superb design freedom, durable parts, pure material properties.
- **Limitations:** Less cost effective for one-off printing. Limited plastic material options
- **Applications:** Fit, form and provides incredible design freedom and the ability to fulfil functional applications.

Binder Jetting

- **Process:** Liquid bonding materials applied on thin layers of powdered ingredients to build part layer upon layer. Metals and ceramic parts typically fired in furnace after printing
- **Strengths:** Fairly accurate and relatively affordable.
- **Limitations:** Gypsum parts can be brittle. Metal parts have mixed material properties.
- **Applications:** Conceptual, potentially full colour models. Prototyping and tooling. Traditional casting applications. The binder jetting is a cost-effective process to achieve multi-colour or meta prints.

Material Jetting

- **Process:** Droplets of material is deposited layer by layer to make parts, in process curing. Objects are produced in a similar manner of an ink jet printer (Bourell et al., 2017).
- **Strengths:** Highly accurate upto 16μ layers. Multi-material printing. On- the- fly curing limits post-processing requirements.
- **Limitations:** Most cost effective at smaller scale. Limited functional application due to jetting requirement and limited durability.

- **Applications:** Typically, smaller part fit and form testing. Material jetting is excellent for small, high-resolution designs, circuit boards, electronic devices, consumer goods, equipment, etc.

Sheet Lamination

- **Process:** Sheets of material stacked and laminated together through a heat source to form an object. Ultrasonic additive manufacturing (UAM) is a type in which ultrasonic attachment is used to seal thermoplastic sheets, while in Laminated object manufacturing (LOM), adhesives are used.
- **Strengths:** Affordable, full colour ink. No support structure is needed. High manufacturing speed
- **Limitations:** Lower fidelity, challenging post-processing
- **Applications:** Conceptual models. Sheet lamination in paper is a cost-effective option for full-colour models.

Material Extrusion

- **Process:** Material is extruded through nozzle in beads which combine into multi-layer parts. The plastic strengthens soon after being extruded and attaches to the sheet.

There are many materials available for this method. Acrylonitrile Butadiene Styrene (ABS) is the most commonly produced polymer, other polymers such as polylactide (PLA), are also used (Kim et al., 2018).

- **Strengths:** Most printer options, lowest material price. Wide breadth of materials for many applications (e.g., PLA to Carbon Fibre)
- **Limitations:** Less accurate, stepped surface finish, supports required.
- **Applications:** Functional testing and tooling. Certain end use applications (e.g., non-load bearing aerospace parts). It is the most common plastic technology with significant material breadth.

Direct Energy Deposition

- **Process:** Powder or wired fed in the melt pool, created on part surface by a focused laser or electron beam (Jiménez et al., 2019).
- **Strengths:** Rapid deposition rate. Capable of building on existing structures, significant material breadth.
- **Limitations:** Not particularly high accuracy, limited material range.
- **Applications:** Repair applications. DED offers fast deposition and the prospect of extending part life.

OPTIMIZATION

Optimization is process or technique involved in a system, design, or decision making in order to make it fully functional, perfect or effective as much as possible. An optimization problem is a mathematical formulation wherein, design parameters or variables needs to be identified or determined to achieve the

Figure 2. Additive Manufacturing Technologies
(Adapted from (Colosimo et al., 2018)

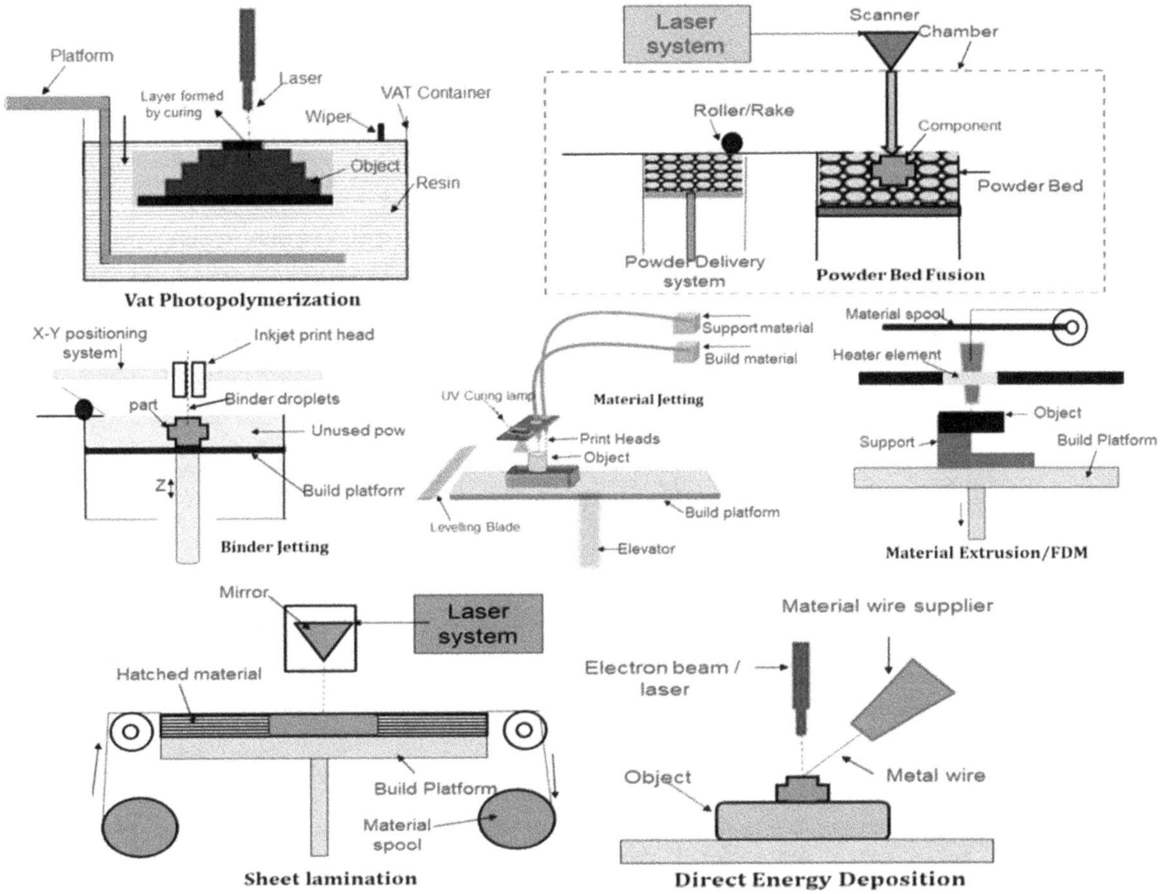

best measurable performance or objective function, by considering the given constraints (Kentli, 2020). Engineering applications of optimization includes in design, planning, control and manufacturing (Attar et al., 2018). In design, optimization is achieved by determining the optimum parameters that lead to best performance. Minimizing manufacturing costs and achieving quality products are applications of optimization in production, planning and control (Aliyi & Lemu, 2019; Junk et al., 2018).

Design for Additive Manufacturing

Design for Additive Manufacturing (DFAM) is design for manufacturability in context to additive manufacturing (AM). Generally, DFAM is a class of design methods or tools whereby functional performance, other key product life-cycle considerations such as manufacturability, reliability, quality and cost can be optimized subjected to the capabilities of additive manufacturing technologies. The concept emerges due to the enormous design freedom and flexibility in AM technologies. Thus, DFAM methods or techniques are required to take maximum advantage of the specific capabilities of AM systems. Typical DFAM tools include topology optimization, design for multi-scale structures, multi-material

Table 1. Classification of Additive Manufacturing methods

S. No.	Category	Principle	Technology	Materials	Benefits	Reference
1	Binder Jetting (BJ)	Liquid binder/s jet are printed on thin layers of powder. By gluing the particles, layer by layer by, the parts are built	• 3D Inkjet	• Polymers • Ceramics • Metals • Composites • Hybrid	• Design freedom • Relatively low cost • Build volume are larger • High print speed	(Kosaraju et al., 2019; Lee et al., 2017; Tofail et al., 2018)
2	Direct Energy Deposition (DED)	During deposition the thermal energy melts the materials	• Laser deposition • Electron beam • Plasma arc melting	• Metals • Hybrid	• Superior grain structure • High quality parts	
3	Material Extrusion (ME)	Material is extruded out through a nozzle	• Fused Deposition Modelling	• Polymers • Composites	• Build functional parts • Inexpensive • Widespread use	
4	Material Jetting (MJ)	The parts are made by depositing, droplets of build materials	• 3D Inkjet Technology	• Polymers • Composites • Hybrid • Ceramics • Biologicals	• High accuracy of deposition • Less wastage • Multi-color parts	
5	Powder Bed Fusion	Thermal energy is fused into a small portion of powder bed in build material	• Electron beam melting (EBM) • Selective Laser Sintering/Melting (SLS/SLM) • Direct Metal Laser Sintering	• Metals • Ceramics • Polymers • Composites • Hybrid	• Powder bed acts Support structure • More material options • Relatively inexpensive • Small footprint	
6	Sheet Lamination (SL)	Bonding of Sheets/foils	• Laminated - object manufacturing • Ultrasound-Additive Manufacturing	• Polymers • Hybrids • Metals • Ceramics	• High speed, • Ease of material handling • Low cost,	
7	Vat Photo-polymerization (VAP)	Liquid polymer is light-cured in a vat	• Stereo Lithography (SLA)	• Polymers • Ceramics	• Excellent accuracy • Excellent surface finish	

design, component reduction, mass modification and other design methods that can use AM-enabled characteristics or features in the software (Silva et al., 2018). Some principles of DFAM include:

- Think Additively: A positive mind-set about deploying DFAM is needed to develop creative ideas and new innovations.

- Design for orientation: requires communication between engineers and operators, running the printers.
- Contour designs: to print just one or two extrusion lines, to create lightweight but strong parts. Example, air craft wing structures.
- Segment and bond parts: Advantages of segmenting parts that are not suitable entirely in the machine. Segmenting of parts can lead to significant time and material savings
- Minimize or eliminate complications: Support structures adds additional time to the printing process. So, by eliminating or minimising the support material, may lead to increase in the quality of the parts.
- Surface treatment is to be provided on the parts

Topology Optimization

Topology optimization (TO) is a type of optimization methods that vary in size and shape optimization. TO differs from shape and size optimization in the sense that design can obtain any shape within the design space, as an alternative of dealing with predefined configurations (Vayre et al., 2012). TO methods solve the problem of material distribution within a design space to obtain an optimum model. TO achieve designs that are not dependent by nature of the initial design. Several algorithms have been developed for TO. Solid Isotropic Material with Penalization (SIMP) and Bi-directional evolutionary structural optimization (BESO) are the most widely used TO algorithms. Both these algorithms are simple and efficient (Kentli, 2020; Meng et al., 2019). The BESO algorithm is used in combination with additive evolutionary structural optimization (AESO) and evolutionary structural optimization (ESO). However, BESO is limited to refinement and coarsening (Aremu et al., 2010). The SIMP algorithm is used to achieve realistic designs to optimize the shape of complex volume fraction (Aremu et al., 2010). A SIMP algorithm is demonstrated with a flow chart in Figure 3. The steps in SIMP algorithm include to compute the finite element analysis; conduct sensitivity analysis; if any improvement is observed over the preceding iteration, then stop the process, update the elemental densities and thus, the final topology is achieved.

Figure 3. SIMP Flow chart

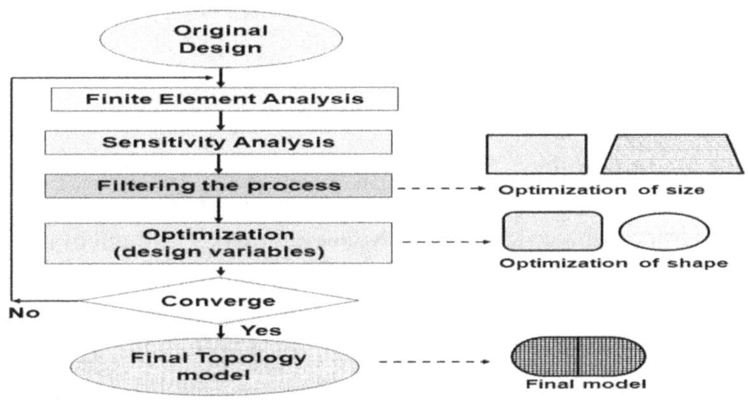

Topology Optimization in Additive Manufacturing

In Today's competitive world, AM provides supreme opportunities for functional parts design and production. Nevertheless, design software, including topology optimization, is needed to completely exploit the design complexity given by the technology (Gao et al., 2015; Jagadish & Bhowmik, 2019; Sun & Hao, 2012). Conventional manufacturing methods are often restricted in the types of designs that can be produced. Designing for AM requires a different approach, given the complexity of design provided by additive manufacturing. TO is a technique that uses mathematical calculations to optimize the geometry of an object (Barbosa & Aroca, 2017). Thus, optimization of topology allows the development of stronger, lighter components. Designers can also optimize the material distribution. Figure 4 demonstrates the steps in a topology optimized design process as: (1) In first step, the original 3D model is drawn in a CAD software. (2) Secondly, the original design is then structurally evaluated with specified boundary conditions to observe the stress distributions and displacement. Based on the results of stress and displacement, TO eliminates material from areas that do not contribute much to the applied loads. (3) In third step, the component is remodeled in CAD software based on TO result. (4) In next step, the modified CAD model is then validated by Finite Element Analysis (FEA) in order to carry the loads and meet design specifications. If the model meets the requirement, the model is tested using some of the prototyping methods, otherwise, remodeling is done again until verification is done. (5) The final design for additive manufacturing is thus achieved.

Case Study - Aircraft Hinge Bracket

TO techniques are generally conducted through the integrated use of the CAD, FEA and various optimization algorithms, considering the different Additive manufacturing techniques (Calignano et al., 2019). In TO, the purpose of CAD is to prepare a rough model of part to be optimized, and purpose of FEA is to evaluate stress distributions and displacements throughout the entire product (Kumar et al., 2019; Yu et al., 2018). Thus, TO of the part is performed so as to eliminate areas that do not support the applied loads appropriately (Zaharin et al., 2018).

Figure 4. Topology Optimized Design Process

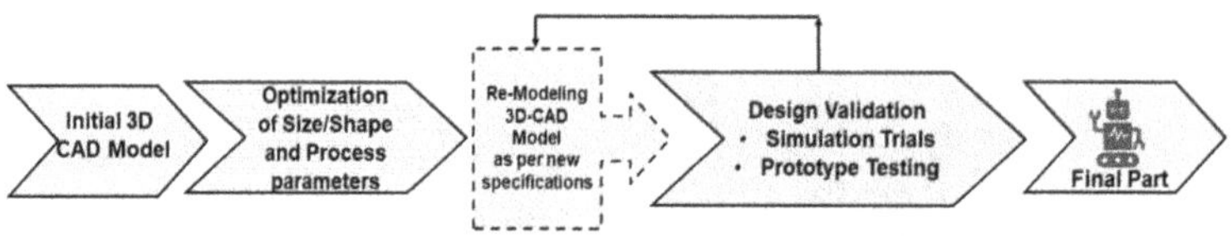

A case study of an aircraft (Airbus, A320) nacelle hinge bracket is presented in Figure 4. In this study it was proved that, as when compared to traditional casting processes AM's Direct Metal Laser Sintering (DMLS) which has achieved benefit of balancing market and environmental sustainability. The flexibility in AM process resulted in topology optimized design, that the CO_2 emission over the entire lifecycle of nacelle hinges weight was reduced by 40% nearly. Further, by eliminating waste in secondary machining, a notable weight saving from 920 g (steel) to 320 g (titanium) which is nearly 35% was achieved through two cycles of optimization. Thus, the final outcome is a lightweight aircraft component while maintaining strength and rigidity as depicted in Figure 5 (Gao et al., 2015; Gebisa & Lemu, 2017; Tofail et al., 2018).

Figure 5. Topology Optimization Process of aircraft hinge bracket
(Adapted from (Gebisa & Lemu, 2017)

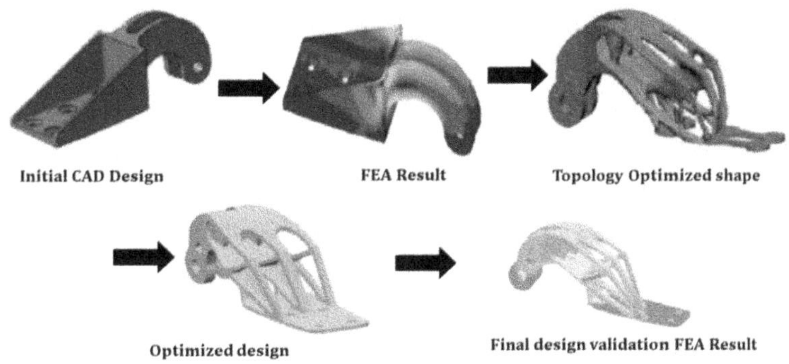

Case Study- Robot Arm

The TO process is employed in a study to develop a robot arm through additive manufacturing. The parts of the robot arm are modelled using Autodesk Fusion360 CAD modelling software. The 3D printing of the parts of the robot was done using an FDM 3D printer (model: Olivetti S2) and the material used was Polylactic Acid (PLA) thermoplastic. The parts are then assembled, and the final assembled robot arm is presented in Figure 6. The integration of TO with AM process resulted in weight reduction, reduced stresses and increased stiffness. This case study showed that nearly 55% weight reduction was achieved and thus concluded that TO design for additive manufacturing results in lightweight designs and decreased lead times.

Figure 6. Robot Arm

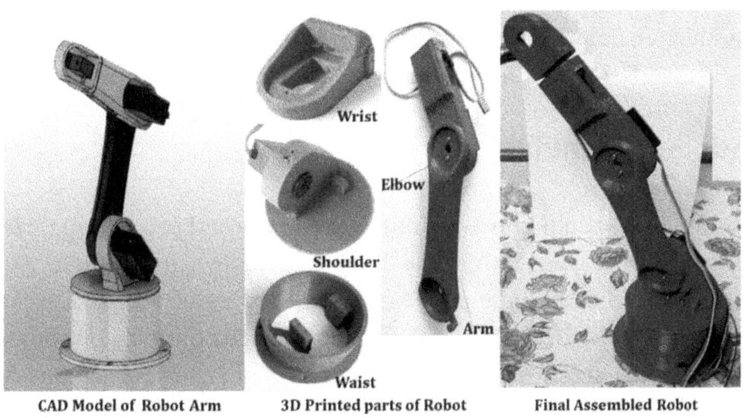

Benefits and Applications of Topology Optimization

- Remarkable freedom of designs (Orme et al., 2018).
- Improved efficiency
- TO pushes the boundaries of design flexibility (Gharbi et al., 2013).
- TO generates several design variations for different applications.
- Objective definitions: e.g. stiffness, weight, material
- Lightweight construction - lighter component with equal or higher stiffness
- Integration of functions
- Manufacturing constraints should be considered during optimization
- Optimized, additively assembled components already support several industries.
- Aerospace industry is the largest adopters of TO designs, due to the advantages of producing light-weight components, decreasing support structures and retaining the strength of the manufactured parts (Gu et al., 2012).
- Satellite and space vehicle organizations extensively use TO and AM assembled components in saving costs (Mower & Long, 2016).
- TO contributes to less wastage of material.
- TO reduces time-to- market.
- Fast calculation and easy to use surfaces without extensive post-processing
- Medical industry/Health is another field benefiting from TO. For example, new opportunities for creating patient-specific, bionic implants with latticed designs are now possible with functionally optimized structures (Negi et al., 2013).
- TO enables additional features such as pore diameter, density, and mechanical properties to be integrated into specific areas of the implantable devices, implants with optimal weight and highly personalized features can be developed.

SIMULATION

Simulation is conducting computer-based experiments with the model and predict the real behavior of the system (Attar, 2011). It is an effective tool in saving time and minimizing the costly trial and errors experiments (Deepak Kumar et al., 2017).

Simulation software replicates the operation of an actual production system using animated, interactive models. Thus, simulation helps companies to visualize the productivity of their production processes and to test technologies safely in order to enhance their performance and profitability.

Simulation of Additive Manufacturing

AM (3D printing) is rapidly expanding into manufacturing, beyond prototyping. However, process repeatability and quality control are the two big challenges that many manufacturers are looking for in AM parts. Achieving good 3D printed parts requires an understanding of the processes involved in design and development (Wiberg et al., 2019). Therefore, planning and optimization techniques for process simulation are critical for AM systems to provide some reliability to specific physical properties such as droplet sizes, shape precision and temperatures (Basu et al., 2019). Simulation optimization may be described as a process which finds the best input variable values among all the possibilities available. The aim of simulation optimization is to reduce resources while optimizing the knowledge obtained in a simulation iteration. AM simulation covers all aspects of the 3D printing process, ranging from the melting of particles in a feedstock to the final toolpath, required to build a part.

Need of Simulation in Additive Manufacturing

Application of simulation tools is an important consideration for cost-effective additive manufacturing. The objective of AM simulation is to addresses all aspects of AM processes, so as to bring end-to-end digital technology. The various aspects include:

- **Functional design:** The first goal is to build an acceptable design that meets the required specifications, and then improve the design through optimization techniques that operate in tandem with simulation (Mourtzis et al., 2014). The optimized functional design can be further enhanced with the use of shape optimizers that work on surface nodes to remove stress hot spots to optimize the AM operation (Bikas et al., 2016).
- **Generate- lattice structure:** The parts produced by AM, generally have a lattice structure, rather than a complete system. One of the goals of AM simulation is to create and optimize a lattice structure using size and shape optimization.
- **Calibrate material:** The material properties of a final component vary significantly from that of the raw material. Hence, the next goal is to monitor the process of phase transformations by conducting multi-scale modelling of the materials.
- **Optimizing AM cycle:** The AM cycle may produce unwanted residual stresses and distortions. To minimize the gap between as-designed and as-manufactured component specifications, it is important to capture physical changes accurately.
- **Performance:** Evaluation as to how the manufactured component can perform with respect to stiffness, fatigue, etc., under real life conditions. For example, Additive simulation takes into account the number of layers, material properties, thermal and structural boundary conditions. Thus, this information may be used to predict residual stresses, distortion and fracture failure.

Steps in Simulation of Additive Manufacturing

Similar to other FEA processes, the sequence of steps in simulation of AM processes includes pre-processor, solver and post-processing. However, some important aspects typical for AM process simulation are to be followed as listed below:

- First a 3D model is created/developed in a CAD software. The models may also be created by means of a 3D object scanner.
- The CAD model is decoded into an STL file. Since, AM is a layered process, the CAD model is to be sliced into layers.
- The STL file is then transported to the 3D printer using the customized machine software.
- The raw materials or consumables are then loaded and the 3D printer is set as per the required specifications for printing. Progressive material addition is a crucial step in the pre-processing stage of AM simulation.

- The 3D model is made, by depositing layer upon layer of the substrate. The number of layers depends on the process parameters such as layer thickness, size of the element, etc. By slicing into digital layers, the 3D shape is formed.
- The part is removed from the build framework and its support system.
- Finally, post-processing might be necessary, such as polishing, cleaning and painting.

Case Study Simulating an Aircraft Bracket

To be able to fully embrace additive manufacturing and produce titanium parts for critical aerospace applications, a thorough understanding of microstructural and mechanical properties is required. Here is an example in which an aircraft bracket made of Aluminum (AlSi10Mg) was simulated. The simulation time is about 20 min. The distortions captured in simulation, is presented in Figure 7, which shows the accuracy of simulation with actual parts.

Figure 7. Simulation of an Aircraft Bracket
(Adapted from MSC Software, 2017)

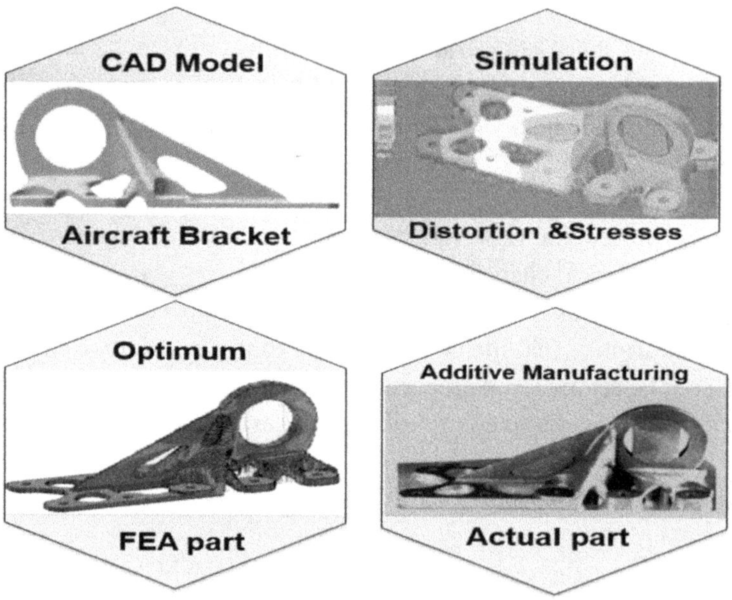

Case Study Titanium Bike Frame Part

Another example is a thin walled bike frame part made of titanium. Additive manufactured thin walled structures have a tendency to have some distortions. The objective of the simulation is to know, how well can the simulation with inherent strain captures the distortions. The simulation results are depicted in Figure 8. The actual bike frame part which is bent inwards is accurately captured in simulation. The failures in the support structures, stress hot spots are captured in the simulation. Further, some bucking defects along the curved line, which is also predicted in the simulation, as shown in Figure 8.

Figure 8. Simulation of a Titanium Bike frame part
(Adapted from MSC Software, 2017)

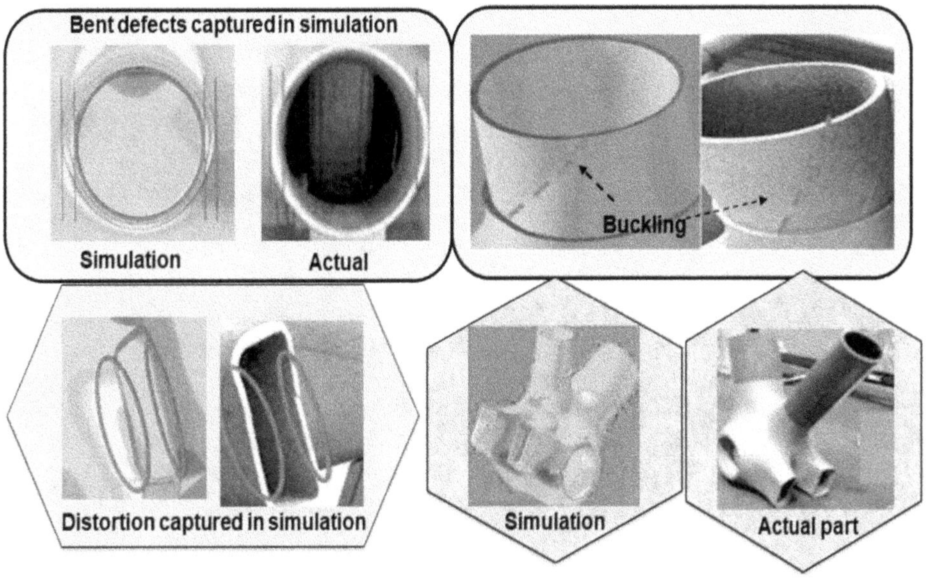

Case Study Impeller Disc

Residual stresses and deformation caused during thermal dilation of the substrate and added structures are one of the issues challenging the AM processes. As an example, Simufact additive simulation enabled us in predicting the residual stresses and distortions of an impeller disc is presented in Figure 9. System control variables have been chosen accurately in Simufact simulation so as to optimize this SLM cycle to reduce printing time and material loss successfully. The simulation guided us to pre-compensate and ensured that the right quality part was produced in right time. This shows how effective is the AM simulation in predicting the stresses and distortions.

Simulation is a thus, critical tool for additive manufacturer's, which enhances the accuracy and quality of AM components (J. Zhang et al., 2019). Therefore, a list of companies providing creative simulation solutions for AM Technology is presented in Table 2.

Benefits of AM Simulation Software

- Greater probability of success is achieved through simulation.
- Simulation aids in determining the design and quality of a components.
- The simulation program helps to prevent failed prints of the part.
- Deeper understanding of AM processes. Simulation of the 3D printing process, makes easier to understand process and their influence on the component's life.
- Production optimization: Because simulation helps to reduce the risk of component deformation, production speed and post-processing can be improved.
- Simulation helps the printing process to be more repeatable.
- Virtual simulation reduces manufacturing costs and reduces material wastage.

Figure 9. Simulation of an Impeller Disc
(Adapted from MSC Software, 2017)

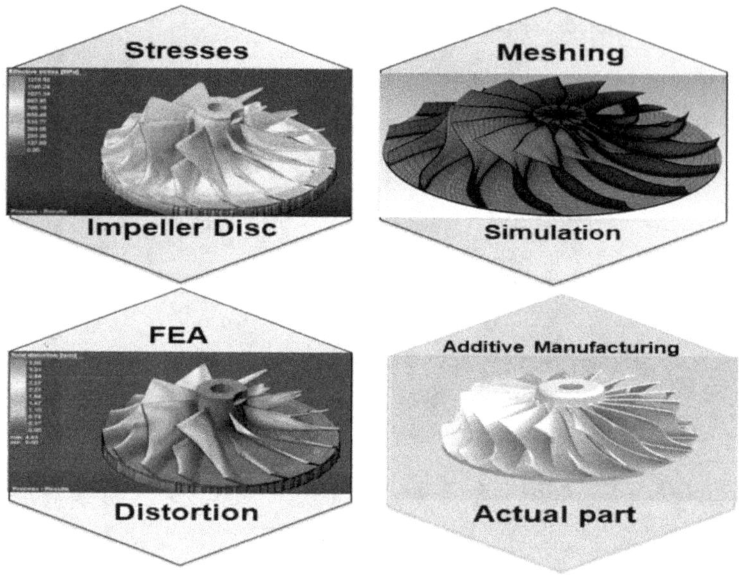

- Simulation provides user assistance in optimizing the Additive Manufacturing Design with its modelling software, and guarantees reduced development and production costs.
- Since incorrect prints can easily become crucial for any budget due to expensive machine running times, it makes sense to use simulations during product development to define the component's key points about the printing process

SUMMARY AND CONCLUSION

In this paper, a brief review on topology optimization and cutting-edge AM simulation technologies is presented. Applications in both these two directions were cited with case study examples. Despite the promising results reported, one must keep in mind of the key challenges in additive manufacturing technology.

Key Challenges

- High machine and material costs
- Limited available materials
- Fluctuating market trends like product life cycle
- Expensive tooling requirements
- Low-volume production
- Reliability and Quality control

Table 2. Companies providing front-line 3D Printing Simulation software

Sl. No.	Simulation Software	Capabilities	Benefits	Reference
1	ANSYS Additive Suite Additive Print	• Simulation suite includes Design for additive manufacturing (DFAM) Modules include design validation, print design, and process simulation. • Topology optimization • Additional materials database, of Al alloy A357, AlSi10Mg, titanium Ti64, etc. are available. • Structural and Thermal analysis	• Reduce physical trial-and-error experiments. • Design geometries for more accurate printing. • Improved robustness for thin-walled structures. • Faster simulation times	(ANSYS® Additive Suite) www.ansys.com
2	FLOW-3D	• Simulates AM processes like directed energy deposition (DED); binder jetting & laser powder bed fusion	• Accurately evaluate thermal distortions, porosity	www.flow3d.com
3	COMSOL Multiphysics®	• Optimize the AM process for both plastics and metals	• Best printing strategy and part geometry	www.comsol.co.in
4	Additive Manufacturing ESI group	Modules include: • Powder coating; • Melt pool shape and dimensions • Material porosity; Surface roughness • Residual stresses; Distortion	• Solutions are more reliable and efficient	ESI Additive manufacturing www.esi-group.com
5	GENOA 3DP AlphaSTAR Corporation's	• GENOA 3DP can predict deformation, fractures, residual stresses and voids	• Best suited for thermoplastics	www.alphastarcorp.com
6	e-Xstream's Digimat-AM	• Simulation module of 3D printing processes in plastics and composites	• Ability to analyze part's performance	www.e-xstream.com
7	Amphyon Amphyon	• Simulate various stages of AM process • Suitable for metal 3D printing processes • Automatic optimization	• Surface quality and shape accuracy can be increased	Amphyon (Altair, 2017) www.altair.com
8	Simufact Additive	• Simufact Additive covers the simulation of Fusion Powder Bed methods like • Direct Metal Laser Sintering and • Selective Laser Melting (SLM)	• Reduce distortions • Optimize build-up orientation and structure • Reduce material and energy costs • Minimize residual stress • Increase productivity	www.simufact.com Simufact Additive (MSC Software, 2017)
9	Autodesk's Netfabb	• Cloud-based capabilities • Predicts thermo-mechanical response during directed energy deposition and metal powder bed fusion of AM parts	• Simulation of complex parts • Innovative products • faster solvers	Netfabb (Autodesk, 2017) www.autodesk.com
10	Dassault Systems SIMULIA	• Optimize the 3D printing process • Predicts distortions • Simulate -microstructure of a material	• Captures material changes during AM process	(3Dsim, 2017), http://3dsim.com/

However, in the current scenario, these limitations, create opportunities for advanced research and developments in additive manufacturing. These opportunities include as below

Opportunities

- Innovative Designs
- Reduced developed times
- Light weight structures
- Customization
- Reduced inventory and reduced product costs
- Recycling
- Implementing quality management techniques
- Modern trainings at Universities/educational Institutes
- New academic programs

Future Directions

The motivation driving the future additive manufacturing implementation relates to the current market strategies and need of complex designs, shorter product life cycle and mass customization. In today's competitive world, AM provides supreme opportunities for functional parts design and production. Thus, advanced topology optimization techniques need to be integrated with AM processes in response to the much needed multi-functional and complex products. Simulation software is a step towards greater reliability and automation. Simulation software will be one of the key factors in accelerating this transformation and building the foundation for smart, highly optimized additive production, in the near future.

CONFLICT OF INTEREST

The authors state no conflicts of Interest.

ACKNOWLEDGMENT

The authors acknowledge the financial funding received from National Project Implementation Unit (NPIU), MHRD, India, under the frame of Collaborative Research Scheme Project grant reference [CRS ID: 1-5732031971].

REFERENCES

Abdulhameed, O., Al-Ahmari, A., Ameen, W., & Mian, S. H. (2019). Additive manufacturing: Challenges, trends, and applications. *Advances in Mechanical Engineering*, *11*(2), 1–27. doi:10.1177/1687814018822880

Aboulkhair, N. T., Simonelli, M., Parry, L., Ashcroft, I., Tuck, C., & Hague, R. (2019). 3D printing of Aluminium alloys: Additive Manufacturing of Aluminium alloys using selective laser melting. *Progress in Materials Science*, *106*, 100578. doi:10.1016/j.pmatsci.2019.100578

Aliyi, A. M., & Lemu, H. G. (2019). Case study on topology optimized design for additive manufacturing. *IOP Conf. Series (IRMES 2019). Materials Science and Engineering, 659*(012020), 1–9. doi:10.1088/1757-899X/659/1/012020

Aremu, A., Ashcroft, I., Hague, R., Wildman, R., & Tuck, C. (2010). Suitability of SIMP and BESO Topology Optimization Algorithms for Additive Manufacture. In *In: 21st Annual International Solid Freeform Fabrication Symposium* (pp. 679–692). University of Texas. http://sffsymposium.engr.utexas.edu/Manuscripts/2010/2010-57-Aremu.pdf

Attar, E. (2011). *Simulation of Selective Electron Beam Melting Processes* (Doctoral dissertation). University of Erlangen, Nurnberg, Germany. https://d-nb.info/1013101472/34

Attar, H., Ehtemam-Haghighi, S., Kent, D., & Dargusch, M. S. (2018). Recent developments and opportunities in additive manufacturing of titanium-based matrix composites: A review. *International Journal of Machine Tools & Manufacture, 133*, 85–102. doi:10.1016/j.ijmachtools.2018.06.003

Barbosa, G. F., & Aroca, R. V. (2017). An IoT-Based Solution for Control and Monitoring of Additive Manufacturing Processes. *Journal of Powder Metallurgy & Mining, 06*(01). Advance online publication. doi:10.4172/2168-9806.1000158

Basu, B., Lundback, A., & Lindgren, L. E. (2019). Simulation of Ti-6Al-4V Additive manufacturing using coupled physically based flow stress and metallurgical model. *Materials, 12*(23). doi:10.3390/ma12233844

Bhardwaj, T., Shukla, M., Paul, C. P., & Bindra, K. S. (2019). Direct Energy Deposition - Laser Additive Manufacturing of Titanium-Molybdenum alloy: Parametric studies, microstructure and mechanical properties. *Journal of Alloys and Compounds, 787*, 1238–1248. doi:10.1016/j.jallcom.2019.02.121

Bikas, H., Stavropoulos, P., & Chryssolouris, G. (2016). Additive manufacturing methods and modelling approaches: A critical review. *International Journal of Advanced Manufacturing Technology, 83*(1–4), 389–405. doi:10.100700170-015-7576-2

Bourell, D., Kruth, J. P., Leu, M., Levy, G., Rosen, D., Beese, A. M., & Clare, A. (2017). Materials for additive manufacturing. *CIRP Annals, 66*(2), 659–681. doi:10.1016/j.cirp.2017.05.009

Calignano, F., Galati, M., Iuliano, L., & Minetola, P. (2019). Design of Additively Manufactured Structures for Biomedical Applications: A Review of the Additive Manufacturing Processes Applied to the Biomedical Sector. *Journal of Healthcare Engineering, 2019*, 1–6. doi:10.1155/2019/9748212 PMID:30992744

Chang, K.-H. (2015). Rapid Prototyping. In K. H. Chang (Ed.), *e-Design: Computer-Aided Engineering Design* (pp. 743–786). Elsevier. doi:10.1016/B978-0-12-382038-9.00014-4

Colosimo, B. M., Huang, Q., Dasgupta, T., & Tsung, F. (2018). Opportunities and challenges of quality engineering for additive manufacturing. *Journal of Quality Technology, 50*(3), 233–252. doi:10.1080/00224065.2018.1487726

Dass, A., & Moridi, A. (2019). State of the Art in Directed Energy Deposition: From Additive Manufacturing to Materials Design. *Coatings, 9*(7), 1–26. doi:10.3390/coatings9070418

Deepak Kumar, S., Karthik, D., Mandal, A., & Pavan Kumar, J. S. R. (2017). Optimization of Thixoforging process parameters of A356 alloy using Taguchi's experimental design and DEFORM Simulation. *Materials Today: Proceedings*, *4*(9), 9987–9991. doi:10.1016/j.matpr.2017.06.307

Ford, S., & Despeisse, M. (2016). Additive manufacturing and sustainability: An exploratory study of the advantages and challenges. *Journal of Cleaner Production*, *137*, 1573–1587. doi:10.1016/j.jclepro.2016.04.150

Frăţilă, D., & Rotaru, H. (2017). Additive manufacturing – a sustainable manufacturing route. MATEC Web of Conferences, 94. doi:10.1051/matecconf/20179403004

Gao, W., Zhang, Y., Ramanujan, D., Ramani, K., Chen, Y., Williams, C. B., Wang, C. C. L., Shin, Y. C., Zhang, S., & Zavattieri, P. D. (2015). The status, challenges, and future of additive manufacturing in engineering. *CAD Computer Aided Design*, *69*, 65–89. doi:10.1016/j.cad.2015.04.001

Gebisa, A. W., & Lemu, H. G. (2017). A case study on topology optimized design for additive manufacturing. *IOP Conference Series. Materials Science and Engineering*, *276*(1), 012026. Advance online publication. doi:10.1088/1757-899X/276/1/012026

Gharbi, M., Peyre, P., Gorny, C., Carin, M., Morville, S., Le Masson, P., Carron, D., & Fabbro, R. (2013). Influence of various process conditions on surface finishes induced by the direct metal deposition laser technique on a Ti–6Al–4V alloy. *Journal of Materials Processing Technology*, *213*(5), 791–800. doi:10.1016/j.jmatprotec.2012.11.015

Gibson, I., Rosen, D. W., & Stucker, B. (2010). Additive manufacturing technologies: Rapid prototyping to direct digital manufacturing. *Additive Manufacturing Technologies: Rapid Prototyping to Direct Digital Manufacturing*. doi:10.1007/978-1-4419-1120-9

Gu, D. D., Meiners, W., Wissenbach, K., & Poprawe, R. (2012). Laser additive manufacturing of metallic components: Materials, processes and mechanisms. *International Materials Reviews*, *50*(3), 133–164. doi:10.1179/1743280411Y.0000000014

Guo, N., & Leu, M. C. (2013). Additive manufacturing: Technology, applications and research needs. *Frontiers of Mechanical Engineering*, *8*(3), 215–243. doi:10.100711465-013-0248-8

Herzog, D., Seyda, V., Wycisk, E., & Emmelmann, C. (2016). Additive manufacturing of metals. *Acta Materialia*, *117*, 371–392. doi:10.1016/j.actamat.2016.07.019

Jagadish, & Bhowmik, S. (2019). Parameters Optimization of FDM for the Quality of Prototypes Using an Integrated MCDM Approach. In K. Kumar, D. Zindani, & J. P. Davim (Eds.), *Additive Manufacturing Technologies From an Optimization Perspective* (pp. 199–220). IGI Global. https://www.igi-global.com/gateway/chapter/230144

Jiménez, M., Romero, L., Domínguez, I. A., Espinosa, M. del M., & Domínguez, M. (2019). Additive Manufacturing Technologies: An Overview about 3D Printing Methods and Future Prospects. *Complexity*, *2019*(9656938), 1–30. doi:10.1155/2019/9656938

Junk, S., Klerch, B., Nasdala, L., & Hochberg, U. (2018). Topology optimization for additive manufacturing using a component of a humanoid robot. *Procedia CIRP*, *70*, 102–107. doi:10.1016/j.procir.2018.03.270

Kalita, H., Zindani, D., & Kumar, K. (2019). Additive Manufacturing: A Tool for Better Education. In K. Kumar, D. Zindani, & J. P. Davim (Eds.), *Additive Manufacturing Technologies From an Optimization Perspective* (pp. 41–76). IGI Global. doi:10.4018/978-1-5225-9167-2.ch003

Kentli, A. (2020). Topology Optimization Applications on Engineering Structures. In A. Kentli (Ed.), *Truss and Frames—Recent Advances and New Perspectives* (pp. 1–23). IntechOpen. doi:10.5772/intechopen.90474

Kim, H., Lin, Y., & Tseng, T.-L. B. (2018). A review on quality control in additive manufacturing. *Rapid Prototyping Journal, 24*(3), 645–669. doi:10.1108/RPJ-03-2017-0048

Kosaraju, S., Krishna Mohan B., & Singh, S. K. (2019). *Recent Advancement in Additive Manufacturing*. doi:10.4018/978-1-5225-9167-2.ch001

Kumar, S. D., Ghose, J., & Mandal, A. (2019). Thixoforming of light-weight alloys and composites: An approach toward sustainable manufacturing. In *Sustainable Engineering Products and Manufacturing Technologies* (pp. 25–43). Elsevier. doi:10.1016/B978-0-12-816564-5.00002-5

Larimian, T., & Borkar, T. (2019). Additive Manufacturing of In Situ Metal Matrix Composites. In B. AlMangour (Ed.), *Additive Manufacturing of Emerging Materials* (pp. 1–28). Springer International Publishing. doi:10.1007/978-3-319-91713-9_1

Lee, J.-Y., An, J., & Chua, C. K. (2017). Fundamentals and applications of 3D printing for novel materials. *Applied Materials Today, 7*, 120–133. doi:10.1016/j.apmt.2017.02.004

Liu, S., & Shin, Y. C. (2019). Additive manufacturing of Ti6Al4V alloy: A review. *Materials & Design, 164*, 107552. doi:10.1016/j.matdes.2018.107552

Manfredi, D., Calignano, F., Krishnan, M., Canali, R., Paola, E., Biamino, S., Ugues, D., Pavese, M., & Fino, P. (2014). Additive Manufacturing of Al Alloys and Aluminium Matrix Composites (AMCs). In W. A. Monteiro (Ed.), *Light Metal Alloys Applications* (pp. 1–33). InTech. doi:10.5772/58534

Martin, J. H., Yahata, B. D., Clough, E. C., Mayer, J. A., Hundley, J. M., & Schaedler, T. A. (2018). Additive manufacturing of metal matrix composites via nanofunctionalization. *MRS Communications, 8*(02), 297–302. doi:10.1557/mrc.2018.95

Mehrpouya, M., Dehghanghadikolaei, A., Fotovvati, B., Vosooghnia, A., Emamian, S. S., & Gisario, A. (2019). The Potential of Additive Manufacturing in the Smart Factory Industrial 4.0: A Review. *Applied Sciences, 9*(18), 3865, 1–34. doi:10.3390/app9183865

Meng, L., Zhang, W., Quan, D., Shi, G., Tang, L., Hou, Y., Breitkopf, P., Zhu, J., & Gao, T. (2019). From Topology Optimization Design to Additive Manufacturing: Today's Success and Tomorrow's Roadmap. *Archives of Computational Methods in Engineering*, 1–26. doi:10.100711831-019-09331-1

Mourtzis, D., Doukas, M., & Bernidaki, D. (2014). Simulation in Manufacturing: Review and Challenges. *Procedia CIRP, 25*, 213–229. doi:10.1016/j.procir.2014.10.032

Mower, T. M., & Long, M. J. (2016). Mechanical behavior of additive manufactured, powder-bed laser-fused materials. *Materials Science and Engineering A, 651*, 198–213. doi:10.1016/j.msea.2015.10.068

Negi, S., Dhiman, S., & Sharma, R. K. (2013). Basics, applications and future of additive manufacturing technologies: A review. *Journal of Manufacturing Technology Research*, *5*(1–2), 75–96.

Ngo, T. D., Kashani, A., Imbalzano, G., Nguyen, K. T. Q., & Hui, D. (2018). Additive manufacturing (3D printing): A review of materials, methods, applications and challenges. *Composites. Part B, Engineering*, *143*, 172–196. doi:10.1016/j.compositesb.2018.02.012

Orme, M., Madera, I., Gschweitl, M., & Ferrari, M. (2018). Topology Optimization for Additive Manufacturing as an Enabler for Light Weight Flight Hardware. *Designs*, *2*(51), 1–22. doi:10.3390/designs2040051

Promakhov, V., Zhukov, A., Ziatdinov, M., Zhukov, I., Schulz, N., Kovalchuk, S., Dubkova, Y., Korsmik, R., Klimova-Korsmik, O., Turichin, G., & Perminov, A. (2019). Inconel 625/TiB2 Metal Matrix Composites by Direct Laser Deposition. *Metals*, *9*(2), 141, 1–22. doi:10.3390/met9020141

Saboori, A., Gallo, D., Biamino, S., Fino, P., & Lombardi, M. (2017). An Overview of Additive Manufacturing of Titanium Components by Directed Energy Deposition: Microstructure and Mechanical Properties. *Applied Sciences*, *7*(9), 883, 1–23. doi:10.3390/app7090883

Silva, F. J. G., Campilho, R. D. S. G., Gouveia, R. M., Pinto, G., & Baptista, A. (2018). A Novel Approach to Optimize the Design of Parts for Additive Manufacturing. *Procedia Manufacturing*, *17*, 53–61. doi:10.1016/j.promfg.2018.10.012

Singh, S., & Ramakrishna, S. (2017). Biomedical applications of additive manufacturing: Present and future. *Current Opinion in Biomedical Engineering*, *2*, 105–115. doi:10.1016/j.cobme.2017.05.006

Sun, Y., & Hao, M. (2012). Statistical analysis and optimization of process parameters in Ti6Al4V laser cladding using Nd:YAG laser. *Optics and Lasers in Engineering*, *50*(7), 985–995. doi:10.1016/j.optlaseng.2012.01.018

Tofail, S. A. M., Koumoulos, E. P., Bandyopadhyay, A., Bose, S., O'Donoghue, L., & Charitidis, C. (2018). Additive manufacturing: Scientific and technological challenges, market uptake and opportunities. *Materials Today*, *21*(1), 22–37. doi:10.1016/j.mattod.2017.07.001

Vayre, B., Vignat, F., & Villeneuve, F. (2012). Designing for Additive Manufacturing. *Procedia CIRP*, *3*, 632–637. doi:10.1016/j.procir.2012.07.108

Wang, X., Jiang, M., Zhou, Z., Gou, J., & Hui, D. (2017). 3D printing of polymer matrix composites: A review and prospective. *Composites. Part B, Engineering*, *110*, 442–458. doi:10.1016/j.compositesb.2016.11.034

Wiberg, A., Persson, J., & Ölvander, J. (2019). Design for additive manufacturing – a review of available design methods and software. *Rapid Prototyping Journal*, *25*(6), 1080–1094. doi:10.1108/RPJ-10-2018-0262

Yang, Y., Song, X., Li, X., Chen, Z., Zhou, C., Zhou, Q., & Chen, Y. (2018). Recent Progress in Biomimetic Additive Manufacturing Technology: From Materials to Functional Structures. *Advanced Materials*, *30*(1706539), 1–34. doi:10.1002/adma.201706539 PMID:29920790

Yasa, E., & Ersoy, K. (2018). Additive Manufacturing of Polymer Matrix Composites. In M. C. Kuşhan (Ed.), *Aircraft Technology* (pp. 147–169). InTech. doi:10.5772/intechopen.75628

Yu, T., Zhao, Y., Sun, J., Chen, Y., & Qu, W. (2018). Process parameters optimization and mechanical properties of forming parts by direct laser fabrication of YCF101 alloy. *Journal of Materials Processing Technology, 262*, 75–84. doi:10.1016/j.jmatprotec.2018.06.023

Zaharin, H. A., Abdul Rani, A. M., Ginta, T. L., & Azam, F. I. (2018). Additive Manufacturing Technology for Biomedical Components: A review. *IOP Conference Series. Materials Science and Engineering, 328*(012003), 1–7. doi:10.1088/1757-899X/328/1/012003

Zhang, C., Chen, F., Huang, Z., Jia, M., Chen, G., Ye, Y., Lin, Y., Liu, W., Chen, B., Shen, Q., Zhang, L., & Lavernia, E. J. (2019). Additive manufacturing of functionally graded materials: A review. *Materials Science and Engineering A, 764*, 138209. doi:10.1016/j.msea.2019.138209

Zhang, J., Li, X., Xu, D., & Yang, R. (2019). Recent progress in the simulation of microstructure evolution in titanium alloys. *Progress in Natural Science: Materials International, 29*(3), 295–304. doi:10.1016/j.pnsc.2019.05.006

KEY TERMS AND DEFINITIONS

Additive Manufacturing: AM also known as Rapid Prototyping or 3D printing is defined as the process of joining materials to produce parts from 3D model data, usually layer by layer, as opposed to subtractive manufacturing and formative manufacturing methodologies as per ISO/ASTM standards.

Additive Simulation: Some of the 3D Printing Simulation software include ANSYS Additive suite, Autodesk's Netfabb, Simufact Additive, etc.

Applications of Additive Manufacturing: The rapid prototyping has been widely used as a powerful tool for product creation in various industries. Major areas of AM applications include: Product design, manufacturing, tool and mold making, automotive, electronics, art, medical, bioengineering, etc.

DFAM: DFAM is design for manufacturability in context to AM. Some of the tools of DFAM include topology optimization.

Optimization: Optimization is a process or technique involved in a system, design, or decision making in order to make it fully functional, perfect or effective as much as possible. An optimization problem is a mathematical formulation wherein, design parameters or variables needs to be identified or determined to achieve the best measurable performance or objective function, by considering the given constraints.

Simulation: Simulation is conducting computer-based experiments with the model and predict the real behavior of the system. It is an effective tool in saving time and minimizing the costly trial and errors experiments. Simulation optimization is a process which finds the best input variable values among all the possibilities available.

Topology Optimization: TO is a type of optimization methods that vary in size and shape optimization. SIMP and BESO are broadly used TO algorithms.

Chapter 11
Simulation Applications for Industrial and Medical Products Additive Manufacturing

Seung Hwan Joo
Inha University, South Korea

Sung Mo Lee
Korea Shipbuilding & Offshore Engineering, South Korea

Jin Ho Yoo
Korea-Additive Manufacturing User Group, South Korea

Hyeon Jin Son
Winforsys, South Korea

Seung Ho Lee
Advanced Technology Research Institute, Korea Shipbuilding & Offshore Engineering, South Korea

ABSTRACT

For 3D printing technology to be used at the manufacturing site, excellent 3D printers, materials, and software are essential. Moreover, in the additive manufacturing (AM) process, software simulation is becoming more important as materials are diversified, and output shapes are more complicated and larger. The goal of the AM process simulation is to prevent build-up failures by predicting the macroscopic distortion and stress of the part. In the AM process simulation, structural deflection or thermal deformation easily occurs in the case where the shape of the additive manufacturing products is large and complex. So, it is necessary to provide more optimized parameters for the build-up process and more precise production of supporters. This chapter is an example of applying AM process simulation to industrial and medical parts to produce excellent products.

DOI: 10.4018/978-1-7998-4054-1.ch011

AUTOMOBILE DIFFERENTIAL GEAR [INDUSTRIAL APPLICATION]

Introduction

For the continuation of the development of the national key industry and new growth industry, the advancement of the casting industry, which is a representative root industry, is critical. In addition to the casting industry, forging, surface preparation, heat treatment, and molding industries are referred to as root technology. These industries have a lower added value and labor-intensive while being the root of automobile, shipbuilding, and machine engine industries.

In particular, the casting industry of Korea has been fulfilling its role as the rear industry. However, over 90% of all companies are small to medium-sized companies, and, as the industry is mainly about multi-product, small quantity production, the technological difference is severe with short product development cycle, leading to the weakness in proprietary technology development capability and quality improvement.

For cast parts, the automobile industry obtains information such as order information and production plan from customers to produce parts required for the assembly of each module, taking the lean production method that is appropriate for securing flexibility of production line with swift response to customer demands. However, in the case of a differential gear case, the delivery is being delayed due to an increased defect rate notwithstanding this production method. As such, the result of an analysis of the causes showed issues in the casting molding process.

According to Kim, K.S (2014), a differential gear case assembled to an automobile differential gear is a case part of the gear combination device that enables rotation speed to be different between the left and right wheels of a differential gear as a part of the automobile differential restriction device. Its form is complex with highly important cleanliness as well as measurement precision of the inner diameter part. Also, it is absolutely necessary to secure mechanical characteristics and suppress defects of case materials for securing the durability of the final product. Cast molding in the process of casting process of hollow case types is mostly conducted as the gas torch method of the heating casting mold, during which casting defect occurs to cause a significant loss due to serious quality issues (p.51-55).

According to Yu, R (2011), in general, the factors that influence the quality of casting mold parts are pressure, speed, location, time, and temperature of the mold, among which the temperature of the mold exerts the greatest influence on the appearance quality of a molded part. Thus, the temperature control of cast is determinative to external appearance quality as well as qualities such as measurement stability (p.2887-2893).

Mid-sized molded parts are manufactured via direct heating with a gas torching method. When it is completed as a final molded product after going through the next processes, a variety of defects such as flashes, fusion defects, measurement defects, and gas defects. According to Yu, R (2011), these defects cause delivery delays and production costs increase due to the lowering of productivity and quality (p.2887-2893). In addition, in the case of a differential case, it takes a significant weight because it is manufactured as solid in the form of casting.

According to Yun, G.B (2013), this research uses the Design for Additive Manufacturing (DfAM) design method for the differential gear case body part with the purpose of increasing fuel efficiency while maximizing structure strength and making it lightweight to solve the above issues and thereby conducted lightweight design (p.1-18). Because it is difficult for a traditional machine process method to solve these issues, the 3D printing process method is applied to enable manufacturing by integrating specific forms internally. In addition, by applying the 3D printing process method with the Powder Bed Fusion (PBF) using metal powder, a support structure is created, which structure should be removed via post-processing. For the post-processing, heat treatment is conducted for surface finishing and material strength increase to improve the additive surface quality.

Manufacturing Method of Automobile Differential Gear Case

Differential Gear Case for Automobiles

A differential case part for automobiles is an automobile difference suppression part and is a case part (Mold) of the gear combination device that allows differentiated gear's left and right wheels to rotate at different speeds. For a vehicle to smoothly rotate without slipping during driving upon turning, the outer wheels must rotate more than the inner wheels. Also, during driving on an uneven surface, the rotation of wheels of both sides must differ. A differential gear device fulfills such a role (Figure 1).

Figure 1. Automobile differential gear
(Source: Metal3D)

Reverse Engineering of Automobile Differential Gear Part

The automobile differential gear part used in this research does not have the drawings as it is a product actually applied to automobiles. Thus, this research uses reverse engineering methods to secure 3D design drawings, based on which the drawings for the 3D printing process are obtained for the automobile differential gear. The 3D measurement device applied to the reverse engineering process is shown in Figure 2 and uses the Scan in a Box-FX device, which is a non-contact and optical shape measurement method, to reverse engineer the automobile differential gear part (Reverse engineering conducted in the precision of ±0.1mm). Based on the coordinate data obtained from measured data, a 3D CAD program was used to modify and design the case surface of the differential gear, using SOLIDWORKS software.

Figure 2. Automobile differential gear case reverse engineering process
(Source: Metal3D)

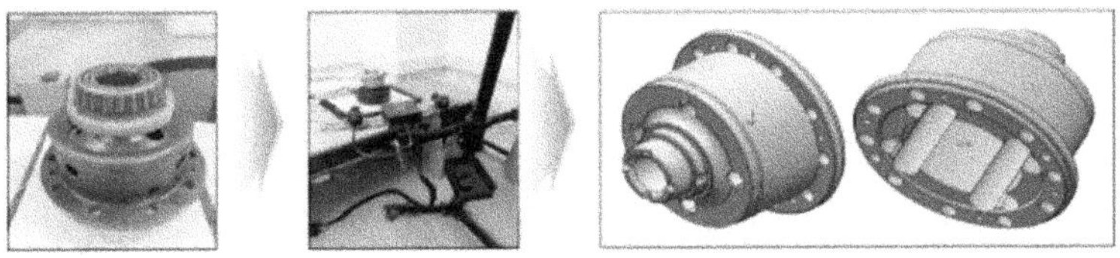

Light Weight Automobile Differential Gear Part

To create a lightweight automobile differential gear part, the lattice structure was applied to the gear case. DfAM is an advanced concept from DfM (Design for Manufacturing) which maximizes the time and cost productivity in product manufacturing through maximizing the best features of the manufacturing method. It is a design approach combined with additive manufacturing technology. It minimizes the manufacturing steps in product development and manufacturing process and maximizes functionality and performance of the product to mass-produce the upgraded products quickly with lower unit cost.

Lattice structure technology has recent 3D printing technology has an excellent form of freedom and can create diverse lightweight structures with a manufacturing method that can accept the diverse demands of the designer. The existing study conducted for applying the lattice structure to the automobile differential gear case. Among lattice structures being provided by the 3-Matic software, this research will proceed with an open-type lattice structure. The characteristics of this structure are large surface, even internal flow, and excellent thermal characteristics, which is why it is widely used.

The structure design was conducted by a total of 8 types of lattice structures that can maintain the shape and ratio of 1:1:1: between X, Y, Z. The specimen was manufactured as a cylindrical porous specimen of a diameter of 10mm and a height of 15mm as specified in the ISO 13314 standard. The lattice thickness was applied as 0.3 mm and precision level 0.3 mm as conditions.

By using the 3-Matic software, lightweight structures are made as follows.

- First, create a quad shape located at the farthest surface and the innermost surface.
- The middle volume is filled with unit cell structure.
- Combine the created quad pillar and the inner lattice structure.
- Finally, extract it as STL file of lightweight structures.

By using the Metalsys 250 3D printer, a total of 8 types of lattice specimens (Figure 3) were manufactured. (A total of 24 specimens were manufactured, 3 for each lattice type). Afterward, lattice specimens were removed from the plate by using the wire electric discharge machining. Then, the residual pow-

Figure 3. Optimum lattice technology development
(Source: Metal3D)

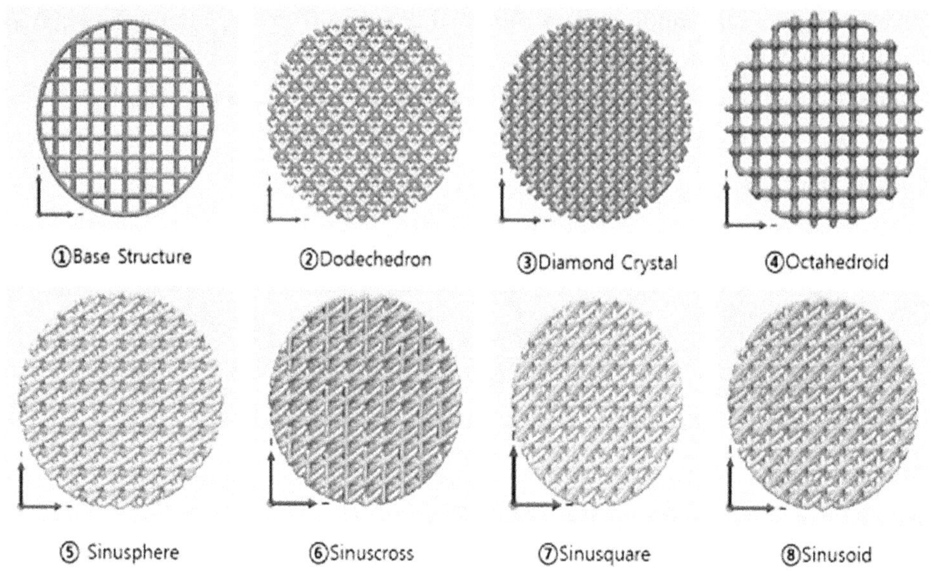

ered inside the specimen was removed by an ultrasonic cleaner. In order to check the strength of lattice specimens, pressure strength testing was conducted with the Instron 5989 (USA) device. The pressure strength measurement results per lattice type are indicated in Table. 1. The result confirmed that the pressure strength of the Dodecahedron type was higher than other lattice types. The Dodecahedron type structure has been selected as the optimum lattice structure. The optimum lattice structure derived by this test result was applied to the automobile differential gear case (Figure 4).

Table 1. Lattice type pressure strength result

	Compressive Stress at Maximum Load [MPa]	Compressive Stress at Yield (Offset 0.2%) [MPa]	Modulus (Automatic Young's) [MPa]
Solid	1056	619	19054
Base Structure	62	44	3449
Dodechedron	560	274	12732
Diamond Crystal	126	87	5081
Octahedroid	99	52	4033
Sinusphere	287	131	7860
Sinuscross	453	212	10688
Sinusquare	399	168	8953
Sinusoid	336	145	7949

Figure 4. Automobile differential gear case output
(Source: Metal3D)

Conclusion

In this research, it was found that an existing automobile differential gear case manufactured via casting causes diverse defects, leading to delivery delays and production cost increase due to the lowering of productivity and quality. In addition, in the case of a differential case, it takes a significant weight because it is manufactured as solid in the form of casting. This research uses the DfAM design method to produce parts by using a metal 3D printer via lightweight design.

It was confirmed that the automobile differential gear part does not have the drawings as it is a product actually applied to automobiles. Thus, this research used reverse engineering methods to secure 3D design drawings.

This research applied the lattice structure to the differential gear case body part with the purpose of increasing fuel efficiency while maximizing structure strength and making it lightweight. A total of 8 types of lattice specimens were manufactured to conduct pressure strength testing, and the Dodecahedron type structure has been selected with excellent strength.

In order to confirm the lightweight of the automobile differential gear case designed as lattice structure, ANSYS SpaceClaim software was used to measure the weight of before and after of the lattice structure. As a result, the volume of the configureuration applied with the lattice structure was confirmed to be reduced. As a result, the reduction was about 54% from the weight of 4.8 kg before applying the lattice structure and to the weight of differential gear case at 2.6 kg after the application of lattice structure.

TIRE MOLD [INDUSTRIAL APPLICATION]

Introduction

According to Chu, C.H (2006), automobile tires play the basic role of driving and braking. Regardless of the road environment such as paved or unpaved, tires must operate properly. In addition, smooth driving also must be guaranteed with an optimized design considering flow, resistance, and friction according to natural environments such as curve, angle of inclination, snow, rain, and wind (p.11-25).

However, the process of manufacturing tires is complex. This is due to thread. Thread is a line-shaped groove on a tire that we as general users simply consider as an aesthetic element. The role of a tire thread is allowing smooth driving by complementing liquidity via a flow of air while driving in a pleasant environment. During rainy days, threads prevent slipping by releasing rainwater through the threads, making the weight of the tire lighter. According to Alkadi, F (2019), during dynamic driving or changing a lane, it also increases handling function, and, when an automobile is put on the brakes, the friction factor increases, minimizing the tire deformation to halt the vehicle while maintaining the grip force evenly (p.211-222).

The important thing is that it is difficult to mold the line patterns of threads on a tire in a spherical shape. The traditional manufacturing process is limited in designing sipe, the threads model implemented on the basic base form and manufacturing it as integrated.

Currently, tire molding is manufactured by part. And only the basic base form is molded by using a milling machine. Sipe form uses a laser cutting method, at which time the thread pattern becomes a complex and non-uniform 3D shape by going through a structural interpretation complementing work.

According to Alkadi, F (2019), the tire mold manufacturing method is conducted via repeated work of folding and bending a thin sheet paper by using a laser cutter, after which a casting work is conducted with rubber resin by covering it over the base mold shape. However, when making tire patterns by pouring the rubber resin into it, often the sipe is either stuck inside the rubber resin or is torn as it is detachable. This is linked to the product defect rate (p.211-222).

3D printer process is a technology of manufacturing by creating 2D cross-section data based on 3D shape data then forming the material as a thin film for additive. In academia, it is well known as Additive Manufacturing (AM). This is a concept contrasted to subtractive, which is producing by cutting or trimming materials and started from the CNC machine tool.

The metal 3D printing process is a method of fusing by melting and injecting laser and beam into powder. Due to the heat, the powder goes through phase transformation as solid-liquid-solid, and, via the repeated process of heat flux and cooling, thermal stress and deformation occur.

According to Oh, J.W (2019), the thermal stress and thermal deformation that can happen in the metal 3D printing process are the most important causes of product defects. In order to solve these issues, numerous trials and errors cause a lot of cost and time. If a virtual simulation of manufacturing process can be conducted by using Computer Aided Engineering (CAE) before conducting 3D printing, the deformation and residual stress that can occur in the overall process of 3D printing including additive process, heat treatment, cutting, and Hot Isostatic Press (HIP) can be predicted and can be used to modify the process and design data by reflecting the information to the process. In other words, you can obtain the 3D printing result value that is based on “Print Right the First Time” (p.1-14).

In this research, experiment and interpretation are used to examine the thermal deformation that occurs while tire mold manufacturing with metal 3D printing of Powder Bed Fusion (PBF) method. This research will compare and validate the interpretation result obtained via Smufact additive simulation and tire mold scan data actually manufactured a 3D printer. In addition, by using the validated interpretation result, this research will propose an optimized 3D printing process method based on the analysis of the tire mold deformation according to the compensation function.

Metal 3D Printing Process Simulation

According to Hwang, I.H (2017), the types of 3D printing process simulation are additive manufacturing interpretation, heat treatment interpretation for removing residual stress, cutting and removal of the support process interpretation from the base plate, and HIP process interpretation for high-density of the output . In addition, by considering the stress and deformation of the output after 3D printing, strength and fatigue interpretations can be conducted (P.161-162).

Metal 3D Printing Process Simulation Process

The process for predicting tire mold deformation via 3D printing process simulation is as follows. First, after modeling a tire mold of the same size, the interpretation is conducted by inputting 3D printing conditions and materials required for the interpretation. Next, by measuring the unique deformation rate of the calibration specimen printed to fit the conditions of the metal 3D printing device, apply it to the simulation. By inputting the optimum process parameter, simulation is conducted. Lastly, after the completion of the 3D printing, the interpretation is conducted in the order of deriving thermal deformation that occurred on the tire mold.

In Advance Process for Accurate Simulation

According to Hwang, I.H (2017), the macro-scale mechanical method interpretation provides a quick interpretation result. However, for accuracy, the unique deformation rate must be reflected. The unique deformation rate varies according to the characteristics of the powder material as well as the laser speed, size, hatching pattern, and other features of the equipment. Thus, it is created by calibration that considers the actual features of the device. Calibration is a method of deriving a unique deformation rate with simulation by using the deformation amount obtained from physical testing. After measuring and inputting the deformation amount (2-3 mm) of the printed and cut the cantilever beam-shaped test specimen, the software calculates the unique deformation rate. By using the calculated unique deformation rate, an accurate process simulation can be conducted by including the actual process variables. This technology has been used in the welding interpretation and is an appropriate method that can increase accuracy due to the similar characteristics of additive manufacturing and welding (p.161-162).

Calibration Interpretation Using Unique Deformation Rate

According to Hwang, I.H (2017), this is the essential part of metal 3D printing process interpretation and is a process of simulating metal 3D printing additive manufacturing. The interpretation method including the mechanical interpretation method of interpreting the unique deformation rate that considers printing process variables by using calibration, a thermal interpretation that considers heat input and loss of the actual printing process without calibration process, and thermos-mechanical interpretation method.

The mechanical interpretation method is a method of using a unique deformation rate interpretation method, which must be advanced by the calibration interpretation before conducting an additive simulation. Calibration interpretation, calculates the unique deformation rate by actually printing a specimen in each direction (0°, 90°), cutting it to measure the deformation amount, and entering the values to the Simufact Additive Print software (p.161-162).

Figure 5. Measurement of deformation amount after cutting the calibration specimen
(Source: Metal3D)

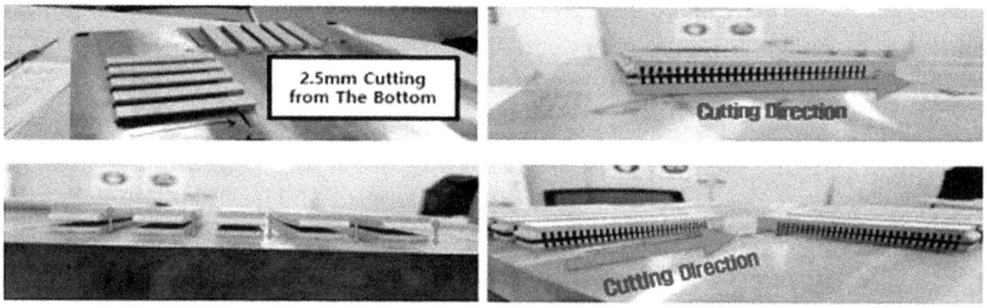

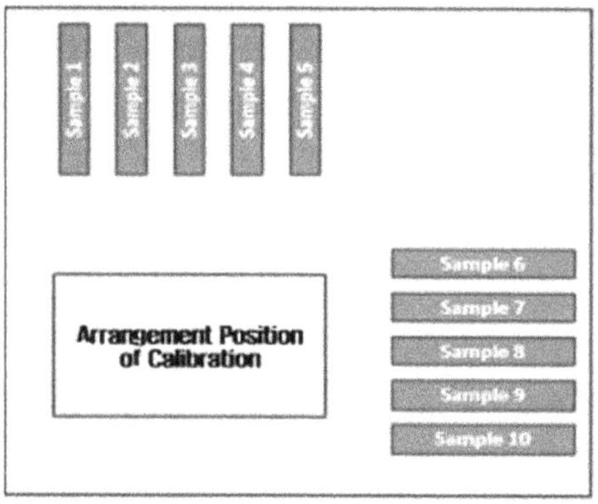

(Unit : mm)

Number	Deformation Value
1	2.67
2	2.67
3	2.66
4	2.65
5	2.65
6	2.65
7	2.72
8	2.68
9	2.64
10	2.67

The unique deformation rate calculated as Figure 5 is a value that includes numerous variables of the actual printing process. By using such a value, additive process simulation can be conducted. By cutting the specimen, the deformation rate is measured, which found the 2.14 mm of deformation on the calibration specimen of 0°, while 2.12 mm was found on 90°.

Tire Mold Finite Element Model Generation

Among various types of finite element interpretation programs, Simufact, a commercial additive interpretation program, was used. In addition, the voxel mesh method, which is most advantageous for additive manufacturing interpretation, was used to conduct the meshwork. In order to confirm tire mold deformation and residual stress, the Finite Element Method (FEM) model was used. For an accurate evaluation, a tire mold of the same size as the actual tire was generated with FEM. Based on the characteristics of the 3D printing process, the part model to which laser is applied experiences a very large thermal change. Thus, the detailed mesh (Solid Type) was used (Figure 6).

Figure 6. FEM model applied to the tire mold
(Source: Metal3D)

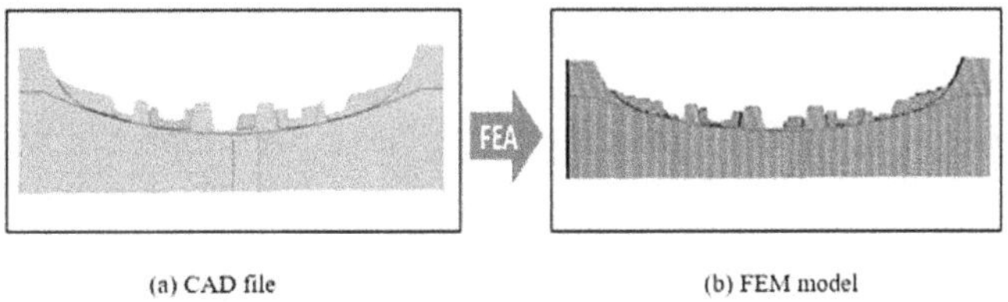

Table 2. SUS 316L chemical composition table (in wt %)

C	Cr	Cu	Mn	Mo	N	Ni	P	S	Si
0.009	16.82	0.31	1.74	2.08	0.029	10.26	0.03	0.024	0.27

(Source: Oerlikon)

Characteristics of the 3D Printing Material Used for Tire Mold Manufacturing

The material used for this interpretation was SUS 316L Steel. In terms of the specific heat, thermal conductivity, thermal expansion coefficient, density, and mechanical characteristics, temperature dependency for changing physical properties according to the temperature has been considered. Thus, in the 3D printing additive interpretation, the material's physical property values according to temperature change must be set. This indicates the material's temperature dependence. Thus, the literature was reviewed to set the physical property values of a material according to the temperature. The physical property data of the material used for 3D printing manufacturing of the tire mold are shown in Table 2 and Table 3. The atmosphere temperature was assumed as 22 °C. The cooling method used after 3D printing was passive cooling based on time. In order to assume that the tire mold is cooled at room temperature enough after 3D printing, the analysis was conducted for 3,600 seconds including the additive and cooling processes. The additive analysis was conducted using the same conditions as the actual 3D printing process.

Table 3. SUS 316L physical·mechanical characteristics

Properties	Values
Density	7.99kg/m^3
Specific Heat	0.45 J/Kg$^{\cdot k}$
Thermal Conductivity	16.2 W/m$^{\cdot k}$
Young's Modulus	205 N/mm^2
Yield Strength	547 N/mm^2
Poisson's Ratio	0.3

(Source: Oerlikon)

Interpretation of tire Mold Deformation via Support Removal

A supporter is the most important element for a 3D printer. It releases heat and fixes so that geometrical change would not occur. Actually, the printing output of the PBF generates a supporter before printing, after which the supporter is cut from the base plate. Next, the supporter must be removed from the output. In this process, deformation or stress may occur due to the cut direction of the supporter. Thus, the deformation or stress was predicted in real-time by interpretation.

Design of Deformation Compensation

In the actual metal 3D printing process, the high heat source (Laser) often causes output to be different from the actual CAD data due to the contraction and expansion deformation. In order to solve this issue, the method of changing diverse process conditions such as additive direction or support change can be used. However, another method is designing deformation compensation. The deformation compensation design is a method of permitting the distorted product toward the opposite direction and predicting the amount of deformation by the simulation to modify the design of the initial shape. With Simufact simulation, a changed CAD file can be generated in the form of STL. By using the generated STL file, simulation is conducted again to proceed to the actual printing.

Result and Discussion of Tire Mold 3D Printing Additive Interpretation

Result of Tire Mold Thermal Deformation Interpretation

The thermal deformation of tire mold was evaluated after removing the support, and Figure 7 shows the comparison of tire mold thermal deformation interpretation results before and after 3D printing. By using the actual 3D printing process conditions, tire mold thermal deformation interpretation was conducted, which confirmed the occurrence in the Z direction on both ends. The maximum deformation was +Z direction at 1.1 mm, the minimum deformation was 0.7 mm in the –Z direction. The comparison of the interpretation result (1.1 mm) and the actual result (1.5 mm) showed about 0.4 mm of the deformation amount difference, while it could be confirmed the deformation occurred at the same location of the actual deformation overall (Figure 7).

The first thing to be confirmed in relation to the printed product via 3D printing additive manufacturing is the size and shape to confirm whether the actual designed original data and the final product coincide after the completion of post-processing on the initial STL file original data and product of additive manufacturing. The measurement may be conducted by using Vernier Calipers that simply measure the length and height. However, in the case of most products manufactured via additive procedure using a 3D printer, the structure is geometrical. Thus, rather than a simple length measuring tool, a 3D scanner is used to validate the size validity. As a result of measuring the actual tire mold printed, the deformation of about 1.5 mm in the Z-axis direction could be confirmed (Figure 7).

Tire Mold Deformation Compensation Interpretation Result

The deformation compensation method was used to conduct thermal deformation interpretation to increase the precision degree of the tire mold. An STL file for tire mold deformation compensation was generated to conduct interpretation after repositioning the support, while the conditions are applied the same as the actual 3D printing process conditions. This research conducted the thermal deformation interpretation according to 3D printing deformation compensation, which confirmed that the thermal deformation was the lowest at 0.04 mm for the method of using the deformation compensation function than the method of building by generating support to the tire mold as shown in Figure 7.

Figure 7. Tire mold thermal deformation interpretation result
(Source: Metal3D)

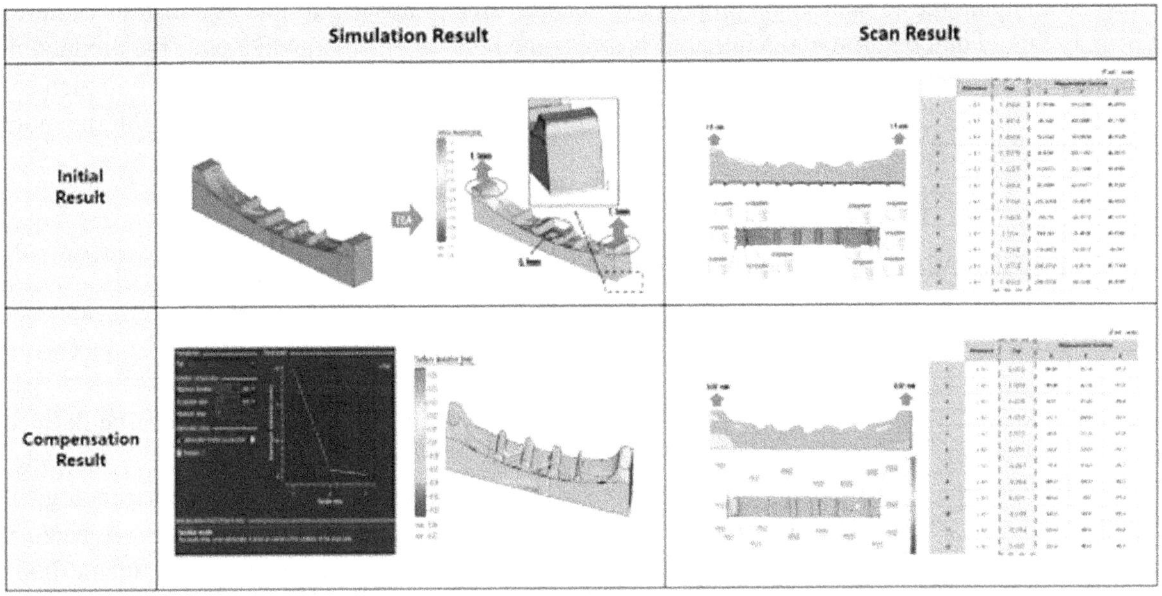

With the same 3D printing build conditions and using compensation design, up to 3.5% of the deformation reduction effect was found, which proves that compensation build function must be used to minimize the deformation. As shown in Figure 11, by using the deformation compensation method, the measurement result of the printed tire mold via scanning confirmed that the maximum was 0.03 mm, which was within the permissible range (±0.1~0.2mm).

Conclusion

This research conducted interpretation to propose the compensation function that can minimize deformation and predict additive interpretation of the deformation due to high heat source (Laser) when manufacturing tire molds with a 3D printer and found the following conclusions.

In order to derive accurate simulation results, a calibration specimen was used to calculate a unique deformation rate. By using the calculated unique deformation rate, a more accurate process simulation was conducted that includes the actual process variables.

In the case of 3D printing output, deformation or stress may occur due to the support cut direction, and, before manufacturing an actual output, interpretation was used to predict.

The thermal deformation interpretation result showed an error of about 0.4 mm when the interpretation result (1.1 mm) and the actual deformation result (1.5 mm) were compared in the case of the tire mold deformation location and amount. Based on this, it was determined that that the deformation occurred in the same location as the actual deformation.

It was confirmed that up to 3.6% of the deformation reduction effect could be derived by using the compensation function than the method of a building by generating support on the tire mold part like the actual 3D printing process.

The result of scan measurement of the tire mold printed via the deformation compensation method, it was confirmed that it fell within the permissible range (±0.1~0.2mm) with a maximum of 0.03 mm.

When applying the compensation proposed by this research to the actual field as a method of minimizing tire mold deformation in the future, it is determined that the method will enable the production of integrated tire molds that are precise and complex in shape and that it will be possible to implement the manufacturing of optimized design while reducing the defect rate of tire production with the mold printed as an integrated type.

CUSTOM CRANIAL IMPLANTS [MEDICAL APPLICATION]

Introduction

According to Alida, M (2009), restoring cranium and replacement of cranium defect can be defined as a medical process of restoring damaged or unhealthy part of a cranium. The main goal is to implement physical shape and shielding of nerves and tissues inside a cranium to achieve the psychological stability of patients. Cranium restoration is conducted on patients suffering from an injury (Accident), infection, or trauma caused by congenital malformation. Skull defects occur due to traumatic bone destruction, brain tumor, or congenital defect and lead to functional and aesthetic defects. Craniofacial reconstruction is a complex surgical process because it includes activation of body parts including the brain, eyes, and other sensory organs within a limited space. Skull damage or malformation influence a large population worldwide and is considered to be the most important and worrisome health issue (p.3186-3192). According to Chen, X (2017), moreover, the recovery of cranial abnormality presents a serious challenge even to very experienced surgeons. This is because the body part subject to surgery is composed of delicate organs such as nerves and soft tissue within the limited area. The best method of treating cranial defects is to use autogenous bone graft because it has a lower risk of complications due to infection compared to implanting other bones. However, the use is limited due to the appropriate donor area, particularly large and complex defect, issue of organ harvesting, the morbidity rate of donor part, and high-cost surgery, thereby limiting their use. For these reasons, recently, the application of this approach has been minimized due to this significant shortcoming. In addition, implant-based on other materials are being demanded (p.4199).

For example, autogenous bone graft includes complex surgical methods, blood loss, and patient pain. Similarly, using a regular implant that requires physical bending or rearrangement to fit the bone curve of a patient is a method that takes a long time, inaccurate, and difficult. In addition, due to the failure and modification of implant caused by dissonance, patients can suffer a lot of psychological stress. Thus, in order to minimize the dissonance, fit the bone curve, and increase the output, a custom implant design concept must be used. Unlike a standard plate, a custom implant design can significantly increase fitting accuracy and operation time.

According to Kullkarni, M (2014), the ultimate goal of cranial bone reconstruction is to alleviate psychological pain due to bone defect, protect the brain, and recover the appearance and psychological stability of the patient. The success of cranial reconstruction depends on implant design, material and manufacturing, a technique of the surgeon, and the defect evaluation before the surgery. Implant with porosity surface is more effective than rough coating. Porosity surface implant provides interface adhesiveness with the bone, providing effective fixation and short recovery time. With enough space for cell adhesion and fluid transportation, a high level of porosity is required. The ideal porous size for bone growth is within the range of 500 ~ 1500 μm. Many researchers identified porous titanium with a porosity of 50% is ideal for bone tissue. Porosity structure that has interconnected and excellent porous causes better implant fixation and significant bone growth. High porous rate and size are beneficial to bone generation, yet the actual increase may decrease the strength of implant (p.111-136).

Thus, based on design and production, the ability to produce a porous structure that controls porosity will be an important element of future clinical success. In the past, a variety of technologies have been used to manufacture porous titanium and alloy thereof that included casting, fabric additive, and powder sintering.

Due to the aging and expansion of the world population, medical demand has rapidly increased. Accordingly, the importance of innovative technology in the medical field is being emphasized. According to Castelan, J (2014), the combination of medical data image, image handling software, and Additive Manufacturing (AM) can generate a complex structure (Customized Implement) of a perfect structure that can make an organized appearance by coinciding with a bone curve. The implementation of combined technology can increase the quality of life of a large population as well as reduce significant costs involved in health care systems. Implant reconstruction may vary according to various factors such as material characteristics, implant design, manufacturing method, and techniques of a doctor (p.1517-3151).

Custom implant design must include the size, shape, and mechanical features of an implant to answer the demands of individual patients. As a result of unique human anatomy, AM is very useful in achieving implant per patient that is perfectly fitted to the bone curve. AM has significantly increased the ability to manufacture a complex and precise geometric structure compared to existing manufacturing methods. With the development of AM technology, the function of preparing parts in accurate shape has improved significantly by using medical image data. The accuracy and process adaptability allows the well-packaged structure to be produced that is graded accordingly. In particular, the application of AM in the medical field is rapidly increasing, which is expected to ignite a revolution in the medical field.

According to Kang, H.G (2018), it can effectively manage medical devices' requirements in terms of product user definition, efficient and cost-effective manufacturing, and short-time required for delivery. Other than the appropriate manufacturing method, related custom implant biometric materials must be implemented for the accurate and relevant requirements of custom implants. Among metal materials that allow the use of titanium alloy (Ti6Al4V ELI), it is considered to be most preferred for custom implants. This can be due to biological compatibility, high weight-to-weight ratio, high erosion resistance, non-magnetic characteristics, high toughness, and mechanical resistance (p.466-477).

In this paper, the comprehensive approach for designing, manufacturing, and validating custom cranial implants has been developed. The main goal of this paper is to introduce a simple method for redesigning and manufacturing cranial defects. By bringing the DICOM (Medical digital image and communication) of CT to a particular software (Mimics), the interest area is reduced based on the segmentation method to produce a 3D model. Before printing out the designed cranial implant with a 3D printer, AM simulation is used to minimize significant time and costs caused by numerous trials and errors via production deformation and stress validation. Furthermore, titanium cranial implants are manufactured with Ti6Al4V ELI powder by using the Powder Bed Fusion (PBF) method. In terms of appearance, accuracy, mechanical properties, and material composition, the final validation is conducted and researched.

3D Printing Methods and Materials for Customized Cranial Implant

According to Khaja Moiduddin (2019), in order to obtain custom cranial implants, a total of 4 main functions are used as the methodology suggested including data collection, 3D modeling production, 3D printing, and specimen validation process. The workflow from data collection to the final test is shown in Figure 8. This method includes segmentation, Standard Tessellation Language (STL) error modification, and thickness reduction, which are aimed at manufacturing reliable cranial implants (p.2513).

Figure 8. Methodology for obtaining custom cranial implants

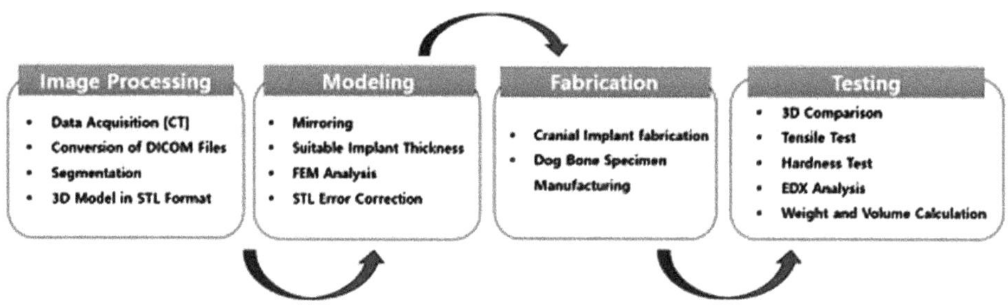

Image Processing

By using the Digital Imaging and Communications in Medicine (DICOM) file, the patient's CT scan data is obtained. Next, the MIMICS® software is used to handle the CT data to generate a CAD model on the necessary anatomical structure. The division of the 2D slice image selects a special image strength (Hounsfield Unit) from the interest areas to process. CT number and Hounsfield device identify tissue characteristics. Specific tissues like bones, skin, and muscles are distinguished with Hounsfield units. Then, by selecting the maximum and minimum threshold values of image strength, the threshold valuation is conducted. In addition, pathological areas are described. Because bones absorb most of the radiation, they have higher Hounsfield value compared to skin and soft tissues.

Customized Implant Design

The process of manufacturing custom implants for patients in this research is shown in Figure 9.

Step 1: 3D CT & MRI data acquisition. By using the digital imaging and communications in medicine (DICOM) file, a patient's computer CT scan data is acquired. Then, MIMICS® software is used to process the CT data and then to create a CAD model on the necessary anatomical structure.

Step 2: Patient custom cranial implant design.

- Design a custom implant for the operation are to fit the defective area of the patient by using a 3D CAD model and 3-Mactic® software.
- According to the opinion on the request document by the operating surgeon, the design insertion fixation part, porous part, and connection holes.

Step 3: Creation of supporter for 3D printing. In order to prevent geometrical change, make a supporter for the custom cranial implant by using MAGICS® software.

Step 4: 3D printing. The Metalsys 150 device which uses PBF method was used to print a custom cranial implant.

Step 5: Process after custom cranial implant output. Conduct post-process including thermal treatment / support removal / bluster (Shot Peening) surface treatment / final polishing / cleaning.

Step 6: Custom cranial implant product validation. Use a 3D scanner for the final validation process including analysis of differences by comparing the design data and measurement data/shape analysis/size measurement.

Step 7: Completion of a custom cranial implant for a patient.

Figure 9. 3D Printing flowchart for custom cranial implant
(Source: Metal3D)

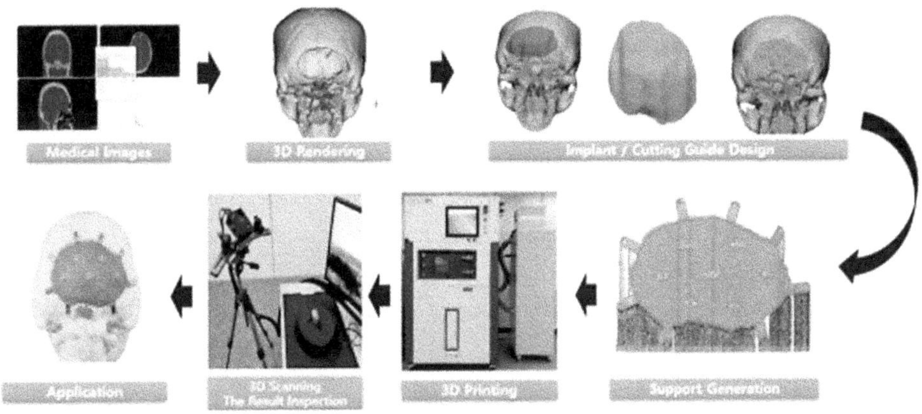

Fabrication of Designed Cylinder (Lattice Structures)

In order to increase the effectiveness of a cranial implant, the implant must be light and has good mechanical strength. Currently, in the case of medical implants, most apply the lattice structure to manufacture a product. The application of the lattice structure has the characteristics of minimized heat generation and thermal effect in addition to the ease of fusion between bones and fluid, which is why it is widely applied to medical implants. According to Khaja, M (2019), The existing study conducted for applying the lattice structure to cranial implants. Among lattice structures provided by 3-Matic software, this is an open-type lattice structure, and the structure design is conducted by a total of 8 types of lattice structures that can maintain the shape and ratio of 1:1:1: between X, Y, Z. The specimen was manufactured as a cylindrical porous specimen of a diameter of 10 mm and a height of 15 mm as specified in the ISO 13314 standard (p.2513).

By using the Metalsys 150 3D printer, a total of 8 types of lattice specimens were manufactured. (A total of 24 specimens were manufactured, 3 for each lattice type). Afterward, lattice specimens were removed from the plate by using the wire electric discharge machining. Then, the residual powered inside the specimen was removed by an ultrasonic cleaner. In order to check the strength of lattice specimens, pressure strength testing was conducted with the Instron 5989 (USA) device. The pressure strength measurement results per lattice type are indicated. The result confirmed that the pressure strength of the Dodecahedron type was higher than other lattice types. The Dodecahedron type structure has been selected as the optimum lattice structure. The optimum Lattice structure derived by this test result was applied to the cranial implant.

Materials

Bio-materials for orthopedic usage must have the following characteristics.

1. Non-toxic
2. Excellent strength
3. High resistance against fatigue
4. Chemically not activated
5. Not corrosive
6. Low elastic modulus similar to bones

The materials used in this research is Ti6Al4V ELI powder (Chemical composition according to ASTM F2924), which is Grade 5 of the titanium powder generally used for cranial implants. The spherical powder is used in metal 3D printing is because its liquidity is superior to square-shaped or irregular powder, making it beneficial for increasing the density and mechanical property of the final structure for the purpose of even coverage or dispensing of the powder layer. As such, the result of the powder shape confirmation with the Scanning Electron Microscope (SEM) showed that spherical powder.

Metal 3D Printing Process Simulation Process

The process for predicting cranial implant deformation via 3D printing process simulation is shown in Figure 10. First, after modeling the cranial implant shape, interpretation is conducted by putting in necessary materials and 3D printing conditions for the interpretation. Next, measure the unique deformation rate of the calibration specimen printed according to the conditions that fit the metal 3D printing device for implementation to apply to the simulation. The simulation proceeds by entering the optimum process parameters. Lastly, the interpretation is conducted in the order of deriving the thermal deformation and stress that occur on the cranial implant.

Creation of the Cranial Implant Finite Element Model

There are many types of finite element interpretation programs. However, the ANSYS Additive Print, the commercial additive interpretation program, is conducted the mesh work with the voxel mesh method,

Figure 10. ANSYS Additive Print simulation process
(Source: Metal3D)

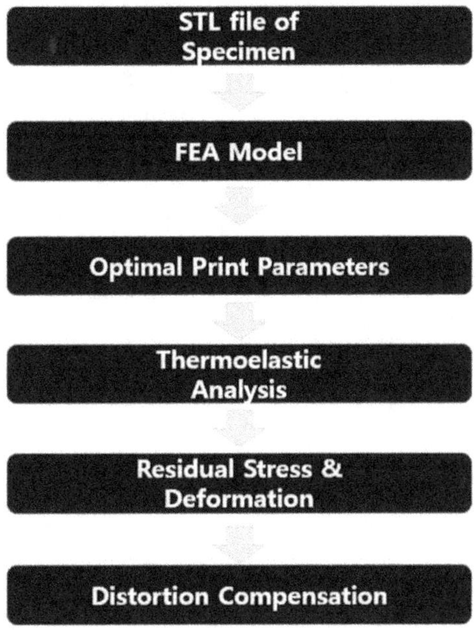

which is most advantageous for additive manufacturing interpretation. In order to confirm the deformation and residual stress of the cranial implant, research was conducted by using a Finite Element Model (FEM). For an accurate evaluation, the cranial implant applied with lattice structure is generated with FEM. Due to the characteristics of the 3D printing process, the part model to which laser is applied goes through an extremely high level of thermal change. Thus, a minute mesh (Solid Type) was used.

Characteristics of 3D Printing Materials for Cranial Implant Production

The material used for this interpretation was Ti6Al4V ELI. In terms of the specific heat, thermal conductivity, thermal expansion coefficient, density, and mechanical characteristics, temperature dependency for changing physical properties according to the temperature has been considered. Thus, in the 3D printing additive interpretation, the material's physical property values according to temperature change must be set. This indicates the material's temperature dependence. Thus, the literature was reviewed to set the physical property values of a material according to the temperature. The physical property data of the material used for 3D printing manufacturing of the cranial implant are shown in Table 4 and Table 5. The cooling method used after 3D printing was passive cooling based on time. The atmosphere

Table 4. Ti6Al4V ELI alloy chemical composition table in wt (%)

Al	V	Fe	O	Ca	Ni	H	Y	Ti
6.35	3.99	0.18	0.11	0.02	<0.01	0.002	<0.001	Bal

(Source: AP&C powder company)

Table 5. Ti6Al4V ELI alloy physical·mechanical characteristics

Properties	Values
Density	4.4 g/cm^3
Specific Heat Capacity	0.5263 J/g-°C
Thermal Conductivity	0.00001358K^{-1}
Young's Modulus	120 GPa
Yield Strength	1100 MPa
Poisson's Ratio	0.3

(Source: AP&C powder company)

temperature was assumed as 22 °C. The additive analysis was conducted using the same conditions as the actual 3D printing process.

Application of Supporter for Cranial Implant Manufacturing

A supporter is the most important element for a 3D printer. It releases heat and fixes so that geometrical change would not occur. Supporter generation significantly depends on experience. For a new or complex for, accurate support is impossible to be created. A wrongly created supporter causes thermal stress, which is the biggest cause of decreasing measurement precision by changing the measurements. In this research, in order to reduce thermal stress and thermal deformation, the supporter angles (45°, 60°, 75°, 90°) are changed to validate via simulation. The supporter was applied the same as the created supporter on the Magics software to conduct interpretation.

Result of Custom Cranial Implant Additive Simulation Interpretation

For cranial implant thermal deformation, the deformation was evaluated after removing the supporter after cooling was completed. By using the actual 3D printing process conditions, thermal deformation interpretation was conducted. First, the supporter angle was generated as 90° for interpretation. The interpretation results confirmed a significant deformation occurred on the cranial implant connection points. When this interpretation result is compared to the initial cranial implant, about 2.1 mm of contraction deformation in the –y direction could be confirmed. After comparing the interpretation result with the data of scanned actual output, a similar thermal deformation phenomenon was confirmed. The result of thermal deformation interpretation according to the supporter angle change in this research confirmed that the cranial implant generated with a supporter of 45° showed the minimum amount of thermal deformation. The result of the supporter angle generated at 45° shows that the highest deformation occurs on the connection part, the same as the result of the supporter generated at 90°. As a result of comparing with the initial cranial implant, contraction deformation occurred around 1.9 mm to the direction of –y and showed about a 0.2 mm reduction effect than creating the supporter at 90°. The result of the comparison between the interpretation result and the actual output confirmed that the thermal deformation occurred similar to the above result (Figure 11).

Among the ANSYS Additive Print functions, there is the deformation compensation function. Compensation is a method of deformation compensation method for increasing the precision of key parts. Based

Figure 11. Cranial implant thermal deformation interpretation and actual output scan data comparison (Source: Metal3D)

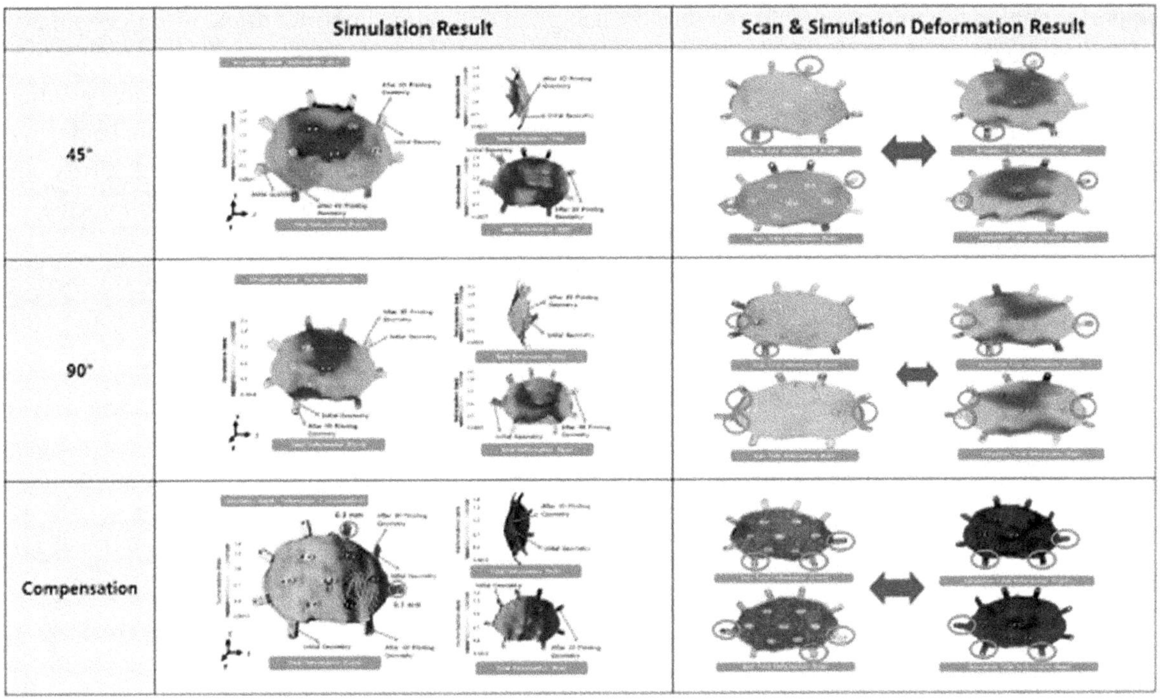

on the thermal deformation interpretation result of the supporter angle 45°, the deformation compensation function was used. The result derived was the same as Figure 16, which could be confirmed to be similar to the preceding result with the occurrence of thermal deformation on the implant fixation part. The result of comparing the interpretation result with the initial cranial implant, almost no deformation was confirmed, while the deformation was about 0.3 mm. By applying the compensation function only, the maximum of 14% of the deformation reduction effect could be confirmed (Figure 11).

Conclusion

This research conducted the interpretation to propose a compensation that can minimize deformation and predict the deformation due to high thermal source (Laser) with additive interpretation when cranial implants are manufactured with a 3D printer. As such, the following conclusions were derived.

The existing restoration operation using bone cement cannot be used for large area operations while causing irregular shape and bacterial infection. In order to solve these issues, cranial implants were manufactured using 3D printing that can reduce time while enabling customized production.

In order to make an optimized cranial implant model, a porous structure was applied. The porous structure has characteristics including ease of fusion between the bone of the skull itself and fluid, cell proliferation, body fluid circulation, weight reduction, minimization of heat generation, and thermal effect. This research manufactured specimens of a total of 8 types of porous structures to conduct pressure strength tests and selected the Dodecahedron structure with excellent strength.

In the process of 3D printing, supporter generation is most critical. Before the printing and according to the angle (45°, 60°, 75°), the influence of displacement and stress was identified to derive the supporter angle with minimum deformation. By using the optimum supporter angle (45°), it could be confirmed that the deformation of actual output was alleviated after interpretation.

The thermal deformation interpretation result shows that the location of the cranial implant deformation and distortion amount was significantly similar to the interpretation result (2.1 mm) and actual deformation result (2.5 mm), which can be determined to occur in the same location as the actual deformation.

To improve the quality of the cranial implant, a simulation was conducted by using the compensated STL file, which showed the deformation reduction effect of as much as 14%.

When applying the compensation function suggested by this research in the actual field as a method of optimizing cranial implant deformation in the future, it is expected to contribute to the product improvement in 3D printing.

ACKNOWLEDGMENT

This research received no specific grant from any funding agency in the public, commercial, or not-for-profit sectors.

REFERENCES

Alida, M. (2009). Direct fabrication through electron beam melting technology of custom cranial implants designed in a PHAN To M-based haptic environment. *Materials & Design, 30*(8), 3186–3192. doi:10.1016/j.matdes.2008.11.013

Alkadi, F., Lee, J.W., Yeo, J.S., Hwang, S.H., & Choi, J.W. (2019). 3D Printing of ground tire rubber composites. *International Journal of Precision Engineering and Manufacturing-Green Technology, 6*, 211-222. doi:10.100740684-019-00023-6

Avraham, T., Franco, P., Brecht, L., Ceradini, D., Saadeh, P., Hirsch, D., & Levine, J. (2014). Functional outcomes of virtually planned free fibula flap reconstruction of the mandible. *Plast. Reconstr. Surg, 134*, 628–634. 0000000 0513 doi:10.1097/ prs.00000

Bhargava, D., Bartlett, P., Russell, J.L., & Liddington, M. (2010). Construction of titanium cranioplasty plate using craniectomy bone flap as template. *Acta Neurochir*, 173–176. 10.1007/s00701-009-0394-2

Bogu, V. P., Kumar, Y.R., & Khanra, A. (2017). Modelling and structural analysis of skull/cranial implant: beyond mid-line deformities. *Acta of Bioengineering and Biomechanics, 19*(1). doi:10.5277/ABB-00547-2016-04

Carry, B. C., & Goswami, T. (2009). Knee implants review of models and biomechanics. *Mater, 30*, 398–413. https:// doi:10.1016/j.matdes.2008.03.032

Castelan, J., Schaeffer, L., Daleffe, A., Fritzen, D., Salvaro, V., & Silva, F. P. (2014). Manufacture of custom-made cranial implants from DICOM® images using 3D printing, CAD/CAM technology and incremental sheet forming. *Revista Brasileira de Engenharia Biomédica, 30*(3), 1517–3151. doi:10.1590/rbeb.2014.024

Chen, X., Xing, L., & Egger, J. (2017). Computer-aided implant design for the re-storation of cranial defects. *Scientific Reports, 4199*. Advance online publication. doi:10.103841598-017-04454-6

Chu, C.H., Song, M.C., & Luo, V.C.S. (2006). Computer aided parametric design for 3D tire mold production. *Computers in Industry, 57*, 11-25. .2005.04.005 doi:10.1016/ j.compind

Gary, H. (2015). Craniofacial reconstruction with poly(Methyl Methacrylate) customized cranial implants. *The Journal of Craniofacial Surgery, 26*(1), 64–70. doi:10.1097/SCS.0000000000001315 PMID:25376145

Halabi, F. E. (2011). Mechanical character-ization and numerical simulation of polyether–ether–ketone (PEEK) cranial implants. *Journal of the Mechanical Behavior of Biomedical Materials, 4*(8), 1819–1832. doi:10.1016/j.jmbbm.2011.05.039 PMID:22098881

Hwang, I. H., Mehmert, P., & Tran, M. (2017). Metal additive manufacturing process simulation for the hinge of the engine hood. *Korean Society for Precision Engineering, 12*, 161-162. Retrieved from http://www.dbpia.co.kr/Journal/articleDetail?nodeId=NODE07300735

Jayanaidu, P. (2013). Analysis of a drive shaft for automobile applications. *Journal of Mechanical and Civil Engineering, 10*(2), 43–46. doi:10.9790/1684-1024346

Kang, H. G., Park, J.W., & Park, D.W. (2018). Three dimensional printing technique and its application to bone tumor surgery. *J Korean Orthop Assoc*, 466-477.

Khaja, M., M, S.H., Umer, U., & Alkhalefah, H. (2019). Fabrication and analysis of a Ti6Al4V implant for cranial restoration. *Applied Sciences, 2513*. Advance online publication. doi:10.3390/app9122513

Kim, B. J., Hong, K.S., Park, K.J., & Kang, S.H. (2012). Customized cranioplasty implants using three-dimensional printers and polymethyl-methacrylate casting. *J Korean Neurosurg*, 541–546. https://doi:10.3340/jkns.2012.52.6.541

Kim, K. S. (2014). Characteristic analysis of carburized quenching processes for bevel gear of automobile driving part. *Current Nanoscience, 10*(1), 51–55. doi:10.2174/157341370966131111230030

Kulkarni, M., Mazare, A., Schmuki, P., & Lglic, A. (2014). Biomaterial surface modification of titanium and titanium alloys for medical applications. *Nanomedicine One Central Press Altrincham*, 111–136. Retrieved from http://www.onecentralpress.com/wp-content/uploads/2014/11/CHAPTER-5-NM-12-LATEST.pdf

Lee, J. W., Fang, J.-J., Chang, L.-R., & Yu, C.-K. (2007). Mandibular defect reconstruction with the help of mirror imaging coupled with laser stereolithographic modeling technique. *Journal of the Formosan Medical Association, 106*(3), 244–250. doi:10.1016/S0929-6646(09)60247-3 PMID:17389170

Markus, G., Xing, L., Xiaojun, C., Dieter, S., & Jan, E. (2016). Computer-aided planning of cranial 3D implants. *Computer Assisted Radiology and Surgery 30th International Congress and Exhibition*. 10.1109/ embc.2016.7590915

Marreiros, M. M., Heuzé, Y., Verius, M., Unterhofer, C., Freysinger, W., & Recheis, W. (2016). Custom implant design for large cranial defects. *International Journal of Computer Assisted Radiology and Surgery, 11*(12), 2217–2230. doi:10.100711548-016-1454-8 PMID:27358081

Mohammadpour, M., Theodossiades, S., Rahnejat, H., & Saunders, T. (2014). Non-newtonian mixed elastohydrodynamics of differential hypoid gears at high loads. *Meccanica, 49*(5), 1115–1138. doi:10.100711012-013-9857-x

Oh, J. W., Na, H.Y., & Choi, H.S. (2017). Technology Trend of the additive manufacturing (AM). *J. Korean Powder Metall Inst, 24*(6), 1-14. https:// doi:10.4150/KPMI.2017.24.6.494

Parthasarathy, J. (2014). 3D modeling, custom implants and its future perspectives in craniofacial surgery. *Ann. Maxillofac, 4*(1), 9–18. doi:10.4103/2231-0746.133065 PMID:24987592

Patil, S. M., Deshpande, A. S., Bhalerao, R. R., Metkari, S. B., & Patil, P. M. (2019). A three-dimensional finite element analysis of the influence of varying implant crest module designs on the stress distribution to the bone. *Dental Research Journal, 16*(3), 145–152. doi:10.4103/1735-3327.255750 PMID:31040869

Singare, S., Lian, Q., Wang, W.P., Wang, J., Liu, Y., Li, D., & Lu, B. (2009). Rapid prototyping assisted surgery planning and custom implant design. *Rapid Prototyp, 19–23*. Advance online publication. doi:10.1108/13552540910925027

Staffa, G., Barbanera, A., Faiola, A., & Fricia, M. (2012). Custom made bio ceramic implants in complex and large cranial re-construction: A two-year follow-up. *Journal of Cranio-Maxillofacial Surgery, 40*(3),65-70. https:// doi:10.1016/j.jcms.2011.04.014

Ting, W., Martin, E., Lorenz, F., & Aleksandra, P. (2006). Anatomically constrained deforma-tion for design of cranial implant: methodology and validation. *International Conference on Medical Image Computing and Computer-Assisted Intervention*, 9-16. 565_210.1007/ 11866

Tun, T., & Lwin, K.T. (2008). Optimizing the microstructure and mechanical properties of austempered ductile iron for automobile differential gear. *Journal of Metals Materials and Minerals, 18*(2), 199–205.

Williams. L.R., Fan, K.F., & Bentley, R.P. (2015). Custom-made titanium cranioplasty: early and late complications of 151 cranioplasties and review of the literature. *Int J Oral Maxillofac Surg, 44*(5), 599-608. doi:10.1016/j.ijom.2014.09.006

Yan, R., Luo, D., Huang, H., Li, R., Yu, N., Liu, C., H, Min., & Rong, Q. (2018). Electron beam melting in the fabrication of three-dimensional mesh titanium mandibular prosthesis scaffold. *Scientific Reports, 750*. Advance online publication. doi:10.103841598-017-15564-6 PMID:29335485

Yoon, H.S., Choi, J.S., & Oh, Y.K. (2007). Thermal deformation of aluminum alloy casting materials for tire mold by numerical analysis. *Transactions of Nonferrous Metals Society of China, 22*(3), 859–863. doi:10.1016/S1003-6326(12)61816-0

Yu, J. Y., Jung, T., Kim, S., & Rhee, S. (2011). Laser welding of cast iron and carburized steel for differential gear. *Journal of Mechanical Science and Technology, 25*(11), 2887–2893. doi:10.100712206-011-0809-x

Yun, G. B. (2013). Study in weight lightening of differential gear and process development of cold forging for electric vehicle. Graduate School, Gyeongsang National University.

ADDITIONAL READING

ANSYS Software Company. (2018). ANSYS releases additive print, suite for metal additive manufacturing. Retrieved from http://www.tenlinks.com/news/ansys-releases-additive-print-suite-for-metal-additive-mfg/

Cooka, P. S., & Murphyb, A. B. (2020). Simulation of melt pool behaviour during additive manufacturing: Underlying physics and progress. *Additive Manufacturing*, *31*, 1–23. doi:10.1016/j.addma.2019.100909

Graf, M., Härtel, S., & Awiszus, B. (2019). Application of numerical simulation for lightweight design. *Journal of Mechanical Engineering*, *8*(1), 176–191. doi:10.3390/met8121009

Joe, J. M. (1999). Validation of computer-assisted manufacture of titanium plates for cranioplasty. *Int. J. Oral Maxillofac*, *28*(4), 309–313. doi:10.1016/S0901-5027(99)80165-9 PMID:10416902

Marta, S. (2019). Simulation and validation of thermography inspection for components produced by additive manufacturing. *Applied Thermal Engineering*, *159*, 113872. doi:10.1016/j.applthermaleng.2019.113872

Pompili, A., Caroli, F., Carpanese, L., Caterino, M., Raus, L., Sestili, G., & Occhipinti, E. (1998). Cranioplasty performed with a new osteoconductive osteoinducing hydroxyapatite-derived material. *Journal of Neurosurgery*, *89*(2), 236–242. doi:10.3171/jns.1998.89.2.0236 PMID:9688118

Qian, X., Zhang, Y., Yin, Y., Zhou, J., Shen, X., & Ji, X. (2019). Optimization on the deformation of circular bosses of a compressor casing wax pattern in selective laser sintering process. *Procedia Manufacturing*, *37*, 578–583. doi:10.1016/j.promfg.2019.12.091

Samman, N., Luk, W. K., Chow, T. W., Cheung, L. K., Tideman, H., & Clark, R. K. (1999). Custom-made titanium mandibular reconstruction tray. *Australian Dental Journal*, *44*(3), 195–199. doi:10.1111/j.1834-7819.1999.tb00221.x PMID:10592564

Shane, E. (2018). Ansys Advantage Explores Simulation for Metal Additive Manufacturing. Retrieved from Lecture Notes Online Web Site: https://www.ansys.com/blog/aa-additive-intro

KEY TERMS AND DEFINITIONS

3D Printing: It is a three-dimensional object from a computer-aided design (CAD) model, usually by successively adding material layer by layer, which is why it is also called additive manufacturing.

3D Scanner: It is the process of analyzing a real-world object or environment to collect errors data on its shape and possibly its appearance.

Additive Manufacturing Simulations: The physics behind the manufacturing process can be accurately recreated in software platforms enabling end to end digitalization – predicting residual stresses, voids, cracks, and so on, factors which will be crucial in the service life of a part.

Design for Additive Manufacturing (DfAM): It is design for manufacturability as applied to additive manufacturing (AM). It is a general type of design methods or tools whereby functional performance and/or other key product life-cycle considerations such as manufacturability, reliability, and cost can be optimized subjected to the capabilities of additive manufacturing technologies.

Differential Gear: It is an automotive part of the power transmission device.

Lattice Structure: A lattice is an ordered array of points describing the arrangement of particles that form a crystal.

Medical Implant: An implant is a medical device manufactured to replace a missing biological structure, support a damaged biological structure, or enhance an existing biological structure.

Powder Bed Fusion (PBF): It is a subset of additive manufacturing (AM) whereby a heat source (e.g., laser, thermal print head) is used to consolidate material in powder form to form three-dimensional (3D) objects.

Tire Mold: It is a precision mold that gives the final shape to the tire.

Topology Optimization: It is a mathematical method that optimizes material layout within a given design space, for a given set of loads, boundary conditions and constraints with the goal of maximizing the performance of the system.

Chapter 12
Multiscale Modeling of the Laser Additive Manufacturing Process

Seshadev Sahoo
Siksha O Anusandhan (Deemed), India

Jyotirmoy Nandy
Siksha O Anusandhan (Deemed), India

ABSTRACT

Additive manufacturing (AM) has emerged as the most versatile process in the manufacturing sector. The advantages of AM such as applicability in a wide range of industries, ease of manufacturing, and reduction in waste production have increased its demand over the past decades. Out of the many techniques under AM, direct metal laser sintering (DMLS) is one of the most efficient manufacturing techniques that uses a high-powered laser beam to sinter metal powders in a layer-by-layer fashion. With the current usage of computational modeling, the prediction of microstructure evolution and other thermo-mechanical properties of different materials have been of great advantage to researchers. Along with a detailed classification of AM techniques, this chapter focuses on the use of continuum, phase field, and atomistic modeling under the DMLS process. The results show that multiscale modeling can be advantageous in gaining deeper insight into various phenomena like diffusion and sintering.

INTRODUCTION

In the recent era, the manufacturing industries face global competition to fabricate their product in reduced time with high quality and low cost. With this aim, Rapid Prototyping technologies have been developed to overcome this issue. In other names, rapid prototyping is also referred to as 3D printing or Additive Manufacturing (AM). Today this AM emerges as one of the smart manufacturing technologies which allow parts to be manufactured through a layer-by-layer build strategy. This technology has been shown its capability in various industrial sectors ranging from the medical industry to the aerospace industry to

DOI: 10.4018/978-1-7998-4054-1.ch012

make near net shape components such that fuel injector nozzle, medical and dental implants, punches, dies, and custom tooling (Guo & Leu, 2013). The components produced by the AM process offers a significant improvement in integration, mechanical properties, lifetime, and energy savings. In comparison with the traditional manufacturing processes, AM processes are a layer-based method that fabricated near net shape components with a predefined digital CAD model without the need for expensive tools. This process utilizes a high-frequency mobile source of heat to produce fully dense components. The advantages of the AM processes over conventional manufacturing methods are design freedom, rapid and cost-effective, ability to produce topologically optimized structure, and also low volume manufacturing of components. The AM processes can permit a high level of flexibility in comparison to the cost of production and time of processing of older traditional processes. Thus, manufacturing industries are exhibiting a huge amount of interest inapplicability of the AM process. The major classifications in currently available in AM technologies are (i) laser sintering (ii) laser melting, (iii) direct energy deposition (Nandy et al., 2019; Frazier, 2014). These classifications consist of various techniques of production based on the type of materials, consolidation method and its mechanism, and the source of energy used.

To meet the requirements for the production of high quality end-use metallic parts, AM processes are been developed with each passing year to make the process more efficient. Although this process can process and manufacture a wide range of metals, AM also excels in the manufacture of high-quality ceramics and composites. The quality of final products manufactured using AM techniques is dependent on internal properties such as the emerging microstructure, amount of pore formation, distribution of stress, rise in internal defects, and other important physical properties. The properties of finished products such as mechanical strength, morphology, and structural growth depend upon initial process parameters. Previous literature on AM techniques demonstrates that several kinds of heat, as well as mass transfer, take place throughout the process, which gives rise to complicated mechanical behavior in the finished specimens. These behaviors are governed by both the physical and chemical properties of the material used. Important characteristics including shape and size of particle, porosity, flowability, composition, and processing conditions play an essential part in the quality of the finished product. As the final microstructure affects both the mechanical and physical aspects of the finished products, research worldwide recommends the computational modeling of microstructures (Mayer, 2005). This helps in predicting the structural changes and final properties of different materials throughout different AM processes. The recent market has shown a huge interest in computational modeling as it helps in qualitative improvement of the samples and reduces material wastage.

This chapter intends to provide a detailed knowledge of multiscale computational modeling used for AM techniques. In this chapter, the classification of AM techniques is shown which is then followed by computational modeling in one of the widely used AM processes called direct metal laser sintering (DMLS). This chapter consists of three different modeling techniques at different length scales. Comprehensive knowledge of microstructure modeling is provided which will help researchers to adopt computational modeling for different AM techniques.

CLASSIFICATION OF AM TECHNIQUES

Laser Sintering

This process is widely used in the area of additive manufacturing. The layer by layer build mechanism allows users to gain speed over production time and improve the mechanical properties of different metal powders and alloys. The two types of laser sintering processes based on technical differences are:

1. Direct Metal Laser Sintering
2. Selective Laser Sintering

The only common feature amongst both these processes is that in both the processes metal powders are partially melted followed by rapid solidification. These processes follow the principle of liquid phase sintering.

DMLS process is a comparatively novel manufacturing technique that was developed by Rapid Product Innovations and EOS GmbH. This process uses a high-powered laser source to shape 3D products using a layer by layer build fashion. Metal in powder form is spread on the build platform in the form of a fine layer. The laser source scans over the metal powder pattern according to the described CAD design. After every scan, a recoating blade spreads a new layer of powder metal on the build platform. After each step, the build platform lowers itself after the re-coater blade completes the spreading of the metal powder. The phenomenon of solid-state sintering in the liquid phase remains the main principle behind the fusing of metal powders in the DMLS process. As compared to the traditional manufacturing process, the DMLS process has a huge advantage over the production of complicated parts due to feedstock flexibility. Other important advantages include the less production of waste products, less expensive as there are no support structures, a high degree of freedom, and completely environment friendly (Gu & Shen, 2008).

The densification of metal powders throughout this process is caused due to overall sintering mechanism. This effect of the process parameters plays a major role in the densification process. Laser power is one of the major parameters where one can control the input laser energy on the powder bed. Scan speed is another major process parameter where one can control the speed of the laser power source scanning the powder bed. Hatch spacing refers to the amount of distance between two consecutive laser scans. Hatch pattern, layer thickness, and laser beam size are other influential process parameters in the DMLS process. The densification of the metal powders also depends on the material properties. Important material properties are chemical composition, the distribution of particles in the powder bed and the size variation of the metal powders.

Selective Laser Sintering (SLS) is similar in operation to the DMLS process. The only difference is the usage of high-power laser inputs in SLS. In this process, polymeric powders materials are used which include amorphous polymers (polycarbonate powders) and semi-crystalline polymers (nylons like polyamide). SLS process gains an advantage over DMLS due to its ability to manufacture metallic components without the use of polymer binder. This process also enables the fabrication of metallic components using a sacrificial polymer binder. This expands the usability of the SLS process but requires post-processing using heat treatment for the removal of the polymer binder.

Amongst the wide range of advantages using laser sintering, there are a few limitations involved in the process. The major limitations are incomplete densification of varied structural and qualitative properties. However, these limitations can be overcome by the usage of post-processing using methods. These methods include hot isostatic pressing, secondary infiltration, and heat treatment using furnace (Mengucci et al., 2016). Also, varying the process parameters in an optimized manner helps to overcome the limitations mentioned above.

Laser Melting

The growing demand for manufacturing highly dense parts in limited time gave rise to the introduction of the laser melting process. The process differentiates itself from the traditional laser sintering process due to the complete/fully melting of metallic powders. The process of complete melting and solidification is accompanied by improved parameters. In comparison to DMLS, this process consists of high-frequency laser power, smaller laser beam spot size, and a very less width of layer thickness. This process has gained higher importance in the manufacturing sector due to its ability to process non-ferrous pure metals (Tolochko et al., 2004). This process faces certain difficulties due to the full melting phenomenon. The unstable melt pool gives rise to a high rate of shrinkage which induces a high amount of stress in the final parts. New microstructures and distinct phases are generated in this process due to the non-equilibrium conditions. Hence, this process has been researched widely to produce high-quality parts.

Direct Metal Deposition

This form of additive manufacturing includes many processes such as:

1. Direct metal deposition (DMD)
2. Direct laser fabrication (DLF)
3. Laser engineered net shaping (LENS)

This form of additive manufacturing includes many processes such as direct metal deposition (DMD), laser engineered net shaping (LENS), and direct laser fabrication (DLF). This process is different from the above-mentioned processes due to its unique coaxial powder delivery system. In this process, a distinctive powder feeder delivers powder in a pressurized gas delivery system. The workpiece moves in the X-Y plane and is guided by a computerized control setup. The usage of all axes processing unit allows this process to coat as well as fabricate new multifaceted parts. Out of the three manufacturing techniques under LMD, the DMD process is different from the other techniques due to the closed-loop feedback system and a coaxial nozzle system. A controller is used in this process to control the various process parameters accompanied by a 6-degree computer-aided manufacturing system. Out of the many metals used in this process, titanium alloys are the most used ones (Corbin et al., 2017; Dinda et al., 2009). The LENS and DLF techniques use a raster scan technique along the X-Y plane. A consistent powder delivery unit introduces powder to the melt pool on the substrate level. After every layer, the laser and powder feed shift according to the Z-direction. Studies in this technique of producing high-quality products have been conducted mostly using nickel-alumina, alumina ceramics, and niobium carbides (Balla et al., 2008; Bandopadhyay et al., 2009; Unocic & DuPont, 2004).

MULTISCALE MODELING OF DMLS PROCESS

This unique technique has been functional throughout the wide segment of single-use components as well as alloys due to which it is hard to understand the overall technical characteristics throughout manufacturing. Specific properties such as the rate of sintering and outcome of variation in process parameters upon final structural and mechanical characteristics of finished parts are some of the major challenges for the researchers. This technique is associated with a wide array of phenomena such as transfer of momentum, mass and heat, and few other chemical reactions which take place during the process. The significance of computational modeling and microstructure evolution has grown over the past few decades. Rising demands in the industries, improvements in their productivity, quality, and design has led to an increased significance of computational modeling.

Continuum Modeling

In continuum modeling, the top surface of the layered bed is scanned using a high-power laser source. The powder particle absorbs the energy source as a concentrated heat flux which causes the powdered metal to melt. The mobile highly powered heat source can be referred to as the Gaussian laser heat source. Rapid cooling causes consolidation of molten particles as the moving heat sources moves from the powder bed. In recent years, many authors have examined and analyzed the heat transfer and fusion process using continuum modeling. Samantaray et al., (2018) performed computational simulations using continuum modeling and studied the sintering behavior of AlSi10Mg particles in the process of DMLS. Panda & Sahoo (2018) have studied the residual stresses during heat transfer in DMLS using this AlSi10Mg alloy. A detailed description of the work can be found elsewhere (Panda & Sahoo, 2019). According to their model, this heat transfer process causes the conduction of heat between the powdered particles and convection of heat amidst the powder bed and the ambient atmosphere. On the other hand, heat radiation arises from a laser source up to the powdered layer due to the generation of heat.

To model the mechanism of heat transfer, the governing equation which is used is,

$$\frac{\partial}{\partial x}\left(k_{\alpha}\frac{\partial T}{\partial x}\right)+\frac{\partial}{\partial y}\left(k_{\beta}\frac{\partial T}{\partial y}\right)+\frac{\partial}{\partial z}\left(k_{\lambda}\frac{\partial T}{\partial z}\right)+\dot{Q}\left(x,y,z,t\right)=\rho c_{p}\frac{\partial T}{\partial t} \tag{1}$$

Here, k_{α}, k_{β}, and k_{λ} represent thermal conductivities in X, Y, and Z directions respectively (W/m K) and T denotes the temperature of the specimen (K). $\dot{Q}\left(x,y,z,t\right)$ represents the rate of the internally generated heat per unit volume (W/m^3), ρ represents the density of metal powder (kg/m^3), C_p denotes the specific heat capacity (J/kg K) and t represents interaction time.

To model the residual stress, the following equation has been used,

$$div\sigma = 0 \tag{2}$$

where σ denotes stress tensor.

The input heat source adheres to a Gaussian distribution in the above continuum modeling.

Mathematically, input heat flux denoted by q has been shown as:

$$q\left(r\right)=\frac{2AP}{\Pi r_0^2}e^{-2r^2/r_0^2} \tag{3}$$

Here, A represents absorptivity of powder system, P denotes input laser power, r_O denotes laser beam radius whereas r represents a distance from the center of spot generated on the top surface in powdered bed and radial distance between the laser beam.

Few assumptions which are taken into consideration for the simulation are as follows:

1. AlSi10Mg powder particles are considered to be spherical.
2. Gaussian distribution of heat source is considered to be the laser energy source which is used to scan the powdered bed perpendicularly in direction.
3. The properties of the powder particles are considered to be homogenous in every direction.
4. The properties of the powder particles are considered to be temperature-dependent.
5. Conduction, convection, and radiation are considered to dominate the heat transfer phenomena.
6. Continuous heat transfer coefficient has been assumed amidst the powder bed and enclosing environment.
7. No such amount of force from the surrounding has been considered to be directed towards a powder bed or substrate.

The governing equations mentioned above are solved using the finite element method. Table 1 shows the thermo-physical properties of AlSi10Mg used for the simulation.

Table 1. Properties of AlSi10Mg powder

Temperature	20°C	100°C	200°C	300°C	400°C
Thermal conductivity(k), W/mK	147	155	159	159	155
Emissivity (ε)	0.3				
Specific heat capacity (c_p), J/kgK	739	755	797	838	922
Density (ρ), g/m^3	2.67				
Heat transfer coefficient (h), W/m^{2}K	80				

Source: Samantaray et al. (2018)

This model considers a coupled heat transfer mechanism in which conduction, convection, and radiation all take place. The simulations are performed and results after each step are considered as input for the next step. The detailed about the simulation procedure and discussion about simulated results are discussed elsewhere (*Samantaray et al., 2018; Panda & Sahoo, 2019).*

Phase-Field Modeling

Additive manufacturing happens to be a fast process due to the rapid solidification/cooling phenomenon. A lot of process parameters like laser power, scanning speed, hatching space, hatching array, and laser spot size affect the overall microstructure. These process parameters engage in playing a vital part in the thermo-physical as well as the mechanical characteristics of the resulting specimen. Evaluation of temperature across the microstructure is a vital assessment in the process as it is directly related to the grain growth pattern.

Two major techniques have been used to model the growth of microstructure over the past few decades which are as follows:

1. Sharp interface modeling
2. Phase-field modeling

Both of these models faced multiple evolutions due to varied requirements over the years. However, the sharp interface modeling faces various challenges due to its incapability in tracking interface, performing complex numerical simulations, and reduces overall time. Contrary to this, phase-field modeling has appeared to be much more efficient in microstructure modeling. It gains its advantages due to the ease of modeling processes such as solidification, grain growth, and martensitic transformations (Ahmed et al., 2016; Denoual & Vattre, 2016; Gandin & Rappaz, 1994). In phase-field modeling, the composition of the input material is described as a few field variables. These field variables are categorized as conserved and non-conserved field variables. Cahn-Hilliard and Cahn-Allen equations define the rate of change of conserved and non-conserved order parameters. These order parameters are expressed as a function of space and time coordinates.

Cahn-Hilliard equation:

The Cahn-Hillard equation is used to model the process of phase separation in specifically binary alloys. In this work, the equation is coupled with the mobility function to study the process of solid and liquid phase formation (Cahn et al., 1996).

$$\frac{\partial \varnothing}{\partial t} = M.\nabla^2 \left[\frac{\partial f}{\partial \varnothing} - \varepsilon_{\varnothing}^2 \nabla_{\varnothing}^2 \right] \tag{4}$$

Where $\varnothing$ denotes order parameter, ∇ represents the Laplacian operator (divergence), f denotes free energy, $\varepsilon_{\varnothing}$ represents gradient energy coefficient for mobility M. This equation is a modified version of Fick's 2nd law of transient diffusion.

Cahn-Allen equation:

This equation is used to study the process of phase formation in multicomponent alloys. It is a reaction-diffusion reaction and is different from the Cahn-Hillard equation in terms of the composition of the alloy. Cahn-Hillard equation deals with the binary alloy system whereas the Cahn-Allen equation deals with the multi-component alloy system (Shen & Yang, 2010).

$$\frac{\partial \varnothing}{\partial t} = -M \left[\frac{\partial f}{\partial \varnothing} - \varepsilon_{\varnothing}^2 \nabla_{\varnothing}^2 \right] \tag{5}$$

This equation is also called the Ginzburg-Landau equation.

Generally, the phase-field model based on the thin interface analysis is used to simulate the microstructure evolution in additive manufacturing. Temperature and diffusion equations are combined with the governing equations. These equations govern the behavior of interface which remains initially smooth but later faces many changes during the process. By using both the order parameters as mentioned earlier, the governing equations which represent the anisotropic phase-field modeling are as follows (Nandy et al., 2017):

$$\tau A^2(\varphi)\frac{\partial\varphi}{\partial t} = W_\varphi^2\nabla.\left(A^2(\varphi)\nabla\varphi\right) - W_\varphi^2\partial_x\left[A(\varphi)A'(\varphi)\partial_y\varphi\right] + W_\varphi^2\partial_y\left[A(\varphi)A'(\varphi)\partial_x\varphi\right] - \frac{d_g(\varphi)}{d\varphi} - \frac{L\left(T_c - T_m\right)}{HT_m}\frac{dp(\varphi)}{d\varphi} + \eta'\left(x,\ T_c\right) \tag{6}$$

$$\frac{\partial T}{\partial t} = \nabla.\left(\alpha(\varphi)\nabla T\right) + \frac{Lh'(\varphi)}{C_p}\frac{\partial\varphi}{\partial t} \tag{7}$$

where τ represents the time scale, φ is the order parameter, $A(\varphi)$ is the anisotropy parameter, $A'(\varphi)$ represents the derivative of $A(\varphi)$ in correspondence to order parameter, Cp represents specific heat capacity at constant pressure, W_φ is the measure of length, L is the latent heat of fusion, Tc denotes the critical temperature, Tm denotes the melting temperature, H is the nucleation barrier, $\eta'(x, T_c)$ denotes the stochastic noise function following thermal fluctuations, $\alpha = k / \rho Cp$ denotes thermal diffusion coefficient and $h'(\varphi)$ is the smooth derivative function in limits h(0)) = 0 and h(1) = 1.

Using this modeling, the microstructure evolution of AlSi10Mg was analyzed in the DMLS process by Nandy et al. (2018). Temperature gradients were extracted using thermal modeling in the DMLS process. The governing equations are solved using the finite-difference technique with a five-point stencil in 2D spaces. The MATLAB software is used to carry out the simulations. The details of the computational domain, simulation procedure, and analysis of the results can be found elsewhere (Satpathy et al., 2018).

Atomistic Modeling

In atomistic modeling, the underlying mechanism behind fusion is studied in a more demonstrated way as compared to phase-field modeling. In atomistic modeling, the solid-state, as well as liquid state sintering, can be studied based on heat flux generated. During solid-state sintering, two major mass transport plays a vital role in the sintering process (bulk and surface). Surface, grain boundary and volume bulk diffusion affect the overall mass transfer from the surface and core of particles. During liquid state sintering, viscous flow is considered to be the major phenomenon as the melting of particles takes place and they fuse onto each other due to surface tension forces. In this way, this modeling allows researchers to study the characteristics of the sintering of nano-scale particles which is a big advantage over phase-field modeling. Various investigations can be carried out using atomistic modeling such a neck growth analysis, calculation of dihedral angle, self-diffusivity, and mean square deviation. Some other important calculations which can be made using this modeling are the radius of gyration and shrinkage ratio. Optimization of process parameters can also be achieved using the results obtained from atomistic modeling. In this modeling technique, the selection of appropriate potentials is considered as an important part. Many researchers have performed simulations to study various phenomena in AM techniques using atomistic modeling. Nandy et al. (2019) performed simulations at the atomistic level through Large Scale Atomic/Molecular Massively Parallel Simulator (LAMMPS) (Plimpton et al., 2007). For setting up the AlSi10Mg nanoparticles, the individual elements are first set up in two spherical atoms of radius 4nm using the 'hybrid/overlay' style. The calculation of forces amongst Aluminum-Aluminum atoms and Magnesium-Magnesium atoms (Type-2) are done using the Embedded Atom Method (EAM) potential. For Silicon-Silicon atoms (Type-3), the interatomic force calculations are done using Tersoff potential. The L-J/cut potential is used to calculate the force between type 1-2, type 1-3, and type 2-3. Each one of them consists of 32,399 particles inside them. A spacing of 0.3nm is maintained between the two particles. The time step size is maintained at 0.001ps for this simulation. The laser power and scan speed are kept the same as mentioned in the previous section. The temperatures are extracted from the thermal model and a canonical ensemble NVT is used where the amount of substance (N), volume (V), and temperature (T) are conserved. After equilibration at 300K, simulations are run for a total time of 450ps, and temperatures are increased at every 100K to 754K.

EAM potential for type 1 and 2 atoms:

$$E_{total} = \sum_{i} F_i\left(\rho_i\right) + \frac{1}{2}\sum_{\substack{i,j \\ i \neq j}} \varnothing_{i,j}\left(r_{i,j}\right) \tag{8}$$

Here, ρ_i represents the electron density which denotes the function common to both atoms i and j and $F_i\left(\rho_i\right)$ shows the embedding energy of atom i with the electron density due to its surrounding atoms. $\varnothing_{i,j}$ denotes the 2-body central potential between atom i and j at separation distance $r_{i,j}$,

Tersoff potential for type 3 atoms:

$$E = \frac{1}{2}\sum_{i}\sum_{j \neq i} V_{ij} \tag{9}$$

where V_{ij} represents bonding energy between atoms

LJ/Cut potential for interaction between type 1-2, 1-3, and 2-3 atoms:

$$U\left(r\right) = 4\varepsilon\left[\left(\frac{\sigma}{r}\right)^{12} - \left(\frac{\sigma}{r}\right)^{6}\right] \tag{10}$$

Here ε denotes the depth of potential well and σ denotes the finite distance at which the inter-particle potential remains zero. Respective values denoting ε and σ are taken as 3.92KJ/mol and 80Å during this simulation study. After the establishment of the model, the nanoparticle pair was first equilibrated at room temperature (300 K). Subsequently, the simulations were performed using various process parameters to examine and analyze the consolidation of kinetics during the process of sintering. As the equilibrium ends, external heat is provided to the system. To ensure the completion of the process, external heat is provided at regular intervals.

CONCLUSION

This chapter focuses on the importance of multiscale modeling in AM processes. Using phase field and atomistic modeling, simulations are performed to understand the necking behavior of AlSi10Mg in one of the AM processes (DMLS). This chapter has disclosed certain research gaps during the DMLS process where additional research is required. Major gaps are as follows:

1. Development of multiscale models of DMLS, as it will provide researchers deeper insight into processes like sintering and diffusion.
2. Studies on the effect of important parameters like hatch spacing, hatching pattern as well as laser beam are neglected over past years which could provide an optimized selection of process parameters.
3. Quantitative analysis of different powders under conditions like lack of fusion can help reducing incomplete melting and morphological disruptions.

ACKNOWLEDGMENT

This research received no specific grant from any funding agency in the public, commercial, or not-for-profit sectors.

REFERENCES

Ahmed, K., Allen, T., & El-Azab, A. (2016). Phase field modeling for grain growth in porous solids. *Journal of Materials Science*, *51*(3), 1261–1277. doi:10.100710853-015-9107-9

Balla, V. K., Bose, S., & Bandyopadhyay, A. (2008). Processing of bulk alumina ceramics using laser engineered net shaping. *International Journal of Applied Ceramic Technology*, *5*(3), 234–242. doi:10.1111/j.1744-7402.2008.02202.x

Bandyopadhyay, A., Krishna, B., Xue, W., & Bose, S. (2009). Application of laser engineered net shaping (LENS) to manufacture porous and functionally graded structures for load-bearing implants. *Journal of Materials Science. Materials in Medicine*, *20*(S1), 29–34. doi:10.100710856-008-3478-2 PMID:18521725

Cahn, J. W., Elliott, C. M., & Novick-Cohen, A. (1996). The Cahn-Hilliard equation with a concentration dependent mobility: Motion by minus the Laplacian of the mean curvature. *European Journal of Applied Mathematics*, *7*(3), 287–301. doi:10.1017/S0956792500002369

Corbin, D. J., Nassar, A. R., Reutzel, E. W., Beese, A. M., & Kistler, N. A. (2017). Effect of directed energy deposition processing parameters on laser deposited Inconel® 718: External morphology. *Journal of Laser Applications*, *29*(2), 022001. doi:10.2351/1.4977476

Denoual, C., & Vattré, A. (2016). A phase field approach with a reaction pathways-based potential to model reconstructive martensitic transformations with a large number of variants. *Journal of the Mechanics and Physics of Solids*, *90*, 91–107. doi:10.1016/j.jmps.2016.02.022

Dinda, G. P., Dasgupta, A. K., & Mazumder, J. (2009). Laser aided direct metal deposition of Inconel 625 superalloy: Microstructural evolution and thermal stability. *Materials Science and Engineering A*, *509*(1-2), 98–104. doi:10.1016/j.msea.2009.01.009

Frazier, W. E. (2014). Metal additive manufacturing: A review. *Journal of Materials Engineering and Performance*, *23*(6), 1917–1928. doi:10.100711665-014-0958-z

Gandin, C. A., & Rappaz, M. (1994). A coupled finite element-cellular automaton model for the prediction of dendritic grain structures in solidification processes. *Acta Metallurgica et Materialia*, *42*(7), 2233–2246. doi:10.1016/0956-7151(94)90302-6

Gu, D., & Shen, Y. (2008). Processing conditions and microstructural features of porous 316L stainless steel components by DMLS. *Applied Surface Science*, *255*(5), 1880–1887. doi:10.1016/j.apsusc.2008.06.118

Guo, N., & Leu, M. C. (2013). Additive manufacturing: Technology, applications, and research needs. *Frontiers of Mechanical Engineering*, *8*(3), 215–243. doi:10.100711465-013-0248-8

Mayer, S. (2005). Optimised mould temperature control procedure using DMLS. *EOS Whitepaper, EOS GmbH Ltd*, 1-10.

Mengucci, P., Barucca, G., Gatto, A., Bassoli, E., Denti, L., Fiori, F., Girardin, E., Bastianoni, P., Rutkowski, B., & Czyrska-Filemonowicz, A. (2016). Effects of thermal treatments on microstructure and mechanical properties of a Co-Cr-Mo-W biomedical alloy produced by laser sintering. *Journal of the Mechanical Behavior of Biomedical Materials*, *60*, 106–117. doi:10.1016/j.jmbbm.2015.12.045 PMID:26803005

Nandy, J., Sarangi, H., & Sahoo, S. (2019). A Review on Direct Metal Laser Sintering: Process Features and Microstructure Modeling. *Lasers in Manufacturing and Materials Processing*, *6*(3), 280–316. doi:10.100740516-019-00094-y

Nandy, J., Yedla, N., Gupta, P., Sarangi, H., & Sahoo, S. (2019). Sintering of AlSi10Mg particles in direct metal laser sintering process: A molecular dynamics simulation study. *Materials Chemistry and Physics*, *236*, 121803. doi:10.1016/j.matchemphys.2019.121803

Panda, B. K., & Sahoo, S. (2018). Numerical simulation of residual stress in laser based additive manufacturing process. *IOP Conference Series. Materials Science and Engineering*, *338*(1), 012030. doi:10.1088/1757-899X/338/1/012030

Panda, B. K., & Sahoo, S. (2019). Thermo-mechanical modeling and validation of stress field during laser powder bed fusion of AlSi10Mg built part. *Results in Physics*, *12*, 1372–1381. doi:10.1016/j.rinp.2019.01.002

Plimpton, S., Paul, C., & Aidan, T. (2007). LAMMPS-large-scale atomic/molecular massivelyparallel simulator. *Sandia National Laboratories*, *18*, 43.

Samantaray, M., Sahoo, S., & Thatoi, D. (2018). Computational modeling of heat transfer and sintering behavior during direct metal laser sintering of AlSi10Mg alloy powder. *Comptes Rendus. Mécanique*, *346*(11), 1043–1054. doi:10.1016/j.crme.2018.08.006

Satpathy, B. B., Nandy, J., & Sahoo, S. (2018). Investigation of consolidation kinetics and microstructure evolution of Al alloys in direct metal laser sintering using phase field simulation. *IOP Conference Series. Materials Science and Engineering*, *338*(1), 012045. doi:10.1088/1757-899X/338/1/012045

Shen, J., & Yang, X. (2010). Numerical approximations of allen-cahn and cahn-hilliard equations. *Discrete and Continuous Dynamic Systems*, *28*(4), 1669–1691. doi:10.3934/dcds.2010.28.1669

Tolochko, N. K., Mozzharov, S. E., Yadroitsev, I. A., Laoui, T., Froyen, L., Titov, V. I., & Ignatiev, M. B. (2004). Balling processes during selective laser treatment of powders. *Rapid Prototyping Journal*, *10*(2), 78–87. doi:10.1108/13552540410526953

Unocic, R. R., & DuPont, J. N. (2004). Process efficiency measurements in the laser engineered net shaping process. *Metallurgical and Materials Transactions. B, Process Metallurgy and Materials Processing Science*, *35*(1), 143–152. doi:10.100711663-004-0104-7

ADDITIONAL READING

Conner, B. P., Manogharan, G. P., Martof, A. N., Rodomsky, L. M., Rodomsky, C. M., Jordan, D. C., & Limperos, J. W. (2014). Making sense of 3-D printing : Creating a map of additive manufacturing products and services. *Addit Manuf*, *1*, 64–76. doi:10.1016/j.addma.2014.08.005

Gu, D. D., Meiners, W., Wissenbach, K., & Poprawe, R. (2012). Laser additive manufacturing of metallic components: Materials, processes, and mechanisms. *International Materials Reviews*, *57*(3), 133–164. doi:10.1179/1743280411Y.0000000014

Gupta, P., Pal, S., & Yedla, N. (2016). Molecular dynamics based cohesive zone modeling of Al (metal)-Cu50Zr50 (metallic glass) interfacial mechanical behavior and investigation of dissipative mechanisms. *Materials & Design*, *105*, 41–50. doi:10.1016/j.matdes.2016.05.054

Nandy, J., Sarangi, H., & Sahoo, S. (2017). Modeling of microstructure evolution in direct metal laser sintering: A phase field approach. *IOP Conference Series. Materials Science and Engineering*, *178*(1), 012028. doi:10.1088/1757-899X/178/1/012028

Nandy, J., Sarangi, H., & Sahoo, S. (2018). Microstructure evolution of Al-Si-10Mg in direct metal laser sintering using phase-field modeling. *Advances in Manufacturing*, *6*(1), 107–117. doi:10.100740436-018-0213-1

Sahoo, S. (2017). Simulation study on rapid solidification of eutectic Al-Cu alloy: A molecular dynamics approach. *Int. J. Computational Materials Science and Surface Engineering*, *7*, 18–25.

Sahoo, S. (2019). An approach toward multiscale modeling of direct metal laser sintering process. *Metal Powder Report*, *74*(2), 72–76. doi:10.1016/j.mprp.2019.01.005

Samantaray, M., Sahoo, S., & Thatoi, D. N. (2019). Modeling of thermal and solidification behavior during laser additive manufacturing of AlSi10Mg alloy powders and its experimental validation. *Journal of Laser Applications*, *31*(3), 032019. doi:10.2351/1.5104318

Sames, W. J., List, F. A., Pannala, S., Dehoff, R. R., & Babu, S. S. (2016). The metallurgy and processing science of metal additive manufacturing. *International Materials Reviews*, *61*(5), 315–360. doi:10.1080/09506608.2015.1116649

KEY TERMS AND DEFINITIONS

Additive Manufacturing: It is an advance manufacturing process which fabricates components in a layered fashion by fusing the powder particles.

Coalescence: The particles agglomerates changes from uneven shapes to more monolithic, spherical shapes.

Direct Metal Laser Sintering: It is powder based additive manufacturing process that has the potential to make the metallic components from powder particles.

LAMMPS: Large scale atomic/molecular massively parallel simulator platform used for atomistic simulation of material systems.

Laser Energy Density: It is defined as laser energy concentrated per unit volume.

Sintering: The process where compaction of powder particles takes place into solid mass by the application of heat and pressure without melting.

Solidification: Solidification, also known as freezing, is a phase change of matter that results in the production of a solid, when the temperature of a liquid is lowered below its freezing point.

APPENDIX

Table 2. Nomenclature

Symbols	Meaning (Units)
k_α, k_β, and k_γ	thermal conductivities in the x, y and z directions (W/m K)
T	Temperature (K)
$\dot{Q}(x,y,z,t)$	rate of the internal heat generation per unit volume (W/m^3)
ρ	density of the metal powder (kg/m^3)
C_p	specific heat capacity (J/kg K)
σ	stress tensor (N/m^2)
q	Heat flux (W/m^2)
A	laser absorptance of the powder system (AU)
P	Laser power (W)
r_0	radius of the laser beam (mm)
r	Radial distance between the laser beam and the center of the spot generated on the top surface of the powder bed (mm)
$\varnothing$	order parameter
∇	Laplacian operator (divergence)
f	bulk free energy
$\varepsilon_\varnothing$	gradient energy coefficient
M	Mobility
τ	time scale
$A(\varphi)$	anisotropy parameter
$A'(\varphi)$	derivative of $A(\varphi)$ with respect to the order parameter
W_φ	length scale
L	latent heat of fusion
T_m	melting temperature (K)
T_c	critical temperature (K)
H	nucleation barrier
$\eta'(x,T_c)$	stochastic noise function emulating thermal fluctuations
α	thermal diffusion coefficient
$h'(\varphi)$	smooth derivative function with limits $h(0)) = 0$ and $h(1) = 1$
C_p	specific heat capacity at constant pressure
ρ_i	electron density
$\varnothing_{i,j}$	two-body central potential between atom i and j
$r_{i,j}$	separation distance (Å)
$F_i(\rho_i)$	embedding energy of atom i with the electron density due to all its neighbors
V_{ij}	bonding energy between the atoms
ε	depth of the potential well (KJ/mol)
σ	finite distance at which the inter-particle potential is zero (Å)

Chapter 13
Powder Bed Fusion Additive Manufacturing of NI-Based Superalloys:
Applications, Characteristics, and Limitations

Evren Yasa
Eskisehir Osmangazi University, Turkey

Ozgur Poyraz
https://orcid.org/0000-0001-9892-5738
Eskişehir Technical University, Turkey

ABSTRACT

Emerging additive manufacturing technologies have been gaining interest from different industries and widened their fields of application among aerospace and defense. The introduction of powder bed fusion processes was one of the significant developments in terms of direct metal part manufacturing of different materials and complex geometries, presenting good properties, and decreasing the need for tooling to allow fast product development as well as small-volume production. In this respect, nickel-based superalloys are one of the most employed material groups for aerospace and defense applications due to their mechanical strength, creep, wear, and oxidation resistance at both ambient and elevated temperatures. Nevertheless, the use of some materials has not become widespread due to several reasons such as processing difficulties, absence of design criteria or material properties. This chapter presents a comprehensive benchmark for powder bed fusion additive manufacturing of nickel-based superalloys considering applications, characteristics, and limitations.

DOI: 10.4018/978-1-7998-4054-1.ch013

INTRODUCTION

Powder bed fusion (PBF) additive manufacturing (AM) technologies, which are categorized as one of the seven main AM groups according to International Standards Organization (2015), include various processes employing different energy sources such as laser, electron or heat. Electron Beam Melting (EBM), Selective Laser Melting (SLM), Selective Laser Sintering (SLS) and Selective Heat Sintering (SHS) are the most commonly known powder bed fusion processes. SLM is also known with different commercial names from different machine vendors such as Direct Metal Laser Melting (DMLM), Direct Metal Laser Sintering (DMLS) and Laser Cusing, Laser-based powder fusion, etc. Like other AM technologies, these processes are used for production of 3 dimensional (3D) parts based on a Computer Aided Design (CAD) model file and can be applied for a wide range of powder materials. Many metallic alloys are included among the wide material range such as aluminum, cobalt, copper, nickel, steel and titanium alloys. With the proper use of PBF AM technologies, many of these alloys can be processed to obtain fully functional parts, prototypes and tools. The technology may offer various advantages such as low material consumption, good material and dimensional properties, design freedom to enable production of complex geometries with improved functionality and internal features, decreased need for tooling, reduced production times and thus easy transition from design to manufacturing and testing. However, it is only possible to benefit from all advantages offered by AM technologies with certain alloys while others still remain as a challenge for research and industrial organizations. Nickel-based superalloys, which are of great interest to aerospace and defense industries, are such an example. Initially developed for conventional manufacturing processes such as casting, forging or welding, several nickel based superalloys are easily adopted to PBF AM technologies, while the scientific or industrial trials still remain unsuccessful for other ones. Diverse reasons lead to problems for adopting nickel-based superalloys for PBF AM technologies such as absence of materials in powder form fulfilling the AM requirements, processing difficulties, material defects and cracks, residual stresses, the need for support structures, lack of knowledge on property enhancement processes (heat treatment, hot isostatic pressing), etc. On top of these, there is a lack of material property – processibility benchmarks, and material selection - design criteria are not fully set for PBF AM of nickel-based superalloys. This chapter presents a comprehensive benchmark for PBF AM of nickel-based superalloys considering applications, characteristics and limitations. The rest of this chapter is arranged as follows: Development of nickel based superalloys and PBF AM technologies are presented in the background section. The subsequent section describes the PBF AM in detail and discusses results of research on various nickel based superalloy types by demonstrating the issues and problems. Following this, the next section provides solutions and recommendations for the encountered problems. Last two sections emphasize the future research directions and conclude the chapter by summarizing the presented information.

BACKGROUND

History, Metallurgy and Types of Nickel Based Superalloys

Materials have a direct impact on the entire lifecycle of an aircraft or engine component, from initial design phase to disposal at the end of its life (Mouritz, 2012). They virtually affect many aspects including cost, performance, safety and operational life. In this regard, new materials and processing methods have been introduced since the beginning of 20th century. As an example, superalloys, still under development, are of great interest both for scientific institutions and industrial companies for being used particularly in jet engines but also in land/marine gas turbines and rockets. Although material science efforts on high temperature applications were initiated in the second decade of 20th century and patents issued for nickel-chrome type alloys with aluminum and/or titanium additions, limited strength capabilities of those historical alloys could not respond to the increased needs of designers (Donachie & Donachie, 2002). As a result of higher demands of Second World War period, the earliest material with the term "superalloy" was used in turbo chargers and aircraft engines (Bowman, 2000). Since then, the utilization of superalloys continuously increased and they are now being used for different component applications such as disks, shafts, cases, blades, vanes, combustors, turbochargers, thrust reversers, exhaust valves, ducts, pumps and many others. Today nearly half of the commercial turbofan engine components are made of nickel based superalloys. For instance, PW4000 of Pratt & Whitney has a superalloy utilization rate of 39% (Paulonis & Schirra, 2001).

The term "superalloy" is currently used for alloys maintaining a good level of mechanical strength at raised temperatures and they are categorized according to various major elements of nickel, nickel-iron and cobalt to have a favorable alliance of mechanical resistance and surface properties at the same time. Chronologically, the nickel-iron based superalloys were the first developed group to contain a high weight ratio (wt. %) of iron (Fe) which behaves as a mutual base material with Nickel (Ni) (Donachie & Donachie, 2002). A typical alloy grade of this category is Incoloy® 800, which was introduced in 1950 as a result of research activities to reduce the use of nickel (Reed, 2006). Most of the Ni-Fe based alloys, commonly used as wrought materials, feature a solid-solution strengthening mechanism and their precipitates have spherical structure. To enable the use at elevated temperature applications, Cobalt (Co) based superalloys utilizes Co as the base material and common strengthening mechanism is having solid solution elements as well as the carbide formation.

Depending on the user requirements, Ni-based superalloys might be used either in cast or wrought forms. Further application of this group of superalloys was developed for powder metallurgy as well as for AM in recent years. Having a face-centered-cubic (FCC) matrix; this group is subcategorized depending on strengthening composition, phases and strengthening mechanism. It is paramount to appreciate these groups in order to analyze any problems which may occur during or after PBF AM. Like in other metallic alloys, chemical composition of the Ni-based superalloys influences the phases and the array of phases influences the microstructure. In this regard, most of the Ni-based superalloys include common elements of Ni, Fe, Cr, C, Nb, Ti, Al, Mo, V and W. However, the weight ratios of these alloying elements differ from each other and this leads to precipitation of different phases in different superalloys. The solid solution austenitic FCC gamma (γ) phase constitutes the matrix of the alloy and it is essential for all Ni-based superalloys independent of the subcategory (Sims, 1984). Together with this, intermetallic gamma prime (γ') phase has also an ordered FCC structure and found in Ni-based superalloys having a considerable amount of Ti or Al, forming Ni_3Ti or Ni_3Al. These alloying elements take the corners of the unit cell, surround Ni atoms, and generate an ordered plane and strength the alloy through thermal process or following heat treatments such as aging. However, Ni-based superalloys may have a wide service temperature range and additional phases are needed for the instability of γ' between 600°C-850°C. In this context, body centered tetragonal gamma double prime (γ'') can enhance the material strength up to 650°Cwhich has Nb or V to form Ni_3Nb or Ni_3V. A typical and common alloy formed through this is Inconel 718 (Reed, 2006).Carbides, which are mostly undesirable phases among other metallic alloys, can form at the grain boundaries of Ni-based superalloys to inhibit grain boundary motion and to stabilize the material at high temperatures (Sabol, 1969).Apart from these, Ni-based superalloys may also be solid solution strengthened through solid solution of elements else typical examples of these alloys include Inconel 625 and Hastelloy-X.

Powder Bed Fusion Additive Manufacturing Technologies

Although the PBF AM technologies are recognized with different commercial terms, general principles of this group of technologies remain the same. This group of technologies produces the parts by fusing a powder layer within a definite cross sectional zone defined by the sliced CAD model of workpiece with the scanning path. A basic classification of this group of technologies starts with the fusion mechanism and classifies the technologies as partial melting and full melting processes. Full melting processes (DMLM/ DMLS /SLM, EBM) demonstrate a sufficient level of physical and mechanical properties comparing to those of conventional materials while the partial melting or sintering processes (SLS, SHS) need additional processes like curing in order to enhance those (Kruth et al., 2005). Owing to this fact, the use of SLS technology is not in the same level of full melting processes and is less common among PBF AM of Ni-based superalloys. Thus, this chapter focuses only on PBF AM technologies based on full melting.

Full melting PBF AM technologies are classified based on the applied energy source and two common energy sources are laser and electron beams. Laser powder bed fusion additive manufacturing (L-PBF AM) mostly uses one or multiple laser beams to melt the powder particles. Nowadays, fiber lasers are the most commonly employed laser type and their power extent can be between 100 W and 1000 W with an average wavelength of 1050 nm to 1100 nm and a beam size of 50 μm – 100 μm at focus. Classified as a low temperature process, L-PBF AM can process the parts in ambient temperatures or below 250°Cjust by heating the base plate (build platform) with the help of embedded resistances for several reasons such as a good connection of the very first consolidated layers to the base plate by removing the moist from metal powder. Another environmental condition of L-PBF AM is the use of gases like argon or nitrogen as a protective atmosphere for the melt pool. The particle size of the metal powder used in L-PBF AM mostly ranges from 15 μm to 45 μm. Having all these easily adjustable conditions, L-PBF AM is a suitable process to have larger machines and today the maximum build volumes of L-PBF AM machines from various machine manufacturers can go up to 400x800x500 mm. Figure 1 displays the schematic of basic principles of the L-PBF AM.

Electron beam powder bed fusion additive manufacturing (E-PBF AM) on the other hand employs very high energy electron beam to melt the metallic powder. The power used in E-PBF AM processes can be as high as 3 kW with a limited part resolution due to the larger focus beam size of 180 μm. The particle size of the metal powders used for E-PBF AM are between 45 μm and105 μm in accordance with the larger minimum focus size and the melt pool width/depth associated with it. Moreover, the powder cost is reduced with coarser powders. The consequences of using high power and large powder particle sizes, brings the risk of spreading the powder and shadowing the electron beam itself. Several measures are taken to pack the powder bed and eliminate this risk such as applying vacuum to building chamber or preheating the powder before melting in order to avoid powder smoke in the chamber. In this regard, the applied vacuum values for E-PBF AM are typically around 5 x10^{-5} mbar and additional to this, 4 x10^{-3} mbar of helium inert gas is introduced to ensure a clean and controlled building chamber. As mentioned within measures, E-PBF AM processes the parts in elevated temperatures as high as 1100°C, depending on the processed material, by heating all the powder bed with the help of electron beam itself which can scan in high speeds going up to 15 m/s. That's why this process is considered as a high temperature

Figure 1. Principles of laser powder bed fusion additive manufacturing
(Poyraz, 2018)

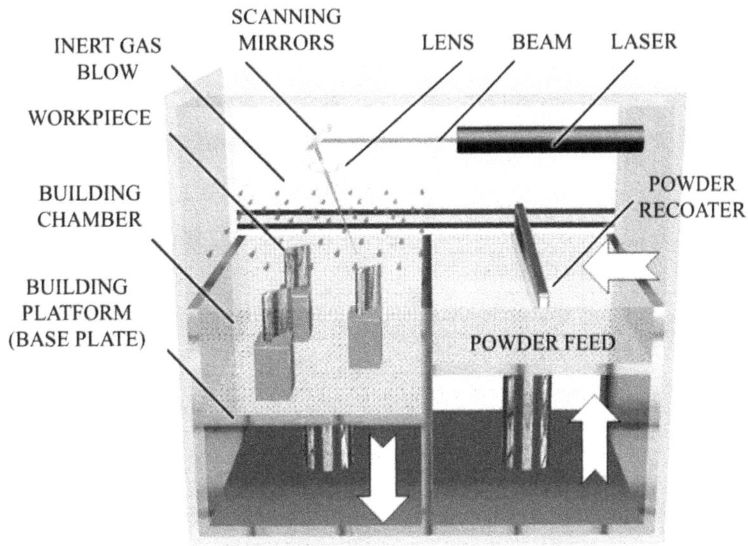

process. The disadvantage of applying counter measures of vacuum and high temperature preheating in E-PBF AM process, leads to smaller machine sizes by influencing the technical and economic feasibility of using larger machines in a negative way. Currently, the E-PBF AM machine sizes to process superalloys are limited to only Ø250 mm which is nearly the half size of the L-PBF AM machines. On the other hand, preheating to high temperatures has its own advantages of decreasing the thermal gradient between the ambient temperature and melting point and thus reducing the residual stresses. Moreover, this reduces the needs of manufacturing supports for heat dissipation.

The major differences of L-PBF AM and E-PBF AM processes are summarized in Table 1. There are plenty of works in the literature comparing L-PBF parts to E-PBF parts in terms of different material properties although both methods use computer controlled melting patterns and a layerwise processing principle (Zhong et al., 2017). For example, for nuclear applications, a comparative study was carried out for AISI 316L stainless steel powder in terms of mechanical properties, engineering problems and manufacturing process (Zhong et al., 2017). The study results clearly show that both processes led to almost fully dense parts. Both at room and high temperatures, the mechanical attributes of SLM parts possess excellent strength and average ductility whereas the EBM parts show better ductility but relatively reduced ultimate tensile strength. In terms of needs for supports, EBM was better whereas the dimensional accuracy and surface quality, SLM was considered as more advantageous. In another study where the compressive properties of parts produced by either SLM or EBM were compared, it was concluded that the SLM printed parts exhibit higher strength, probably due to geometric irregularities of EBM printed parts, and the EBM printed parts were more sensitive to the specimen size (Xiao et al. 2019). Regarding biomedical applications, various aspects such as the adequacy to manufacture small details in the form of lattice structures and surface quality as well as microstructure, are evaluated in different studies (Harun et al., 2018). Fatigue properties of SLM and EBM parts, which are critical for structural applications where a cyclic loading is present, are also addressed in the literature. In a study where high cycle fatigue was the focus, it is concluded that taking the roughness into account SLM parts gave longer

Table 1. Comparison of laser and electron beam powder bed fusion additive manufacturing processes

	L-PBF	E-PBF
Energy Source	Laser	Electron beam
Preheating	Up to 250°C	Up to 1100°C
Processing Atmosphere	Inert gas (Argon or Nitrogen)	Vacuum (Supported by Helium)
Applicable Materials	Various alloys of aluminum, cobalt -chromium, ferrous, nickel based superalloys, steels, titanium alloys, precious metals.	Various alloys of cobalt -chromium, nickel based superalloys, titanium alloys and titanium aluminates.
Powder Size	15 µm - 45 µm	45 µm - 105 µm
Layer Thickness	20 µm - 50 µm	50 µm - 100 µm
Dimensional Accuracy	100 µm - 200 µm	400 µm - 500 µm
Surface Quality	R_a 5 µm – R_a 10 µm	R_a 25 µm – R_a 35 µm
Geometric Complexity	High	Intermediate
Melting Rates	7 cm^3/h - 70 cm^3/h	55 cm^3/h - 80 cm^3/h
Maximum Available Machine Sizes to be Used for Superalloys	Up to 500 mm by 500 mm or 400 mm by 800 mm	Up to Ø 250 mm

(Poyraz & Kushan, 2018)

lives whereas when machined, they are comparable but lower than hot rolled specimens (Vayssette et al., 2018). The results are also consistent with another study (Nicoletto et al., 2018). Thus, depending on the application, material and technical specifications, each process offers various advantages and limitations. The selection between these two powder bed AM processes shall depend on the expected results.

CHALLENGES ON POWDER BED FUSION ADDITIVE MANUFACTURING OF SUPERALLOYS

Appropriate part selection or design, together with a detailed process planning and the use of proper process parameters are vital for the success of powder bed fusion additive manufacturing of superalloys which have high material unit prices, elevated machine investment budgets and relatively more expensive engineering and labor costs due to employment of qualified personnel. The first point to be considered in the selection of target parts for powder bed fusion additive manufacturing of superalloys are their technological fit and feasibility. The criteria to be considered for technological feasibility are mainly the part size, the part material, the mechanical-metallurgical requirements of the part and the geometric-surface quality requirements provided by final product design. The estimated cost of PBF parts, including pre-processing, build and post-processing, is also a critical factor to be taken into account as well as the added value when the product portfolio is screened. Since the part sizes are mostly limited by the available machines in the market, one should focus on mechanical and metallurgical requirements of the part. Following material properties, dimensional and surface quality requirements shall also be taken into account. However, there are more opportunities to alter the obtained dimensional properties with the selection of appropriate post-processes at a compromise of increased production time. The altering may also be very challenging when intricate geometries with internal features are considered.

Therefore, the following sub-sections are targeted towards the most critical challenges encountered in the above-mentioned areas.

Processability of Materials, Metallurgical and Mechanical Properties

Processability studies for PBF AM most importantly deal with developing a proper microstructure without any pores or cracks to produce parts at or close to full density of their conventionally produced counterparts. Just after achieving the proper microstructure and density, minor improvements can be obtained by utilizing post processes such as heat treatment or hot isostatic pressing (HIP). Although proper microstructure can vary between L-PBF AM and E-PBF AM for different materials, it is essential to have microstructural integrity and thus sufficient overlap between two consecutive scans and layers. Two important influencing factors on the microstructural integrity are the energy density and scanning strategy. The energy density can be calculated as volumetric, surface or linear density for each process. The parameters used in those calculations are mainly layer thickness (μm), focus diameter (μm), hatch distance (μm), laser power (W) and scanning velocity (mm/s) for L-PBF AM and layer thickness (μm), focus offset (μm), beam current (mA) and scan speed (mm/s) for E-PBF AM. Figure 2 shows the schematic of volumetric energy density parameters for L-PBF AM.

Figure 2. Schematics of laser powder bed fusion additive manufacturing volumetric energy density parameters

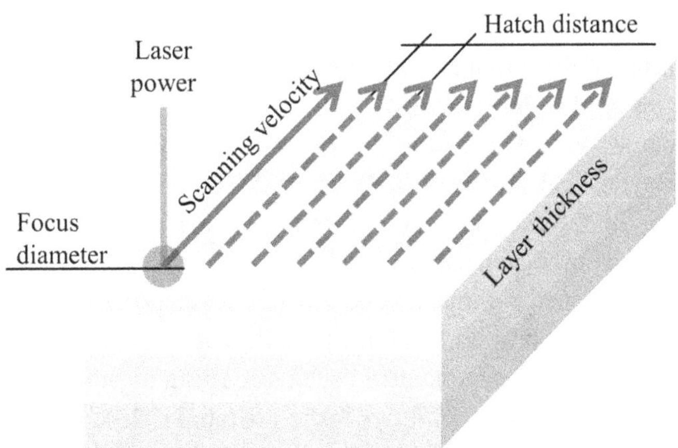

Among these parameters some of them are constraint by the used material or machine specifications. For example, the layer thickness is constrained by the powder size distribution, and focus diameter in L-PBF or the focus offset is governed by the machine specifications limiting the use of larger hatch distance in E-PBF. In this regard, most research is done for developing the optimum linear energy density ratio by changing scanning velocity/speed and focus diameter/offset. Alteration of these two values considerably affects the microstructure and mechanical attributes of metal alloys. In addition to these, various scanning strategies are applied to utilize a uniform temperature distribution along part surface and through previous layers. The applied scanning strategies include but are not limited to line scanning

Figure 3. Various scanning strategies applied to powder bed fusion additive manufacturing (Poyraz, 2018)

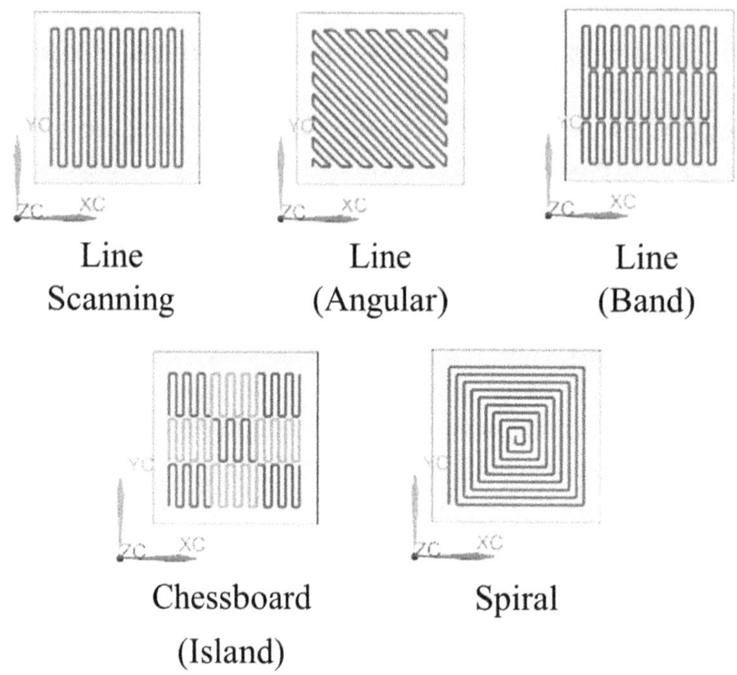

with constant or alternating angles, island (chess board) scanning and spiral scanning. Figure 3 shows various scanning strategies applied in PBF -AM.

Within the current state-of-the art, available superalloys materials for PBF AM include CM 247 LC, CMSX-4, Hastelloy X, Haynes 230, Haynes 282, Inconel 625, Inconel 713 LC, Inconel 718, Inconel 738 LC, Inconel 939, Nimonic 263, Rene 41, Rene 80, Rene 142 and Waspalloy. Table 2 shows the major alloying elements of PBF AM produced superalloys.

Although most of the mentioned parameters and process development approaches can be applicable for the listed superalloys, certain properties of these materials may reveal challenges for PBF AM. Initially developed for conventional manufacturing such as casting or billet casting and forging, one of the major problems of many Ni-based superalloys is the potential risk of weld cracking and this risk is directly related to PBF AM, analogous to welding process. The recent studies relate the weld cracking susceptibility of superalloys to high fractions of Gamma Prime (γ'). Aluminum (Al) and Titanium (Ti) are used to define an empirical limit for weld cracking (Donachie & Donachie, 2002). A widened approach includes Chromium (Cr) and Cobalt (Co) elements on top of Al and Ti. (Ott et al., 2005). Figure 4 shows a reproduction (modified from Ott et al., 2005) of weldability assessment diagram including superalloy grades available for PBF AM.

This type of cracking is defined as strain age, reheat or post weld heat treatment cracking and it is a special type of solid-state cracking to be seen among precipitation hardened nickel based superalloys. It occurs due to residual stress relief during reheating of the material and it coincides with the precipitation of hardening phases (Hanning & Andersson, 2016). In PBF AM, reheating is done by the consecutive scans and layers neighboring to the completed scan line and leads to reheat cracking. Figure 5 shows

Table 2. Major alloying elements (in weight %) of powder bed fusion additively manufactured superalloys

Alloy	Cr	Co	Mo	W	Al	Ti	Ta	Fe	Nb	Re	Hf	Zr	Ni
CM 247 LC	8.0	9.3	0.5	9.5	5.6	0.7	3.2	-	-	-	1.4	-	Bal.
CMSX-4	6.5	9.6	0.6	6.4	5.6	1.0	6.5	-	-	3.0	0.1	-	Bal.
HASTELLOY X	22.0	1.5	9.0	0.6	0.25	-	-	18.5	-	-	-	-	Bal.
HAYNES 230	22.0	-	2.0	14.0	0.3	-	-	-	-	-	-	-	Bal.
HAYNES 282	20.0	10.0	8.5	-	1.5	2.1	-	0.7	-	-	-	-	Bal.
INCONEL 625	21.5	-	9.0	-	0.2	0.2	-	2.5	3.6	-	-	-	Bal.
INCONEL 713LC	12.0	-	4.5	-	5.9	0.6	-	-	2.0	-	-	0.1	Bal.
INCONEL 718	19.0	-	3.0	-	0.5	0.9	-	18.5	5.1	-	-	-	Bal.
INCONEL 738LC	16.0	8.5	1.75	2.6	3.4	3.4	1.75	-	0.9	-	-	-	Bal.
INCONEL 939	22.4	19.0	-	2.0	1.9	3.7	-	-	1.0	-	-	0.1	Bal.
NIMONIC 263	20.0	20.0	5.9	-	0.5	2.1	-	-	-	-	-	-	Bal.
RENE 41	19.0	11.0	1.0	-	1.5	3.1	-	-	-	-	-	-	Bal.
RENE 80	14.0	9.0	4.0	6.0	3.0	4.7	-	-	-	-	0.8		Bal.
RENE 142	6.8	12.0	1.5	4.9	6.15	-	6.35	-	-	2.8	1.5	-	Bal.
WASPALLOY	19.5	13.5	4.5	-	1.3	3.0	-	-	-	-	-		Bal.

(Reeds, 2006)

typical cracks for two sets of process parameters on CM 247 LC material fabricated by L-PBF AM (Carter, 2013).

While the cracking susceptibility related to chemical composition and phases is defined as strain age cracking (categorized among solid state cracking), other cracking mechanisms of heat affected zone

Figure 4. Weldability assessment of superalloys used in PBF AM
(Adapted from Ott et al., 2005).

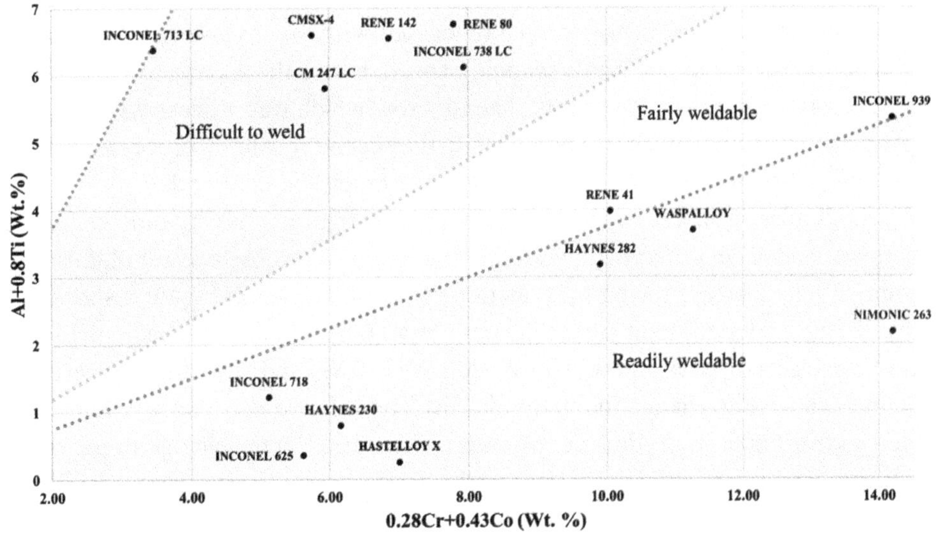

Figure 5. Crack formations on L-PBF AM fabricated samples of CM 247 LC material (Carter, 2013)

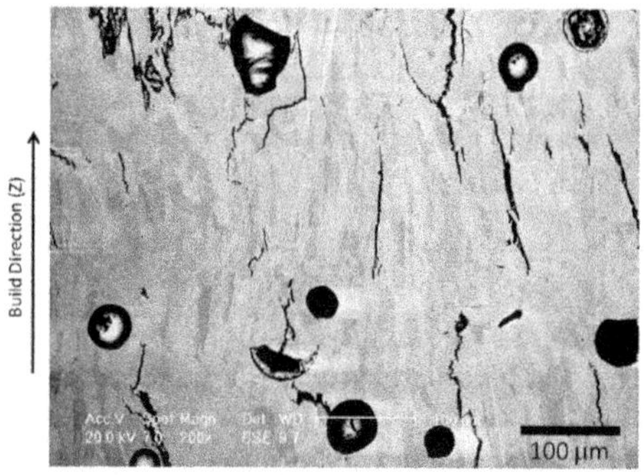

Laser power 200 W, scanning speed 400 mm/s

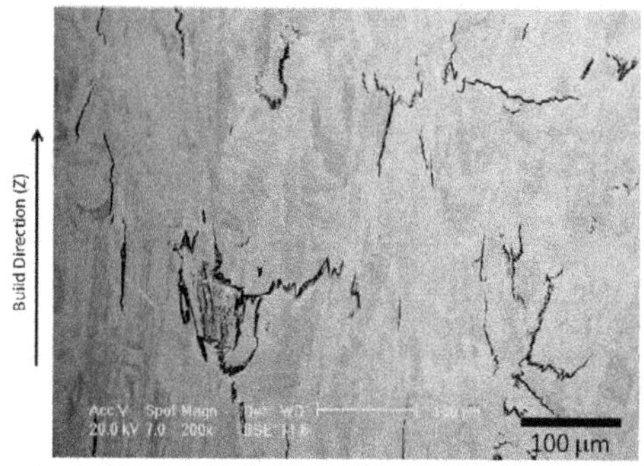

Laser power 150 W, scanning speed 600 mm/s

liquation cracking and solidification cracking may also be applied to PBF AM of superalloys. In this respect, liquation cracking takes place afters the material is heated to an elevated temperature below the melting point with a high heating rate and some phases are not fast enough to get into solid solution but remain as low melting point phases at grain boundaries (Henderson et al., 2004). Later on, they can melt and enervate the grain boundaries to propagate cracks under service conditions. On the other hand, solidification cracking, which was reported to be seen under low welding speeds (Dye et al., 2001) and high effective power upon cooling of two phase solid-liquid region as a result of dendrite interlocking and semi-continuous grain boundary is less likely to be seen in PBF AM which utilizes high scanning speeds.

The superalloy types which are considered as readily weldable (For example Inconel 625) alloys mostly suffer from pores instead of cracks. For this problem, which is relatively easy to solve, it is necessary to select and optimize the mentioned energy input parameters correctly. Independent of the different energy density approaches such as volumetric, surface or liner, the selected parameters should facilitate adequate level of energy for melting the material. While insufficient energy may reveal un-melted zones to act as pores, the use of excessive energy can lead to uncontrolled melt pools or unnecessary power consumption. Figure 6 shows light optical microstructure photos of Inconel 625 L-PBF AM sample under 500X zoom. The figure includes two directions as perpendicular and parallel to build direction. First row of photos shows the condition of the sample processed with 40 µm layer thickness, 110 µm hatch distance, 100 W laser power and 600 mm/s scanning speed, where the second row has different power and speed parameters of 350 W and 860 mm/s. The first row which was process with an energy density of 37.9 J/mm^3 has visually distinguishable pores while the second row to be processed with 92.5 J/mm^3 is nearly fully dense. It is also important to consider that empirical energy density optimization may not guarantee overall microstructure of the part and minor changes in porosity ratio can be experienced with same energy densities but different power/scanning speed values. As a result of these, unnecessary power consumption may increase the part production costs. The graph in Figure 6 shows the changing of relative part density under constant layer thickness of 40 µm and hatch distance of 110 µm. Various combinations of three sets of laser power (100 W, 200 W, 350 W) and scanning speed (600 mm/s, 860 mm/s, 1060 mm/s) shows the change in energy density and relative part density. Through the graph, it can be perceived that under a certain level of laser power (100 W) the change in the scanning speed and thus the energy density drastically influences the relative part density. However, above a certain

Figure 6. Energy density parameters and influences on part porosity/density for L-PBF AM Inconel 625 samples.

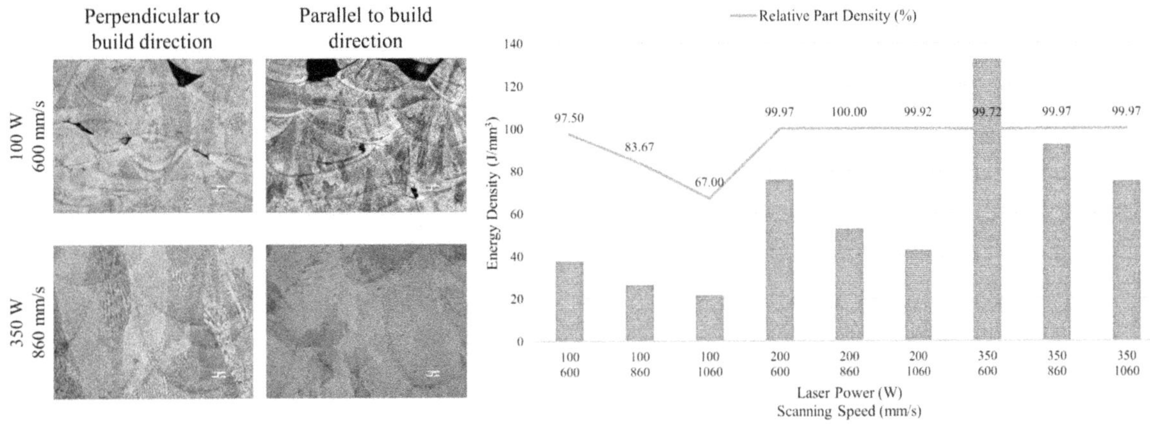

laser power (200 W), the change in the scanning speed and thus the energy density does not have much influence on the final part density.

The energy density values for core (in-skin) regions of various readily weldable superalloys such as Inconel 625 (Yasa et al., 2016), Inconel 718 (Poyraz & Kushan, 2018), Hastelloy X (Han et al., 2018), Haynes 230 (Bauer et al., 2015) and Haynes 282 (Ramakrishnan & Dinda, 2019) have a certain range changing between 55 J/mm^3 – 70 J/mm^3 and within this range the materials can be processed to an acceptable level of porosity (less than 0.5%), a fine microstructure and thus high mechanical properties. Higher energy density levels can also be used for these alloys which does not represent any risk of cracking. Fairly weldable alloys such as Inconel 939 or less difficult to weld alloys such as Inconel 738 LC can be processed with slightly higher energy density of 70 J/mm^3 - 75 J/mm^3, while maintaining these values by using the power or hatch distance levels different to those of readily weldable alloys (Shaikh, 2018, Perevoshchikova et al., 2017). The situation with difficult to weld alloys such as CM 247 LC or Inconel 713 LC is not straight forward like these. Power input and scanning speed was found to be the greatest influencing parameters for these alloys and for a crack free process it is recommended to utilize low powers with high scanning speeds. But this approach for optimum crack free parameters are contrary to the maximum part density target and further research is needed on difficult to weld alloys (Carter, 2013, Harrison, 2016). Current major research directions to overcome these challenges are and benefit from lower thermal gradients to decrease risk of cracks due to micro residual stresses. Figure 7 shows two types of fractal scan strategies used in L-PBF AM processing of difficult to weld CM 247 LC superalloy.

After overcoming challenges on porosity and microcracks, further targets are followed such as maintaining the required mechanical properties. In this regard, careful consideration is needed for superalloys which are expected to service under difficult conditions like high temperatures. For this reason, the evaluation should include room and high temperature mechanical attributes of elasticity modulus, yield strength, tensile strength and creep resistance. Since most of these alloys are suitable for precipitation hardening, necessary post processing (heat treatment and/or HIP) should be conducted at this step. Typical heat treatments applied to superalloys after PBF AM includes stress relieving, solution treatment, annealing and aging. These heat treatments can be used as single process steps and/or a combination of two to four according to requirements. Even though the subject superalloy is not a

Figure 7. Fractal scan strategies used in L-PBF AM processing of difficult to weld CM 247 LC superalloy (Adapted from Catchpole-Smith et al., 2017)

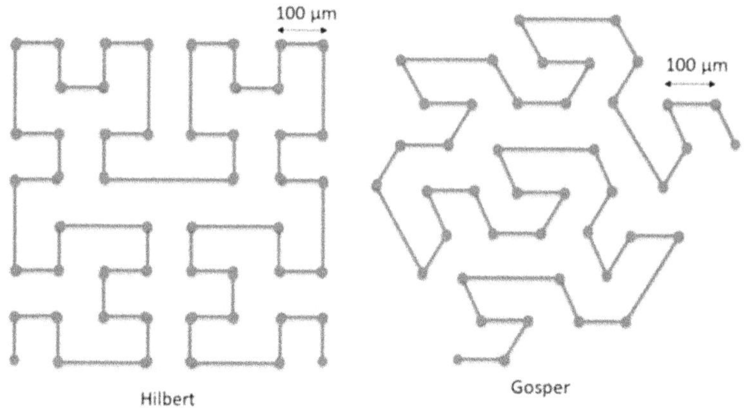

precipitation hardenable alloy, it is recommended to apply stress relieving to minimize residual stresses after PBF AM processes which induce higher thermal gradients to material. Another advantage of stress relieving treatment is the temperature which is usually below annealing or recrystallization and requires simple heat treatment furnaces to be used close to PBF AM facilities. Solution treatment is mostly used prior to aging and it intends to homogenize microstructure, dissolved secondary phases and recrystallize material. Annealing of superalloys, which have similar heating cycles to solution treating for hardenable superalloys, is mostly used to improve ductility to enhance better forming and machining properties. Precipitation treatments are used to strengthen age-hardenable alloys by causing the precipitation of phases from the supersaturated matrix that is developed by solution treating and retained by rapid cooling from the solution treating temperature (Donachie & Donachie, 2002). The standard heat treatment cycles for conventionally produced superalloys may also be applied to PBF AM produced counterparts and enhance material properties. A typical example to this is Inconel 718 material and standard heat treatment cycle defined in AMS 5662, 5663 and 5664. Both L-PBF and E-PBF AM produced parts may have benefit by applying similar heat treatment cycles (solution treating and aging) as defined in the standard.Table 3 shows the enhancement in room temperature mechanical atrributes of Inconel 718 material manufactured by L-PBF and E-PBF AM.

Table 3. Comparison of mechanical properties of powder bed fusion additive manufactured Inconel 718.

	Laser Powder Bed Fusion	**Electron Beam Powder Bed Fusion**
Reference	EOS, 2016	Balachandramurthi et al., 2019
Yield Strength - MPa (As Built)	710	920
Tensile Strength - MPa (As Built)	1040	1075
Yield Strength - MPa (Solution Treated and Aged)	1200	1096
Tensile Strength - MPa (Solution Treated and Aged)	1470	1172

Micro-cracking due to thermal factors cannot always be wiped out only by optimizing the AM process for some of the difficult-to-weld nickel-based superalloys and this may present a significant challenge. In order to overcome this challenge, one of the most widely used methods to heal the internal cracks and provide a route to eliminate or minimize the porosities in powder bed fusion AM technologies is the hot isostatic pressing (HIP). With typical pressures from 400 to 2000 bar and temperatures up to 2,000°C, HIPing can improve the ductility and fatigue properties of additively manufactured parts for high-performance applications. Yet, it should be noted that defects that are seemingly eliminated during HIP may re-open during subsequent heat treatments or even during service life (Tammas-Williams et al., 2016). The reason is considered as the use of non-soluble gases with the used material. For example, argon is non-soluble in metals and the trapped gas exerts enough pressure to re-open the cracks in post-HIP treatments or during service. The HIPing conditions, temperature and pressure, are chosen so that the pressure is above the reduced yield stress at elevated temperature allowing plastic flow and creep to eliminate the porosity. The HIP pressure depends on the size and content of the pores to be closed whereas the temperature to reduce the yield stress is usually above 0.7 times the melting temperature which is also high enough for creep and diffusion. For nickel superalloys, the temperature needs to be greater than the gamma prime solvus temperature to allow for creep. The elevated temperature and the applied isostatic pressure in the HIP process lead to variations in the microstructure as well as grain size. This does not only change the mechanical properties including fatigue but also the scatter in the mechanical properties is affected. It should be noted that HIPing can only close non-surface pores and cracks.

In a study by Wang, it is shown that applying a HIP treatment after SLM for Hastelloy® X consolidates the microstructure and eliminates the defects leading to good mechanical properties including fatigue and acceptable finish (Wang, 2011). In a different study by Murr et al., the HIP treatment was applied on Inconel 625 samples produced by E-PBF. The as-built specimens displayed a columnar architecture parallel to the build axis of thin, crystallographically coincident γ^2 precipitates with grains of 20 μm in diameter and 500 μm in length. The HIP treatment dissolved the γ^2 precipitates and led to an equiaxed grain structure with an average grain diameter of 50 μm. As expected, the post-HIP samples experienced a roughly 17% decrease in hardness and 20% decrease in yield stress from the as-built state while the elongation of the samples increased by 57% which was explained by the dissolution of Ni3Nb γ^2 precipitate columnar grain structure and the development of a more equiaxed grain structure during HIP treatment (Murr et al., 2011). The HIP treatment was also applied on Inconel 718 specimens produced by L-PBF by different research groups. Amato et al. showed that in the as-built specimens, an unusual columnar microstructural architecture composed of primarily γ^2 phase precipitate columns within directionally solidified and similarly textured grains was observed. Upon HIPing under argon atmosphere at a temperature of 1163°C for 4 h prefaced by a stress relief annealing (at a temperature of 982°C for half an hour), the micro hardness (Vickers) rose to 5.5 GPa from 3.8 GPa. A variance of 5–6% was observed in the strength of cylinders. This is attributed to the orientation of the columnar γ^2 phase architecture: parallel to the z-axis cylinder axis (and also parallel to the tensile axis) and perpendicular to the x-axis cylinder axis (and also perpendicular to the tensile axis) (Amato et al. 2012). Mostafa et al. applied a different heat treatment sequence consisting of homogenization at 1100 °C for 1 h followed by furnace cooling, then HIP at 1160 °C for 4 h under a pressure of 100 MPa, also followed by furnace cooling. Because of the HIP treatment, it was observed that the γ^2 phase was fully dissolved and the further growth of carbides was promoted. Moreover, the average grain size magnified by ~50 μm to ~150 μm around diameter after the HIP treatment (Mostafa et al. 2017). Moussaoui et al. showed that the amount of porosity was considerably decreased to less than 0.01% owing to the heat treatment being an HIP (1160 °C, 102 MPa, 3 h and air cooling) followed by solution treatment (980 °C, 1 h and air cooling) and double ageing (720 °C, 8 h, furnace cooling + 620 °C, 8 h and air cooling). This heat treatment also disappeared the dendritic microstructure and equiaxial and mainly columnar grains were observed. Ruttert et al. has investigated the influence of HIP on the CMSX-4 built by E-PBF. They showed that the resulting directionally solidified microstructures is about two orders of magnitude finer than in cast form. Moreover, in comparison to cast material, these specimens include less undesired defects like porosity, cracks, segregation, and brittle phases. In consistency with other studies, it was shown that HIP temperature should be above the γ¢-solvus temperature. Otherwise, even kept for short holding durations, γ¢-coarsening occurred along the columnar grain boundaries. Thus, a standard HIP recipe for cast materials is not recommended. Moreover, it is mentioned that with the latest HIP equipment with a quenching capability, it is possible to finish the heat treatment and HIP in the same sequence (Ruttert et al., 2016).

Geometrical Accuracy and Surface Quality

Thermal stresses are a critical aspect in PBF since they lead to distortion. They arise from notably high changes of temperature with respect to time in the vicinity of the melt pool and by the volume shrinkage during phase change and the cooling in the solid state. One of the routes to minimize the residual thermal stresses is the change of process parameters that can be easily controlled. For example, increasing the scan speed leads to larger thermal gradients leading to a more pronounced residual stresses and deformations. Increasing the laser power, on the other hand, leads to increased melt pool sizes and slightly reduced distortions. The scan strategy is a very dominant factor affecting the thermal stresses. Scan strategies with limited scan vector lengths and random scanning improves the geometrical accuracy. As one of the process parameters, preheating may affect the thermal gradients. Moreover, in comparison to E-PBF, higher thermal gradients in L-PBF induce high transient stresses which cause increased residual stresses in a range close to materials yield strength and deformations (Risse, 2019).

In order to allow the excessive heat to be dissipated by conduction to the base plate, the overhang surfaces need to be supported. Furthermore, these support structures are used to anchor the part to the base plate in order to resist any forces. Since the preheating in E-PBF sinter the powder material and these sintered powder particles act as supports, the design of the support structures is more critical in L-PBF. On one hand, it is important to have a sufficient amount of supports to avoid distortions and cracking while on the other hand, it is critical to have minimum supports so that removing them after the process is easy and does not involve too much efforts. There are different approaches to minimize the supports in the L-PBF process. Some of the studies in the literature aimed for reducing the supports by re-orienting the part (Cheng and To, 2019; Das, et al. 2015). As shown in Figure 8, different orientations of the same T part lead to different amounts of support structures. Although changing the part orientation is one of the simplest measures to eliminate the adverse effect of overhangs, generally for a complex geometry, the part orientation needs to be optimized for more than one objective such as maximum build volume, build time, surface quality, etc. (Moroni et al., 2015). Additionally, in a complex part, there are generally more overhang surfaces than only one conflicting each other regarding the optimized part orientation. Thus, by only changing the build orientation, it is generally very difficult to eliminate the need for supports. Yet, different algorithms can be used to minimize them.

If not completely eliminated, then the design of support structures becomes an important topic. Poyraz et al. has examined the effect of different parameters of the block support structures which are widely

Figure 8. T part needs different levels of supports when oriented in different directions

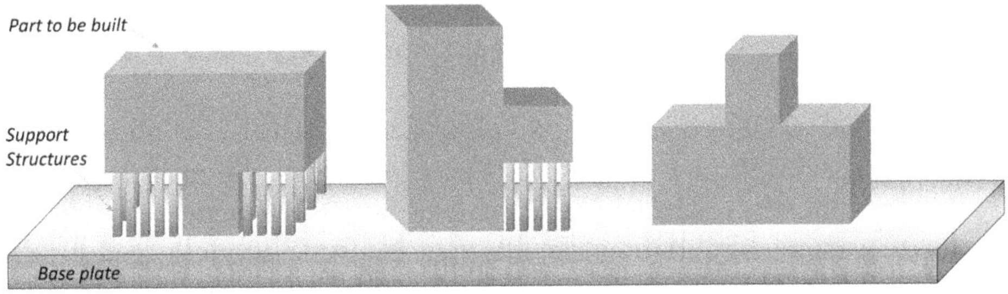

used (Poyraz et al., 2015). For the SLM of Inconel 625, the influence of hatch distance, fragmentation, top tooth length and Z-offset (see Figure 9) on the thermally induced deformations was studied. The experimental results conclude that, in comparison to tooth parameters of supports, hatch distance has a greater influence on the deformations. As the hatch distance is increased, the distortion measured on the edge of the part becomes more evident. After a certain level, when the residual stresses outreached the yield stress, a separation of support structures from the part is observed. Regarding the ease of removing the supports, higher hatch distance and fragmentation interval resulted in detaching support from the part easily (Poyraz et al. 2015).

Figure 9. Block support design parameters
(Poyraz et al. 2015)

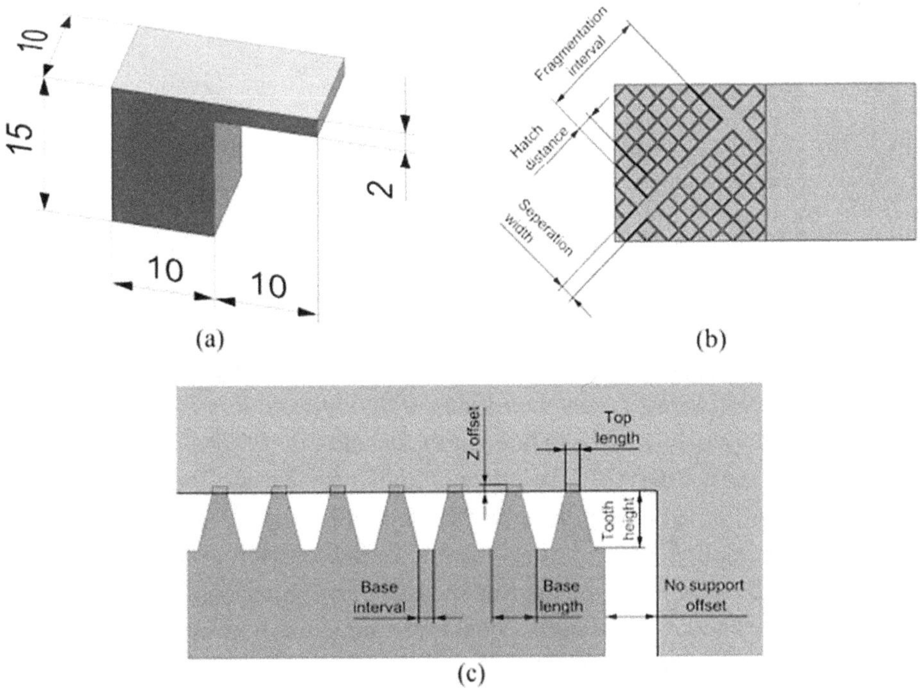

Hussein et al.'s study described an application of lightweight support structures where they concluded that the support type, the volume fraction, and the size of the cells are the major aspects that control the manufacturability, support volume and thus the total processing time (Hussein et al., 2013). In another study carried out with Hastelloy X, the process parameters were selected as a laser power of 80-90 W and a scan speed of 400-500 mm/s in comparison to part hatch parameters of higher laser power and scan speeds. It was concluded that the most critical factors to minimize the part distortion as a result of residual stresses by optimally conducting excessive heat away is the proper selection of the type of supports and process parameters (Calignano and Minetola, 2019). The typical support structures used for L-PBF of nickel superalloys are block type structures exhibited in Figure 9. The relevant parameters are provided in Table 4 modified from the study of Carter et al.

Table 4. Typical block type support parameters used for L-PBF of nickel superalloys

Support Parameters	
Hatch distance:	0.5 mm with a rotation of 45°
Fragmentation interval:	4 mm
Separation:	0.5 mm
Teeth geometry (see Figure 9):	Height: 1.5 mm Top length: 0.6 mm Base length: 1.5 mm Base interval: 0.2 mm
Borders:	Borders on with teeth geometry No fragmentation No perforations

(Modified from Carter,2013)

Surface quality is an important aspect for critical parts made of superalloys since it may affect final product's aerodynamic performance in terms of losses or shocks, mechanical integrity on fatigue strength (Konečná et al., 2016) or surface integrity on wear performance (Winter et al., 2013) (Tascioglu et al., 2019). For a proper and successful implementation of PBF AM, surface quality should be considered during design and production phases. The low surface quality obtained by PBF AM is one of the limitations of the process and the reason for this is the layered nature of the process like other AM processes. In this regard, stair stepping effect, which is seen as a texture on part surfaces, occurs due to geometric differences between consecutive layers' outer and inner contours, and it is more noticeable in inclined surfaces. As mentioned previously, it leads to poor surface quality of $_{Ra}$ 5 µm – R_a 10 µm for L-PBF AM and R_a 25 µm – R_a 35 µm for E-PBF AM. The cause of higher surface roughness values of E-PBF AM relative to L-PBF AM is affiliated to the metal powder with a large particle size and thus bigger layer thickness. According to modeling and experimental studies, the second factor to influence the surface roughness as a result of PBF AM is the inclination angle of parts' outer and inner surfaces (Yasa et al., 2016). However, in contrary to layer thickness, inclination angle has a reverse proportionate influence on stair stepping effect and the surface roughness decrease with the increasing inclination angle. Both the layer thickness and inclination angle may change surface roughness and more importantly surface waviness, since the waviness is the measurement of the more widely spaced texture of surface or is the measurement the irregularities whose spacing is higher than sampling length used in roughness measurements (Oberg et al. 2004). Figure 10 shows the model for the relation between surface waviness, layer thickness and inclination angle (Yasa et al., 2016). It also shows experimental specimens produced by L-PBF AM of Inconel 625 and the benchmark between modeling and experiments. Although the use of inclined test artifacts together with contact measurement techniques are common in literature, curved test artifacts together with non-contact measurement techniques may also be applied for a detailed and comprehensive characterization of as built PBF AM surfaces of superalloys (Poyraz et al., 2019).

Figure 10. The model for the relation between surface waviness, layer thickness and inclination angle as well as the tested specimens and obtained results in comparison to predicted waviness
(Yasa et al., 2016)

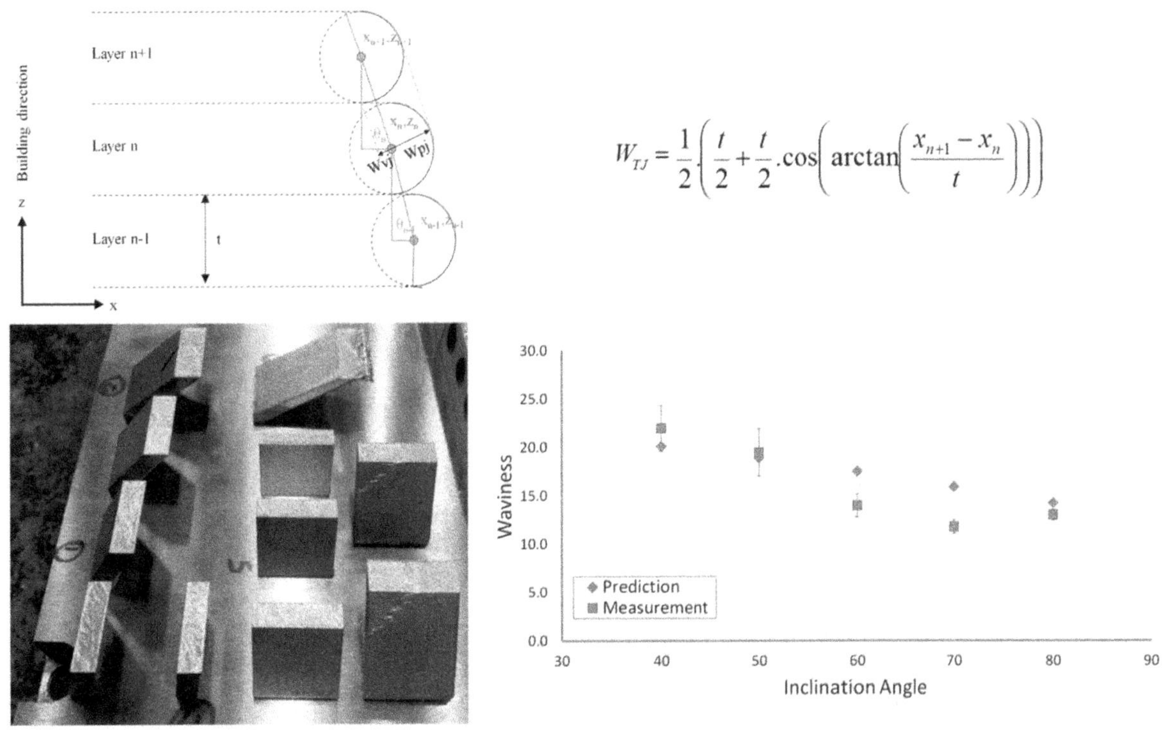

FUTURE RESEARCH DIRECTIONS

Additive manufacturing of full density components from nickel superalloys is gaining more importance because nickel based superalloys are frequently employed in a broad range of industries from aerospace, power generation, oil and gas to nuclear reactors. Their superior mechanical and chemical properties even at elevated temperatures make these alloys a right candidate for several demanding applications. In order to enhance the technological level of powder bed fusion, AM technologies have recently been studied for a diverse range of applications regarding different aspects. New materials, especially the ones that are difficult to weld with high ratios of Al+Ti%, are being developed while inherent problems of additive manufacturing such as residual stresses and leading deformations, surface quality, and anisotropic material properties. Moreover, components with a ductile core and hard, corrosion and abrasion resistance periphery from the same uniformly mixed powder blend are very one of the interesting future research directions. The demand for larger parts is always in place in order to increase the potential to integrate several parts into a single part eliminating the need of joining processes and simplifying the supply chain. Another important field of study is the software development for AM processes. Currently, the CAD/CAE/CAM software is mainly targeted towards conventional manufacturing processes. For example, more accurate and faster simulation tools with integrated material database for AM materials are needed for process modeling of complex PBF AM processes in order to eliminate the number of trial-and-errors. Moreover, advanced design tools like topology optimization and generative design software to enable design for AM and to explore the real potential of these processes while minimizing the need for supports shall be developed and integrated in the CAD software. When the design of any part takes process limitations into account covering the whole process chain such as stair stepping effect, overhang surfaces, need for supports, post-processing requirements, etc., the potential of the AM parts become much higher overcoming many challenges such as limited surface quality or dimensional accuracy.

CONCLUSION

Additive manufacturing of full density components from nickel superalloys has been gaining more importance especially in the last decade because nickel based superalloys are frequently employed in various industries such as aerospace, marine and nuclear reactors as well as chemical industries due to their superior mechanical and chemical properties at even elevated temperatures. Additionally, the advantages of having a low buy-to-fly ratio, decreased lead times from design to testing, suitability for low volume production and almost infinite design complexity make powder bed fusion additive manufacturing processes an appealing manufacturing route for highly demanding applications. In order to enhance the technological level of these technologies, different aspects have recently been studied for a diverse range of applications.However, there are still some critical challenges to widen the range of these AM processes including the limited processability of nickel based superalloys, obtained mechanical and metallurgical properties as well as the surface quality and dimensional accuracy. To conclude, it is important to realize that powder bed AM is a new technology in comparison to well established traditional manufacturing routes and it is getting more mature every day. Moreover, it is important to note that due to the inherent nature of the processes, additive manufacturing may lead to new process and supply chains as well as new manufacturability problems related to high residual stresses and leading deformations, inferior surface quality and dimensional accuracy in contrast to machining,anisotropic material properties, etc. These problems necessitate different solution routes rather than conventional processes. For example, heat treatments developed for cast or wrought materials may not give the optimum results for additively manufactured parts. Thus, as new technologies and solutions emerge, AM of nickel superalloys will become a more reliable and mature technology for applications with demanding and strict technical requirements.

ACKNOWLEDGMENT

This research received no specific grant from any funding agency in the public, commercial, or not-for-profit sectors.

REFERENCES

Amato, K. N., Gaytan, S. M., Murr, L. E., Martinez, E., Shindo, P. W., Hernandez, J., & Medina, F. (2012). Microstructures and mechanical behavior of Inconel 718 fabricated by selective laser melting. *Acta Materialia*, *60*(5), 2229–2239. doi:10.1016/j.actamat.2011.12.032

Balachandramurthi, A., Moverare, J., Mahade, S., & Pederson, R. (2019). Additive Manufacturing of Alloy 718 via Electron Beam Melting: Effect of Post-Treatment on the Microstructure and the Mechanical Properties. *Materials (Basel)*, *12*(1), 68. doi:10.3390/ma12010068 PMID:30585242

Bauer, T., Dawson, K., Spierings, A. B., & Wegener, K. (2015, August). Microstructure and mechanical characterisation of SLM processed Haynes® 230®. In *Proceedings of the 26th annual international Solid Freeform Fabrication Symposium*, (pp. 813-822). Academic Press.

Bowman, R. (2000). *Superalloys: A primer and history*. The Minerals, Metals & Materials Society.

Calignano, F., & Minetola, P. (2019). Influence of process parameters on the porosity, accuracy, roughness and support structures of Hastelloy X produced by laser powder bed fusion. *Materials (Basel)*, *12*(19), 3178. doi:10.3390/ma12193178 PMID:31569803

Carter, L. N. (2013). *Selective laser melting of nickel superalloys for high temperature applications* (Doctoral dissertation). University of Birmingham.

Catchpole-Smith, S., Aboulkhair, N., Parry, L., Tuck, C., Ashcroft, I. A., & Clare, A. (2017). Fractal scan strategies for selective laser melting of 'unweldable' nickel superalloys. *Additive Manufacturing*, *15*, 113–122. doi:10.1016/j.addma.2017.02.002

Cheng, B., Shrestha, S., & Chou, K. (2016). Stress and deformation evaluations of scanning strategy effect in selective laser melting. *Additive Manufacturing*, *12*, 240–251. doi:10.1016/j.addma.2016.05.007

Cheng, L., & To, A. (2019). Part-scale build orientation optimization for minimizing residual stress and support volume for metal additive manufacturing: Theory and experimental validation. *Computer Aided Design*, *113*, 1–23. doi:10.1016/j.cad.2019.03.004

Das, P., Chandran, R., Samant, R., & Anand, S. (2015). Optimum Part Build Orientation in Additive Manufacturing for Minimizing Part Errors and Support Structures. *Procedia Manufacturing*, *1*, 343–354. doi:10.1016/j.promfg.2015.09.041

Donachie, M. J., & Donachie, S. J. (2002). *Superalloys: A technical guide*. ASTM International. doi:10.31399/asm.tb.stg2.9781627082679

Dye, D., Hunziker, O., & Reed, R. C. (2001). Numerical analysis of the weldability of superalloys. *Acta Materialia*, *49*(4), 683–697. doi:10.1016/S1359-6454(00)00361-X

EOS. (2016). *EOS NickelAlloy IN718 Material Data Sheet.* Developed for EOS M400 SF.

Han, Q., Mertens, R., Montero-Sistiaga, M. L., Yang, S., Setchi, R., Vanmeensel, K., & Fan, H. (2018). Laser powder bed fusion of Hastelloy X: Effects of hot isostatic pressing and the hot cracking mechanism. *Materials Science and Engineering A, 732*, 228–239. doi:10.1016/j.msea.2018.07.008

Hanning, F., & Andersson, J. (2016, October). A Review of Strain Age Cracking in Nickel Based Superalloys. In *7th International Swedish Production Symposium, SPS16*, (pp. 1-7). Academic Press.

Harrison, N. J. (2016). *Selective Laser Melting of Nickel Superalloys: solidification, microstructure and material response* (Doctoral dissertation). University of Sheffield.

Harun, W. S. W., Manam, N. S., Kamariah, M. S. I. N., Sharif, S., Zulkifly, A. H., Ahmad, I., & Miura, H. (2018). A review of powdered additive manufacturing techniques for Ti-6al-4v biomedical applications. *Powder Technology, 331*, 74–97. doi:10.1016/j.powtec.2018.03.010

Henderson, M. B., Arrell, D., Larsson, R., Heobel, M., & Marchant, G. (2004). Nickel based superalloy welding practices for industrial gas turbine applications. *Science and Technology of Welding and Joining, 9*(1), 13–21. doi:10.1179/136217104225017099

Hussein, A., Hao, L., Yan, C., Everson, R., & Young, P. (2013). Advanced lattice support structures for metal additive manufacturing. *Journal of Materials Processing Technology, 213*(7), 1019–1026. doi:10.1016/j.jmatprotec.2013.01.020

International Standards Organization / American Society for Testing and Materials. (2015). *Standard Terminology for Additive Manufacturing – General Principles – Terminology* (ISO / ASTM52900 – 15). Author.

Konečná, R., Nicoletto, G., Kunz, L., & Bača, A. (2016). Microstructure and directional fatigue behavior of Inconel 718 produced by selective laser melting. *Procedia Structural Integrity, 2*, 2381–2388. doi:10.1016/j.prostr.2016.06.298

Kruth, J. P., Vandenbroucke, B., Vaerenbergh, J. V., & Mercelis, P. (2005, April). Benchmarking of Different SLS/SLM Processes as Rapid Manufacturing Techniques. In *Proceedings of International Conference of Polymers & Moulds Innovations (PMI)*. Gent, Belgium: Academic Press.

Moroni, G., Syam, W. P., & Petrò, S. (2015, March). Functionality-based part orientation for additive manufacturing. In *Proceedings of 25th CIRP Design Conference*. Haifa, Israel: Academic Press.

Mostafa, A., Picazo Rubio, I., Brailovski, V., Jahazi, M., & Medraj, M. (2017). Structure, Texture and Phases in 3D Printed IN718 Alloy Subjected to Homogenization and HIP Treatments. *Metals, 7*(196), 1–23. doi:10.3390/met7060196

Mouritz, A. P. (2012). *Introduction to aerospace materials*. Elsevier.

Murr, L. E., Martinez, E., Gaytan, S. M., Ramirez, D. A., Machado, B. I., Shindo, P. W., Martinez, J. L., Medina, F., Wooten, J., Ciscel, D., Ackelid, U., & Wicker, R. B. (2011). Microstructural Architecture, Microstructures, and Mechanical Properties for a Nickel-Base Superalloy Fabricated by Electron Beam Melting. *Metallurgical and Materials Transactions. A, Physical Metallurgy and Materials Science, 42*(11), 3491–3508. doi:10.100711661-011-0748-2

Nicoletto, G., Konečná, R., Frkáň, M., & Riva, E. (2018). Surface roughness and directional fatigue behavior of as-built EBM and DMLS Ti6Al4V. *International Journal of Fatigue*, *116*, 140–148. doi:10.1016/j.ijfatigue.2018.06.011

Oberg, E., Jones, F. D., Horton, H. L., Ryffel, H. H., McCauley, C., Heald, R., & Hussain, M. (2004). Machinery's Handbook: A Reference Book for the Mechanical Engineer, Designer. *Manufacturing Engineer, Draftsman, Toolmaker, and Machinist, 26.*

Ott, E. A., Groh, J., & Sizek, H. (2005). Metals Affordability Initiative: Application of Allvac alloy 718Plus® for aircraft engine static structural components. *Superalloys*, *718*, 625–706. doi:10.7449/2005/Superalloys_2005_35_45

Paulonis, D.F. & Schirra, J.J. (2001). Alloy 718 at Pratt & Whitney- Historical Perspective and Future Challenges. *TMS Superalloys 718. 625. 706 and Various Derivatives.*

Perevoshchikova, N., Rigaud, J., Sha, Y., Heilmaier, M., Finnin, B., Labelle, E., & Wu, X. (2017). Optimisation of selective laser melting parameters for the Ni-based superalloy IN-738 LC using Doehlert's design. *Rapid Prototyping Journal*, *23*(5), 881–892. doi:10.1108/RPJ-04-2016-0063

Poyraz, O. (2018). *Investigation of the Effects of Process Parameters for Laser Additive Manufacturing by Using Modeling and Simulation Methods* (Doctoral dissertation). Eskisehir Osmangazi University, Eskisehir, Turkey.

Poyraz, Ö., & Kuşhan, M. C. (2018). Investigation of the effect of different process parameters for laser additive manufacturing of metals. *Journal of the Faculty of Engineering and Architecture of Gazi University*, *33*(2), 729–742.

Poyraz, Ö., & Kuşhan, M. C. (2019). Residual Stress-induced Distortions in Laser Powder Bed Additive Manufacturing of Nickel-based Superalloys. *Strojniski Vestnik. Jixie Gongcheng Xuebao*, *65*(6).

Poyraz, Ö., Solakoğlu, E. U., Ören, S., Tüzemen, C., & Akbulut, G. (2019). Surface texture and form characterization for powder bed additive manufacturing. *Journal of the Faculty of Engineering and Architecture of Gazi University*, *34*(3), 1653–1664.

Poyraz, Ö., Yasa, E., Akbulut, G., Orhangül, A., & Pilatin, S. (2015, August) Investigation of support structures for direct metal laser sintering of IN625 parts. In *Proceedings of Solid Freeform Fabrication Symposium*. Austin, TX: Academic Press.

Ramakrishnan, A., & Dinda, G. P. (2019). Microstructure and mechanical properties of direct laser metal deposited Haynes 282 superalloy. *Materials Science and Engineering A*, *748*, 347–356. doi:10.1016/j.msea.2019.01.101

Reed, R. C. (2006). *The Superalloys: Fundamentals and Applications.* Cambridge. doi:10.1017/CBO9780511541285

Risse, J. (2019). *Additive Manufacturing of Nickel-Base Superalloy IN738LC by Laser Powder Bed Fusion* (Doctoral Dissertation). Rheinisch-Westfälischen Technischen Hochschule Aachen, Aachen, Germany.

Ruttert, B., Ramsperger, M., Mujica Roncery, L., Lopez-Galilea, I., Körner, C., & Theisen, W. (2016). Impact of hot isostatic pressing on microstructures of CMSX-4 Ni-base superalloy fabricated by selective electron beam melting. *Materials & Design*, *110*, 720–727. doi:10.1016/j.matdes.2016.08.041

Sabol, G. P., & Stickler, R. (1969). Microstructure of Nickel-Based Superalloys. *Physica Status Solidi. B, Basic Research*, *35*(1), 11–52. doi:10.1002/pssb.19690350102

Shaikh, A. S. (2018). *Development of γ'Precipitation Hardening Ni-Base Superalloy for Additive Manufacturing* (Master's Thesis). Chalmers University.

Sims, C. T. (1984). *A history of superalloy metallurgy for superalloy metallurgists*. Superalloys. doi:10.7449/1984/Superalloys_1984_399_419

Tammas-Williams, S., Withers, P. J., Todd, I., & Prangnell, P. B. (2016). The Effectiveness of Hot Isostatic Pressing for Closing Porosity in Titanium Parts Manufactured by Selective Electron Beam Melting. *Metallurgical and Materials Transactions. A, Physical Metallurgy and Materials Science*, *47*(5), 1939–1946. doi:10.100711661-016-3429-3

Tascioglu, E., Kaynak, Y., Poyraz, Ö., Orhangül, A., & Ören, S. (2019, September). The Effect of Finish-Milling Operation on Surface Quality and Wear Resistance of Inconel 625 Produced by Selective Laser Melting Additive Manufacturing. In *Proceeding of International Conference on Advanced Surface Enhancement* (pp. 263-272). Springer.

Vayssette, B., Saintier, N., Brugger, C., Elmay, M., & Pessard, E. (2018). Surface roughness of Ti-6Al-4V parts obtained by SLM and EBM: Effect on the High Cycle Fatigue life. *Procedia Engineering*, *213*, 89–97. doi:10.1016/j.proeng.2018.02.010

Wang, F. (2011). Mechanical property study on rapid additive layer manufacture Hastelloy® X alloy by selective laser melting technology. *International Journal of Advanced Manufacturing Technology*, *58*(5-8), 545–551. doi:10.100700170-011-3423-2

Winter, K., Hartmann, J., Jeschke, P., & Lahmer, M. (2013). Experimental and numerical investigation of streamwise surface waviness on axial compressor blades. In *ASME Turbo Expo 2013: Turbine Technical Conference and Exposition*. American Society of Mechanical Engineers Digital Collection. 10.1115/GT2013-95983

Xiao, L., Song, W., Hu, M., & Li, P. (2019). Compressive properties and micro-structural characteristics of Ti–6Al–4V fabricated by electron beam melting and selective laser melting. *Materials Science and Engineering A*, *764*, 138204. doi:10.1016/j.msea.2019.138204

Yasa, E., Poyraz, O., Solakoglu, E. U., Akbulut, G., & Oren, S. (2016). A study on the stair stepping effect in direct metal laser sintering of a nickel-based superalloy. *Procedia CIRP*, *45*, 175–178. doi:10.1016/j.procir.2016.02.068

Zhong, Y., Rännar, L.-E., Wikman, S., Koptyug, A., Liu, L., Cui, D., & Shen, Z. (2017). Additive manufacturing of ITER first wall panel parts by two approaches: Selective laser melting and electron beam melting. *Fusion Engineering and Design*, *116*, 24–33. doi:10.1016/j.fusengdes.2017.01.032

Chapter 14
Laser Additive Manufacturing in Industry 4.0:
Overview, Applications, and Scenario in Developing Economies

Christ P. Paul
Raja Ramanna Centre for Advanced Technology, India & Homi Bhabha National Institute, Mumbai, India

Arackal N. Jinoop
https://orcid.org/0000-0002-1885-6427
Raja Ramanna Centre for Advanced Technology, India & Homi Bhabha National Institute, Mumbai, India

Saurav K. Nayak
Raja Ramanna Centre for Advanced Technology, India & Homi Bhabha National Institute, Mumbai, India

Alini C. Paul
https://orcid.org/0000-0003-4474-7731
Nitte Meenakshi Institute of Technology, India

ABSTRACT

Additive manufacturing is one of the nine technologies fuelling the fourth industrial revolution (Industry 4.0). High power lasers augmented with allied digital technologies is changing the entire manufacturing scenario through metal additive manufacturing by providing feature-based design and manufacturing with the technology called laser additive manufacturing (LAM). It enables the fabrication of customized components having complex and lightweight designs with high performance in a short period. The chapter compiles the evolution and global status of LAM technology highlighting its advantages and freedoms for various industrial applications. It discusses how LAM is contributing to Industry 4.0 for the fabrication of customized engineering and prosthetic components through case studies. It compiles research, development, and deployment scenarios of this new technology in developing economies along with the future scope of the technology.

DOI: 10.4018/978-1-7998-4054-1.ch014

INTRODUCTION

Industrial revolution witnesses a marked era, where innovations or inventions change the way products and services are produced for faster delivery with increased profitability at a lower cost. The first (1760-1840), second (1870-1914) and third (1969 onwards) industrial revolutions are characterized by massive changes in industrial operations mainly due to the introduction of innovations in the steam engine, electricity and automation, respectively (Horn, Rosenband, & Smith, 2010), (Reisman, 1996), (Engelman, 2020), (Meak, 2020)]. These three industrial revolutions were centred on mass production for increasing profits and the market was extraordinarily driven by sellers. Subsequently, the market took a U-turn and shifted from sellers' driven market to buyers' market. It pushed industries to adopt the policy of supplying the best quality products and services at the lowest possible price immediately meeting all possible consumer's expectations. Industries coming with more innovations and better after-sales support started dominating the market. The trend is evident nowadays too; it is recognized as the commencement of the fourth industrial revolution or Industry 4.0. In this period, industries are toiling to bring "first product in the market" and there was not much market share for "me-too-products" (a product that is designed with same design philosophy/ feature as another existing product in the market) in this unending race. In this way, the period of mass production is translating into mass customization with a fine balance between customer satisfaction and mass production. It is expected that autonomous there will be a breakthrough in the field of robotics, nanotechnology, quantum computing, internet of things, biotechnology, artificial intelligence (AI), autonomous vehicles and additive manufacturing (AM) (Quintanilla, Hope, Darnton, & Hunter, 2019). As per literature, nine different sectors are recognized as pillars of Industry 4.0 (Cheng, Liu, Qiang, & Liu, 2016) (Vaidya, Ambad, & Bhosle, 2018):

1. Big Data: It deals with the analysis of large volume, velocity, veracity, variety and value (Tao, Tang, Zou, & Qi, 2019) of data being collected over time from different operations in a company/institution/population. This analysis would lead to useful insights into the input parameters, processes and output by identifying trends, patterns, and relationships among them. These insights would be later useful in adjusting parameters to improve production processes across different platforms. In the period of smart manufacturing supported by big data, researchers are trying to assimilate various technologies and stimulate new ideas (Tao, Tang, Zou, & Qi, 2019).
2. Augmented Reality: This technology will enable showcasing a product in the natural environment without creating an actual physical copy of the product (augmented imagery in a real-world) and running the product and its augmented image interactively in real-time). It will allow manufacturers to indicate how their products would look like in a real environment to their prospective customers and allow controlling the product by using augmented image (Palmarini, Erkoyuncu, Roy, & Torabmostaedi, 2018).
3. Simulation: It means playing out a scenario of an actual situation, process or environment. 3D simulation of component development, material development, and production processes are used extensively (Vieira, Dias, Santos, Pereira, & Oliveira, 2018) (Mordor Intelligence LLP, 2020). It will obtain the real-time data to replicate the real world in a virtual model, which includes machines, process, products, and humans.
4. Internet of Things (IoT): It is the network of physical devices with electronics, software, sensors, actuators and internet connectivity enabling data exchange for direct integration of the real world into computer-based systems for monitoring and control ensuing improved efficiency, cost-effectiveness,

and lower human efforts. Industrial IoT (IIoT) possesses machinery connected via the internet to send/ receive the information/ data enabling the innovative analytics platforms to analyze it to make appropriate decisions.

5. Systems Integration: Through Industry 4.0, the entire organization and subsequent companies will be interconnected and interacting in a closed system. System integration is achieved when all the computer systems in a manufacturing unit are interconnected allowing for actual communication and data transfer among them with the help of coordinating software.
6. Autonomous Robots: This technology allows systems to sense, think, act, and confirm the execution of assignment autonomously helping in an increase in the company's profitability, productiveness, and competitiveness. Now, robots can deliver wider variety of services and are more autonomous, flexible, co-operative, act together and co-work with humans with safety.
7. Cloud Computing: It involves a shared network of configurable computing resources for storing, managing and processing data that conveniently provides on-demand computer access and requires hardly any management effort and service provider interaction. This trend would help in faster production of goods and their introduction to the market (Xu, 2012).
8. Cyber Security: With the introduction of IoT, Cloud Computing and Systems integration, the production efficiency increases manifold, but so does the risk of hacking and data theft. Hence, robust cybersecurity systems must be developed to protect valuable data on customers, their products and intellectual property.
9. Additive Manufacturing (AM): ASTM defines AM as "a process of joining materials to make objects from 3D model data, usually layer upon layer, as opposed to subtractive manufacturing methodologies" (Standard Terminology for Additive Manufacturing Technologies, 2019) (Paul, Bhargava, Kumar, Pathak, & Kukreja, 2012). This technology aims towards fabricating final products layer-by-layer by providing the solid model of the product as an input. For fabricating metallic components via. AM; plasma, electron beam or lasers are being used as the heating source. The processes using lasers to manufacture components by AM are termed as "Laser Additive Manufacturing (LAM)" (Paul, Jinoop, & Bindra, Metal Additive Manufacturing using lasers, 2018).

This chapter presents a comprehensive review of how LAM is contributing to Industry 4.0. It introduces LAM processes, their evolution, and their advantages. It also presents the relation of LAM with other pillars of Industry 4.0. It summarizes the global scenario of LAM technology with a specific mention of developing economies.

EXTENSION OF AM TO LAM

Chuck Hull developed the first commercial AM process known as stereolithography in 1983. The process developed three-dimensional objects by curing of photopolymers layers using ultraviolet lasers/ light. At this stage, AM used polymers as raw material for building requisite 3D shapes and its application was limited to prototyping. Even though AM started with polymers, it leapt, when applied to metals opening new possibilities in the manufacturing processes. Subsequently, AM of metals came to be known as Metal Additive Manufacturing (MAM). Most of the commercial MAM processes deploy high energy density sources, like – lasers, plasma and electron beams to melt metallic materials. Fig. 1 presents the most commonly used MAM systems that deploy high energy sources.

Figure 1. Most Commonly used MAM Systems that deploy high energy sources

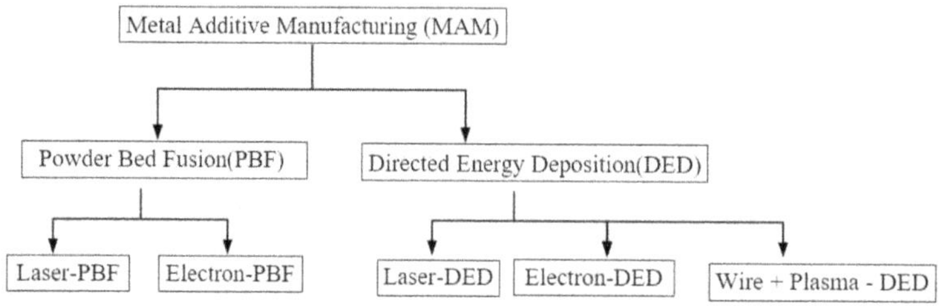

The AM revolution for metals started when Carl Deckard and Joe Beaman from the University of Texas patented Selective Laser Sintering (SLS) process in 1986. In 1995, EOSINT M250 was launched for enabling rapid manufacturing of metallic tools (Bhavar, et al., 2014). Selective Laser Melting (SLM) process was invented in 1995 by Dr. Wilhelm Meiners and Dr. Konrad Wissenbach of the Fraunhofer Institute for Laser Technology ILT in Aachen, Germany, and Dr. Dieter Schwarze and Dr. Matthias Fockele from the Fockele & Schwarze (F&S) Stereolithographietechnik GmbH (Germany Patent No. DE19649865C1, 1998). In the same year, Sandia National Laboratory reported the development of "Laser Engineered Net Shaping (LENS^{TM})" technology using dynamic blowing technique. LENS^{TM} technology was used by Optomech to develop its first AM system and introduce it to the market in 1998 (Sandia National Laboratories, 2017). In 2000, Andersson and Larsson patented the Electron Beam Melting (EBM) process, which was authorised by Arcam AB. Even though the electron beam is a high energy source, it can only be used to process electrically conductive materials, whereas lasers can be used with any material that can absorb energy at its wavelength. Electron beams also run the risk of powder expulsion due to the accumulation of excess negative charge creating a more diffuse beam. Arc is also used as a heat source and has higher build rates as compared to electron and lasers as a heat source, but it suffers from low dimensional accuracy. Thus, by considering the advantages of lasers over other high energy density sources, MAM is most commonly performed using laser-based technologies and it is termed as LAM. Complex shapes and structures that were initially difficult to manufacture using conventional processes can now easily be manufactured using LAM without greatly compromising on the structural properties of the components. LAM processes can be of two types, i.e., Laser Powder Bed Fusion (LAM-PBF) and Laser Directed Energy Deposition (LAM-DED) (Paul, Jinoop, & Bindra, Metal Additive Manufacturing using lasers, 2018) (Nayak, Mishra, Paul, Jinoop, & Bindra, 2020). In LAM-PBF technology, focused laser beams are deployed for selectively melting a thin layer of powder pre-placed by spreading over the build plate. The 3D component is built by melting metal powder layers one upon another as directed by design (Bajaj, Wright, Todd, & Jägle, 2019). One of the major advantages of the technology is the unlimited design freedom with lower surface roughness as compared to other high energy AM counterparts (Ding, Pan, Cuiuri, & Li, 2015). Fig. 2 (a) presents the schematic of a LAM-PBF system. Thus, LAM-PBF involves the pre-spreading of powders. On the other hand, LAM-DED involves feedstock feeding mechanism. In LAM-DED, the focused laser beam is used to create a melt pool, into which the feedstock is added for material deposition. Depending on the type of feedstock material DED can be classified into (i) Powder Fed Deposition (PFD) (Fig. 2(b)) (ii) Wire Fed Deposition (WFD) (Fig. 2(c)) (Sandia National Laboratories, 2017). In PFD, feedstock material in powder form is carried by the carrier gas to the point of deposition using a coaxial nozzle and deposits the material as per the requisite geometry (Tan, Pang, Kaminski, & Pepin, 2019). In WFD, the wire is fed through a

Figure 2. Schematic of (a) LAM-PBF (b) PFD Based LAM-DED (c) WFD based LAM-DED

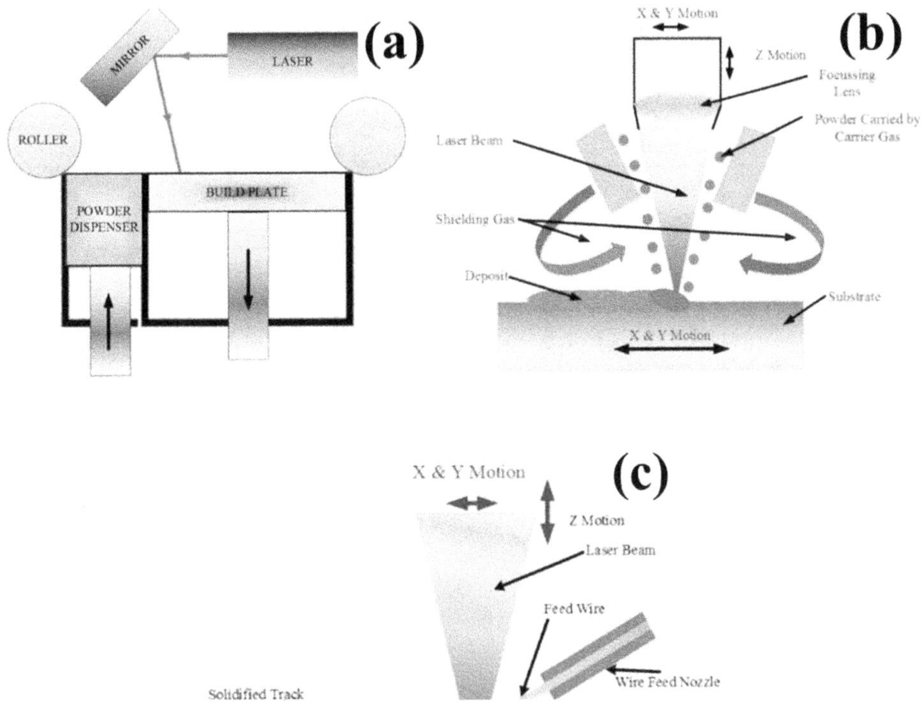

co-axial or lateral feeding system and melted using a high power laser source. The major advantage of WFD over PFD is the 100% utilization of raw-material making it a greener route as compared to PFD. The advantage of co-axial wire feeding over lateral feeding is the omnidirectionally it offers allowing fabrication of complex engineering components (Motta, Demir, & Previtali, 2018). Table 1 compares the characteristics, features of LAM-PBF, PFD Based LAM-DED and WFD based LAM-DED.

GLOBAL SCENARIO OF LAM TECHNOLOGY

There are many published studies on present status, growth rate and future forecast for AM technology in various application domains. Technavio, in its report, predicts the Compound Annual Growth Rate (CAGR) greater than 21% by 2022 based on application in aerospace, health care, tooling, academic institutions, and automobile (Technavio, 2018). It further forecasts that the major end-users of MAM are service bureaus, research institutions, and the aerospace industry. It is expected that the market growth in the aerospace section (29%) and automotive section is mainly due to huge demand for engine and system components and the requirement for lightweight, faster, and fuel-efficient vehicles respectively. A 15% growth rate is predicted for AM (Compound Annual Growth Rate (CAGR), 2015–2025) and Asia-Pacific region will see a growth of 18.6% (CAGR 2015–2025), with China having greater than 70% growth (Frost and Sullivan, 2016). Technavio identifies the following firms as the key players in the global LAM market: EOS, Renishaw, and SLM Solutions (The Fabricator, 2016). Table 2 summarizes the information regarding major LAM system manufacturers along with model, laser details and build volume (Paul, Jinoop, & Bindra, Metal Additive Manufacturing using lasers, 2018) (Optomec, 2018)

Table 1. Comparison between LAM-PBF, PFD Based LAM-DED and WFD based LAM-DED

Property	LAM-PBF	PFD based LAM-DED	WFD based LAM-DED
Laser power	Low	High	High
Spot size	microns	mm range	mm range
Material addition	Preplaced powder	In-situ powder feeding	In-situ wire feeding
Multi-material freedom	Lower	High	High
Roughness	Lower	High	High
Deposition rate	Low	High	High
Complex geometry	Unlimited	Limited	Limited
Layer thickness	Low	High	High
Dimensional Accuracy	High	Low	Low

(DM3D Technology LLC, 2008) (EOS GmbH, n.d.) (SLM Solutions-Machines, n.d.) (GE Additive, 2016) (Renishaw plc., 2015).

LAM AND INDUSTRY 4.0

LAM is one of the potential game-changing technologies with the inherent capability of changing the complete manufacturing sequence from "mass production to mass customization". It is enabling 'Just in time' manufacturing, 'mass customization' and 'onsite manufacturing' of components in industries. LAM is a digital technology contributing to Industry 4.0 as one of the major pillars delineating the future. It allows industries to congregate and interpret data across machines, which allows swift, flexible and efficient processes to develop components at superior quality and reduced cost. LAM has the following four special freedoms (Paul, Jinoop, & Bindra, Metal Additive Manufacturing using lasers, 2018):

1. Material design freedom: LAM enables the freedom of adding different materials within a component at various locations. It is possible to join metallurgical incompatible materials by introducing compatible buffer or graded layers. Further, the introduction of graded layers provides a smoother transition and better interface strength across dissimilar materials (Banait, et al., 2017).
2. Shape design freedom: LAM opens new opportunities in design and manufacturing through "feature-based design and manufacturing", where complex geometries (such as undercuts and internal spaces) and lightweight structures with equally good performance (through superior material usage) can be built. In contrast to conventional manufacturing, increased complexity in LAM does not lead to higher cost (Pinkerton, 2016), which makes it attractive for high-value materials. With the help computational software, stress flow fields could be identified that led to re-design and manufacture of components that although had relatively complex shapes but utilized less material and as a result, were lighter than before. Fig. 3 presents the comparison between AM and conventional manufacturing techniques concerning complexity, cost per part and number of parts.

Table 2. LAM technologies with commercial machine details

Technology	Manufacturer	Model	Laser	Build Volume	Scan Speed (m/s)
LAM-DED	Optomec	LENS™ 450	Fiber Laser, 400W	100 mm x 100 mm x 100 mm	0.06
LAM-DED	Optomec	LENS™ MR-7	Fiber Laser, 500W, 1 kW or 2 kW	300 mm x 300 mm x 300 mm	0.06
LAM-DED	Optomec	LENS™ 850-R	Fiber Laser, 1000 W or 2000 W	900 mm x 1500 mm x 900 mm	0.06
LAM-DED	DMD3D	DMD 105D	Fiber coupled laser- Fiber/ Diode / Disc, 1 kW	2D: 800 mm x800 mm x3000 mm 3D:300 mm x300 mm x300 mm	0.03
LAM-DED	DMD3D	DMD 503D/505D	1 to 5kW Fiber coupled laser - Diode/ Disc/Fiber Laser	2D: 1219 mm x1219 mm x 600 mm 3D: 673 mm x 673 mm x 474 mm	0.03
LAM-DED	DMD3D	DMD IC106	1kW Fiber coupled laser – Diode or Disc Laser	300 mm x 300 mm x 300 mm	0.016
LAM-DED	DMD3D	DMD 44R/66R	1 to 5kW Fiber-coupled Diode/disc/ fiber Laser	44R: 1425 mm x 1020 mm x 1020 mm 66R: 2330 mm x 1670 mm x 1670 mm	0.041
LAM-PBF	EOS	EOS M 100	Fiber laser, 200 W	Ø 100 mm x 95 mm	7
LAM-PBF	EOS	EOS M 290	Fiber laser, 400 W	250 mm x 250 mm x 325 mm	7
LAM-PBF	EOS	EOSINT M 280	Fiber laser, 200 W or 400 W	250 mm x 250 mm x 325 mm	7
LAM-PBF	EOS	EOS M400	Fiber laser, 1 kW	400 mm x 400 mm x 400 mm	7
LAM-PBF	SLM Solutions	SLM 500	Fiber Laser, Twin: 2×(400 W / 700 W) Quad: 4×(400 W / 700 W)	500 mm x 280 mm x 365 mm	up to 10
LAM-PBF	SLM Solutions	SLM 280 2.0	Single (1x 400 W), Twin (2x 400 W), Single (1x 700 W), Twin (2x 700 W), Dual (1x 700 W and 1x 1000 W) IPG fiber laser	280 mm x 280 mm x 365 mm	10
LAM-PBF	SLM Solutions	SLM 125	Fiber laser, 400 W	125 mm x 125 mm x 125 mm	10
LAM-PBF	Concept Lasers GmbH	Mlab Cusing	Fibre laser, 100 W (CW)	50 mm x 50 mm x 80 mm 70 mm x 70 mm x 80 mm (x, y) 90 mm x 90 mm x 80 mm (x, y)	7
LAM-PBF	Concept Lasers GmbH	Mlab cusing 200R	Fibre laser, 200 W (CW)	100 mm x 100 mm x 100 mm (x, y, z) 70 mm x 70 mm x 80 mm (x, y, z) 90 mm x 90 mm x 80 mm (x, y, z) 50 mm x 50 mm x 80 mm (x, y, z)	7
LAM-PBF	Concept Lasers GmbH	M1 cusing	Fibre laser, 200 W or 400 W	250 mm x 250 mm x 250 mm	7
LAM-PBF	Concept Lasers GmbH	M Line Factory	Fibre laser, max 4×1.000 watt	400 mm x 400 mm x up to 425 mm	Max. 5
LAM-PBF	Concept Lasers GmbH	M2 cusing	2 x 200 W (CW), optional 2 x 400 W (CW)	250 mm x 250 mm x 280 mm	7 4,5 m/s for variable focus move
LAM-PBF	Concept Lasers GmbH	XLINE 2000R	Fibre laser, 2×1 kW CW	800 mm x 400 mm x 500 mm	7
LAM-PBF	Renishaw	AM250	250 W laser	250 mm × 250 mm × 300 mm.	*
LAM-PBF	Renishaw	AM400	400 W laser	250 mm × 250 mm × 300 mm	*
LAM-PBF	Renishaw	RenAM 500M	500 W laser	250 mm × 250 mm × 350 mm	*
LAM-PBF	Renishaw	RenAM 500Q	4 X 500 W laser	250 mm × 250 mm × 350 mm	*

* Data unavailable in commercial websites

Figure 3. Unit manufacturing cost for additive and conventional manufacturing; (a) with volume of production; (b) with product complexity
(Adapted from (Pinkerton, 2016))

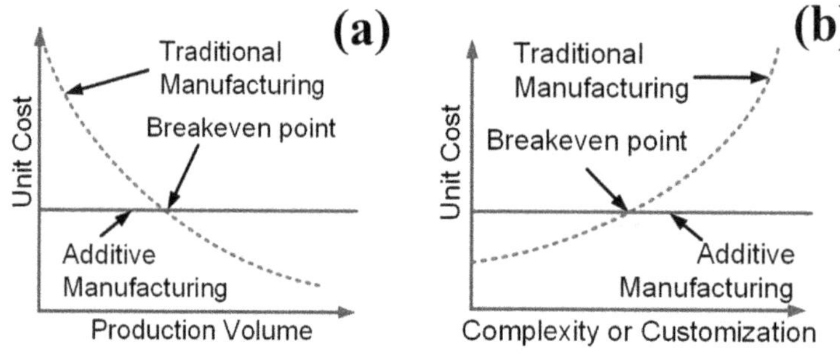

3. Logistics freedom: The gap between manufacturer and consumer is reduced by taking input directly from the customer for mass customization. Design, fabrication and performance evaluation of the parts can be performed at far and remote locations through cyber technologies. This reduces the supply-chain management by decreasing the number of stages involved in the fabrication of parts (Pour, Zanardini, Bacchetti, & Zanoni, 2016), (Silva & Rezende, 2013).
4. Post-processing freedom: The strategy for the fabrication of components can be developed to achieve material properties/ geometry incorporating post-processing requirements. Post-processing can be done on components for deriving required material properties and geometry. Recently, laser-based post-processing techniques such as laser annealing (Shiva, Palani, Paul, & Singh, 2016), laser peening (Jinoop, Investigation on laser shock processing of direct metal sintered Inconel 718 (M.Tech Thesis), 2016), laser polishing (Rosa, Mognol, & Hascoët, 2014) are also applied to LAM samples for improving the properties and surface finish.

The above mentioned exclusive freedoms offered by LAM allow for fabrication as per functional needs by providing the finest solution in terms of material, shape, and processing. Further, LAM is significantly contributing to advanced manufacturing in association with the other eight pillars of Industry 4.0. The relation of LAM with the other eight pillars of Industry 4.0 are described below:

1. Autonomous Robots are devices that perform tasks with minimal human intervention by learning from their surroundings and making independent decisions. Autonomous robots will strengthen

the LAM technology and benefit the manufacturing sector by reducing error, improved efficiency, and productivity with less human health hazards "and therefore consistently improving the quality and productivity of the work (Volini, Berg, & Moradian, 2017)". With this technology, automation of build plate replacement, and post-processing material handling is possible, which can improve productivity by reducing the cycle time. GE Aviation recently began using robotic AM processes for the Advanced Turboprop engine. The major aim was to reduce waste and part numbers. GE could reduce 855 separate parts to 12. It can also be noted that one-third of the entire engine is produced by AM (Robotic Industries Association (RIA), 2018), (Dusen, 2017).

2. Simulations of the process, material and product development are another pillar for Industry 4.0. The 3D simulations are the source for imitation of the process and use real-time data to develop a virtual model incorporating 3M (man, machines and materials) interaction. Simulation translates the testing process from real to virtual world in computers providing the advantage of reduced numbers of experimental iterations and ultimately less expensive. A combination of LAM simulation and experiments is bringing essential dynamics to the industries allowing process optimizations even before the start of production trials. As LAM is a complex process with multi-physics involved in it, simulation of the process is being used to predict distortion and residual stresses (Li, Liu, & Guo, 2016). This is helping the industries to estimate post-processing allowance accurately and improve the quality of the products thereby fabricating components in the first go itself. Another aspect of simulation especially for LAM is the generative design that involves generating multiple design alternatives for a set of input design goals (Autodesk Inc., 2020).
3. Big Data technology is also complimenting LAM through data visualization and real-time process control. Deployment of advanced data processing tools is yielding data close-to-real estimation and enabling improved visualization of the LAM process. The published report shows that General Electric's (GE) Aviation sector integrates big data and LAM to gather and evaluate data on their built components to predict flaws using sensors placed strategically (Chin, 2017).
4. Cloud-based manufacturing has two distinct applications. The first is within an industry, where cloud manufacturing facilitates is used to connect and use various manufacturing processes by appropriate and optimized linking of LAM with conventional subtractive machinery yielding hybrid manufacturing. The second is for LAM product users by transferring the product information (3D model, material and duty conditions) to the potential vendor through the internet. Cloud-based manufacturing provides the logistics freedom for LAM processes and reduces the steps in supply chain management for the part fabrication across the globe.
5. Internet of things (IoT) and AM are changing the way the world thinks, sees and understands things. IoT is the network of devices rooted with the ability to sense, collect, analyze and communicate data. These smart devices take many forms, ranging from wearables to high-value, engineering components. LAM techniques can potentially develop such functional components by reducing manufacturing costs and material wastage. The technology is stepping towards the Digital Twins - a digital replica of real world (processes, people, places, systems, and devices) providing the system health and process dynamics throughout its life cycle. Optomech developed technology for fabricating conformal electronics like antennas and sensors for enabling IoT applications using AM. The developed technology enables the fabrication of smart devices with a reduced cost of manufacturing using a variety of materials (Optomec Inc., 2018).
6. Augmented reality is an assortment of technologies that allows instantaneous mixing of contents generated by the computer to be superimposed over a live camera view of the physical world. It

can be considered as any system, which combines virtual and real and is communicating in real-time (Furht, 2011). Augmented reality could be used as a preview function of LAM. Just as LAM speeds up the process of manufacturing engineering components, augmented reality can simulate the model before sending for fabrication. Augmented reality is an ideal technique to provide enriched information, experience and potentially fabricate parts using LAM.

7. Cybersecurity has a similar threat for LAM as other technology involving digital data transfer with or without wire. Digital data transfer between AM machines is prone to cyber-attacks and the most vulnerable at any stage. For instance, in the pre-processing stage, the STL file is commonly generated in all LAM machines and these files are easy to edit due to universality they possess. Another investigation on STL shows that void can be placed inside the component through illegal access of these files which can reduce the yield strength by 14% for a tensile sample (Sturm, Williams, Camelio, White, & Parker, 2014). As LAM is mostly used for fabricating customized medical implants and devices, the changes made through illegal access in the cyber world can lead to multiple implications. Thus, cyber-attack is a serious threat and the protection of associated elements (data and manufacturing systems) is an ever-evolving issue. Advanced encryption of data, system identity, and machine-access management systems are being used to deliver safe and consistent transfer of information (Sturm, Williams, Camelio, White, & Parker, 2014).
8. Horizontal and vertical system integration is another pillar for Industry 4.0. The major concern for industries from the early days is the lack of interaction in horizontal and vertical partners. In horizontal partnership, real-time monitoring of activity can be used to schedule the tasks across the production line to adhere time schedules at a critical path. Computer-based manufacturing can aid in faster decisions to minimize the idle time between the fabrication of components in LAM systems. LAM allows customer interaction by providing the opportunity of real-time feedback, thereby improving customer satisfaction and provides vertical integration of the system. The design of the product can be seized online as per the customer feedback through simulations and 3D visualizations. Firms, divisions, activities, and abilities will become much more unified which enables automatic value chains through cross-company and worldwide integration of data.

LAM IN ENGINEERING AND PROSTHETIC APPLICATIONS

The journey to Industry 4.0 is a journey from a customized product era to of mass-customization era. In reality, mass customization is possible in Industry 4.0 through a combination of nine pillars of Industry 4.0. But, the major contributing factor towards "mass customization" is advanced manufacturing techniques, like - LAM, which provides the ability to personalize a product to an individual consumer's requirement with economic viability. The global scenario of LAM is presented in the following section for various engineering and medical sectors.

Aerospace

The first commercial AM machine, initially referred to as Rapid Prototyping, was developed in the late 1980s and Pratt & Whitney, the American aerospace manufacturers were among the first customers for the technology. Thus, the aerospace applications of AM started long back and the saga continued to LAM. The following

Factors attributed to the deployment of LAM in aerospace industries:

1. Components with a complex geometry: Most of the aerospace components have complex geometry because it requires parts with integrated functions. This can be an aerofoil or gas turbine blades with a cooling passage embedded in it. The aerospace sector also requires thin-walled structures, which can be fabricated expeditiously using LAM in comparison to that using conventional manufacturing processes (Paul, et al., 2012).
2. High-performance materials: Aerospace sectors use high-performance materials (such as – titanium based alloys, nickel based superalloys, ceramics, etc.). These materials are not only challenging, but also costly, and time-consuming to process using conventional manufacturing processes. However, LAM involves melting and deposition of the material in layer-by-layer fashion and thereby reduces the tedious machining process. The process window once developed can be used for fabricating an infinite number of parts (Jinoop, Paul, & Bindra, 2019).
3. Low buy to fly ratio: The buy to fly ratio refers to the weight of raw material acquired for a component and the final weight of the component. LAM can reduce this ratio significantly as powders and wire (wire has almost zero buy-to-fly ratios) are the most common raw material for fabricating components instead of ingots.
4. Quick turnaround time: For commercial aircraft, life expectancy is now more than 30 years, circumventing the necessity to preserve and substitute old tooling is a prominent inventory price benefit for companies. Through LAM, the manufacturing time for the test/replacement parts can now be as low as two weeks according to Airbus. The components can be promptly transported and mounted in a damaged aircraft to get it back into the air (Brandt, 2016).

LAM-PBF is suitable due to the ability to fabricate complex shapes, while LAM-DED provides an advantage in multi-axis fabrication and multi-material capability for aerospace applications. Table 3 summarises some examples of LAM built engineering components and their characteristics.

Table 3. Examples of LAM applications in engineering applications

S. No	Component	Material	AM Machine Utilized	Advantages	References
1	Primary flight control hydraulic component for Airbus A380	Titanium alloy	EOS M 290	35% Weight Reduction and fewerparts	(EOS GmbH, 2019)
2	Satellite Components	Al-Si-10Mg	EOS M400	40% Weight Reduction. Fabrication time of 80 hours	(EOS GmbH, n.d.)
3	Fuel Nozzle by General Electric	Nickel alloy	-	25% weight reduction and higher mechanical strength (5 times than conventional)	(Brandt, 2016) (Kellner, 2017)

Medical

In the medical industry, customized products are required to provide anatomy specific and case-specific solutions for prostheses as there is a difference in shape, size and mechanical properties of the human body parts (Paul, Jinoop, & Bindra, Metal Additive Manufacturing using lasers, 2018). On the other hand, case-specific solutions are required to address the orthopaedic abnormality and accidental injury issues, as it is almost impossible to have an idea before the event occurs. These two solutions are of great interest in the medical sector and require tailoring of the components. Commercially, various metals, like - cobalt chrome alloy, commercially pure titanium, Ti-6Al-4V, Nitinol, Tantalum, Magnesium, and alloys, etc. are used for fabricating implants (Malyala, Manmadhachary, Kumar, & Alwala, 2017). The envisaged deployment of LAM for the fabrication of prosthetics is presented in Fig. 4.

Figure 4. Deployment of LAM for the fabrication of prosthetics

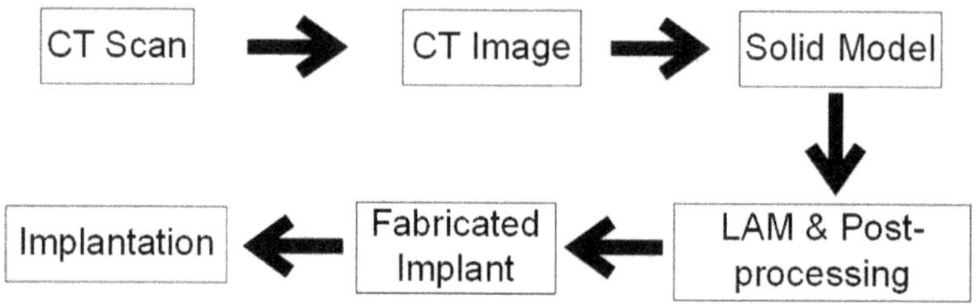

It starts with image acquisition using advanced imaging methods, such as - computed tomography (CT) and magnetic resonance imaging (MRI) to get the data cloud having requisite bone information. CT and MRI data are most suitable for modelling of bone structures and soft tissues, respectively. CT acquires 2D data where the slice thickness can be controlled during scanning. The slice thickness has an inverse relationship with the quality of the CT. Digital Imaging and Communications in Medicine (DICOM) images are created in this initial pre-processing stage to produce AM medical models that in turn are used to generate a 3D CAD model (Malyala, Manmadhachary, Kumar, & Alwala, 2017). Once bone data is produced in 3D, the required operations are performed using medical software such as region growing, cut, split, merge, etc. An STL file of the selected tissues is then produced and the anatomic parts (as defined in the STL file) can be LAMed, if required. The final post-processing of the tangible 3D-printed model is required before clinical use. The typical techniques used for post-processing also includes sterilization, ultrasonic cleaning, shot peening, etc. depending on design, process, and material.

Table 4. Examples of LAM applications in medical applications

Sl No.	Component	Material	AM Machine Utilized	Advantages	References
1	Hip implant	Titanium alloy	EOSINT M 280	Incorporated a large number of cavities	(Metal Powder Report, 2015)
2	Sternum	Titanium	EOSINT M280	Light and biocompatible making it an ideal prosthesis	(Dzian, 2018)

Heat treatment is usually performed to reduce stresses within the material. Partially melted powders should also be removed from the cellular structures (Brandt, 2016). In some cases, the surgeons will perform a mock surgery on the medical prototype first for necessary planning and practice for the actual surgery. Table 4 summarises some examples of LAM built medical prosthetics and their characteristics.

Other Engineering Applications

Table 5 summarises some other examples of LAM built engineering components and their characteristics.

Table 5. Examples of LAM applications in other engineering applications

Sl No.	Component	Material	AM Machine Utilized	Weight Reduction From Conventional Methods	Advantages	References
1	Suspension system (TransFIORmers)	Titanium	RensihawAM250	40%	Lighter, more rigid than conventional parts. Reduces un-sprung mass of the bike	(Renishaw plc., 2016)
2	Bicycle frame (Empire cycles)	Titanium	RensihawAM250	33%	Extremely strong and has high corrosion resistance	(Renishaw plc., 2019)
3	Conformal cooling channels (Salcomp)	-	EOSINT M 270	-	Production increased by more than 56,000 units. Cooling time reduced from 14 to 8 seconds. Rejection rate reduced from 2-1.4%	(EOS GmbH, 2013)

AM IN DEVELOPING ECONOMIES

One of the common modes of increasing gross domestic production (GDP) for developing economies has been by providing low-cost products to developed nations through a major reduction in labour costs (Tiger model economy). Recently, this is challenged with a reduction in labour costs in developed nations through automation. Now, the option available with developing economies to either restart the export of natural resources or adapt newer technologies where automation is not explicitly possible due to the involvement of creativity and human jurisdiction. As mass customization is possible with AM technologies, it is seen as one of the newer technologies to be adopted by developing economies (D'Aveni, 2019). BRICS (Brazil, Russia, India, China, and South Africa) nations are on lead to take this opportunity to their advantage.

AM in Brazil

AM is a rapidly growing market in Brazil. Editora Aranda's survey revealed that 54% of Brazillian plastic product manufacturing companies use AM for validating their designs or fabricating short runs of assembly parts. Other Industrial sectors such as architecture, medicine, jewellery, dentistry, and Do-It-Yourself (DIY) are also increasingly using AM. The rising application of AM has motivated the production of local printer manufacturing and development of a whole range of filaments with different technical features. The excitement around AM has also led to many AM related start-up companies and services for industries across a broad range (Souza, 2019)

AM in Russia

AM development in Russia is at nascent stages although it has mastered equipment and raw materials production for AM (json.tv, 2019). Tomsk Electronic Technologies is the first in the country to develop the electron beam metal AM machine in 2017 (Jackson B., 2018). Tomsk Polytechnic University has setups for electron-based AM, laser-based AM and fabrication of components with reinforced composites, etc. The centre is also involved in the development of AM related apparatus and software (Teslenko, Digilina, & Abdullaev, 2019). All-Russian Research Institute of Aviation Materials is involved in research for the manufacturing of AM consumables for prototyping applications in the field of metallurgy, aviation and space industry (Teslenko, Digilina, & Abdullaev, 2019). Many leading industries in Russia have begun shifting from prototyping to directly producing and repairing functional equipment. Tikhvin Freight Car Building Plant (UWC - United Wagon Company) is the first company in Russia to apply additive technologies for creating foundry model tooling (Teslenko, Digilina, & Abdullaev, 2019). AAC (Aresenyev Aviation Company) of Russia underwent modernization and is now using AM for producing moulds for components of Ka series of helicopters (Jackson B., 2018). Russian aircraft engine builder Aviadvigatel received $1.13 billion from the state for research and development of the next PD-35 engine that involves additive technology. Aviadvigatel, an aircraft engine builder uses all version of laser additive manufacturing processes for the cast part production on burned-out models, part repair and cultivating metal parts respectively. Russia's first AM built satellite launched by cosmonaut Fyodor Yurchikhin in 2017 (Jackson B., 2018). The "Luch" Design Bureau of Perm Motor Plant presented an unmanned aircraft with all the parts built by AM in 30 hours (Teslenko, Digilina, & Abdullaev, 2019). Experts predict that Russia will be developing AM technology in the field of powder materials, usage of carbon fibers for improving the strength of components, increasing production accuracy, service development by leasing AM systems and graphene usage for production of metal fibers and batteries (Teslenko, Digilina, & Abdullaev, 2019).

AM in China

China is heavily investing in AM technologies and prioritizing its development to compete with the US and Europe in the manufacturing sector. According to a forecast by the China Industry Information Institute, the projected output for China's AM industry is $7.68 billion, or one-third of the global market, by 2020 (Wei, 2019).

There are extensive works by universities and research institutes to develop indigenous AM capabilities with augmentation of machine learning (ML) and neural network (NN) algorithms (Qi, Chen, Li, Chen, & Li, 2019). The list of institutes includes Tsinghua University, Huazhong University of Science and Technology, Xi'an Jiaotong University, and South China University of Technology (Lin, Zhang, Zhang, Wang, & Zhang, 2012), (Scott, 2016), (Wang, et al., 2019), (Glover, 2017). Tsinghua University has developed AM systems for bio-manufacturing/bio-printing such as Rapid Ice Prototyping (RIP) process for fabricating porous tissue engineering scaffolds at low temperatures, Aerodynamically assisted tip-pen direct writing (TPDW) system capable of fabricating complex 3D scaffolds with nearly 50 μm feature size. Tsinghua University has a laser micro cladding deposition manufacturing equipment that has a digitalized micro-fluid technology for building fully dense parts. Huazhong University of Science and Technology developed technology by combining features of traditional forging and casting to the MAM process to produce components free of porosity and cracks (Scott, 2016). Researchers from

Xi'an Jiaotong University developed the hydrogel AM for bioprinting by incorporating the capacitor edge effect in the process which allows for patterning a large number of hydrogel solutions consisting of different chemical and physical properties (Wang, et al., 2019). The South China University of Technology printed a beak of Titanium for a red-crowned crane whose beak was injured in a fight rendering it unable to eat. The red-crowned crane was able to eat shortly after the transplant (Glover, 2017). Xi'an ZhiRong a Chinese company built a new electron beam additive manufacturing (EBAM) system named ZcompleX that has metal wire as feedstock (Clarke, 2017). Farsoon Technologies, of China, specializes in prototyping services, equipment for laser sintering, AM related materials and technical support (O'Neal, 2019). Nanfang Additive Manufacturing Technology Co. and Tubular Goods Research Institute (TGRI) of China National Petroleum Corporation (CNPC) signed a contract to explore the feasibility of electron beam manufacturing (EBM) for making oil and gas pipelines components. (Jackson B., 2018).

AM in South Africa

SA's Department of Science and Technology in 2016 presented a strategy to position SA strongly in the future AM market (Campbell, Beer, & Pei, 2011). More than 50 business operators are providing AM related services, like - service providers for consulting and design, suppliers for AM machines and related technologies. Currently AM in SA is focused on jewellery, tooling, and prototyping applications, but the government strategy also wants to apply AM in the fields of military and aerospace, automotive, medical, dental, conventional manufacturing (casting, tooling, refurbishment), materials development (titanium) (Abisourour, 2018). Aerosud, a SA aeronautical engineering, and manufacturing company based in Pretoria, together with Council for Scientific and Industrial Research (CSIR) has developed one of the world's biggest MAM machines as a part of project Aeroswift. In 2014, Doctors in SA used AM parts for successfully conducting jawbone replacement surgery. SA is also capable of indigenously developing its own AM machine named Robobeast possessing world-class quality. SA is hoping to give a push to an entrepreneurial environment hopes to generate a large number of jobs with the help of Robobeast (Mashambanhaka, 2019).

AM in India

National Institution for Transforming India (NITI) Aayog, the highest policy-making permanent commission, has identified 15 disruptive technologies for the future (National Institution for Transforming India, 2016). It includes AM as one of the technologies for India to adopt for long term results. All India Council for Technical Education (AICTE) introduced a National Doctoral Fellowship in India from the year 2018 and identified "Digital Manufacturing" as one of the thrust areas for research in India (AICTE National Doctoral Fellowship Scheme, 2018).

LAM technologies are growing in India from medical to manufacturing sector. In India, LAM technologies are used in academic and research institutes along with industries. Some of the major companies in India catering to the industry needs are Wipro, INTECH-DMLS, Incredible AM Pvt. Ltd., Objectify Technologies Pvt. Ltd, RAPID DMLS Inc., Renishaw, etc. They primarily use the LAM-PBF system to cater to various engineering and medical requirements.

The academic institutes in India, like - Indian Institute of Technologies (IIT's) are also venturing into the area of MAM. IIT Kharagpur, IIT Kanpur, IIT Palakkad have LAM-PBF systems installed ((Sarkar, Kumar, & Nath, 2017), (Kumar, et al., 2012), (Panchagnula & Simhambhatla, 2018), (Jhavar,

Jain, & Paul, 2014) (Chouhan, Sharma, Gulhane, Palani, & Lad, 2018), (Inauguration of Metal Additive Manufacturing (3D Metal Printing) Facility at IIT Palakkad, 2018)). IIT Bombay has a Hybrid MAM that uses Gas Metal Arc based deposition system and co-axial wire-based LAM-DED system. IIT Indore has micro-plasma and Metal Inert Gas (MIG) based deposition systems ((Kumar, et al., 2012) (Jhavar, Jain, & Paul, 2014) (Chouhan, Sharma, Gulhane, Palani, & Lad, 2018)).

Research Institutes in India working in LAM are Centre Manufacturing Technology Institute Bangalore (CMTI), Defence Metallurgical Research Laboratory Hyderabad (DMRL), International Advanced Research Centre for Powder Metallurgy and New Materials Hyderabad (ARCI), Central Mechanical Engineering Research Institute Durgapur (CMERI), Central Glass & Ceramic Research Institute (CGCRI) and Raja Ramanna Centre for Advanced Technology (RRCAT), etc. LAM-PBF machines are available in CMTI, RRCAT, and ARCI, while the LAM-DED machine is available in DMRL and CGCRI. RRCAT and CMERI have their own indigenously developed LAM-DED set-up, while RRCAT also has an indigenously developed LAM-PBF system.

Figure 5. Scheme of work at the authors' lab as per Industry 4.0

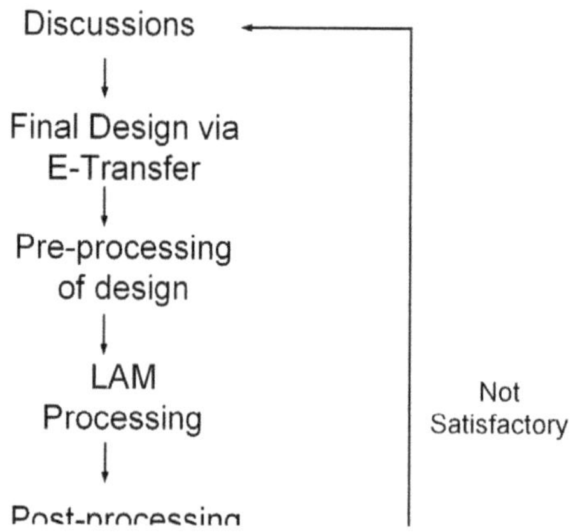

RRCAT is one of the premier National Laboratories in India that is working from LAM system development to process development until the component qualification for specific applications. It is the only

Table 6. Indigenously built LAM facilities at RRCAT

S. No.	Technology	Material	Build Size	Laser Power	Part/ Beam Manipulator
1	LAM-DED (Fig. 6(a))	Powder	250 mm x 250 mm x 250 mm	2 kW	CNC
2	LAM-DED (Fig. 6(b))	Powder	size: 1.6 m diameter x 2.1 m height	2 kW	Robotic arm
3	LAM-DED (Fig. 6(c))	Wire	250 mm x 250 mm x 250 mm	2 kW	CNC
4	LAM-PBF (Fig. 6(e))	Powder	250 mm x 250 mm x 250 mm	500 W	Galvano scanner

Figure 6. LAM Systems developed at RRCAT a) PFD based LAM-DED with processing volume of 250 mm X 250 mm X 250 mm b) PFD based LAM-DED with 1.6 m diameter and 2.1 m height and robotic head c) co-axial WFD based LAM-DED d) Architecture of LAM-DED systems e) LAM-PBF with build volume 250 mm x 250 mm x 250 mm
(Adapted from Nayak et al., 2020)

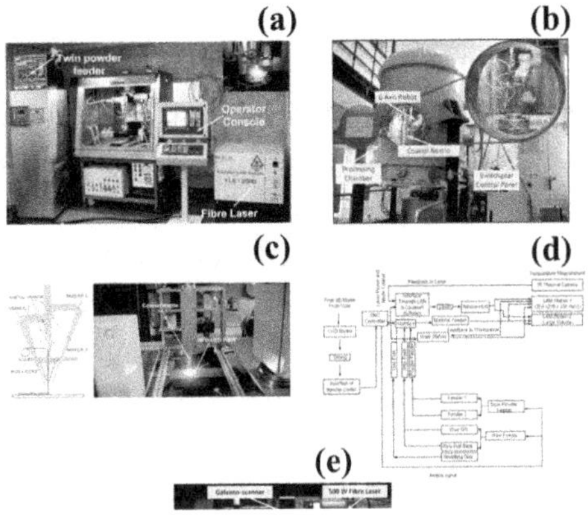

AM laboratory in the country to have both indigenously developed LAM-PBF and LAM-DED systems. The scheme of work at the author's lab is presented in Fig. 5. RRCAT has four indigenously developed LAM systems in which one is the LAM-PBF system with three systems under the LAM-DED category (Fig. 6). Table 6 summarises the LAM systems developed at RRCAT. Fig. 6(d) presents the architecture of LAM-DED systems developed at RRCAT.

FUTURE SCOPE

AM has revolutionized manufacturing by providing freedom in shape design, materials, logistics, and post-processing. It has also reduced the environmental impact concerning material consumption and water usage (Huang, Liu, & Mokasdar, 2013). Research and Development in LAM are being pursued to overcome inherent challenges and explore new possibilities. Some of the recent research focuses are:

- Development of higher power lasers (> 6 kW) possessing improved efficiency and process control for improving process repeatability
- Combining additive with subtractive and formative technologies i.e. developing hybrid technologies (Li, Ramakrishna, & Singh, 2018) for improving component quality (Paul, Jinoop, & Bindra, Metal Additive Manufacturing using lasers, 2018)
- Development of in-situ sensing technologies for monitoring temperature, cooling rate, etc. for eventually predicting microstructure and residual stresses
- System development for building large size engineering components that finds an application in rapid tooling and end-use components in the aerospace sector and nuclear industries (Huang, Leu, Mazumder, & Donmez, 2015).

CONCLUSION

The compendium of the global scenario of LAM and some of our above-described works presents a small glimpse of an immense area of enormous development in the domain. This glimpse fortifies the outlook that LAM is poised to solve imminent industrial problems for providing superior components and to work as a strong thrust force to bring revolution called "Industry 4.0". Ever-growing innovative technologies, entry of more competitors and compulsion to work towards adverse duty conditions is transforming the market to a customer-centric market working for better human life in terms of comfort and longevity. LAM is repositioning itself as game-changing technology and reforming the complete manufacturing sequence from "mass production to mass customization" and from "centralized to decentralized" manufacturing. The scope of LAM is extended to design and fabricate complex multi-functional parts with accurate control during fabrication. These advantages are enabling LAM to contribute to Industry 4.0 in a big way through the following five special freedoms - material design, shape design, quality, logistics, and post-processing. LAM is significantly contributing to advanced manufacturing in association with the other eight pillars (autonomous robots, simulation, cloud technology, cybersecurity, augmented reality, cybersecurity, system integration, and big data) of Industry 4.0. LAM is deployed worldwide for fabricating customized engineering and prosthetic components which is presented through typical case-studies.

FUNDING INFORMATION

This research received no specific grant from any funding agency in the public, commercial, or not-for-profit sectors.

ACKNOWLEDGMENT

The authors express their sincere gratitude to Mr. Debashis Das, Director RRCAT for his constant support and encouragement. Thanks are due to Mr. S V Nakhe, Director – Laser Group and Dr. K S Bindra, Associate Director – Laser Group for constant encouragement in this evolving program at RRCAT. The authors are also thankful to our collaborators and colleagues at Laser Additive Manufacturing Laboratory for their support.

REFERENCES

Abisourour, M. (2017). *Could 3D printing revolutionize manufacturing in Africa and the Middle East.* Retrieved from https://infomineo.com/additive-manufacturing-africa-middle-east

Additive, G. E. (n.d.). *Concept lasers-metal 3D printers.* Retrieved from https://www.ge.com/additive/de/who-we-are/concept-laser

AICTE. (2018). *AICTE National Doctoral Fellowship Scheme.* Retrieved from https://www.aicte-india.org/sites/default/files/AICTE%20NDF%20guidelines.pdf

Autodesk Inc. (2020). *Generative Design*. Retrieved from https://www.autodesk.com/solutions/generative-design

Bajaj, P., Wright, J., Todd, I., & Jägle, E. A. (2019). Predictive process parameter selection for Selective Laser Melting Manufacturing: Applications to high thermal conductivity alloys. *Additive Manufacturing*, *27*, 246–258. doi:10.1016/j.addma.2018.12.003

Banait, S. M., Jinoop, A. N., Paul, C. P., Kumar, H., Pawade, P. S., & Bindra, K. S. (2017). Investigating Laser Additive Manufacturing of Functionally Graded Ni-Cr-B-Si and SS 316L. In *International Conference on Advances in Thermal Systems, Materials and Design Engineering* (pp. 1-6). Mumbai: SSRN Electronic Journal. 10.2139srn.3101590

Bhavar, V., Kattire, P., Patil, V., Khot, S., Gujar, K., & Singh, R. (2017). A review on powder bed fusion technology of metal additive manufacturing. In A. B. Badiru, V. V. Valencia, & D. Liu (Eds.), *Additive manufacturing handbook: Product development for the defense industry* (pp. 251–261). Taylor and Francis. doi:10.1201/9781315119106-15

Brandt, M. (Ed.). (2016). *Laser Additive Manufacturing: Materials, design, technologies, and applications*. Woodhead Publishing.

Campbell, R. I., Beer, D. J., & Pei, E. (2011). Additive manufacturing in South Africa: Building on the foundations. *Rapid Prototyping Journal*, *17*(2), 156–162. doi:10.1108/13552541111113907

Cheng, G., Liu, L., Qiang, X., & Liu, Y. (2016). Industry 4.0 Development and Application of Intelligent Manufacturing. In *2016 International Conference on Information System and Artificial Intelligence* (pp. 407-410). Hong Kong: IEEE. 10.1109/ISAI.2016.0092

Chin, B. (2017). *The Perfect Combo: 3D Printing & Big Data*. Retrieved from https://publish.illinois.edu/digitalmaking2017/the-perfect-combo-3d-printing-big-data/

Chouhan, P., Sharma, S., Gulhane, S., Palani, I. A., & Lad, B. K. (2018). Investigation on influence of Nitinol wire made tailored structure manufactured using MIG additive technique for machine tool operations. *Materials Today: Proceedings*, *5*, 21350–21354.

Clarke, C. (2017). *China develops ZcompleX, an electron beam additive manufacturing machine to rival sciaky*. Retrieved from https://www.3dprintpulse.com/additive-manufacturing/china/?open-article-id=6396751&article-title=china-develops-zcomplex--an-electron-beam-additive-manufacturing-machine-to-rival-sciaky&blog-domain=3dprintingindustry.com&blog-title=3d-printing-industry

D'Aveni, R. A. (2019). *How 3-D printing can jumpstart developing economies*. Retrieved from https://www.forbes.com/sites/richarddaveni/2019/03/19/how-3d-printing-can-jumpstart-developing-economies/#3d6fdeeb4400

DM3D Technology LLC. (2008). *DMD machines solutions*. Retrieved from http://www.pomgroup.com

Deloitte. (2017). *Using autonomous robots to drive supply chain innovation*. Retrieved from https://www2.deloitte.com/in/en.html

Ding, D., Pan, Z., Cuiuri, D., & Li, H. (2015). Wire-feed additive manufacturing of metal components: Technologies, developments and future interests. *International Journal of Advanced Manufacturing Technology*, *81*(1-4), 465–481. doi:10.100700170-015-7077-3

Dusen, M. V. (2017). *GE's 3D-printed airplane engine will run this year*. Retrieved from https://www.ge.com/reports/mad-props-3d-printed-airplane-engine-will-run-year/

Dzian, A. Ž., Živčák, J., Penciak, R., & Hudák, R. (2018). Implantation of a 3D-printed titanium sternum in a patient with a sternal tumor. *World Journal of Surgical Oncology*, *16*(1), 1–4. doi:10.118612957-018-1315-8 PMID:29334989

Engelman, R. (n.d.). *The second industrial revolution,1870-1914*. Retrieved from http://ushistoryscene.com/article/second-industrial-revolution

EOS GmbH. (n.d.a). *Tooling: Salcomp - Additive manufacturing permits optimized cooling for maximum production efficiency*. Retrieved from https://www.eos.info/press/customer_case_studies/salcomp

EOS GmbH. (n.d.b). *First metal 3D printed primary flight control hydraulic component flies on an Airbus A380, EOS: Customer case studies*. Retrieved from https://www.eos.info/press/case_studies/first-3d-printed-hydraulic-component-flies-on-airbus-a380

EOS GmbH. (n.d.c). *Additive manufacturing of satellite components, EOS: Customer case studies*. Retrieved from https://www.eos.info/case_studies/additive-manufacturing-of-antenna-bracket-for-satellite

EOS GmbH. (n.d.d). *EOS systems and equipment for metal manufacturing*. Retrieved from https://www.eos.info

Frost and Sullivan. (n.d.). *Research and analysis*. Retrieved from https://ww2.frost.com

Furht, B. (Ed.). (2011). *Handbook of Augmented Reality*. Springer. doi:10.1007/978-1-4614-0064-6

Glover, S. (2017). *Additive manufacturing & applications in China*. Retrieved from https://www.laserchirp.com/2017/01/additive-manufacturing-applications-in-china

Government of India. (2016). *National Institution for Transforming India*. Retrieved from http://niti.gov.in

Horn, J., Rosenband, L., & Smith, R. (Eds.). (2010). *Reconceptualizing the industrial revolution*. MIT Press. doi:10.7551/mitpress/8585.001.0001

Huang, S. H., Liu, P., Mokasdar, A., & Hou, L. (2013). Additive manufacturing and its societal impact: A literature review. *International Journal of Advanced Manufacturing Technology*, *67*(5-8), 1191–1203. doi:10.100700170-012-4558-5

Huang, Y., Leu, M. C., Mazumder, J., & Donmez, A. (2015). Additive manufacturing: Current state, future potential, gaps and needs, and recommendations. *Journal of Manufacturing Science and Engineering*, *137*(1), 014001–014010. doi:10.1115/1.4028725

ISO. (2019). *ISO/ASTM 52900:2015 Additive manufacturing — General principles — Terminology*. Retrieved from https://www.iso.org/standard/69669.html

J'son & Partners™. (2019). *The 3D printing market in Russia and in the world in 2018 (Additive Manufacturing, AM)*. Retrieved from http://json.tv/en/ict_telecom_analytics_view/the-3d-printing-market-in-russia-and-in-the-world-in-2018-additive-manufacturing-am

Jackson, B. (2018). *Nanfang additive manufacturing confirms contract for 3D printed oil and gas pipelines*. Retrieved from https://3dprintingindustry.com/news/nanfang-additive-manufacturing-confirms-contract-3d-printed-oil-gas-pipelines-133504

Jackson, B. (2018). *Russian helicopters backs additive manufacturing for future aviation*. Retrieved from https://3dprintingindustry.com/news/russian-helicopters-backs-additive-manufacturing-future-aviation-133704

Jhavar, S., Jain, N. K., & Paul, C. P. (2014). Development of micro-plasma transferred arc (μ-PTA) wire deposition process for additive layer manufacturing applications. *Journal of Materials Processing Technology*, *214*(5), 1102–1110. doi:10.1016/j.jmatprotec.2013.12.016

Jinoop, A. N. (2016). *Investigation on laser shock processing of direct metal sintered Inconel 718* (Unpublished M.Tech Thesis). National Institute of Technology Warangal, Warangal, Telangana, India.

Jinoop, A. N., Paul, C. P., & Bindra, K. S. (2019). Laser assisted direct energy deposition of Hastelloy-X. *Optics & Laser Technology*, *109*, 14–49. doi:10.1016/j.optlastec.2018.07.037

Kellner, T. (2017). *Mind Meld: how GE and a 3D-printing visionary joined forces*. Retrieved from https://www.ge.com/reports/mind-meld-ge-3d-printing-visionary-joined-forces

Kumar, B. Y., Billo, T., Murmu, A., Choube, H., Karunakaran, K. P., & Bernard, A. (2012). Hybrid layered manufacturing using gas metal arc welding. *International Journal of Advanced Manufacturing Systems*, *14*(1), 11–18.

Li, C., Liu, J. F., & Guo, Y. B. (2016). Prediction of residual stress and part distortion in Selective Laser Melting. *Procedia CIRP*, *45*, 171–174. doi:10.1016/j.procir.2016.02.058

Li, C., Ramakrishna, S., & Singh, S. (2018). A review of digital manufacturing-based hybrid additive manufacturing processes. *International Journal of Advanced Manufacturing Technology*, *95*, 2281–2300. doi:10.100700170-017-1041-3

Lin, F., Zhang, L., Zhang, T., Wang, J., & Zhang, R. (2012). Innovative education in additive manufacturing in China. In *23rd Annual International Solid Freeform Fabrication Symposium* (pp. 14-44). uTexas: SFF.

Malyala, S. K., Manmadhachary, A., Kumar, Y. R., & Alwala, A. (2017). Manufacturing of patient specific additive manufacturing medical models for complex surgeries. *Materials Today: Proceedings*, *4*(2), 1134–1139.

Mashambanhaka, F. (2019). *3D printing Africa: South Africa's 3D printing sector*. Retrieved from https://3dprint.com/242448/3d-printing-africa-south-africas-3d-printing-sector

Meak, L. (2020). *The fourth industrial revolution: Changing how we live and work*. Retrieved from https://newsroom.cisco.com/feature-content?type=webcontent&articleId=1999882

Meiners, W., Wissenbach, K. D., & Gasser, A. D. (1998). *Germany Patent No. DE19649865C1*.

Metal Powder Report. (2015). *Titanium hip implant printed with EOS technology*. Retrieved from https://www.materialstoday.com/additive-manufacturing/news/titanium-hip-implant-printed-with-eos-technology/

Mordor Intelligence, L. L. P. (2020). *Simulation software market - growth, trends, and forecast (2020 - 2025)*. Retrieved from https://www.mordorintelligence.com/: https://www.mordorintelligence.com/industry-reports/simulation-software-market

Motta, M., Demir, A. G., & Previtali, B. (2018). High-speed imaging and process characterization of co-axial laser metal wire deposition. *Additive Manufacturing*, *22*, 497–507. doi:10.1016/j.addma.2018.05.043

Nayak, S. K., Mishra, S. K., Paul, C. P., Jinoop, A. N., & Bindra, K. S. (2020). Effect of energy density on laser powder bed fusion built single tracks and thin wall structures with 100 μm preplaced powder layer thickness. *Optics & Laser Technology*, *125*, 106016. doi:10.1016/j.optlastec.2019.106016

O'Neal, B. (2019). *Airbus & Farsoon Technologies in Collaboration to Develop High-Performance Polymers for Additive Manufacturing*. Retrieved from https://www.3dprintpulse.com/additive-manufacturing/china/?open-article-id=9819375&article-title=airbus---farsoon-technologies-in-collaboration-to-develop-high-performance-polymers-for-additive-manufacturing&blog-domain=3dprint.com&blog-title=3dprint-com

Optomec Inc. (2018). *The internet of things*. Retrieved from https://optomec.com/printed-electronics/aerosol-jet-core-applications/internet-of-things/

Palakkad, I. I. T. (2019). *Inauguration of metal additive manufacturing (3D metal printing) facility at IIT Palakkad*. Retrieved from https://iitpkd.ac.in/news/inauguration-metal-additive-manufacturing-3d-metal-printing-facility-iit-palakkad

Palmarini, P., Erkoyuncu, J. A., Roy, R., & Torabmostaedi, H. (2018). A systematic review of augmented reality applications in maintenance. *Robotics and Computer-integrated Manufacturing*, *49*, 215–228. doi:10.1016/j.rcim.2017.06.002

Panchagnula, J. S., & Simhambhatla, S. (2018). Manufacture of complex thin-walled metallic objects using weld-deposition based additive manufacturing. *Robotics and Computer-integrated Manufacturing*, *49*, 194–203. doi:10.1016/j.rcim.2017.06.003

Paul, C. P., Bhargava, P., Kumar, A., Pathak, A. K., & Kukreja, L. M. (2012). Laser Rapid Manufacturing: technology, applications, modeling and future prospects. In J. P. Davim (Ed.), *Lasers in Manufacturing* (pp. 1–60). ISTE-Wiley.

Paul, C. P., Jinoop, A. N., & Bindra, K. S. (2018). Metal Additive Manufacturing using lasers. In R. Singh & J. P. Davim (Eds.), *Additive Manufacturing: Applications and innovations* (pp. 37–94). CRC Press. doi:10.1201/b22179-2

Paul, C. P., Mishra, S. K., Premsingh, C. H., Bhargava, P., Tiwari, P., & Kukreja, L. M. (2012). Studies on laser rapid manufacturing of cross-thin-walled porous structures of Inconel 625. *International Journal of Advanced Manufacturing Technology*, *61*(5-8), 757–770. doi:10.100700170-011-3742-3

Pinkerton, A. J. (2016). Lasers in additive manufacturing. *Optics & Laser Technology*, *78*, 25–32. doi:10.1016/j.optlastec.2015.09.025

Pour, M. A., Zanardini, M., Bacchetti, A., & Zanoni, S. (2016). Additive Manufacturing Impacts on Productions and Logistics Systems. *IFAC-PapersOnLine*, *49*(12), 1679–1684. doi:10.1016/j.ifacol.2016.07.822

Qi, X., Chen, G., Li, Y., Chen, X., & Li, C. (2019). Applying neural-network-based machine learning to additive manufacturing: Current applications, challenges, and future perspectives. *Engineering*, *5*(4), 721–729. doi:10.1016/j.eng.2019.04.012

Quintanilla, P., Hope, D. A., Darnton, H., & Hunter, A. (2019). *Responsible use of technology*. Retrieved from https://www.weforum.org/whitepapers/responsible-use-of-technology

Reisman, G. (1996). *Capitalism: A Complete and Integrated Understanding of the Nature and Value of Human Economic Life*. Jameson Books.

Renishaw plc. (2016). *Metal 3D printing pushes the boundaries in Moto2™ through defiant innovation*. Retrieved from http://www.renishaw.com/en/metal-3d-printing-pushes-the-boundaries-in-moto2-through-defiant-innovation--38935

Renishaw plc. (2019). *Resource centre - Case study: First metal 3D printed bicycle frame manufactured by Renishaw for Empire cycles*. Retrieved from https://resources.renishaw.com/en/details/case-study-first-metal-3d-printed-bicycle-frame-manufactured-by-renishaw-for-empire-cycles--57197

Robotic Industries Association (RIA). (2018). *Collaborative robot market growing*. Retrieved from https://www.controleng.com/articles/collaborative-robot-market-growing/

Rosa, B., Mognol, P., & Hascoët, J. Y. (2014). Topography modeling of laser polishing on AISI 316L. *Mechanics & Industry*, *15*(1), 51–61. doi:10.1051/meca/2014003

Sandia National Laboratories. (2017). *Laser Engineered Net Shaping™*. Retrieved from https://www.sandia.gov/mst/technologies/net-shaping.html

Sarkar, S., Kumar, C. S., & Nath, A. K. (2017). Effect of Different Heat Treatments on Mechanical Properties of Laser Sintered Additive Manufactured Parts. *Journal of Manufacturing Science and Engineering*, *139*(11), 1–11. doi:10.1115/1.4037437

Scott, C. (2016). *China develops new metal 3D printing technology, Combining old and new manufacturing techniques*. Retrieved from https://3dprint.com/143613/china-metal-3d-printing-tech/

Shiva, S., Palani, I. A., Paul, C. P., & Singh, B. (2016). Laser annealing of laser additive– manufactured Ni-Ti structures: An experimental–numerical investigation. *Proceedings of the Institution of Mechanical Engineers. Part B, Journal of Engineering Manufacture*, *232*(6), 1054–1067. doi:10.1177/0954405416661582

Silva, J. V., & Rezende, R. A. (2013). Additive manufacturing and its future impact in logistics. *IFAC Proceedings Volumes, 46*(24), 277-282.

Solutions, S. L. M. (2018). *SLM® machines*. Retrieved from https://slm-solutions.com

Souza, H. (2019). *3D printing: a growing market in Brazil*. Retrieved from https://3dprint.com/236228/3d-printing-a-growing-market-in-brazil/

Sturm, L. D., Williams, C. B., Camelio, J. A., White, J., & Parker, R. (2017). Cyber-physical vulnerabilities in additive manufacturing systems: A case study attack on the. STL file with human subjects. *Journal of Manufacturing Systems*, *44*, 154–164. doi:10.1016/j.jmsy.2017.05.007

Tan, Z. E., Pang, J. H., Kaminski, J., & Pepin, H. (2019). Characterisation of porosity, density, and microstructure of directed energy deposited stainless steel AISI 316L. *Additive Manufacturing*, *25*, 286–296. doi:10.1016/j.addma.2018.11.014

Tao, F., Tang, Y., Zou, X., & Qi, Q. (2019). A field programmable gate array implemented fibre channel switch for big data communication towards smart manufacturing. *Robotics and Computer-integrated Manufacturing*, *57*, 166–181. doi:10.1016/j.rcim.2018.12.005

Technavio. (2020). *Metal additive manufacturing market by application and geography - forecast and analysis 2020-2024*. Retrieved from https://www.technavio.com/report/global-metal-additive-manufacturing-market-analysis

Teslenko, I. B., Digilina, O. B., & Abdullaev, N. V. (2019). Status and perspectives for the use of additive technologies in various branches of Russian industry. *IOP Conference Series. Materials Science and Engineering*, *483*, 012093. doi:10.1088/1757-899X/483/1/012093

The Fabricator. (2016). *Aerospace 3-D printing market to grow at 55.85 percent CAGR to 2020*. Retrieved from https://www.thefabricator.com/thefabricator/news/shopmanagement/aerospace-3-d-printing-market-to-grow-at-55-85-percent-cagr-to-2020

Vaidya, S., Ambad, P., & Bhosle, S. (2018). Industry 4.0 – A Glimpse. *Procedia Manufacturing*, *20*, 233–238. doi:10.1016/j.promfg.2018.02.034

Vieira, A. A., Dias, L. M., Santos, M. Y., Pereira, G. A., & Oliveira, J. A. (2018). Setting an industry 4.0 research and development agenda for simulation - a literature review. *International Journal of Simulation Modelling*, *17*(3), 377–390. doi:10.2507/IJSIMM17(3)429

Wang, W., Lu, T., Yang, M., Sun, D., Xia, Y., & Wang, T. (2019). Hydrogel 3D printing with the capacitor edge effect. *Science Advances*, *5*, 1–7. PMID:30915393

Wei, H. (2019). *Enabler of digital transformation*. Retrieved from http://global.chinadaily.com.cn/a/201904/15/WS5cb3dda3a3104842260b62af.html

Xu, X. (2012). From cloud computing to cloud manufacturing. *Robotics and Computer-integrated Manufacturing*, *28*(1), 75–86. doi:10.1016/j.rcim.2011.07.002

ADDITIONAL READING

Gibson, I., Rosen, D. W., & Stucker, B. (2015). *Additive manufacturing technologies: Rapid prototyping to direct digital manufacturing*. Springer-Verlag. doi:10.1007/978-1-4939-2113-3

Gilchrist, A. (2016). *Industry 4.0: The Industrial Internet of Things*. Apress. doi:10.1007/978-1-4842-2047-4

Gu, D. (2015). *Laser Additive Manufacturing of High-Performance Materials*. Springer-Verlag. doi:10.1007/978-3-662-46089-4

KEY TERMS AND DEFINITIONS

Additive Manufacturing: A process of joining materials to make objects from 3D model data, usually layer upon layer, as opposed to subtractive and formative manufacturing methodologies.

Directed Energy Deposition: An additive manufacturing process in which focused thermal energy is used to fuse materials by melting as they are being deposited.

Industry 4.0: A name for the current trend of automation and data exchange in manufacturing technologies, including cyber-physical systems, the Internet of things, cloud computing and cognitive computing and creating the smart factory.

Laser: A laser is a device that emits light through a process of optical amplification based on the stimulated emission of electromagnetic radiation.

Manufacturing: Manufacturing is the processing of raw materials or parts into finished goods using tools, human labour, machinery, and chemical processing.

Mass Customization: Mass customization is a marketing and manufacturing technique which combines the flexibility and personalization of custom-made products with the low unit costs associated with mass production.

Powder Bed Fusion: An additive manufacturing process in which thermal energy selectively fuses regions of a powder bed.

Compilation of References

3D . Printing in Aerospace & Aviation | GE Additive. (n.d.). Retrieved February 25, 2020, from https://www.ge.com/additive/additive-manufacturing/industries/aviation-aerospace

3D . Systems Company. (2017). *3DXpert software for metal additive manufacturing*. Retrieved from Lecture Notes Online Web Site: https://www.3dsystems.com/software/3dxpert

3. Dsim Company. (2017). *Additive manufacturing simulation*. Retrieved from Lecture Notes Online Web Site: http://3dsim.com

Abbaszadeh, M., Hönnige, J. R., Martina, F., Neto, L., Kashaev, N., Colegrove, P., & Klusemann, B. (2019). Numerical Investigation of the Effect of Rolling on the Localized Stress and Strain Induction for Wire + Arc Additive Manufactured Structures. *Journal of Materials Engineering and Performance*, *28*(8), 4931–4942. doi:10.100711665-019-04249-y

Abdulhameed, O., Al-Ahmari, A., Ameen, W., & Mian, S. H. (2019). Additive manufacturing: Challenges, trends, and applications. *Advances in Mechanical Engineering*, *11*(2), 1–27. doi:10.1177/1687814018822880

Abe, T., & Sasahara, H. (2016). Dissimilar metal deposition with a stainless steel and nickel-based alloy using wire and arc-based additive manufacturing. *Precision Engineering*, *45*, 387–395. doi:10.1016/j.precisioneng.2016.03.016

Abioye, T. E., Folkes, J., & Clare, A. T. (2013). A parametric study of Inconel 625 wire laser deposition. *Journal of Materials Processing Technology*, *213*(12), 2145–2151. doi:10.1016/j.jmatprotec.2013.06.007

Abisourour, M. (2017). *Could 3D printing revolutionize manufacturing in Africa and the Middle East*. Retrieved from https://infomineo.com/additive-manufacturing-africa-middle-east

Aboulkhair, N. T., Everitt, N. M., Ashcroft, I., & Tuck, C. (2014). Reducing porosity in AlSi10Mg parts processed by selective laser melting. *Additive Manufacturing*, *1*, 77–86. doi:10.1016/j.addma.2014.08.001

Aboulkhair, N. T., Simonelli, M., Parry, L., Ashcroft, I., Tuck, C., & Hague, R. (2019). Progress in Materials Science 3D printing of Aluminium alloys : Additive Manufacturing of Aluminium alloys using selective laser melting. *Progress in Materials Science*, *106*(July), 100578. doi:10.1016/j.pmatsci.2019.100578

Acharya, R., Sharon, J. A., & Staroselsky, A. (2017). Prediction of microstructure in laser powder bed fusion process. *Acta Materialia*, *124*, 360–371. doi:10.1016/j.actamat.2016.11.018

Additive Manufacturing & 3D Printing Processes | GE Additive. (n.d.). Retrieved February 25, 2020, from https://www.ge.com/additive/additive-manufacturing/information/additive-manufacturing-processes

Additive Manufacturing for the Aerospace Industry. (2019). Elsevier.

Additive manufacturing in aerospace is growing - 3Dnatives. (n.d.). Retrieved February 25, 2020, from https://www.3dnatives.com/en/additive-manufacturing-aerospace-growing-061220184/

Additive Manufacturing in the Aerospace Industry > ENGINEERING.com. (n.d.). Retrieved February 25, 2020, from https://www.engineering.com/AdvancedManufacturing/ArticleID/14218/Additive-Manufacturing-in-the-Aerospace-Industry.aspx

Additive Manufacturing Industry overview | GE Additive. (n.d.). Retrieved February 25, 2020, from https://www.ge.com/additive/industry-overview

Additive, G. E. (n.d.). *Concept lasers-metal 3D printers.* Retrieved from https://www.ge.com/additive/de/who-we-are/concept-laser

Aerospace 3D printing Applications | 3D Hubs. (n.d.). Retrieved February 25, 2020, from https://www.3dhubs.com/knowledge-base/aerospace-3d-printing-applications/

Agarwala, M., Bourell, D., Beaman, J., Marcus, H., & Barlow, J. (1995). Direct selective laser sintering of metals. *Rapid Prototyping Journal, 1*(1), 26–36. doi:10.1108/13552549510078113

Agustí, J. I. (2017). Environmental design guidelines for digital fabrication. *Journal of Cleaner Production, 142,* 2780–2791. doi:10.1016/j.jclepro.2016.10.190

Ahmad, B., van der Veen, S. O., Fitzpatrick, M. E., & Guo, H. (2018). Residual stress evaluation in selective-laser-melting additively manufactured titanium (Ti-6Al-4V) and inconel 718 using the contour method and numerical simulation. *Additive Manufacturing, 22,* 571–582. doi:10.1016/j.addma.2018.06.002

Ahmed, K., Allen, T., & El-Azab, A. (2016). Phase field modeling for grain growth in porous solids. *Journal of Materials Science, 51*(3), 1261–1277. doi:10.100710853-015-9107-9

AICTE. (2018). *AICTE National Doctoral Fellowship Scheme.* Retrieved from https://www.aicte-india.org/sites/default/files/AICTE%20NDF%20guidelines.pdf

Airbus Gets 1st Production LEAP-1A Jet Engines - GE Reports. (n.d.). Retrieved February 25, 2020, from https://www.ge.com/reports/airbus-gets-1st-production-jet-engines-with-3d-printed-parts-from-cfm/

Akbari, M., & Kovacevic, R. (2018). An investigation on mechanical and microstructural properties of 316LSi parts fabricated by a robotized laser / wire direct metal deposition system. *Additive Manufacturing, 23*(August), 487–497. doi:10.1016/j.addma.2018.08.031

Akca, E., & Gursel, A. (2015). A Review on Superalloys and IN718 Nickel-Based INCONEL Superalloy. *Periodicals of Engineering and Natural Sciences, 3*(1), 15–27. doi:10.21533/pen.v3i1.43

Al-Dabbagh, J. B., Al-Faluji, I. K., & Hashim, Y. B. (2010). *Negative Thermal Expansion in Ferromagnetic Fe-Ni Invar Alloys.* Paper presented at the Proceedings of ICSSST2010: 3rd International Conference on Solid State Science & Technology, Kuching, Sarawak, Malaysia.

Alida, M. (2009). Direct fabrication through electron beam melting technology of custom cranial implants designed in a PHAN To M-based haptic environment. *Materials & Design, 30*(8), 3186–3192. doi:10.1016/j.matdes.2008.11.013

Aliyi, A. M., & Lemu, H. G. (2019). Case study on topology optimized design for additive manufacturing. *IOP Conf. Series (IRMES 2019). Materials Science and Engineering, 659*(012020), 1–9. doi:10.1088/1757-899X/659/1/012020

Alkadi, F., Lee, J.W., Yeo, J.S., Hwang, S.H., & Choi, J.W. (2019). 3D Printing of ground tire rubber composites. *International Journal of Precision Engineering and Manufacturing-Green Technology, 6,* 211-222. doi:10.100740684-019-00023-6

Amato, K. N., Gaytan, S. M., Murr, L. E., Martinez, E., Shindo, P. W., Hernandez, J., & Medina, F. (2012). Microstructures and mechanical behavior of Inconel 718 fabricated by selective laser melting. *Acta Materialia*, *60*(5), 2229–2239. doi:10.1016/j.actamat.2011.12.032

Ambrogio, G., Gagliardi, F., Muzzupappa, M., & Filice, L. (2019). Additive-incremental forming hybrid manufacturing technique to improve customised part performance. *Journal of Manufacturing Processes, 37*(November), 386–391. doi:10.1016/j.jmapro.2018.12.008

Anon Software. (2017). *Topology optimization guide.* Retrieved from Lecture Notes Online Web Site: http:// www.topology-opt.com/software-list/

ANSYS Software Company. (2019). *Ansys additive suite.* Retrieved from https:// www. ansys. com /products/struct-ures/ansys-additive-suite

Anton, W., Johan, P., & Johan, O. (2018). An optimization framework for additive manufacturing given topology optimization results. *Toolsand Methods for Competitive Engineering (TMCE) 2018 Proceedings,* 533-542.

Aremu, A., Ashcroft, I., Hague, R., Wildman, R., & Tuck, C. (2010). Suitability of SIMP and BESO Topology Optimization Algorithms for Additive Manufacture. In *In: 21st Annual International Solid Freeform Fabrication Symposium* (pp. 679–692). University of Texas. http://sffsymposium.engr.utexas.edu/Manuscripts/2010/2010-57-Aremu.pdf

Arısoy, Y. M., Criales, L. E., & Özel, T. (2019). Modeling and simulation of thermal field and solidification in laser powder bed fusion of nickel alloy IN625. *Optics & Laser Technology*, *109*, 278–292. doi:10.1016/j.optlastec.2018.08.016

Asgari, H., Salarian, M., Ma, H., Olubamiji, A., & Vlasea, M. (2018). On thermal expansion behavior of invar alloy fabricated by modulated laser powder bed fusion. *Materials & Design*, *160*, 895–905. doi:10.1016/j.matdes.2018.10.025

Asiabanpour, B., Mokhtar, A., & Houshmand, M. (2008). Rapid manufacturing. In *Collaborative Engineering*. Theory and Practice. doi:10.1007/978-0-387-47321-5_7

ASTM International. (2014). *F3049 - 14 Standard guide for characterizing properties of metal powders used for additive manufacturing processes*. Author.

ASTM International. (2015). *ISO/ASTM 52900 Standard Terminology for Additive Manufacturing - General Principles - Terminology*. ASTM International.

ASTM International. (2017). *B822-17 Standard Test Method for Particle Size Distribution of Metal Powders and Related Compounds by Light Scattering*. Author.

Attar, E. (2011). *Simulation of Selective Electron Beam Melting Processes* (Doctoral dissertation). University of Erlangen, Nurnberg, Germany. https://d-nb.info/1013101472/34

Attar, H., Bermingham, M. J., Ehtemam-Haghighi, S., Dehghan-Manshadi, A., Kent, D., & Dargusch, M. S. (2019). Evaluation of the mechanical and wear properties of titanium produced by three different additive manufacturing methods for biomedical application. *Materials Science and Engineering A*, *760*, 339–345. doi:10.1016/j.msea.2019.06.024

Attar, H., Ehtemam-Haghighi, S., Kent, D., & Dargusch, M. S. (2018). Recent developments and opportunities in additive manufacturing of titanium-based matrix composites: A review. *International Journal of Machine Tools & Manufacture*, *133*, 85–102. doi:10.1016/j.ijmachtools.2018.06.003

Atzeni, E., Barletta, M., Calignano, F., Iuliano, L., Rubino, G., & Tagliaferri, V. (2016). Abrasive Fluidized Bed (AFB) finishing of AlSi10Mg substrates manufactured by Direct Metal Laser Sintering (DMLS). *Additive Manufacturing*, *10*, 15–23. doi:10.1016/j.addma.2016.01.005

Autodesk Inc. (2020). *Generative Design*. Retrieved from https://www.autodesk.com/solutions/generative-design

Avraham, T., Franco, P., Brecht, L., Ceradini, D., Saadeh, P., Hirsch, D., & Levine, J. (2014). Functional outcomes of virtually planned free fibula flap reconstruction of the mandible. *Plast. Reconstr. Surg, 134*, 628–634. 0000000 0513 doi:10.1097/ prs.00000

Ayarkwa, K. F., Williams, S. W., & Ding, J. (2017). Assessing the effect of TIG alternating current time cycle on aluminium wire + arc additive manufacture. *Additive Manufacturing, 18*, 186–193. doi:10.1016/j.addma.2017.10.005

Aydinöz, M. E., Brenne, F., Schaper, M., Schaak, C., Tillmann, W., Nellesen, J., & Niendorf, T. (2016). On the microstructural and mechanical properties of post-treated additively manufactured Inconel 718 superalloy under quasi-static and cyclic loading. *Materials Science and Engineering A, 669*, 246–258. doi:10.1016/j.msea.2016.05.089

Bai, Y., Yang, Y., Wang, D., & Zhang, M. (2017). Influence mechanism of parameters process and mechanical properties evolution mechanism of maraging steel 300 by selective laser melting. *Materials Science and Engineering A, 703*, 116–123. doi:10.1016/j.msea.2017.06.033

Bajaj, P., Wright, J., Todd, I., & Jägle, E. A. (2019). Predictive process parameter selection for Selective Laser Melting Manufacturing: Applications to high thermal conductivity alloys. *Additive Manufacturing, 27*, 246–258. doi:10.1016/j.addma.2018.12.003

Balachandramurthi, A. R., Moverare, J., Dixit, N., Deng, D., & Pederson, R. (2019). Microstructural influence on fatigue crack propagation during high cycle fatigue testing of additively manufactured Alloy 718. *Materials Characterization, 149*, 82–94. doi:10.1016/j.matchar.2019.01.018

Balachandramurthi, A., Moverare, J., Mahade, S., & Pederson, R. (2019). Additive Manufacturing of Alloy 718 via Electron Beam Melting: Effect of Post-Treatment on the Microstructure and the Mechanical Properties. *Materials (Basel), 12*(1), 68. doi:10.3390/ma12010068 PMID:30585242

Balla, V. K., Bose, S., & Bandyopadhyay, A. (2008). Processing of bulk alumina ceramics using laser engineered net shaping. *International Journal of Applied Ceramic Technology, 5*(3), 234–242. doi:10.1111/j.1744-7402.2008.02202.x

Banait, S. M., Jinoop, A. N., Paul, C. P., Kumar, H., Pawade, P. S., & Bindra, K. S. (2017). Investigating Laser Additive Manufacturing of Functionally Graded Ni-Cr-B-Si and SS 316L. In *International Conference on Advances in Thermal Systems, Materials and Design Engineering* (pp. 1-6). Mumbai: SSRN Electronic Journal. 10.2139srn.3101590

Bandyopadhyay, A., & Heer, B. (2018). Additive manufacturing of multi-material structures. *Materials Science and Engineering R Reports, 129*, 1–16. doi:10.1016/j.mser.2018.04.001

Bandyopadhyay, A., Krishna, B., Xue, W., & Bose, S. (2009). Application of laser engineered net shaping (LENS) to manufacture porous and functionally graded structures for load-bearing implants. *Journal of Materials Science. Materials in Medicine, 20*(S1), 29–34. doi:10.100710856-008-3478-2 PMID:18521725

Barazanchi, A., Li, K. C., Al-Amleh, B., Lyons, K., & Waddell, J. N. (2017). Additive Technology: Update on Current Materials and Applications in Dentistry. *Journal of Prosthodontics, 26*(2), 156–163. doi:10.1111/jopr.12510 PMID:27662423

Barbosa, G. F., & Aroca, R. V. (2017). An IoT-Based Solution for Control and Monitoring of Additive Manufacturing Processes. *Journal of Powder Metallurgy & Mining, 06*(01). Advance online publication. doi:10.4172/2168-9806.1000158

Barros, R., Silva, F. J. G., Gouveia, R. M., Saboori, A., Marchese, G., Biamino, S., Salmi, A., & Atzeni, E. (2019). *Stress Analysis Before and After Heat Treatment*. Academic Press.

Bartlett, J. L., & Li, X. (2019). An overview of residual stresses in metal powder bed fusion. *Additive Manufacturing, 27*, 131–149. doi:10.1016/j.addma.2019.02.020

Bassoli, E., Sola, A., Celesti, M., Calcagnile, S., & Cavallini, C. (2018). *Development of Laser-Based Powder Bed Fusion Process Parameters and Scanning Strategy for New Metal Alloy Grades : A Holistic Method Formulation*. doi:10.3390/ma11122356

Basu, B., Lundback, A., & Lindgren, L. E. (2019). Simulation of Ti-6Al-4V Additive manufacturing using coupled physically based flow stress and metallurgical model. *Materials, 12*(23). doi:10.3390/ma12233844

Bauer, T., Dawson, K., Spierings, A. B., & Wegener, K. (2015, August). Microstructure and mechanical characterisation of SLM processed Haynes® 230®. In *Proceedings of the 26th annual international Solid Freeform Fabrication Symposium*, (pp. 813-822). Academic Press.

Baufeld, B., Brandl, E., & Van Der Biest, O. (2011). Wire based additive layer manufacturing: Comparison of microstructure and mechanical properties of Ti-6Al-4V components fabricated by laser-beam deposition and shaped metal deposition. *Journal of Materials Processing Technology*, *211*(6), 1146–1158. doi:10.1016/j.jmatprotec.2011.01.018

Bekker, A. C. M., Verlinden, J. C., & Galimberti, G. (2016). *Challenges in Assessing the Sustainability of Wire + Arc Additive Manufacturing for Large Structures*. Academic Press.

Berdine, G. G., DiPaola, M., & Weinberg, M. (2019). Economic and Regulatory Perspectives on Additive Manufacturing. In M. Dipaola & F. M. Wodajo (Eds.), 3D Printing in Orthopaedic Surgery (pp. 41-48). Elsevier. doi:10.1016/B978-0-323-58118-9.00004-X

Bhardwaj, T., Shukla, M., Paul, C. P., & Bindra, K. S. (2019). Direct Energy Deposition - Laser Additive Manufacturing of Titanium-Molybdenum alloy: Parametric studies, microstructure and mechanical properties. *Journal of Alloys and Compounds*, *787*, 1238–1248. doi:10.1016/j.jallcom.2019.02.121

Bhargava, D., Bartlett, P., Russell, J.L., & Liddington, M. (2010). Construction of titanium cranioplasty plate using craniectomy bone flap as template. *Acta Neurochir*, 173–176. 10.1007/s00701-009-0394-2

Bhavar, V., Kattire, P., Patil, V., Khot, S., Gujar, K., & Singh, R. (2017). A review on powder bed fusion technology of metal additive manufacturing. In A. B. Badiru, V. V. Valencia, & D. Liu (Eds.), *Additive manufacturing handbook: Product development for the defense industry* (pp. 251–261). Taylor and Francis. doi:10.1201/9781315119106-15

Bikas, H., Stavropoulos, P., & Chryssolouris, G. (2016). Additive manufacturing methods and modelling approaches: A critical review. *International Journal of Advanced Manufacturing Technology*, *83*(1-4), 389–405. doi:10.100700170-015-7576-2

Birnbaum, A. J., Vikelic, S., & Lawrence Yao, Y. (2010). Advances in laser-induced plastic deformation processes. In J. Lawrence, J. Pou, D. K. Y. Low, & E. Toyserkani (Eds.), Advances in Laser Materials Processing (pp. 535-572). Woodhead Publishing.

Boddeti, N., Ding, Z., Kaijima, S., Maute, K., & Dunn, M. L. (2018). Simultaneous digital design and additive manufacture of structures and materials. *Scientific Reports*, *8*(1), 15560. doi:10.103841598-018-33454-3 PMID:30348943

Bogers, M., Hadar, R., & Bilberg, A. (2016). Additive manufacturing for consumer-centric business models: Implications for supply chains in consumer goods manufacturing. *Technological Forecasting and Social Change*, *102*, 225–239. doi:10.1016/j.techfore.2015.07.024

Bogu, V. P., Kumar, Y.R., & Khanra, A. (2017). Modelling and structural analysis of skull/cranial implant: beyond midline deformities. *Acta of Bioengineering and Biomechanics, 19*(1). doi:10.5277/ABB-00547-2016-04

Bourell, D. L. (2016). Perspectives on Additive Manufacturing. *Annual Review of Materials Research*, *46*(1), 1–18. doi:10.1146/annurev-matsci-070115-031606

Bourell, D., Kruth, J. P., Leu, M., Levy, G., Rosen, D., Beese, A. M., & Clare, A. (2017). Materials for additive manufacturing. *CIRP Annals*, *66*(2), 659–681. doi:10.1016/j.cirp.2017.05.009

Bowman, R. (2000). *Superalloys: A primer and history*. The Minerals, Metals & Materials Society.

Brackett, D., Ashcroft, L., & Hague, R. J. M. (2011). Topology optimization for additive manufacturing. *Wolfson School of Mechanical and Manufacturing Engineering*, 348-362. Retrieved from https://sffsymposium.engr.utexas.edu/Manuscripts/2011/2011-27-Brackett.pdf

Brandt, M. (Ed.). (2016). *Laser Additive Manufacturing: Materials, design, technologies, and applications*. Woodhead Publishing.

Brezinová, J., Hudák, R., Guzanová, A., Draganovská, D., Ižaríková, G., & Koncz, J. (2016). Direct Metal Laser Sintering of Ti6Al4V for Biomedical Applications: Microstructure, Corrosion Properties, and Mechanical Treatment of Implants. *Metals*, *6*(7), 171. doi:10.3390/met6070171

Butscher, A., Bohner, M., Roth, C., Ernstberger, A., Heuberger, R., Doebelin, N., Von Rohr, P. R., & Müller, R. (2012). Printability of calcium phosphate powders for three-dimensional printing of tissue engineering scaffolds. *Acta Biomaterialia*, *8*(1), 373–385. doi:10.1016/j.actbio.2011.08.027 PMID:21925623

Cabrini, M., Lorenzi, S., Pastore, T., Pellegrini, S., Manfredi, D., Fino, P., Biamino, S., & Badini, C. (2016). Evaluation of corrosion resistance of Al–10Si–Mg alloy obtained by means of Direct Metal Laser Sintering. *Journal of Materials Processing Technology*, *231*, 326–335. doi:10.1016/j.jmatprotec.2015.12.033

Cahn, J. W., Elliott, C. M., & Novick-Cohen, A. (1996). The Cahn-Hilliard equation with a concentration dependent mobility: Motion by minus the Laplacian of the mean curvature. *European Journal of Applied Mathematics*, *7*(3), 287–301. doi:10.1017/S0956792500002369

Caiazzo, F. (2018). Laser-aided Directed Metal Deposition of Ni-based superalloy powder. *Optics & Laser Technology*, *103*, 193–198. doi:10.1016/j.optlastec.2018.01.042

Calignano, F. (2018). Investigation of the accuracy and roughness in the laser powder bed fusion process. *Virtual and Physical Prototyping*, *13*(2), 97–104. doi:10.1080/17452759.2018.1426368

Calignano, F., Galati, M., Iuliano, L., & Minetola, P. (2019). Design of Additively Manufactured Structures for Biomedical Applications: A Review of the Additive Manufacturing Processes Applied to the Biomedical Sector. *Journal of Healthcare Engineering*, *2019*, 1–6. doi:10.1155/2019/9748212 PMID:30992744

Calignano, F., & Minetola, P. (2019). Influence of process parameters on the porosity, accuracy, roughness and support structures of Hastelloy X produced by laser powder bed fusion. *Materials (Basel)*, *12*(19), 3178. doi:10.3390/ma12193178 PMID:31569803

Campbell, R. I., Beer, D. J., & Pei, E. (2011). Additive manufacturing in South Africa: Building on the foundations. *Rapid Prototyping Journal*, *17*(2), 156–162. doi:10.1108/13552541111113907

Carroll, B. E., Otis, R. A., Paul, J., Suh, J., Dillon, R. P., Shapiro, A. A., Hofmann, D. C., Liu, Z., & Beese, A. M. (2016). Functionally graded material of 304L stainless steel and inconel 625 fabricated by directed energy deposition : Characterization and thermodynamic modeling. *Acta Materialia*, *108*, 46–54. doi:10.1016/j.actamat.2016.02.019

Carry, B. C., & Goswami, T. (2009). Knee implants review of models and biomechanics. *Mater*, *30*, 398–413. https://doi:10.1016/j.matdes.2008.03.032

Carter, L. N. (2013). *Selective laser melting of nickel superalloys for high temperature applications* (Doctoral dissertation). University of Birmingham.

Carter, L. N., Martin, C., Withers, P. J., & Attallah, M. M. (2014). The influence of the laser scan strategy on grain structure and cracking behaviour in SLM powder-bed fabricated nickel superalloy. *Journal of Alloys and Compounds*, *615*, 338–347. doi:10.1016/j.jallcom.2014.06.172

Casalino, G., Mortello, M., & Campanelli, S. L. (2015). Ytterbium fiber laser welding of Ti6Al4V alloy. *Journal of Manufacturing Processes*, *20*, 250–256. doi:10.1016/j.jmapro.2015.07.003

Castelan, J., Schaeffer, L., Daleffe, A., Fritzen, D., Salvaro, V., & Silva, F. P. (2014). Manufacture of custom-made cranial implants from DICOM® images using 3D printing, CAD/CAM technology and incremental sheet forming. *Revista Brasileira de Engenharia Biomédica*, *30*(3), 1517–3151. doi:10.1590/rbeb.2014.024

Catchpole-Smith, S., Aboulkhair, N., Parry, L., Tuck, C., Ashcroft, I. A., & Clare, A. (2017). Fractal scan strategies for selective laser melting of 'unweldable' nickel superalloys. *Additive Manufacturing*, *15*, 113–122. doi:10.1016/j.addma.2017.02.002

Chai, X., Chai, H., Wang, X., Yang, J., Li, J., Zhao, Y., Cai, W., Tao, T., & Xiang, X. (2017). Fused deposition modeling (FDM) 3D printed tablets for intragastric floating delivery of domperidone. *Scientific Reports*, *7*(1), 1–9. doi:10.103841598-017-03097-x PMID:28588251

Chang, J., He, J., Mao, M., Zhou, W., Lei, Q., Li, X., Li, D., Chua, C.-K., & Zhao, X. (2018). Advanced Material Strategies for Next-Generation Additive Manufacturing. *Materials (Basel)*, *11*(1), 166. doi:10.3390/ma11010166 PMID:29361754

Chang, K.-H. (2015). Rapid Prototyping. In K. H. Chang (Ed.), *e-Design: Computer-Aided Engineering Design* (pp. 743–786). Elsevier. doi:10.1016/B978-0-12-382038-9.00014-4

Chaudhuri, B., Mondal, B., Ray, S. K., & Sarkar, S. C. (2016). A novel biocompatible conducting polyvinyl alcohol (PVA)-polyvinylpyrrolidone (PVP)-hydroxyapatite (HAP) composite scaffolds for probable biological application. *Colloids and Surfaces. B, Biointerfaces*, *143*, 71–80. doi:10.1016/j.colsurfb.2016.03.027 PMID:26998868

Cheepu, M., Ashfaq, M., & Muthupandi, V. (2017). A new approach for using interlayer and analysis of the friction welding of titanium to stainless steel. *Transactions of the Indian Institute of Metals*, *70*(10), 2591–2600. doi:10.100712666-017-1114-x

Chen, B., Su, Y., Xie, Z., Tan, C., & Feng, J. (2020). *Development and characterization of 316L/Inconel625 functionally graded material fabricated by laser direct metal deposition.* doi:10.1016/j.optlastec.2019.105916

Chen, B., Yao, Y., Song, X., Tan, C., Cao, L., & Feng, J. (2018). *Microstructure and mechanical properties of additive manufacturing AlSi10Mg alloy using direct metal deposition.* doi:10.1080/00150193.2018.1392147

Chen, X., Li, J., Cheng, X., Wang, H., & Huang, Z. (2018). Effect of heat treatment on microstructure, mechanical and corrosion properties of austenitic stainless steel 316L using arc additive manufacturing. *Materials Science and Engineering A*, *715*(October), 307–314. doi:10.1016/j.msea.2017.10.002

Chen, X., Ware, H. O. T., Baker, E., Chu, W., Hu, J., & Sun, C. (2017). The Development of an All-polymer-based Piezoelectric Photocurable Resin for Additive Manufacturing. *Procedia CIRP*, *65*, 157–162.

Chen, D., & Zheng, X. (2018). Multi-material additive manufacturing of metamaterials with giant, tailorable negative Poisson's ratios. *Scientific Reports*, *8*(1), 9139. doi:10.103841598-018-26980-7 PMID:29904093

Cheng, B., Shrestha, S., & Chou, K. (2016). Stress and deformation evaluations of scanning strategy effect in selective laser melting. *Additive Manufacturing*, *12*, 240–251. doi:10.1016/j.addma.2016.05.007

Cheng, G., Liu, L., Qiang, X., & Liu, Y. (2016). Industry 4.0 Development and Application of Intelligent Manufacturing. In *2016 International Conference on Information System and Artificial Intelligence* (pp. 407-410). Hong Kong: IEEE. 10.1109/ISAI.2016.0092

Cheng, L., & To, A. (2019). Part-scale build orientation optimization for minimizing residual stress and support volume for metal additive manufacturing: Theory and experimental validation. *Computer Aided Design, 113*, 1–23. doi:10.1016/j.cad.2019.03.004

Chen, J., Li, K., Wang, Y., Xing, L., Yu, C., Liu, H., Ma, J., Liu, W., & Shen, Z. (2020). The effect of hot isostatic pressing on thermal conductivity of additively manufactured pure tungsten. *International Journal of Refractory Metals & Hard Materials, 87*, 105135. doi:10.1016/j.ijrmhm.2019.105135

Chen, X., Xing, L., & Egger, J. (2017). Computer-aided implant design for the re-storation of cranial defects. *Scientific Reports, 4199*. Advance online publication. doi:10.103841598-017-04454-6

Chen, X., Li, J., Cheng, X., He, B., Wang, H., & Huang, Z. (2017). Microstructure and mechanical properties of the austenitic stainless steel 316L fabricated by gas metal arc additive manufacturing. *Materials Science and Engineering A, 703*(January), 567–577. doi:10.1016/j.msea.2017.05.024

Chen, Z., Ye, H., & Xu, H. (2018). Distortion control in a wire-fed electron-beam thin-walled Ti-6Al-4V freeform. *Journal of Materials Processing Technology, 258*(March), 286–295. doi:10.1016/j.jmatprotec.2018.04.008

Chern, A. H., Nandwana, P., Yuan, T., Kirka, M. M., Dehoff, R. R., Liaw, P. K., & Duty, C. E. (2019). A review on the fatigue behavior of Ti-6Al-4V fabricated by electron beam melting additive manufacturing. *International Journal of Fatigue, 119*(August), 173–184. doi:10.1016/j.ijfatigue.2018.09.022

Chin, B. (2017). *The Perfect Combo: 3D Printing & Big Data*. Retrieved from https://publish.illinois.edu/digitalmaking2017/the-perfect-combo-3d-printing-big-data/

Chlebus, E., Gruber, K., Kuźnicka, B., Kurzac, J., & Kurzynowski, T. (2015). Effect of heat treatment on the microstructure and mechanical properties of Inconel 718 processed by selective laser melting. *Materials Science and Engineering A, 639*, 647–655. doi:10.1016/j.msea.2015.05.035

Cho, H., Erdeniz, D., Sharp, K. W., & Dunand, D. C. (2018). Experimental and modeling study of compressive creep in 3D-woven Ni-based superalloys. *Acta Materialia, 155*, 236–244. doi:10.1016/j.actamat.2018.05.067

Choi, J.-P., Shin, G.-H., Yang, S., Yang, D.-Y., Lee, J.-S., Brochu, M., & Yu, J.-H. (2017). Densification and microstructural investigation of Inconel 718 parts fabricated by selective laser melting. *Powder Technology, 310*, 60–66. doi:10.1016/j.powtec.2017.01.030

Chouhan, P., Sharma, S., Gulhane, S., Palani, I. A., & Lad, B. K. (2018). Investigation on influence of Nitinol wire made tailored structure manufactured using MIG additive technique for machine tool operations. *Materials Today: Proceedings, 5*, 21350–21354.

Chu, C.H., Song, M.C., & Luo, V.C.S. (2006). Computer aided parametric design for 3D tire mold production. *Computers in Industry, 57*, 11-25. .2005.04.005 doi:10.1016/ j.compind

Clarke, C. (2017). *China develops ZcompleX, an electron beam additive manufacturing machine to rival sciaky*. Retrieved from https://www.3dprintpulse.com/additive-manufacturing/china/?open-article-id=6396751&article-title=china-develops-zcomplex--an-electron-beam-additive-manufacturing-machine-to-rival-sciaky&blog-domain=3dprintingindustry.com&blog-title=3d-printing-industry

Cloots, M., Uggowitzer, P. J., & Wegener, K. (2016). Investigations on the microstructure and crack formation of IN738LC samples processed by selective laser melting using Gaussian and doughnut profiles. *Materials & Design*, *89*, 770–784. doi:10.1016/j.matdes.2015.10.027

Cobbs, T., Lucas, B., & Crandall, J. L. (2019). *How 3D Metal Printing Saves Time and Lowers Costs: DED for Repair of Industrial Components from Optomec Inc.* https://optomec.com/wp-content/uploads/2019/03/DEDWebinar-slides.pdf

Collins, P., Brice, D., Samimi, P., Ghamarian, I., & Fraser, H. (2016). Microstructural control of additively manufactured metallic materials. *Annual Review of Materials Research*, *46*(1), 63–91. doi:10.1146/annurev-matsci-070115-031816

Collur, M. M., Paul, A., & Debroy, T. (1987). Mechanism of alloying element vaporization during laser welding. *Metallurgical Transactions. B, Process Metallurgy*, *18*(4), 733–740. doi:10.1007/BF02672891

Colosimo, B. M., Huang, Q., Dasgupta, T., & Tsung, F. (2018). Opportunities and challenges of quality engineering for additive manufacturing. *Journal of Quality Technology*, *50*(3), 233–252. doi:10.1080/00224065.2018.1487726

Compton, O. C., & Nguyen, S. T. (2010). Graphene Oxide, Highly Reduced Graphene Oxide, and Graphene: Versatile Building Blocks for Carbon-Based Materials. *Small*, *6*(6), 711–723. doi:10.1002mll.200901934 PMID:20225186

Conner, B. P., Manogharan, G. P., Martof, A. N., Rodomsky, L. M., Rodomsky, C. M., Jordan, D. C., & Limperos, J. W. (2014a). Making sense of 3-D printing: Creating a map of additive manufacturing products and services. *Additive Manufacturing*, *1–4*, 64–76. doi:10.1016/j.addma.2014.08.005

Cooper, A., Kang, S., Kietzman, J., Prinz, F., Lombardi, J., & Weiss, L. (1999). Automated fabrication of complex molded parts using Mold Shape Deposition Manufacturing. *Materials & Design*, *20*(2-3), 83–89. doi:10.1016/S0261-3069(99)00013-8

Cooper, D. E., Stanford, M., Kibble, K. A., & Gibbons, G. J. (2012). Additive Manufacturing for product improvement at Red Bull Technology. *Materials & Design*, *41*, 226–230. doi:10.1016/j.matdes.2012.05.017

Corbacho, J. L., Suárez, J. C., & Molleda, F. (1998). Welding of invar Fe-36Ni alloy for tooling of composite materials. *Welding International*, *12*(12), 966–971. doi:10.1080/09507119809448543

Corbin, D. J., Nassar, A. R., Reutzel, E. W., Beese, A. M., & Kistler, N. A. (2017). Effect of directed energy deposition processing parameters on laser deposited Inconel® 718: External morphology. *Journal of Laser Applications*, *29*(2), 022001. doi:10.2351/1.4977476

Cordova, L., Campos, M., & Tinga, T. (2019). Revealing the Effects of Powder Reuse for Selective Laser Melting by Powder Characterization. *JOM*, *71*(3), 1062–1072. doi:10.100711837-018-3305-2

Creative Commons, B.-N.-S. (2018). *Direct metal deposition (DMD) – OpenLearn*. Open University.

Criales, L. E., Arısoy, Y. M., Lane, B., Moylan, S., Donmez, A., & Özel, T. (2017a). Predictive modeling and optimization of multi-track processing for laser powder bed fusion of nickel alloy 625. *Additive Manufacturing*, *13*, 14–36. doi:10.1016/j.addma.2016.11.004

Criales, L. E., Arısoy, Y. M., Lane, B., Moylan, S., Donmez, A., & Özel, T. (2017b). Laser powder bed fusion of nickel alloy 625: Experimental investigations of effects of process parameters on melt pool size and shape with spatter analysis. *International Journal of Machine Tools & Manufacture*, *121*, 22–36. doi:10.1016/j.ijmachtools.2017.03.004

Cunningham, C. R., Flynn, J. M., Shokrani, A., Dhokia, V., & Newman, S. T. (2018). Invited review article: Strategies and processes for high quality wire arc additive manufacturing. *Additive Manufacturing*, *22*(June), 672–686. doi:10.1016/j.addma.2018.06.020

D'Aveni, R. A. (2019). *How 3-D printing can jumpstart developing economies.* Retrieved from https://www.forbes.com/sites/richarddaveni/2019/03/19/how-3d-printing-can-jumpstart-developing-economies/#3d6fdeeb4400

Das, P., Chandran, R., Samant, R., & Anand, S. (2015). Optimum Part Build Orientation in Additive Manufacturing for Minimizing Part Errors and Support Structures. *Procedia Manufacturing, 1*, 343–354. doi:10.1016/j.promfg.2015.09.041

Dass, A., & Moridi, A. (2019). State of the Art in Directed Energy Deposition: From Additive Manufacturing to Materials Design. *Coatings, 9*(7), 1–26. doi:10.3390/coatings9070418

Davis, A. E., Breheny, C. I., Fellowes, J., Nwankpa, U., Martina, F., Ding, J., & Prangnell, P. B. (2019). Mechanical performance and microstructural characterisation of titanium alloy-alloy composites built by wire-arc additive manufacture. *Materials Science and Engineering A, 765*(July), 138289. doi:10.1016/j.msea.2019.138289

DebRoy, T., Wei, H. L., Zuback, J. S., Mukherjee, T., Elmer, J. W., Milewski, J. O., Beese, A. M., Wilson-Heid, A., De, A., & Zhang, W. (2018). Additive manufacturing of metallic components – Process, structure and properties. *Progress in Materials Science, 92*, 112–224. doi:10.1016/j.pmatsci.2017.10.001

Deepak Kumar, S., Karthik, D., Mandal, A., & Pavan Kumar, J. S. R. (2017). Optimization of Thixoforging process parameters of A356 alloy using Taguchi's experimental design and DEFORM Simulation. *Materials Today: Proceedings, 4*(9), 9987–9991. doi:10.1016/j.matpr.2017.06.307

Delgadillo, I., Gollisch, H., & Feder, R. (1994). Lattice-vibration and spin-fluctuation effects on photoemission from ferromagnetic Ni. *Physical Review. B, 50*(21), 15808–15813. doi:10.1103/PhysRevB.50.15808 PMID:9975948

Delgado Camacho, D., Clayton, P., O'Brien, W. J., Seepersad, C., Juenger, M., Ferron, R., & Salamone, S. (2018). Applications of additive manufacturing in the construction industry – A forward-looking review. *Automation in Construction, 89*, 110–119. doi:10.1016/j.autcon.2017.12.031

Deloitte. (2017). *Using autonomous robots to drive supply chain innovation.* Retrieved from https://www2.deloitte.com/in/en.html

Deng, Y., Li, X., Wu, L., Yang, Q., & Chen, Y. (2019). Microstructure and Performance of WAAM TiB2-Reinforced Al–Si-Based Composites. In Physics and Engineering of Metallic Materials (Vol. 217, pp. 321–328). doi:10.1007/978-981-13-5944-6_31

Deng, D., Peng, R. L., Brodin, H., & Moverare, J. (2018). Microstructure and mechanical properties of Inconel 718 produced by selective laser melting: Sample orientation dependence and effects of post heat treatments. *Materials Science and Engineering A, 713*, 294–306. doi:10.1016/j.msea.2017.12.043

Denoual, C., & Vattré, A. (2016). A phase field approach with a reaction pathways-based potential to model reconstructive martensitic transformations with a large number of variants. *Journal of the Mechanics and Physics of Solids, 90*, 91–107. doi:10.1016/j.jmps.2016.02.022

Derekar, K. S. (2018). A review of wire arc additive manufacturing and advances in wire arc additive manufacturing of aluminium. *Materials Science and Technology (United Kingdom), 34*(8), 895–916. doi:10.1080/02670836.2018.1455012

Dharmendra, C., Hadadzadeh, A., Amirkhiz, B. S., Janaki Ram, G. D., & Mohammadi, M. (2019). Microstructural evolution and mechanical behavior of nickel aluminum bronze Cu-9Al-4Fe-4Ni-1Mn fabricated through wire-arc additive manufacturing. *Additive Manufacturing, 30*(March), 100872. doi:10.1016/j.addma.2019.100872

Dhinakaran, V., Ajith, J., Yasin, A. F., Jagadeesha, T., Sathish, T., & Stalin, B. (2019). Materials Today : Proceedings Wire Arc Additive Manufacturing (WAAM) process of nickel based superalloys – A review. *Materials Today: Proceedings.* doi:10.1016/j.matpr.2019.08.159

Digital Engineering. (2018). *Altair Debuts inspire 2018 software*. Retrieved from https://www.digital-engineering247.com/article/altair-debuts-inspire-2018-software/

Dinda, G. P., Dasgupta, A. K., & Mazumder, J. (2009). Laser aided direct metal deposition of Inconel 625 superalloy: Microstructural evolution and thermal stability. *Materials Science and Engineering A*, *509*(1-2), 98–104. doi:10.1016/j.msea.2009.01.009

Dinda, G. P., Dasgupta, A. K., & Mazumder, J. (2012). Evolution of microstructure in laser deposited Al–11.28%Si alloy. *Surface and Coatings Technology*, *206*(8–9), 2152–2160. doi:10.1016/j.surfcoat.2011.09.051

Ding, D., Pan, Z., Cuiuri, D., & Li, H. (2015). Wire-feed additive manufacturing of metal components: Technologies, developments and future interests. *International Journal of Advanced Manufacturing Technology*, *81*(1–4), 465–481. doi:10.100700170-015-7077-3

Dini, F., Ghaffari, S. A., Jafar, J., Hamidreza, R., & Marjan, S. (2019). A review of binder jet process parameters; powder, binder, printing and sintering condition. *Metal Powder Report.*

Dinovitzer, M., Chen, X., Laliberte, J., Huang, X., & Frei, H. (2019). Effect of wire and arc additive manufacturing (WAAM) process parameters on bead geometry and microstructure. *Additive Manufacturing, 26*(October), 138–146. doi:10.1016/j.addma.2018.12.013

Direct Metal Laser Melting (DMLM) | GE Additive. (n.d.). Retrieved February 25, 2020, from https://www.ge.com/additive/additive-manufacturing/information/direct-metal-laser-melting-technology

Divya, V. D., Muñoz-Moreno, R., Messé, O. M. D. M., Barnard, J. S., Baker, S., Illston, T., & Stone, H. J. (2016). Microstructure of selective laser melted CM247LC nickel-based superalloy and its evolution through heat treatment. *Materials Characterization*, *114*, 62–74. doi:10.1016/j.matchar.2016.02.004

DM3D Technology LLC. (2008). *DMD machines solutions*. Retrieved from http://www.pomgroup.com

Donachie, M. J., & Donachie, S. J. (2002). *Superalloys: A technical guide*. ASTM International. doi:10.31399/asm.tb.stg2.9781627082679

Du, D., Dong, A., Shu, D., Zhu, G., Sun, B., Li, X., & Lavernia, E. (2019). Influence of build orientation on microstructure, mechanical and corrosion behavior of Inconel 718 processed by selective laser melting. *Materials Science and Engineering A*, *760*, 469–480. doi:10.1016/j.msea.2019.05.013

Dusen, M. V. (2017). *GE's 3D-printed airplane engine will run this year*. Retrieved from https://www.ge.com/reports/mad-props-3d-printed-airplane-engine-will-run-year/

Dutta, B., & Froes, F. H. (2015). The additive manufacturing (AM) of titanium alloys. In Titanium Powder Metallurgy (pp. 447–468). Elsevier.

Dutta, B., & Froes, F. H. S. (2017). The Additive Manufacturing (AM) of titanium alloys. *Metal Powder Report*, *72*(2), 96–106. doi:10.1016/j.mprp.2016.12.062

Dye, D., Hunziker, O., & Reed, R. C. (2001). Numerical analysis of the weldability of superalloys. *Acta Materialia*, *49*(4), 683–697. doi:10.1016/S1359-6454(00)00361-X

Dzian, A. Ž., Živčák, J., Penciak, R., & Hudák, R. (2018). Implantation of a 3D-printed titanium sternum in a patient with a sternal tumor. *World Journal of Surgical Oncology*, *16*(1), 1–4. doi:10.118612957-018-1315-8 PMID:29334989

Edgar, J., & Tint, S. (2015). Additive manufacturing technologies: 3D printing, rapid prototyping, and direct digital manufacturing. *Johnson Matthey Technology Review*, *59*(3), 193–198. doi:10.1595/205651315X688406

Eliseeva, O. V., Kirk, T., Samimi, P., Malak, R., Arróyave, R., Elwany, A., & Karaman, I. (2019). Functionally Graded Materials through robotics-inspired path planning. *Materials & Design*, *182*, 107975. doi:10.1016/j.matdes.2019.107975

Endoh, Y. (1979). Lattice dynamics in ferromagnetic invar alloys. *Journal of Magnetism and Magnetic Materials*, *10*(2), 177–182. doi:10.1016/0304-8853(79)90173-2

Engeli, R., Etter, T., Hövel, S., & Wegener, K. (2016). Processability of different IN738LC powder batches by selective laser melting. *Journal of Materials Processing Technology*, *229*, 484–491. doi:10.1016/j.jmatprotec.2015.09.046

Engelman, R. (n.d.). *The second industrial revolution,1870-1914*. Retrieved from http://ushistoryscene.com/article/second-industrial-revolution

Engineering and manufacturing - great.gov.uk international. (n.d.). Retrieved February 25, 2020, from https://www.great.gov.uk/international/content/about-uk/industries/engineering-and-manufacturing/

EOS GmbH. (n.d.a). *Tooling: Salcomp - Additive manufacturing permits optimized cooling for maximum production efficiency*. Retrieved from https://www.eos.info/press/customer_case_studies/salcomp

EOS GmbH. (n.d.b). *First metal 3D printed primary flight control hydraulic component flies on an Airbus A380, EOS: Customer case studies*. Retrieved from https://www.eos.info/press/case_studies/first-3d-printed-hydraulic-component-flies-on-airbus-a380

EOS GmbH. (n.d.c). *Additive manufacturing of satellite components, EOS: Customer case studies*. Retrieved from https://www.eos.info/case_studies/additive-manufacturing-of-antenna-bracket-for-satellite

EOS GmbH. (n.d.d). *EOS systems and equipment for metal manufacturing*. Retrieved from https://www.eos.info

EOS. (2016). *EOS NickelAlloy IN718 Material Data Sheet*. Developed for EOS M400 SF.

Ewald, S., Kies, F., Hermsen, S., Voshage, M., Haase, C., & Schleifenbaum, J. H. (2019). Rapid Alloy Development of Extremely High-Alloyed Metals Using Powder Blends in Laser Powder Bed Fusion. *Materials (Basel)*, *12*(10), 1706. doi:10.3390/ma12101706 PMID:31130684

ExOne | Binder Jetting Technology. (n.d.). Retrieved February 25, 2020, from https://www.exone.com/en-US/case-studies/what-is-binder-jetting

Eyers, D. R., & Potter, A. T. (2017). Industrial Additive Manufacturing: A manufacturing systems perspective. *Computers in Industry*, *92-93*, 208–218. doi:10.1016/j.compind.2017.08.002

Farber, B., Small, K. A., Allen, C., Causton, R. J., Nichols, A., Simbolick, J., & Taheri, M. L. (2018). Correlation of mechanical properties to microstructure in Inconel 718 fabricated by Direct Metal Laser Sintering. *Materials Science and Engineering A*, *712*, 539–547. doi:10.1016/j.msea.2017.11.125

Fathi, S., & Dickens, P. (2013). Challenges in drop-on-drop deposition of reactive molten nylon materials for additive manufacturing. *Journal of Materials Processing Technology*, *213*(1), 84–93. doi:10.1016/j.jmatprotec.2012.08.006

Fera, M., Macchiaroli, R., Fruggiero, F., & Lambiase, A. (2018). A new perspective for production process analysis using additive manufacturing—complexity vs production volume. *International Journal of Advanced Manufacturing Technology*, *95*(1), 673–685. doi:10.100700170-017-1221-1

Fereiduni, E., Yakout, M., & Elbestawi, M. (2019). Laser-Based Additive Manufacturing of Lightweight Metal Matrix Composites. In B. AlMangour (Ed.), *Additive Manufacturing of Emerging Materials* (pp. 55–109). Springer International Publishing. doi:10.1007/978-3-319-91713-9_3

Fixter, J., Gu, J., Ding, J., Williams, S. W., & Prangnell, P. B. (2017). Preliminary investigation into the suitability of 2xxx alloys for Wire-Arc Additive Manufacturing. *Materials Science Forum, 877*, 611–616. . doi:10.4028/www.scientific.net/MSF.877.611

Flowers, P. F., Reyes, C., Ye, S., Kim, M. J., & Wiley, B. J. (2017). 3D printing electronic components and circuits with conductive thermoplastic filament. *Additive Manufacturing, 18*, 156–163. doi:10.1016/j.addma.2017.10.002

Ford, S., & Despeisse, M. (2016). Additive manufacturing and sustainability: An exploratory study of the advantages and challenges. *Journal of Cleaner Production, 137*, 1573–1587. doi:10.1016/j.jclepro.2016.04.150

Frăţilă, D., & Rotaru, H. (2017). Additive manufacturing – a sustainable manufacturing route. MATEC Web of Conferences, 94. doi:10.1051/matecconf/20179403004

Frazier, W. E. (2014). Metal Additive Manufacturing: A Review. *Journal of Materials Engineering and Performance, 23*(6), 1917–1928. doi:10.100711665-014-0958-z

Frost and Sullivan. (n.d.). *Research and analysis*. Retrieved from https://ww2.frost.com

Furht, B. (Ed.). (2011). *Handbook of Augmented Reality*. Springer. doi:10.1007/978-1-4614-0064-6

Galati, M., Snis, A., & Iuliano, L. (2019). Powder bed properties modelling and 3D thermo-mechanical simulation of the additive manufacturing Electron Beam Melting process. *Additive Manufacturing, 30*, 100897. doi:10.1016/j.addma.2019.100897

Gallmeyer, T. G., Moorthy, S., Kappes, B. B., Mills, M. J., Amin-Ahmadi, B., & Stebner, A. P. (2020). Knowledge of process-structure-property relationships to engineer better heat treatments for laser powder bed fusion additive manufactured Inconel 718. *Additive Manufacturing, 31*, 100977. doi:10.1016/j.addma.2019.100977

Gandin, C. A., & Rappaz, M. (1994). A coupled finite element-cellular automaton model for the prediction of dendritic grain structures in solidification processes. *Acta Metallurgica et Materialia, 42*(7), 2233–2246. doi:10.1016/0956-7151(94)90302-6

Gao, W., Zhang, Y., Ramanujan, D., Ramani, K., Chen, Y., Williams, C. B., & Zavattieri, P. D. (2015). The status, challenges, and future of additive manufacturing in engineering. *Computer Aided Design, 69*, 65–89. doi:10.1016/j.cad.2015.04.001

Gary, H. (2015). Craniofacial reconstruction with poly(Methyl Methacrylate) customized cranial implants. *The Journal of Craniofacial Surgery, 26*(1), 64–70. doi:10.1097/SCS.0000000000001315 PMID:25376145

Gault, B., Raabe, D., Chauvet, E., Kontis, P., Eric, A. J., Vayre, B., Tassin, C., & Blandin, J. (2018). Hot cracking mechanism affecting a non-weldable Ni-based superalloy produced by selective electron Beam Melting. *Acta Materialia, 142*, 82–94. doi:10.1016/j.actamat.2017.09.047

Gaytan, S. M., Cadena, M. A., Karim, H., Delfin, D., Lin, Y., Espalin, D., MacDonald, E., & Wicker, R. B. (2015). Fabrication of barium titanate by binder jetting additive manufacturing technology. *Ceramics International, 41*(5), 6610–6619. doi:10.1016/j.ceramint.2015.01.108

GE Aviation celebrates 30,000th 3D printed fuel nozzle - 3D Printing Industry. (n.d.). Retrieved February 25, 2020, from https://3dprintingindustry.com/news/ge-aviation-celebrates-30000th-3d-printed-fuel-nozzle-141165/

GE. (2018). *Electron Beam Melting (EBM)*. GE Additive.

Gebisa, A. W., & Lemu, H. G. (2017). A case study on topology optimized design for additive manufacturing. *IOP Conference Series. Materials Science and Engineering, 276*(1), 012026. Advance online publication. doi:10.1088/1757-899X/276/1/012026

Geekie, L. & Jamie, J. (2018). Including simulation in the additive manufacturing work flow aa. *Industrlal Equipment*, 14-18. Retrieved from https://www.ansys.com/-/media/ ansys/ corporate/ resourcelibrary/article/including-simulation-in-the-additive-manufacturing-workflow-aa-v12-i3.pdf

Ge, J., Ma, T., Han, W., Yuan, T., Jin, T., Fu, H., Xiao, R., Lei, Y., & Lin, J. (2019). Thermal-induced microstructural evolution and defect distribution of wire-arc additive manufacturing 2Cr13 part: Numerical simulation and experimental characterization. *Applied Thermal Engineering*, *163*(100), 114335. doi:10.1016/j.applthermaleng.2019.114335

Geng, H., Li, J., Xiong, J., Lin, X., & Zhang, F. (2017). Optimization of wire feed for GTAW based additive manufacturing. *Journal of Materials Processing Technology*, *243*, 40–47. doi:10.1016/j.jmatprotec.2016.11.027

Gharbi, M., Peyre, P., Gorny, C., Carin, M., Morville, S., Le Masson, P., Carron, D., & Fabbro, R. (2013). Influence of various process conditions on surface finishes induced by the direct metal deposition laser technique on a Ti–6Al–4V alloy. *Journal of Materials Processing Technology*, *213*(5), 791–800. doi:10.1016/j.jmatprotec.2012.11.015

Gibson, I., Rosen, D. W., & Stucker, B. (2010). Additive manufacturing technologies: Rapid prototyping to direct digital manufacturing. *Additive Manufacturing Technologies: Rapid Prototyping to Direct Digital Manufacturing*. doi:10.1007/978-1-4419-1120-9

Gibson, I., Rosen, D., & Stucker, B. (2015). Additive manufacturing technologies: 3D printing, rapid prototyping, and direct digital manufacturing. In *Additive Manufacturing Technologies: 3D* (2nd ed.). Printing, Rapid Prototyping, and Direct Digital Manufacturing. doi:10.1007/978-1-4939-2113-3

Glover, S. (2017). *Additive manufacturing & applications in China*. Retrieved from https://www.laserchirp.com/2017/01/additive-manufacturing-applications-in-china

Gockel, J., Beuth, J., & Taminger, K. (2014). Integrated control of solidification microstructure and melt pool dimensions in electron beam wire feed additive manufacturing of Ti-6Al-4V. *Additive Manufacturing*, *1–4*, 119–126. doi:10.1016/j.addma.2014.09.004

Gokuldoss, P. K., Kolla, S., Eckert, J., Gokuldoss, P. K., Kolla, S., & Eckert, J. (2017). Additive Manufacturing Processes: Selective Laser Melting, Electron Beam Melting and Binder Jetting—Selection Guidelines. *Materials (Basel)*, *10*(6), 672. doi:10.3390/ma10060672 PMID:28773031

Gonzalez, J. A., Mireles, J., Stafford, S. W., Perez, M. A., Terrazas, C. A., & Wicker, R. B. (2019). Characterization of Inconel 625 fabricated using powder-bed-based additive manufacturing technologies. *Journal of Materials Processing Technology*, *264*, 200–210. doi:10.1016/j.jmatprotec.2018.08.031

Goodridge, R., & Ziegelmeier, S. (2017). 7 - Powder bed fusion of polymers. Laser Additive Manufacturing. doi:10.1016/B978-0-08-100433-3.00007-5

Gorsse, S., Hutchinson, C., Gouné, M., & Banerjee, R. (2017). Additive manufacturing of metals: A brief review of the characteristic microstructures and properties of steels, Ti-6Al-4V and high-entropy alloys. *Science and Technology of Advanced Materials*, *18*(1), 584–610. doi:10.1080/14686996.2017.1361305 PMID:28970868

Gou, J., Shen, J., Hu, S., Tian, Y., & Liang, Y. (2019). Microstructure and mechanical properties of as-built and heat-treated Ti- 6Al-4V alloy prepared by cold metal transfer additive manufacturing. *Journal of Manufacturing Processes*, *42*(October), 41–50. doi:10.1016/j.jmapro.2019.04.012

Govekara, E., Kuznetsova, A., Kotara, M., & Kondo, M. (2018). Annular laser beam based direct metal deposition A new methodology to analyze the functional and physical architecture of assembly oriented product family Japan identification. *Procedia CIRP*, *74*, 222–227. doi:10.1016/j.procir.2018.08.099

Government of India. (2016). *National Institution for Transforming India*. Retrieved from http://niti.gov.in

Graf, B., Marko, A., Petrat, T., Gumenyuk, A., & Rethmeier, M. (2018). 3D laser metal deposition: Process steps for additive manufacturing. *Welding in the World*, *62*(4), 877–883. doi:10.100740194-018-0590-x

Gross, B. C., Erkal, J. L., Lockwood, S. Y., Chen, C., & Spence, D. M. (2014). Evaluation of 3D Printing and Its Potential Impact on Biotechnology and the Chemical Sciences. *Analytical Chemistry*, *86*(7), 3240–3253. doi:10.1021/ac403397r PMID:24432804

Gu, J., Wang, X., Bai, J., Ding, J., Williams, S., Zhai, Y., & Liu, K. (2018). Deformation microstructures and strengthening mechanisms for the wire+arc additively manufactured Al-Mg4.5Mn alloy with inter-layer rolling. *Materials Science and Engineering A*, *712*(November), 292–301. doi:10.1016/j.msea.2017.11.113

Gu, D. (2015). Laser Additive Manufacturing (AM): Classification, Processing Philosophy, and Metallurgical Mechanisms. In *Laser Additive Manufacturing of High-Performance Materials* (pp. 15–71). Springer Berlin Heidelberg. doi:10.1007/978-3-662-46089-4_2

Gu, D. D., Meiners, W., Wissenbach, K., & Poprawe, R. (2012). Laser additive manufacturing of metallic components: Materials, processes and mechanisms. *International Materials Reviews*, *50*(3), 133–164. doi:10.1179/1743280411Y.0000000014

Gu, D., & Shen, Y. (2008). Processing conditions and microstructural features of porous 316L stainless steel components by DMLS. *Applied Surface Science*, *255*(5), 1880–1887. doi:10.1016/j.apsusc.2008.06.118

Guillaume, C. E. (1904). Invar and Its Applications. *Nature*, *71*(1832), 134-139.

Gu, J. L., Ding, J. L., Cong, B. Q., Bai, J., Gu, H. M., Williams, S. W., & Zhai, Y. C. (2014). The Influence of Wire Properties on the Quality and Performance of Wire+Arc Additive Manufactured Aluminium Parts. *Advanced Materials Research*, *1081*, 210–214. doi:10.4028/www.scientific.net/amr.1081.210

Gu, J., Ding, J., Williams, S. W., Gu, H., Bai, J., Zhai, Y., & Ma, P. (2016). The strengthening effect of inter-layer cold working and post-deposition heat treatment on the additively manufactured Al–6.3Cu alloy. *Materials Science and Engineering A*, *651*, 18–26. doi:10.1016/j.msea.2015.10.101

Gu, J., Ding, J., Williams, S. W., Gu, H., Ma, P., & Zhai, Y. (2016). The effect of inter-layer cold working and post-deposition heat treatment on porosity in additively manufactured aluminum alloys. *Journal of Materials Processing Technology*, *230*, 26–34. doi:10.1016/j.jmatprotec.2015.11.006

Gu, J., Gao, M., Yang, S., Bai, J., Ding, J., & Fang, X. (2019). Pore formation and evolution in wire + arc additively manufactured 2319 Al alloy. *Additive Manufacturing*, *30*, 100900. doi:10.1016/j.addma.2019.100900

Gu, J., Gao, M., Yang, S., Bai, J., Zhai, Y., & Ding, J. (2020). Microstructure, defects, and mechanical properties of wire + arc additively manufactured Al[sbnd]Cu4.3-Mg1.5 alloy. *Materials & Design*, *186*, 108357. doi:10.1016/j.matdes.2019.108357

Guo, M., Jia, C., Zhou, J., Liu, W., & Wu, C. (2020). Investigating the generation process of molten droplets and arc plasma in the confined space during compulsively constricted WAAM. *Journal of Materials Processing Technology*, *275*(February), 116355. doi:10.1016/j.jmatprotec.2019.116355

Guo, N., & Leu, M. C. (2013). Additive manufacturing: Technology, applications and research needs. *Frontiers of Mechanical Engineering*, *8*(3), 215–243. doi:10.100711465-013-0248-8

Haase, C., Tang, F., Wilms, M. B., Weisheit, A., & Hallstedt, B. (2017). Combining thermodynamic modeling and 3D printing of elemental powder blends for high-throughput investigation of high-entropy alloys – Towards rapid alloy screening and design. *Materials Science and Engineering A*, *688*, 180–189. doi:10.1016/j.msea.2017.01.099

Hack, H., Link, R., Knudsen, E., Baker, B., & Olig, S. (2017). Mechanical properties of additive manufactured nickel alloy 625. *Additive Manufacturing*, *14*, 105–115. doi:10.1016/j.addma.2017.02.004

Halabi, F. E. (2011). Mechanical character-ization and numerical simulation of polyether–ether–ketone (PEEK) cranial implants. *Journal of the Mechanical Behavior of Biomedical Materials*, *4*(8), 1819–1832. doi:10.1016/j.jmbbm.2011.05.039 PMID:22098881

Han, J., Yang, J., Yu, H., Yin, J., Gao, M., Wang, Z., & Zeng, X. (2017). Microstructure and mechanical property of selective laser melted Ti6Al4V dependence on laser energy density. *Rapid Prototyping Journal*, *23*(2), 217–226. doi:10.1108/RPJ-12-2015-0193

Hanning, F., & Andersson, J. (2016, October). A Review of Strain Age Cracking in Nickel Based Superalloys. In 7th *International Swedish Production Symposium, SPS16*, (pp. 1-7). Academic Press.

Han, Q., Mertens, R., Montero-Sistiaga, M. L., Yang, S., Setchi, R., Vanmeensel, K., & Fan, H. (2018). Laser powder bed fusion of Hastelloy X: Effects of hot isostatic pressing and the hot cracking mechanism. *Materials Science and Engineering A*, *732*, 228–239. doi:10.1016/j.msea.2018.07.008

Harrison, N. J. (2016). *Selective Laser Melting of Nickel Superalloys: solidification, microstructure and material response* (Doctoral dissertation). University of Sheffield.

Harrison, N. J., Todd, I., & Mumtaz, K. (2015). Reduction of micro-cracking in nickel superalloys processed by Selective Laser Melting: A fundamental alloy design approach. *Acta Materialia*, *94*, 59–68. doi:10.1016/j.actamat.2015.04.035

Harrison, N. J., Todd, I., & Mumtaz, K. (2017). Thermal expansion coefficients in Invar processed by selective laser melting. *Journal of Materials Science*, *52*(17), 10517–10525. doi:10.100710853-017-1169-4 PMID:32025047

Harun, W. S. W., Manam, N. S., Kamariah, M. S. I. N., Sharif, S., Zulkifly, A. H., Ahmad, I., & Miura, H. (2018). A review of powdered additive manufacturing techniques for Ti-6al-4v biomedical applications. *Powder Technology*, *331*, 74–97. doi:10.1016/j.powtec.2018.03.010

Henderson, M. B., Arrell, D., Larsson, R., Heobel, M., & Marchant, G. (2004). Nickel based superalloy welding practices for industrial gas turbine applications. *Science and Technology of Welding and Joining*, *9*(1), 13–21. doi:10.1179/136217104225017099

Hengsbach, F., Koppa, P., Holzweissig, M. J., Aydinöz, M. E., Taube, A., Hoyer, K. P., Starykov, O., Tonn, B., Niendorf, T., Tröster, T., & Schaper, M. (2018). Inline additively manufactured functionally graded multi-materials: Microstructural and mechanical characterization of 316L parts with H13 layers. *Progress in Additive Manufacturing*, *3*(4), 221–231. doi:10.100740964-018-0044-4

Herzog, D., Seyda, V., Wycisk, E., & Emmelmann, C. (2016). Additive manufacturing of metals. *Acta Materialia*, *117*, 371–392. doi:10.1016/j.actamat.2016.07.019

He, X., DebRoy, T., & Fuerschbach, P. W. (2003). Alloying element vaporization during laser spot welding of stainless steel. *Journal of Physics. D, Applied Physics*, *36*(23), 3079–3088. doi:10.1088/0022-3727/36/23/033

Hilaire, A., Andrieu, E., & Wu, X. (2019). High-temperature mechanical properties of alloy 718 produced by laser powder bed fusion with different processing parameters. *Additive Manufacturing*, *26*, 147–160. doi:10.1016/j.addma.2019.01.012

Hitzler, L., Hirsch, J., Heine, B., Merkel, M., Hall, W., & Öchsner, A. (2017). On the Anisotropic Mechanical Properties of Selective Laser-Melted Stainless Steel. *Materials (Basel), 10*(10), 1136. doi:10.3390/ma10101136 PMID:28954426

Holemans, W., & Data, R. U. S. A. (2015). *(12) United States Patent. 2.*

Hönnige, J. R., Williams, S., Roy, M. J., Colegrove, P., & Ganguly, S. (2017). Residual Stress Characterization and Control in the Additive Manufacture of Large Scale Metal Structures. In Residual Stresses 2016 (Vol. 2, pp. 455–460). doi:10.21741/9781945291173-77

Hönnige, J. R., Colegrove, P. A., Ganguly, S., Eimer, E., Kabra, S., & Williams, S. (2018). Control of residual stress and distortion in aluminium wire + arc additive manufacture with rolling. *Additive Manufacturing, 22*, 775–783. doi:10.1016/j.addma.2018.06.015

Hooper, P. A. (2018). Melt pool temperature and cooling rates in laser powder bed fusion. *Additive Manufacturing, 22*(May), 548–559. doi:10.1016/j.addma.2018.05.032

Horgar, A., Fostervoll, H., Nyhus, B., Ren, X., Eriksson, M., & Akselsen, O. M. (2018). Additive manufacturing using WAAM with AA5183 wire. *Journal of Materials Processing Technology, 259*(March), 68–74. doi:10.1016/j.jmatprotec.2018.04.014

Horn, J., Rosenband, L., & Smith, R. (Eds.). (2010). *Reconceptualizing the industrial revolution.* MIT Press. doi:10.7551/mitpress/8585.001.0001

Huang, S. H., Liu, P., Mokasdar, A., & Hou, L. (2013). Additive manufacturing and its societal impact: A literature review. *International Journal of Advanced Manufacturing Technology, 67*(5-8), 1191–1203. doi:10.100700170-012-4558-5

Huang, Y., Leu, M. C., Mazumder, J., & Donmez, A. (2015). Additive Manufacturing: Current State, Future Potential, Gaps and Needs, and Recommendations. *Journal of Manufacturing Science and Engineering, 137*(1), 014001. Advance online publication. doi:10.1115/1.4028725

Hunt, J., Derguti, F., & Todd, I. (2014). *Ironmaking & Steelmaking Processes, Products and Applications Selection of steels suitable for additive layer manufacturing Selection of steels suitable for additive layer manufacturing.* doi:10.1179/0301923314Z.000000000269

Hussein, A., Hao, L., Yan, C., Everson, R., & Young, P. (2013). Advanced lattice support structures for metal additive manufacturing. *Journal of Materials Processing Technology, 213*(7), 1019–1026. doi:10.1016/j.jmatprotec.2013.01.020

Hu, Y. L., Lin, X., Yu, X. B., Xu, J. J., Lei, M., & Huang, W. D. (2017). Effect of Ti addition on cracking and microhardness of Inconel 625 during the laser solid forming processing. *Journal of Alloys and Compounds, 711*, 267–277. doi:10.1016/j.jallcom.2017.03.355

Hwang, I. H., Mehmert, P., & Tran, M. (2017). Metal additive manufacturing process simulation for the hinge of the engine hood. *Korean Society for Precision Engineering, 12*, 161-162. Retrieved from http://www.dbpia.co.kr/Journal/articleDetail?nodeId=NODE07300735

Ian, G. D., & Brent, S. (2015). Design for additive manufacturing. *Johnson Matthey Technology Review, 59*(3), 107–145. doi:10.1007/978-1-4939-2113-3_17

Ilogebe, A. B., Waters, C. K., Elliot, A. M., & Shackleford, C. (2019). Morphology of Binder-Jet Additive Manufactured Structural Amorphous Metal Matrix Composites. *International Journal of Engineering Science, 8*, 15–24.

In, T., III. How A Walnut-Sized Part Changed The Way GE Aviation Builds Jet Engines - GE Reports. (n.d.). Retrieved February 25, 2020, from https://www.ge.com/reports/transformation-3d-walnut-sized-part-changed-way-ge-aviation-builds-jet-engines/

Inovar Communications Ltd. (n.d.). *Metal Additive Manufacturing processes.* Author.

International Standards Organization / American Society for Testing and Materials. (2015). *Standard Terminology for Additive Manufacturing – General Principles – Terminology* (ISO / ASTM52900 – 15). Author.

Irwin, J., Reutzel, E. W., Michaleris, P., Keist, J., & Nassar, A. R. (2016). Predicting Microstructure From Thermal History During Additive Manufacturing for Ti-6Al-4V. *Journal of Manufacturing Science and Engineering, 138*(11), 111007. Advance online publication. doi:10.1115/1.4033525

ISO. (2019). *ISO/ASTM 52900:2015 Additive manufacturing — General principles — Terminology.* Retrieved from https://www.iso.org/standard/69669.html

J'son & Partners™. (2019). *The 3D printing market in Russia and in the world in 2018 (Additive Manufacturing, AM).* Retrieved from http://json.tv/en/ict_telecom_analytics_view/the-3d-printing-market-in-russia-and-in-the-world-in-2018-additive-manufacturing-am

Jackson, B. (2018). *Nanfang additive manufacturing confirms contract for 3D printed oil and gas pipelines.* Retrieved from https://3dprintingindustry.com/news/nanfang-additive-manufacturing-confirms-contract-3d-printed-oil-gas-pipelines-133504

Jackson, B. (2018). *Russian helicopters backs additive manufacturing for future aviation.* Retrieved from https://3dprintingindustry.com/news/russian-helicopters-backs-additive-manufacturing-future-aviation-133704

Jagadish, & Bhowmik, S. (2019). Parameters Optimization of FDM for the Quality of Prototypes Using an Integrated MCDM Approach. In K. Kumar, D. Zindani, & J. P. Davim (Eds.), *Additive Manufacturing Technologies From an Optimization Perspective* (pp. 199–220). IGI Global. https://www.igi-global.com/gateway/chapter/230144

Jared, B. H., Aguilo, M. A., Beghini, L. L., Boyce, B. L., Clark, B. W., Cook, A., Kaehr, B. J., & Robbins, J. (2017). Additive manufacturing: Toward holistic design. *Scripta Materialia, 135,* 141–147. doi:10.1016/j.scriptamat.2017.02.029

Järvinen, J.-P., Matilainen, V., Li, X., Piili, H., Salminen, A., Mäkelä, I., & Nyrhilä, O. (2014). Characterization of Effect of Support Structures in Laser Additive Manufacturing of Stainless Steel. *Physics Procedia, 56,* 72–81. doi:10.1016/j.phpro.2014.08.099

Jayanaidu, P. (2013). Analysis of a drive shaft for automobile applications. *Journal of Mechanical and Civil Engineering, 10*(2), 43–46. doi:10.9790/1684-1024346

Jhavar, S., Jain, N. K., & Paul, C. P. (2014). Development of micro-plasma transferred arc (μ-PTA) wire deposition process for additive layer manufacturing applications. *Journal of Materials Processing Technology, 214*(5), 1102–1110. doi:10.1016/j.jmatprotec.2013.12.016

Jia, Q., & Gu, D. (2014a). Selective laser melting additive manufacturing of Inconel 718 superalloy parts: Densification, microstructure and properties. *Journal of Alloys and Compounds, 585,* 713–721. doi:10.1016/j.jallcom.2013.09.171

Jia, Q., & Gu, D. (2014b). Selective laser melting additive manufactured Inconel 718 superalloy parts: High-temperature oxidation property and its mechanisms. *Optics & Laser Technology, 62,* 161–171. doi:10.1016/j.optlastec.2014.03.008

Jiménez, M., Romero, L., Domínguez, I. A., Espinosa, M. del M., & Domínguez, M. (2019). Additive Manufacturing Technologies: An Overview about 3D Printing Methods and Future Prospects. *Complexity, 2019,* 1–30. doi:10.1155/2019/9656938

Jinoop, A. N. (2016). *Investigation on laser shock processing of direct metal sintered Inconel 718* (Unpublished M.Tech Thesis). National Institute of Technology Warangal, Warangal, Telangana, India.

Jinoop, A. N., Paul, C. P., & Bindra, K. S. (2019). Laser assisted direct energy deposition of Hastelloy-X. *Optics & Laser Technology, 109,* 14–49. doi:10.1016/j.optlastec.2018.07.037

Johannes Karl Fink. (2018). *3D Industrial Printing with Polymers*. John Wiley & Sons.

Jones, G. (2019). *Direct Metal Laser Sintering (DMLS) – Simply Explained*. All3DP.

Joseph, J. (2016). *Study of direct laser fabricated high entropy alloys. (PhD Thesis)*. Deakin University. http://dro.deakin.edu.au/view/DU:30088804

Joseph, J., Hodgson, P., Jarvis, T., Wu, X., Stanford, N., & Fabijanic, D. M. (2018). Effect of hot isostatic pressing on the microstructure and mechanical properties of additive manufactured AlxCoCrFeNi high entropy alloys. *Materials Science and Engineering A*, *733*, 59–70. doi:10.1016/j.msea.2018.07.036

Joseph, J., Jarvis, T., Wu, X., Stanford, N., Hodgson, P., & Fabijanic, D. M. (2015). Comparative study of the microstructures and mechanical properties of direct laser fabricated and arc-melted AlxCoCrFeNi high entropy alloys. *Materials Science and Engineering A*, *633*, 184–193. doi:10.1016/j.msea.2015.02.072

Junk, S., Klerch, B., Nasdala, L., & Hochberg, U. (2018). Topology optimization for additive manufacturing using a component of a humanoid robot. *Procedia CIRP*, *70*, 102–107. doi:10.1016/j.procir.2018.03.270

Kalentics, N., Sohrabi, N., Tabasi, H. G., Griffiths, S., Jhabvala, J., Leinenbach, C., Burn, A., & Logé, R. E. (2019). Healing cracks in selective laser melting by 3D laser shock peening. *Additive Manufacturing*, *30*(April), 100881. doi:10.1016/j.addma.2019.100881

Kalita, H., Zindani, D., & Kumar, K. (2019). Additive Manufacturing: A Tool for Better Education. In K. Kumar, D. Zindani, & J. P. Davim (Eds.), *Additive Manufacturing Technologies From an Optimization Perspective* (pp. 41–76). IGI Global. doi:10.4018/978-1-5225-9167-2.ch003

Kang, H. G., Park, J.W., & Park, D.W. (2018). Three dimensional printing technique and its application to bone tumor surgery. *J Korean Orthop Assoc*, 466-477.

Kan, W., Chen, B., Peng, H., Liang, Y., & Lin, J. (2020). Fabrication of nano-TiC reinforced high Nb-TiAl nanocomposites by electron beam melting. *Materials Letters*, *259*, 126856. doi:10.1016/j.matlet.2019.126856

Kapil, S., Legesse, F., Kumar, R., & Karunakaran, K. P. (2017). ScienceDirect Hybrid Layered Manufacturing of Turbine Blades. *Materials Today: Proceedings*, *4*(8), 8837–8847. doi:10.1016/j.matpr.2017.07.234

Karunakaran, K. P., Suryakumar, S., Pushpa, V., & Akula, S. (2010). Robotics and Computer-Integrated Manufacturing Low cost integration of additive and subtractive processes for hybrid layered manufacturing. *Robotics and Computer-integrated Manufacturing*, *26*(5), 490–499. doi:10.1016/j.rcim.2010.03.008

Katti, D. R., Sharma, A., & Katti, K. S. (2017). Predictive Methodologies for Design of Bone Tissue Engineering Scaffolds. In *Materials for Bone Disorders* (pp. 453–492). Elsevier. doi:10.1016/B978-0-12-802792-9.00010-0

Kay, A. L., & Couper, P. (2004). *Junkers aircraft and engines, 1913-1945*. Putnam Aeronautical Books.

Kaynak, Y., & Kitay, O. (2019). The effect of post-processing operations on surface characteristics of 316L stainless steel produced by selective laser melting. *Additive Manufacturing*, *26*, 84–93. doi:10.1016/j.addma.2018.12.021

Keichep, D. M., & Smugeresky, J. E. (n.d.). *Using the Laser Engineered Net Shaping (LENS) Process to Produce Complex Components from a CAD Solid Model*. Academic Press.

Kellens, K., Baumers, M., Gutowski, T. G., Flanagan, W., Lifset, R., & Duflou, J. R. (2017). Environmental dimensions of additive manufacturing: Mapping application domains and their environmental implications. *Journal of Industrial Ecology*, *21*(S1), S49–S68. doi:10.1111/jiec.12629

Keller, T., Lindwall, G., Ghosh, S., Ma, L., Lane, B. M., Zhang, F., Kattner, U. R., Lass, E. A., Heigel, J. C., Idell, Y., Williams, M. E., Allen, A. J., Guyer, J. E., & Levine, L. E. (2017). Application of finite element, phase-field, and CALPHAD-based methods to additive manufacturing of Ni-based superalloys. *Acta Materialia*, *139*, 244–253. doi:10.1016/j.actamat.2017.05.003 PMID:29230094

Kellner, T. (2017). *Mind Meld: how GE and a 3D-printing visionary joined forces.* Retrieved from https://www.ge.com/reports/mind-meld-ge-3d-printing-visionary-joined-forces

Kentli, A. (2020). Topology Optimization Applications on Engineering Structures. In A. Kentli (Ed.), *Truss and Frames—Recent Advances and New Perspectives* (pp. 1–23). IntechOpen. doi:10.5772/intechopen.90474

Khaing, M., Fuh, J., & Lu, L. (2001). Direct metal laser sintering for rapid tooling: Processing and characterization of EOS parts. *Journal of Materials Processing Technology*, *113*(1), 269–272. doi:10.1016/S0924-0136(01)00584-2

Khairallah, S. A., Anderson, A. T., Rubenchik, A., & King, W. E. (2016). Laser powder-bed fusion additive manufacturing: Physics of complex melt flow and formation mechanisms of pores, spatter, and denudation zones. *Acta Materialia*, *108*, 36–45. doi:10.1016/j.actamat.2016.02.014

Khaja, M., M, S.H., Umer, U., & Alkhalefah, H. (2019). Fabrication and analysis of a Ti6Al4V implant for cranial restoration. *Applied Sciences*, *2513*. Advance online publication. doi:10.3390/app9122513

Khan, M., & Dickens, P. (2010). Selective Laser Melting (SLM) of pure gold. *Gold Bulletin*, *43*(2), 114–121. doi:10.1007/BF03214976

Khanna, N., Gandhi, A., Nakum, B., & Srivastava, A. (2018). Optimization and analysis of surface roughness for INVAR-36 in end milling operations. *Materials Today: Proceedings*, *5*(2, Part 1), 5281–5288. doi:10.1016/j.matpr.2017.12.111

Khanna, N., Mistry, S., Rahman Rashid, R. A., & Gupta, M. K. (2019). Investigations on density and surface roughness characteristics during selective laser sintering of Invar-36 alloy. *Materials Research Express*, *6*(8), 086541. doi:10.1088/2053-1591/ab18bd

Khan, P. A. A., & Debroy, T. (1984). Alloying element vaporization and weld pool temperature during laser welding of AlSl 202 stainless steel. *Metallurgical Transactions. B, Process Metallurgy*, *15*(4), 641–644. doi:10.1007/BF02657284

Khodabakhshi, F., Farshidianfar, M. H., Bakhshivash, S., Gerlich, A. P., & Khajepour, A. (2019). Dissimilar metals deposition by directed energy based on powder-fed laser additive manufacturing. *Journal of Manufacturing Processes*, *43*(May), 83–97. doi:10.1016/j.jmapro.2019.05.018

Kim, B. J., Hong, K.S., Park, K.J., & Kang, S.H. (2012). Customized cranioplasty implants using three-dimensional printers and polymethyl-methacrylate casting. *J Korean Neurosurg*, 541–546. https:// doi:10.3340/jkns.2012.52.6.541

Kim, S. G. (2018). *Materialize to unveil metal 3D printing simulation module*. Retrieved from https://www.zdnet.co.kr/view/?no=20181001104213

Kim, D. C., Shin, S., Chung, W., Shin, C., & Lim, K. (2014). Application of 3D printing technology in seismic physical modeling. *J. Korean Soc. Miner. Energy Resour. Eng*, *56*(3), 260–269. doi:10.32390/ksmer.2019.56.3.260

Kim, H., Lin, Y., & Tseng, T.-L. B. (2018). A review on quality control in additive manufacturing. *Rapid Prototyping Journal*, *24*(3), 645–669. doi:10.1108/RPJ-03-2017-0048

Kim, K. S. (2014). Characteristic analysis of carburized quenching processes for bevel gear of automobile driving part. *Current Nanoscience*, *10*(1), 51–55. doi:10.2174/15734137096661311112300030

Kim, S. B., Kim, N. H., Kim, J. H., & Moon, H. S. (2018). Evaluation of the fit of metal copings fabricated using stereolithography. *The Journal of Prosthetic Dentistry*, *120*(5), 693–698. doi:10.1016/j.prosdent.2018.01.012 PMID:29807735

Kim, Y.S., Choi, S.W., & Yang, S.Y. (2019). Investigation to metal 3D printing additive manufacturing (AM) process simulation technology. *Journal of Drive and Control*, *16*, 51–58. doi:10.7839/ksfc.2019.16.3.051

Klobčar, D., Lindič, M., & Bušić, M. (2018). Wire arc additive manufacturing of mild steel. *Materials and Geoenvironment*, *65*(4), 179–186. doi:10.2478/rmzmag-2018-0015

Klosterman, D., Chartoff, R., Osborne, N., & Graves, G. (n.d.). *Laminated Object Manufacturing, a New Process for the Direct Manufacture of Monolithic Ceramics and Continuous Fiber CMCs*. Academic Press.

Kobryn, P. A., & Semiatin, S. L. (2003). Microstructure and texture evolution during solidification processing of Ti–6Al–4V. *Journal of Materials Processing Technology*, *135*(2–3), 330–339. doi:10.1016/S0924-0136(02)00865-8

Kok, Y., Tan, X. P., Wang, P., Nai, M. L. S., Loh, N. H., Liu, E., & Tor, S. B. (2018). Anisotropy and heterogeneity of microstructure and mechanical properties in metal additive manufacturing: A critical review. *Materials & Design*, *139*, 565–586. doi:10.1016/j.matdes.2017.11.021

Konečná, R., Kunz, L., Nicoletto, G., & Bača, A. (2016). Long fatigue crack growth in Inconel 718 produced by selective laser melting. *International Journal of Fatigue*, *92*, 499–506. doi:10.1016/j.ijfatigue.2016.03.012

Konečná, R., Nicoletto, G., Kunz, L., & Bača, A. (2016). Microstructure and directional fatigue behavior of Inconel 718 produced by selective laser melting. *Procedia Structural Integrity*, *2*, 2381–2388. doi:10.1016/j.prostr.2016.06.298

Kontis, P., Kostka, A., Raabe, D., & Gault, B. (2019). Influence of composition and precipitation evolution on damage at grain boundaries in a crept polycrystalline Ni-based superalloy. *Acta Materialia*, *166*, 158–167. doi:10.1016/j.actamat.2018.12.039

Koren, Y., Heisel, U., Jovane, F., Moriwaki, T., Pritschow, G., Ulsoy, G., & Van Brussel, H. (1999). Reconfigurable Manufacturing Systems. *CIRP Annals*, *48*(2), 527–540. doi:10.1016/S0007-8506(07)63232-6

Kosaraju, S., Krishna Mohan B., & Singh, S. K. (2019). *Recent Advancement in Additive Manufacturing*. doi:10.4018/978-1-5225-9167-2.ch001

Koutny, D., Palousek, D., Pantelejev, L., Hoeller, C., Pichler, R., Tesicky, L., & Kaiser, J. (2018). Influence of Scanning Strategies on Processing of Aluminum Alloy EN AW 2618 Using Selective Laser Melting. *Materials (Basel, Switzerland)*, *11*(2), 298. doi:10.3390/ma11020298 PMID:29443912

Kreitcberg, A., Brailovski, V., & Turenne, S. (2017a). Effect of heat treatment and hot isostatic pressing on the microstructure and mechanical properties of Inconel 625 alloy processed by laser powder bed fusion. *Materials Science and Engineering A*, *689*, 1–10. doi:10.1016/j.msea.2017.02.038

Kreitcberg, A., Brailovski, V., & Turenne, S. (2017b). Elevated temperature mechanical behavior of IN625 alloy processed by laser powder-bed fusion. *Materials Science and Engineering A*, *700*, 540–553. doi:10.1016/j.msea.2017.06.045

Krishnan, M., Atzeni, E., Canali, R., Calignano, F., Manfredi, D., Ambrosio Elisa, P., & Iuliano, L. (2014). On the effect of process parameters on properties of AlSi10Mg parts produced by DMLS. *Rapid Prototyping Journal*, *20*(6), 449–458. doi:10.1108/RPJ-03-2013-0028

Królikowski, M. A., & Krawczyk, M. B. (2019). Does Metal Additive Manufacturing in Industry 4.0 Reinforce the Role of Substractive Machining? In Lecture Notes in Mechanical Engineering (Vol. 1, pp. 150–164). doi:10.1007/978-3-030-18715-6_13

Kromm, A., Cabeza, S., Mishurova, T., Nadammal, N., Thiede, T., & Bruno, G. (2018).. . *Residual Stresses in Selective Laser Melted Samples of a Nickel Based Superalloy.*, *6*, 259–264. doi:10.21741/9781945291890-41

Kruth, J. P., Vandenbroucke, B., Vaerenbergh, J. V., & Mercelis, P. (2005, April). Benchmarking of Different SLS/SLM Processes as Rapid Manufacturing Techniques. In *Proceedings of International Conference of Polymers & Moulds Innovations (PMI)*. Gent, Belgium: Academic Press.

Kruth, J. P. (1991). Material Incress Manufacturing by Rapid Prototyping Techniques. *CIRP Annals*, *40*(2), 603–614. doi:10.1016/S0007-8506(07)61136-6

Kruth, J.-P., Deckers, J., Yasa, E., & Wauthlé, R. (2012). Assessing and comparing influencing factors of residual stresses in selective laser melting using a novel analysis method. *Proceedings of the Institution of Mechanical Engineers. Part B, Journal of Engineering Manufacture*, *226*(6), 980–991. doi:10.1177/0954405412437085

Kruth, J.-P., Wang, X., Laoui, T., & Froyen, L. (2003). Lasers and materials in selective laser sintering. *Assembly Automation*, *23*(4), 357–371. doi:10.1108/01445150310698652

Kulkarni, M., Mazare, A., Schmuki, P., & Lglic, A. (2014). Biomaterial surface modification of titanium and titanium alloys for medical applications. *Nanomedicine One Central Press Altrincham*, 111–136. Retrieved from http://www.onecentralpress.com/wp-content/uploads/2014/11/CHAPTER-5-NM-12-LATEST.pdf

Kumar, B. Y., Billo, T., Murmu, A., Choube, H., Karunakaran, K. P., & Bernard, A. (2012). Hybrid layered manufacturing using gas metal arc welding. *International Journal of Advanced Manufacturing Systems*, *14*(1), 11–18.

Kumar, S. D., Ghose, J., & Mandal, A. (2019). Thixoforming of light-weight alloys and composites: An approach toward sustainable manufacturing. In *Sustainable Engineering Products and Manufacturing Technologies* (pp. 25–43). Elsevier. doi:10.1016/B978-0-12-816564-5.00002-5

Kundin, J., Mushongera, L., & Emmerich, H. (2015). Phase-field modeling of microstructure formation during rapid solidification in Inconel 718 superalloy. *Acta Materialia*, *95*, 343–356. doi:10.1016/j.actamat.2015.05.052

Kunze, K., Etter, T., Grässlin, J., & Shklover, V. (2015). Texture, anisotropy in microstructure and mechanical properties of IN738LC alloy processed by selective laser melting (SLM). *Materials Science and Engineering A*, *620*, 213–222. doi:10.1016/j.msea.2014.10.003

Kurian & Arivazhagan. (2014). *Numerical and experimental investigations on laser melting of stainless steel 316L metal powders _ Elsevier Enhanced Reader.pdf*. Elsevier.

Kurz, W., & Trivedi, R. (1994). Rapid solidification processing and microstructure formation. *Materials Science and Engineering A*, *179-180*, 46–51. doi:10.1016/0921-5093(94)90162-7

Ladewig, A., Schlick, G., Fisser, M., Schulze, V., & Glatzel, U. (2016). Influence of the shielding gas flow on the removal of process by-products in the selective laser melting process. *Additive Manufacturing*, *10*, 1–9. doi:10.1016/j.addma.2016.01.004

LaMonica, M. (2013). Additive Manufacturing-Innovations, Advances, and Applications. MIT Technology Review, 116. doi:10.1201/b19360

Larimian, T., & Borkar, T. (2019). Additive Manufacturing of In Situ Metal Matrix Composites. In B. AlMangour (Ed.), *Additive Manufacturing of Emerging Materials* (pp. 1–28). Springer International Publishing. doi:10.1007/978-3-319-91713-9_1

Lasers, S. (n.d.). *Direct Metal Laser Sintering | 3D Selective Laser Melting*. Academic Press.

Lee, Y. S., & Zhang, W. (2019). *Laser Powder Bed Additive Manufacturing*. Retrieved from https://www.flow3d.co.kr/3d-%ED%94%84%EB%A6%B0%ED%8C%85-%ED%8C%8C%EC%9A%B0%EB%8D%94-%EB%B2%A0%EB%93%9C-%EC% 88% 98% EC%B9% 98% ED% 95%B4%EC%84%9D/

Lee, J. W., Fang, J.-J., Chang, L.-R., & Yu, C.-K. (2007). Mandibular defect reconstruction with the help of mirror imaging coupled with laser stereolithographic modeling technique. *Journal of the Formosan Medical Association*, *106*(3), 244–250. doi:10.1016/S0929-6646(09)60247-3 PMID:17389170

Lee, J.-Y., An, J., & Chua, C. K. (2017). Fundamentals and applications of 3D printing for novel materials. *Applied Materials Today*, *7*, 120–133. doi:10.1016/j.apmt.2017.02.004

Leung, C. L. A., Marussi, S., Atwood, R. C., Towrie, M., Withers, P. J., & Lee, P. D. (2018). In situ X-ray imaging of defect and molten pool dynamics in laser additive manufacturing. *Nature Communications*, *9*(1), 1355. doi:10.103841467-018-03734-7 PMID:29636443

Lewis, G. K., & Schlienger, E. (2000). Practical considerations and capabilities for laser assisted direct metal deposition. *Materials & Design*, *21*(4), 417–423. doi:10.1016/S0261-3069(99)00078-3

Li, Y., Xiong, J., & Yin, Z. (2019). Molten pool stability of thin-wall parts in robotic GMA-based additive manufacturing with various position depositions. *Robotics and Computer Integrated Manufacturing*, *56*(August), 1–11. doi:10.1016/j.rcim.2018.08.002

Li, C., Guo, Y. B., & Zhao, J. B. (2017). Interfacial phenomena and characteristics between the deposited material and substrate in selective laser melting Inconel 625. *Journal of Materials Processing Technology*, *243*, 269–281. doi:10.1016/j.jmatprotec.2016.12.033

Li, C., Liu, J. F., & Guo, Y. B. (2016). Prediction of residual stress and part distortion in Selective Laser Melting. *Procedia CIRP*, *45*, 171–174. doi:10.1016/j.procir.2016.02.058

Li, C., Liu, Z. Y., Fang, X. Y., & Guo, Y. B. (2018). Residual Stress in Metal Additive Manufacturing. *Procedia CIRP*, *71*, 348–353. doi:10.1016/j.procir.2018.05.039

Li, C., Ramakrishna, S., & Singh, S. (2018). A review of digital manufacturing-based hybrid additive manufacturing processes. *International Journal of Advanced Manufacturing Technology*, *95*, 2281–2300. doi:10.100700170-017-1041-3

Li, C., White, R., Fang, X. Y., Weaver, M., & Guo, Y. B. (2017). Microstructure evolution characteristics of Inconel 625 alloy from selective laser melting to heat treatment. *Materials Science and Engineering A*, *705*, 20–31. doi:10.1016/j.msea.2017.08.058

Li, G., Gao, M., Chen, C., Zhang, C., & Zeng, X. Y. (2014). Characterisation comparison of laser and laser–arc hybrid welding of Invar 36 alloy. *Science and Technology of Welding and Joining*, *19*(1), 30–37. doi:10.1179/1362171813Y.0000000159

Li, H., Chen, B., Tan, C., Song, X., & Feng, J. (2020). Microstructure evolution and mechanical properties of laser metal deposition of Invar 36 alloy. *Optics & Laser Technology*, *125*, 106037. doi:10.1016/j.optlastec.2019.106037

Li, H., Feng, S., Li, J., & Gong, J. (2018). Effect of heat treatment on the δ phase distribution and corrosion resistance of selective laser melting manufactured Inconel 718 superalloy. *Materials and Corrosion*, *69*(10), 1350–1354. doi:10.1002/maco.201810159

Li, J. L. Z., Alkahari, M. R., Rosli, N. A. B., Hasan, R., Sudin, M. N., & Ramli, F. R. (2019). Review of wire arc additive manufacturing for 3d metal printing. *International Journal of Automotive Technology*, *13*(3), 346–353. doi:10.20965/ijat.2019.p0346

Li, J. Q., Sun, W. A., Ao, W. Q., Gu, K. M., & Xiao, P. (2007). Al2O3–FeCrAl composites and functionally graded materials fabricated by reactive hot pressing. *Composites. Part A, Applied Science and Manufacturing*, *38*(2), 615–620. doi:10.1016/j.compositesa.2006.02.012

Lim, C. W. J., Le, K. Q., Lu, Q., & Wong, C. H. (2016). An Overview of 3-D Printing in Manufacturing, Aerospace, and Automotive Industries. *IEEE Potentials*, *35*(4), 18–22. doi:10.1109/MPOT.2016.2540098

Lin, F., Zhang, L., Zhang, T., Wang, J., & Zhang, R. (2012). Innovative education in additive manufacturing in China. In *23rd Annual International Solid Freeform Fabrication Symposium* (pp. 14-44). uTexas: SFF.

Lipton, J. I., Cutler, M., Nigl, F., Cohen, D., & Lipson, H. (2015). Additive manufacturing for the food industry. *Trends in Food Science & Technology*, *43*(1), 114–123. doi:10.1016/j.tifs.2015.02.004

Li, S., Wei, Q., Shi, Y., Zhu, Z., & Zhang, D. (2015). Microstructure Characteristics of Inconel 625 Superalloy Manufactured by Selective Laser Melting. *Journal of Materials Science and Technology*, *31*(9), 946–952. doi:10.1016/j.jmst.2014.09.020

Liu, R., Wang, Z., Sparks, T., Liou, F., & Newkirk, J. (2017). Aerospace applications of laser additive manufacturing. In *Laser Additive Manufacturing* (pp. 351–371). Elsevier. doi:10.1016/B978-0-08-100433-3.00013-0

Liu, S., & Shin, Y. C. (2019). Additive manufacturing of Ti6Al4V alloy: A review. *Materials & Design*, *164*, 107552. doi:10.1016/j.matdes.2018.107552

Liu, W., Jia, C., Guo, M., Gao, J., & Wu, C. (2019). Compulsively constricted WAAM with arc plasma and droplets ejected from a narrow space. *Additive Manufacturing*, *27*(March), 109–117. doi:10.1016/j.addma.2019.03.003

Liu, Y. J., Wang, H. L., Li, S. J., Wang, S. G., Wang, W. J., Hou, W. T., Hao, Y. L., Yang, R., & Zhang, L. C. (2017). Compressive and fatigue behavior of beta-type titanium porous structures fabricated by electron beam melting. *Acta Materialia*, *126*, 58–66. doi:10.1016/j.actamat.2016.12.052

Li, W., Karnati, S., Kriewall, C., Liou, F., Newkirk, J., Taminger, K. M. B., & Seufzer, W. J. (2017). Fabrication and characterization of a functionally graded material from Ti-6Al-4V to SS316 by laser metal deposition. *Additive Manufacturing*, *14*, 95–104. doi:10.1016/j.addma.2016.12.006

Li, Y., & Gu, D. (2014). Parametric analysis of thermal behavior during selective laser melting additive manufacturing of aluminum alloy powder. *Materials & Design*, *63*, 856–867. doi:10.1016/j.matdes.2014.07.006

Li, Y., Hu, Y., Cong, W., Zhi, L., & Guo, Z. (2017). Additive manufacturing of alumina using laser engineered net shaping: Effects of deposition variables. *Ceramics International*, *43*(10), 7768–7775. doi:10.1016/j.ceramint.2017.03.085

Lockwoodanthony, J. (2016). *Inspire, a true topological optimization tool that goes beyond its name!* Retrieved from https://blog.altair.co.kr/34946

Loughborough University. (n.d.). *The 7 categories of Additive Manufacturing*. Additive Manufacturing Research Group | Loughborough University.

Lowa, A. (2018). A design framework for additive manufacturing based on the integration of axiomatic design approach, inverse problem-solving and an additive manufacturing database. Iowa State University.

Lu, X., Zhou, Y. F., Xing, X. L., Shao, L. Y., Yang, Q. X., & Gao, S. Y. (2017). Open-source wire and arc additive manufacturing system: Formability, microstructures, and mechanical properties. *International Journal of Advanced Manufacturing Technology*, *93*(5–8), 2145–2154. doi:10.100700170-017-0636-z

Lu, Y., Wu, S., Gan, Y., Huang, T., Yang, C., Junjie, L., & Lin, J. (2015). Study on the microstructure, mechanical property and residual stress of SLM Inconel-718 alloy manufactured by differing island scanning strategy. *Optics & Laser Technology*, *75*, 197–206. doi:10.1016/j.optlastec.2015.07.009

Ma, G., Zhao, G., Li, Z., Yang, M., & Xiao, W. (2019). Optimization strategies for robotic additive and subtractive manufacturing of large and high thin-walled aluminum structures. *International Journal of Advanced Manufacturing Technology*, *101*(5–8), 1275–1292. doi:10.100700170-018-3009-3

Mahamood, R., Akinlabi, E., Shukla, M., & Pityana, S. (2012). Functionally Graded Material: An Overview. In *Proceedings of the World Congress on Engineering* (Vol. 3, p. 2065). Retrieved from http://researchspace.csir.co.za/dspace/handle/10204/6548

Mahamood, R. M., Akinlabi, E. T., Shukla, M., & Pityana, S. (2013). Scanning velocity influence on microstructure, microhardness and wear resistance performance of laser deposited Ti6Al4V/TiC composite. *Materials & Design*, *50*, 656–666. doi:10.1016/j.matdes.2013.03.049

Maharubin, S., Hu, Y., Sooriyaarachchi, D., Cong, W., & Tan, G. Z. (2019). Laser engineered net shaping of antimicrobial and biocompatible titanium-silver alloys. *Materials Science and Engineering C*, *105*, 110059. doi:10.1016/j.msec.2019.110059 PMID:31546421

Mahmoudi, M., Tapia, G., Franco, B., Ma, J., Arroyave, R., Karaman, I., & Elwany, A. (2018). On the printability and transformation behavior of nickel-titanium shape memory alloys fabricated using laser powder-bed fusion additive manufacturing. *Journal of Manufacturing Processes*, *35*, 672–680. doi:10.1016/j.jmapro.2018.08.037

Malyala, S. K., Manmadhachary, A., Kumar, Y. R., & Alwala, A. (2017). Manufacturing of patient specific additive manufacturing medical models for complex surgeries. *Materials Today: Proceedings*, *4*(2), 1134–1139.

Manfredi, D., Calignano, F., Krishnan, M., Canali, R., Paola, E., Biamino, S., … Fino, P. (2014). Additive Manufacturing of Al Alloys and Aluminium Matrix Composites (AMCs). In Light Metal Alloys Applications. InTech.

Manfredi, D., Calignano, F., Krishnan, M., Canali, R., Paola, E., Biamino, S., Ugues, D., Pavese, M., & Fino, P. (2014). Additive Manufacturing of Al Alloys and Aluminium Matrix Composites (AMCs). In W. A. Monteiro (Ed.), *Light Metal Alloys Applications* (pp. 1–33). InTech. doi:10.5772/58534

Manogharan, G., Wysk, R., Harrysson, O., & Aman, R. (2015). AIMS- a Metal Additive-Hybrid Manufacturing System : System Architecture and Attributes. *Procedia Manufacturing*, *1*, 273–286. doi:10.1016/j.promfg.2015.09.021

Manyika, J., Chui, M., Bughin, J., Dobbs, R., Bisson, P., & Marrs, A. (2013). *Disruptive technologies: Advances that will transform life, business, and the global economy* (Vol. 180). McKinsey Global Institute. http://www.mckinsey.com/insights/business_technology/disruptive_technologies

Markus, G., Xing, L., Xiaojun, C., Dieter, S., & Jan, E. (2016). Computer-aided planning of cranial 3D implants. *Computer Assisted Radiology and Surgery 30th International Congress and Exhibition*. 10.1109/ embc.2016.7590915

Marreiros, M. M., Heuzé, Y., Verius, M., Unterhofer, C., Freysinger, W., & Recheis, W. (2016). Custom implant design for large cranial defects. *International Journal of Computer Assisted Radiology and Surgery*, *11*(12), 2217–2230. doi:10.100711548-016-1454-8 PMID:27358081

Martinezmartin, A. (2016). *Phase optimization + 3D printer = 48% mass reduction*. Retrieved from https://blog.altair.co.kr/34757

Martin, J. H., Yahata, B. D., Clough, E. C., Mayer, J. A., Hundley, J. M., & Schaedler, T. A. (2018). Additive manufacturing of metal matrix composites via nanofunctionalization. *MRS Communications*, *8*(02), 297–302. doi:10.1557/mrc.2018.95

Mashambanhaka, F. (2019). *3D printing Africa: South Africa's 3D printing sector*. Retrieved from https://3dprint.com/242448/3d-printing-africa-south-africas-3d-printing-sector

Materialise Software Company. (2020). *Materialise 3-Matic*. Retrieved from https:// www. Mater -ialise.com /en/ soft -ware/ magics/ modules

Materials, H., Konyashin, I., Hinners, H., Ries, B., Kirchner, A., Klöden, B., ... Sidorenko, D. (2019). International Journal of Refractory Metals Additive manufacturing of WC-13% Co by selective electron beam melting : Achievements and challenges. *International Journal of Refractory Metals & Hard Materials*, *84*(June), 105028. doi:10.1016/j.ijrmhm.2019.105028

Ma, Y., Cuiuri, D., Li, H., Pan, Z., & Shen, C. (2016). The effect of postproduction heat treatment on γ-TiAl alloys produced by the GTAW-based additive manufacturing process. *Materials Science and Engineering A*, *657*, 86–95. doi:10.1016/j.msea.2016.01.060

Mayer, S. (2005). Optimised mould temperature control procedure using DMLS. *EOS Whitepaper, EOS GmbH Ltd*, 1-10.

McCue, T. (2019). *STL Files: What They Are and How to Use Them*. Academic Press.

Meak, L. (2020). *The fourth industrial revolution: Changing how we live and work*. Retrieved from https://newsroom.cisco.com/feature-content?type=webcontent&articleId=1999882

Mehrpouya, M., Dehghanghadikolaei, A., Fotovvati, B., Vosooghnia, A., Emamian, S. S., & Gisario, A. (2019). The Potential of Additive Manufacturing in the Smart Factory Industrial 4.0: A Review. *Applied Sciences*, *9*(18), 3865, 1–34. doi:10.3390/app9183865

Meiners, W., Wissenbach, K. D., & Gasser, A. D. (1998). *Germany Patent No. DE19649865C1*.

Meng, L., Zhang, W., Quan, D., Shi, G., Tang, L., Hou, Y., Breitkopf, P., Zhu, J., & Gao, T. (2019). From Topology Optimization Design to Additive Manufacturing: Today's Success and Tomorrow's Roadmap. *Archives of Computational Methods in Engineering*, 1–26. doi:10.100711831-019-09331-1

Mengucci, P., Barucca, G., Gatto, A., Bassoli, E., Denti, L., Fiori, F., Girardin, E., Bastianoni, P., Rutkowski, B., & Czyrska-Filemonowicz, A. (2016). Effects of thermal treatments on microstructure and mechanical properties of a Co-Cr-Mo-W biomedical alloy produced by laser sintering. *Journal of the Mechanical Behavior of Biomedical Materials*, *60*, 106–117. doi:10.1016/j.jmbbm.2015.12.045 PMID:26803005

Mercelis, P., & Kruth, J. P. (2006). Residual stresses in selective laser sintering and selective laser melting. *Rapid Prototyping Journal*, *12*(5), 254–265. doi:10.1108/13552540610707013

Mertens, R., Vrancken, B., Holmstock, N., Kinds, Y., Kruth, J. P., & Van Humbeeck, J. (2016). Influence of Powder Bed Preheating on Microstructure and Mechanical Properties of H13 Tool Steel SLM Parts. *Physics Procedia*, *83*, 882–890. doi:10.1016/j.phpro.2016.08.092

Metal Powder Report. (2015). *Titanium hip implant printed with EOS technology*. Retrieved from https://www.materialstoday.com/additive-manufacturing/news/titanium-hip-implant-printed-with-eos-technology/

Mfusi, B. J., Tshabalala, L. C., Popoola, A. P. I., & Mathe, N. R. (2018). The effect of selective laser melting build orientation on the mechanical properties of AlSi10Mg parts. *IOP Conference Series. Materials Science and Engineering*, *430*, 012028. doi:10.1088/1757-899X/430/1/012028

Miao, Q., Wu, D., Chai, D., Zhan, Y., Bi, G., Niu, F., & Ma, G. (2020). Comparative study of microstructure evaluation and mechanical properties of 4043 aluminum alloy fabricated by wire-based additive manufacturing. *Materials & Design*, *186*, 108205. doi:10.1016/j.matdes.2019.108205

Ministry of Food and Drug Safety(MFDS). (2019). *GMP complaint guide by 3D printing customized medical device manufacturing process*. Academic Press.

Mishurova, T., Cabeza, S., Thiede, T., Nadammal, N., Kromm, A., Klaus, M., Genzel, C., Haberland, C., & Bruno, G. (2018). The Influence of the Support Structure on Residual Stress and Distortion in SLM Inconel 718 Parts. *Metallurgical and Materials Transactions. A, Physical Metallurgy and Materials Science*, *49*(7), 3038–3046. Advance online publication. doi:10.100711661-018-4653-9

Moat, R. J., Pinkerton, A. J., Li, L., Withers, P. J., & Preuss, M. (2011). Residual stresses in laser direct metal deposited Waspaloy. *Materials Science and Engineering A*, *528*(6), 2288–2298. doi:10.1016/j.msea.2010.12.010

Mohamed, O. A., Masood, S. H., & Bhowmik, J. L. (2017). Experimental investigation of time-dependent mechanical properties of PC-ABS prototypes processed by FDM additive manufacturing process. *Materials Letters*, *193*, 58–62. doi:10.1016/j.matlet.2017.01.104

Mohammadizadeh, M., Fidan, I., Allen, M., & Imeri, A. (2018). Creep behavior analysis of additively manufactured fiber-reinforced components. *International Journal of Advanced Manufacturing Technology*, *99*(5-8), 1225–1234. doi:10.100700170-018-2539-z

Mohammadpour, M., Theodossiades, S., Rahnejat, H., & Saunders, T. (2014). Non-newtonian mixed elastohydrodynamics of differential hypoid gears at high loads. *Meccanica*, *49*(5), 1115–1138. doi:10.100711012-013-9857-x

Moorehead, M., Bertsch, K., Niezgoda, M., Parkin, C., Elbakhshwan, M., Sridharan, K., Zhang, C., Thoma, D., & Couet, A. (2020). High-throughput synthesis of Mo-Nb-Ta-W high-entropy alloys via additive manufacturing. *Materials & Design*, *187*, 108358. doi:10.1016/j.matdes.2019.108358

Mordor Intelligence, L. L. P. (2020). *Simulation software market - growth, trends, andforecast (2020 - 2025)*. Retrieved from https://www.mordorintelligence.com/: https://www.mordorintelligence.com/industry-reports/simulation-software-market

Moroni, G., Syam, W. P., & Petrò, S. (2015, March). Functionality-based part orientation for additive manufacturing. In *Proceedings of 25th CIRP Design Conference*. Haifa, Israel: Academic Press.

Mostafa, A., Picazo Rubio, I., Brailovski, V., Jahazi, M., & Medraj, M. (2017). Structure, Texture and Phases in 3D Printed IN718 Alloy Subjected to Homogenization and HIP Treatments. *Metals*, *7*(196), 1–23. doi:10.3390/met7060196

Mostafaei, A., Kimes, K. A., Stevens, E. L., Toman, J., Krimer, Y. L., Ullakko, K., & Chmielus, M. (2017). Microstructural evolution and magnetic properties of binder jet additive manufactured Ni-Mn-Ga magnetic shape memory alloy foam. *Acta Materialia*, *131*, 482–490. doi:10.1016/j.actamat.2017.04.010

Motta, M., Demir, A. G., & Previtali, B. (2018). High-speed imaging and process characterization of coaxial laser metal wire deposition. *Additive Manufacturing*, *22*, 497–507. doi:10.1016/j.addma.2018.05.043

Mouritz, A. P. (2012). *Introduction to aerospace materials*. Elsevier.

Mourtzis, D., Doukas, M., & Bernidaki, D. (2014). Simulation in Manufacturing: Review and Challenges. *Procedia CIRP*, *25*, 213–229. doi:10.1016/j.procir.2014.10.032

Mower, T. M., & Long, M. J. (2016). Mechanical behavior of additive manufactured, powder-bed laser-fused materials. *Materials Science and Engineering A*, *651*, 198–213. doi:10.1016/j.msea.2015.10.068

MSC Software Company. (2018). *Simulation software for metal 3D printing*. Retrieved from https://www.mscsoftware.com/sites/default/files/simufact_additive_brochure_kr.pdf

Muguruza, A., Bo, J. B., Gómez, A., Minguella-Canela, J., Fernandes, J., Ramos, F., & Cirera, A. (2017). Development of a multi-material additive manufacturing process for electronic devices. *Procedia Manufacturing*, *13*, 746–753. doi:10.1016/j.promfg.2017.09.180

Muñiz-Lerma, A. J., Nommeots-Nomm, A., Waters, E. K., & Brochu, M. (2018). A Comprehensive Approach to Powder Feedstock Characterization for Powder Bed Fusion Additive Manufacturing: A Case Study on AlSi7Mg. *Materials (Basel)*, *11*(12), 2386. doi:10.3390/ma11122386 PMID:30486411

Murphy, J. (2019). *Selective Laser Melting (SLM) – 3D Printing Simply Explained*. All3DP.

Murr, L. E., Martinez, E., Amato, K. N., Gaytan, S. M., Hernandez, J., Ramirez, D. A., Shindo, P.W., Medina, F. & Wicker, R. B. (2012). Fabrication of metal and alloy components by additive manufacturing: examples of 3D materials science. *Journal of Materials Research and Technology, 1*(1), 42-54.

Murr, L. E., Gaytan, S. M., Ceylan, A., Martinez, E., Martinez, J. L., Hernandez, D. H., Machado, B. I., Ramirez, D. A., Medina, F., & Collins, S. (2010). Characterization of titanium aluminide alloy components fabricated by additive manufacturing using electron beam melting. *Acta Materialia*, *58*(5), 1887–1894. doi:10.1016/j.actamat.2009.11.032

Murr, L. E., Martinez, E., Amato, K. N., Gaytan, S. M., Hernandez, J., Ramirez, D. A., Shindo, P. W., Medina, F., & Wicker, R. B. (2012). Fabrication of Metal and Alloy Components by Additive Manufacturing: Examples of 3D Materials Science. *Journal of Materials Research and Technology*, *1*(1), 42–54. doi:10.1016/S2238-7854(12)70009-1

Murr, L. E., Martinez, E., Gaytan, S. M., Ramirez, D. A., Machado, B. I., Shindo, P. W., Martinez, J. L., Medina, F., Wooten, J., Ciscel, D., Ackelid, U., & Wicker, R. B. (2011). Microstructural Architecture, Microstructures, and Mechanical Properties for a Nickel-Base Superalloy Fabricated by Electron Beam Melting. *Metallurgical and Materials Transactions. A, Physical Metallurgy and Materials Science*, *42*(11), 3491–3508. doi:10.100711661-011-0748-2

Mu, X., Bertron, T., Dunn, C., Qiao, H., Wu, J., Zhao, Z., Saldana, C., & Qi, H. J. (2017). Porous polymeric materials by 3D printing of photocurable resin. *Materials Horizons*, *4*(3), 442–449. doi:10.1039/C7MH00084G

Nadammal, N., Cabeza, S., Mishurova, T., Thiede, T., Kromm, A., Seyfert, C., Farahbod, L., Haberland, C., Schneider, J. A., Portella, P. D., & Bruno, G. (2017). Effect of hatch length on the development of microstructure, texture and residual stresses in selective laser melted superalloy Inconel 718. *Materials & Design*, *134*, 139–150. doi:10.1016/j.matdes.2017.08.049

Nakamura, Y. (1976). The invar problem. *IEEE Transactions on Magnetics*, *12*(4), 278–291. doi:10.1109/TMAG.1976.1059049

Nandy, J., Sarangi, H., & Sahoo, S. (2019). A Review on Direct Metal Laser Sintering: Process Features and Microstructure Modeling. *Lasers in Manufacturing and Materials Processing*, *6*(3), 280–316. doi:10.100740516-019-00094-y

Nandy, J., Yedla, N., Gupta, P., Sarangi, H., & Sahoo, S. (2019). Sintering of AlSi10Mg particles in direct metal laser sintering process: A molecular dynamics simulation study. *Materials Chemistry and Physics*, *236*(March), 121803. doi:10.1016/j.matchemphys.2019.121803

NASA 3-D Prints First Full-Scale Copper Rocket Engine Part | NASA. (n.d.). Retrieved February 25, 2020, from https://www.nasa.gov/marshall/news/nasa-3-D-prints-first-full-scale-copper-rocket-engine-part.html

NASA develops first 3-D printed copper rocket engine part - The Economic Times. (n.d.). Retrieved February 25, 2020, from https://economictimes.indiatimes.com/news/science/nasa-develops-first-3-d-printed-copper-rocket-engine-part/articleshow/47013482.cms?from=mdr

Nayak, S. K., Mishra, S. K., Paul, C. P., Jinoop, A. N., & Bindra, K. S. (2020). Effect of energy density on laser powder bed fusion built single tracks and thin wall structures with 100 μm preplaced powder layer thickness. *Optics & Laser Technology*, *125*, 106016. doi:10.1016/j.optlastec.2019.106016

Negi, S., Dhiman, S., & Sharma, R. K. (2013). Basics, applications and future of additive manufacturing technologies: A review. *Journal of Manufacturing Technology Research*, *5*(1–2), 75–96.

Ngo, T. D., Kashani, A., Imbalzano, G., Nguyen, K. T., & Hui, D. (2018). Additive manufacturing (3D printing): A review of materials, methods, applications and challenges. *Composites. Part B, Engineering*, *143*, 172–196. doi:10.1016/j.compositesb.2018.02.012

Nguyen, Q. B., Nai, M. L. S., Zhu, Z., Sun, C. N., Wei, J., & Zhou, W. (2017). Characteristics of Inconel Powders for Powder-Bed Additive Manufacturing. *Engineering*, *3*(5), 695–700. doi:10.1016/J.ENG.2017.05.012

Nicoletto, G., Konečná, R., Frkáň, M., & Riva, E. (2018). Surface roughness and directional fatigue behavior of as-built EBM and DMLS Ti6Al4V. *International Journal of Fatigue*, *116*, 140–148. doi:10.1016/j.ijfatigue.2018.06.011

Nie, P., Ojo, O. A., & Li, Z. (2014). Numerical modeling of microstructure evolution during laser additive manufacturing of a nickel-based superalloy. *Acta Materialia*, *77*, 85–95. doi:10.1016/j.actamat.2014.05.039

Ni, M., Chen, C., Wang, X., Wang, P., Li, R., Zhang, X., & Zhou, K. (2017). Anisotropic tensile behavior of in situ precipitation strengthened Inconel 718 fabricated by additive manufacturing. *Materials Science and Engineering A*, *701*, 344–351. doi:10.1016/j.msea.2017.06.098

O'Neal, B. (2019). *Airbus & Farsoon Technologies in Collaboration to Develop High-Performance Polymers for Additive Manufacturing*. Retrieved from https://www.3dprintpulse.com/additive-manufacturing/china/?open-article-id=9819375&article-title=airbus---farsoon-technologies-in-collaboration-to-develop-high-performance-polymers-for-additive-manufacturing&blog-domain=3dprint.com&blog-title=3dprint-com

Oberg, E., Jones, F. D., Horton, H. L., Ryffel, H. H., McCauley, C., Heald, R., & Hussain, M. (2004). Machinery's Handbook: A Reference Book for the Mechanical Engineer, Designer. *Manufacturing Engineer, Draftsman, Toolmaker, and Machinist, 26.*

Obidigbo, C., Tatman, E.-P., & Gockel, J. (2019). Processing parameter and transient effects on melt pool geometry in additive manufacturing of Invar 36. *International Journal of Advanced Manufacturing Technology*, *104*(5), 3139–3146. doi:10.100700170-019-04229-5

Ogawa, T., & Koseki, T. (1986). Weldability of Invar (Fe-36%Ni) and its Large-Diameter Pipe. *Welding Journal*, *65*(28), 35–44.

Oh, J. W., Na, H., & Choi, H. (2017). Technology trend of the additive manufacturing (AM). *J. Korean Powder Metall*, *24*(6), 1–14. doi:10.4150/KPMI.2017.24.6.494

Optomec Inc. (2018). *The internet of things*. Retrieved from https://optomec.com/printed-electronics/aerosol-jet-core-applications/internet-of-things/

Orme, M., Madera, I., Gschweitl, M., & Ferrari, M. (2018). Topology Optimization for Additive Manufacturing as an Enabler for Light Weight Flight Hardware. *Designs*, *2*(51), 1–22. doi:10.3390/designs2040051

Ota, H., Emaminejad, S., Gao, Y., Zhao, A., Wu, E., Challa, S., Chen, K., Fahad, H. M., Jha, A. K., Kiriya, D., Gao, W., Shiraki, H., Morioka, K., Ferguson, A. R., Healy, K. E., Davis, R. W., & Javey, A. (2016). Application of 3D Printing for Smart Objects with Embedded Electronic Sensors and Systems. *Advanced Materials Technologies*, *1*(1), 1600013. doi:10.1002/admt.201600013

Ott, E. A., Groh, J., & Sizek, H. (2005). Metals Affordability Initiative: Application of Allvac alloy 718Plus® for aircraft engine static structural components. *Superalloys*, *718*, 625–706. doi:10.7449/2005/Superalloys_2005_35_45

Özel, T., Shaurya, A., Altay, A., & Yang, L. (2018). Process monitoring of meltpool and spatter for temporal-spatial modeling of laser powder bed fusion process. *Procedia CIRP*, *74*, 102–106. doi:10.1016/j.procir.2018.08.049

Palakkad, I. I. T. (2019). *Inauguration of metal additive manufacturing (3D metal printing) facility at IIT Palakkad*. Retrieved from https://iitpkd.ac.in/news/inauguration-metal-additive-manufacturing-3d-metal-printing-facility-iit-palakkad

Palermo, E. (2013). *What is Selective Laser Sintering?* Live Science.

Palmarini, P., Erkoyuncu, J. A., Roy, R., & Torabmostaedi, H. (2018). A systematic review of augmented reality applications in maintenance. *Robotics and Computer-integrated Manufacturing*, *49*, 215–228. doi:10.1016/j.rcim.2017.06.002

Pal, S., Lojen, G., Kokol, V., & Drstvensek, I. (2018). Evolution of metallurgical properties of Ti-6Al-4V alloy fabricated in different energy densities in the Selective Laser Melting technique. *Journal of Manufacturing Processes*, *35*, 538–546. doi:10.1016/j.jmapro.2018.09.012

Pan, Z., Ding, D., Wu, B., Cuiuri, D., Li, H., & Norrish, J. (2018). Arc Welding Processes for Additive Manufacturing: A Review. In Transactions on Intelligent Welding Manufacturing, Transactions on Intelligent Welding Manufacturing (pp. 3–24). doi:10.1007/978-981-10-5355-9_1

Panchagnula, J. S., & Simhambhatla, S. (2018). Manufacture of complex thin-walled metallic objects using weld-deposition based additive manufacturing. *Robotics and Computer-integrated Manufacturing*, *49*, 194–203. doi:10.1016/j.rcim.2017.06.003

Panda, B. K., & Sahoo, S. (2018). Numerical simulation of residual stress in laser based additive manufacturing process. *IOP Conference Series. Materials Science and Engineering*, *338*(1), 012030. doi:10.1088/1757-899X/338/1/012030

Panda, B. K., & Sahoo, S. (2019). Thermo-mechanical modeling and validation of stress field during laser powder bed fusion of AlSi10Mg built part. *Results in Physics*, *12*, 1372–1381. doi:10.1016/j.rinp.2019.01.002

Paolini, A., Kollmannsberger, S., & Rank, E. (2019). Additive manufacturing in construction: A review on processes, applications, and digital planning methods. *Additive Manufacturing*, *30*(October), 100894. doi:10.1016/j.addma.2019.100894

Parthasarathy, J. (2014). 3D modeling, custom implants and its future perspectives in craniofacial surgery. *Ann. Maxillofac*, *4*(1), 9–18. doi:10.4103/2231-0746.133065 PMID:24987592

Patil, S. M., Deshpande, A. S., Bhalerao, R. R., Metkari, S. B., & Patil, P. M. (2019). A three-dimensional finite element analysis of the influence of varying implant crest module designs on the stress distribution to the bone. *Dental Research Journal*, *16*(3), 145–152. doi:10.4103/1735-3327.255750 PMID:31040869

Paul, C. P., Bhargava, P., Kumar, A., Pathak, A. K., & Kukreja, L. M. (2012). Laser Rapid Manufacturing: technology, applications, modeling and future prospects. In J. P. Davim (Ed.), *Lasers in Manufacturing* (pp. 1–60). ISTE-Wiley.

Paul, C. P., Jinoop, A. N., & Bindra, K. S. (2018). Metal Additive Manufacturing using lasers. In R. Singh & J. P. Davim (Eds.), *Additive Manufacturing: Applications and innovations* (pp. 37–94). CRC Press. doi:10.1201/b22179-2

Paul, C. P., Mishra, S. K., Premsingh, C. H., Bhargava, P., Tiwari, P., & Kukreja, L. M. (2012). Studies on laser rapid manufacturing of cross-thin-walled porous structures of Inconel 625. *International Journal of Advanced Manufacturing Technology*, *61*(5-8), 757–770. doi:10.100700170-011-3742-3

Paulonis, D.F. & Schirra, J.J. (2001). Alloy 718 at Pratt & Whitney- Historical Perspective and Future Challenges. *TMS Superalloys 718. 625. 706 and Various Derivatives.*

Pei, E., Loh, G. H., Harrison, D., de Amorim Almeida, H., Verona, M. D. M., & Paz, R. (2017). A study of 4D printing and functionally graded additive manufacturing. *Assembly Automation*, *37*(2), 147–153. doi:10.1108/AA-01-2017-012

Peng, H., Liu, C., Guo, H., Yuan, Y., Gong, S., & Xu, H. (2016). Fabrication of WCp/NiBSi metal matrix composite by electron beam melting. *Materials Science and Engineering A*, *666*, 320–323. doi:10.1016/j.msea.2016.04.079

Perevoshchikova, N., Rigaud, J., Sha, Y., Heilmaier, M., Finnin, B., Labelle, E., & Wu, X. (2017). Optimisation of selective laser melting parameters for the Ni-based superalloy IN-738 LC using Doehlert's design. *Rapid Prototyping Journal*, *23*(5), 881–892. doi:10.1108/RPJ-04-2016-0063

Pham, D., & Dimov, S. S. (2012). *Rapid manufacturing: the technologies and applications of rapid prototyping and rapid tooling*. Springer Science & Business Media.

Pinkerton, A. J. (2016). Lasers in additive manufacturing. *Optics & Laser Technology*, *78*, 25–32. doi:10.1016/j.optlastec.2015.09.025

Plimpton, S., Paul, C., & Aidan, T. (2007). LAMMPS-large-scale atomic/molecular massivelyparallel simulator. *Sandia National Laboratories*, *18*, 43.

Pollock, T. M., Rene, N., Rene, N., & Rene, N. (2006). Nickel-Based Superalloys for Advanced Turbine Engines : Chemistry, Microstructure, and Properties. *Journal of Propulsion and Power*, *22*(2), 361–374. doi:10.2514/1.18239

Popovich, A. A., Sufiiarov, Vs., Borisov, E., Polozov, I., & Masaylo, D. (2018). Design and manufacturing of tailored microstructure with selecive laser melting. *Materials Physics and Mechanics*, *38*, 1–10. doi:10.18720/MPM.3812018_1

Popovich, V. A., Borisov, E. V., Popovich, A. A., Sufiiarov, V. S., Masaylo, D. V., & Alzina, L. (2017a). Functionally graded Inconel 718 processed by additive manufacturing: Crystallographic texture, anisotropy of microstructure and mechanical properties. *Materials & Design*, *114*, 441–449. doi:10.1016/j.matdes.2016.10.075

Popovich, V. A., Borisov, E. V., Popovich, A. A., Sufiiarov, V. S., Masaylo, D. V., & Alzina, L. (2017b). Impact of heat treatment on mechanical behaviour of Inconel 718 processed with tailored microstructure by selective laser melting. *Materials & Design*, *131*, 12–22. doi:10.1016/j.matdes.2017.05.065

Pour, M. A., Zanardini, M., Bacchetti, A., & Zanoni, S. (2016). Additive Manufacturing Impacts on Productions and Logistics Systems. *IFAC-PapersOnLine*, *49*(12), 1679–1684. doi:10.1016/j.ifacol.2016.07.822

Poyraz, O. (2018). *Investigation of the Effects of Process Parameters for Laser Additive Manufacturing by Using Modeling and Simulation Methods* (Doctoral dissertation). Eskisehir Osmangazi University, Eskisehir, Turkey.

Poyraz, Ö., & Kuşhan, M. C. (2018). Investigation of the effect of different process parameters for laser additive manufacturing of metals. *Journal of the Faculty of Engineering and Architecture of Gazi University*, *33*(2), 729–742.

Poyraz, Ö., & Kuşhan, M. C. (2019). Residual Stress-induced Distortions in Laser Powder Bed Additive Manufacturing of Nickel-based Superalloys. *Strojniski Vestnik. Jixie Gongcheng Xuebao*, *65*(6).

Poyraz, Ö., Solakoğlu, E. U., Ören, S., Tüzemen, C., & Akbulut, G. (2019). Surface texture and form characterization for powder bed additive manufacturing. *Journal of the Faculty of Engineering and Architecture of Gazi University*, *34*(3), 1653–1664.

Poyraz, Ö., Yasa, E., Akbulut, G., Orhangül, A., & Pilatin, S. (2015, August) Investigation of support structures for direct metal laser sintering of IN625 parts. In *Proceedings of Solid Freeform Fabrication Symposium*. Austin, TX: Academic Press.

Prashanth, K. G., Scudino, S., Maity, T., Das, J., & Eckert, J. (2017). Is the energy density a reliable parameter for materials synthesis by selective laser melting? *Materials Research Letters*, *5*(6), 386–390. doi:10.1080/21663831.2017.1299808

Pröbstle, M., Neumeier, S., Hopfenmüller, J., Freund, L. P., Niendorf, T., Schwarze, D., & Göken, M. (2016). Superior creep strength of a nickel-based superalloy produced by selective laser melting. *Materials Science and Engineering A*, *674*, 299–307. doi:10.1016/j.msea.2016.07.061

Promakhov, V., Zhukov, A., Ziatdinov, M., Zhukov, I., Schulz, N., Kovalchuk, S., Dubkova, Y., Korsmik, R., Klimova-Korsmik, O., Turichin, G., & Perminov, A. (2019). Inconel 625/TiB2 Metal Matrix Composites by Direct Laser Deposition. *Metals, 9*(2), 141, 1–22. doi:10.3390/met9020141

Qi, Z., Cong, B., Qi, B., Sun, H., Zhao, G., & Ding, J. (2018). Microstructure and mechanical properties of double-wire + arc additively manufactured Al-Cu-Mg alloys. *Journal of Materials Processing Tech., 255*(December), 347–353. doi:10.1016/j.jmatprotec.2017.12.019

Qi, Z., Qi, B., Cong, B., Sun, H., Zhao, G., & Ding, J. (2019). Microstructure and mechanical properties of wire + arc additively manufactured 2024 aluminum alloy components: As-deposited and post heat-treated. *Journal of Manufacturing Processes, 40*(May), 27–36. doi:10.1016/j.jmapro.2019.03.003

Qin, Y., Qi, Q., Scott, P. J., & Jiang, X. (2019). Status, comparison, and future of the representations of additive manufacturing data. *Computer Aided Design*, *111*, 44–64. doi:10.1016/j.cad.2019.02.004

Qiu, C., Adkins, N. J. E., & Attallah, M. M. (2016). Selective laser melting of Invar 36: Microstructure and properties. *Acta Materialia*, *103*, 382–395. doi:10.1016/j.actamat.2015.10.020

Qi, X., Chen, G., Li, Y., Chen, X., & Li, C. (2019). Applying neural-network-based machine learning to additive manufacturing: Current applications, challenges, and future perspectives. *Engineering*, *5*(4), 721–729. doi:10.1016/j.eng.2019.04.012

Qi, Z., Cong, B., Qi, B., Zhao, G., & Ding, J. (2018). Properties of wire + arc additively manufactured 2024 aluminum alloy with different solution treatment temperature. *Materials Letters*, *230*, 275–278. doi:10.1016/j.matlet.2018.07.144

Qi, Z., Qi, B., Cong, B., & Zhang, R. (2018). Microstructure and mechanical properties of wire + arc additively manufactured Al-Mg-Si aluminum alloy. *Materials Letters*, *233*, 348–350. doi:10.1016/j.matlet.2018.09.048

Quan, Z., Larimore, Z., Wu, A., Yu, J., Qin, X., Mirotznik, M., Suhr, J., Byun, J.-H., Oh, Y., & Chou, T.-W. (2016). Microstructural design and additive manufacturing and characterization of 3D orthogonal short carbon fiber/acrylonitrile-butadiene-styrene preform and composite. *Composites Science and Technology*, *126*, 139–148. doi:10.1016/j.compscitech.2016.02.021

Quintanilla, P., Hope, D. A., Darnton, H., & Hunter, A. (2019). *Responsible use of technology*. Retrieved from https://www.weforum.org/whitepapers/responsible-use-of-technology

Rafieazad, M., Ghaffari, M., Vahedi Nemani, A., & Nasiri, A. (2019). Microstructural evolution and mechanical properties of a low-carbon low-alloy steel produced by wire arc additive manufacturing. *International Journal of Advanced Manufacturing Technology*, *105*(5-6), 2121–2134. doi:10.100700170-019-04393-8

Rafieazad, M., Mohammadi, M., & Nasiri, A. M. (2019). On microstructure and early stage corrosion performance of heat treated direct metal laser sintered AlSi10Mg. *Additive Manufacturing*, *28*(April), 107–119. doi:10.1016/j.addma.2019.04.023

Raghavan, N., Simunovic, S., Dehoff, R., Plotkowski, A., Turner, J., Kirka, M., & Babu, S. (2017). Localized melt-scan strategy for site speci fi c control of grain size and primary dendrite arm spacing in electron beam additive. *Acta Materialia*, *140*, 375–387. doi:10.1016/j.actamat.2017.08.038

Rahman, Z., Barakh Ali, S. F., Ozkan, T., Charoo, N. A., Reddy, I. K., & Khan, M. A. (2018). Additive Manufacturing with 3D Printing: Progress from Bench to Bedside. *The AAPS Journal*, *20*(6), 101. doi:10.120812248-018-0225-6 PMID:30209693

Ramakrishnan, A., & Dinda, G. P. (2019). Microstructure and mechanical properties of direct laser metal deposited Haynes 282 superalloy. *Materials Science and Engineering A*, *748*, 347–356. doi:10.1016/j.msea.2019.01.101

Ramirez, A. J., Sowards, J. W., & Lippold, J. C. (2006). Improving the ductility-dip cracking resistance of Ni-base alloys. *Journal of Materials Processing Technology*, *179*(1–3), 212–218. doi:10.1016/j.jmatprotec.2006.03.095

Rao, C. S., & Prasad, K. S. (2012). Advances in Plasma Arc Welding : A Review. *Journal of Mechanical Engineering and Technology*, *4*(1), 35–60.

Raus, A. A., Wahab, M. S., Ibrahim, M., Kamarudin, K., Ahmed, A., & Shamsudin, S. (2017, April). Mechanical and physical properties of AlSi10Mg processed through selective laser melting. In AIP Conference Proceedings (Vol. 1831, No. 1, p. 020027). AIP Publishing LLC. doi:10.1063/1.4981168

Reed, R. C. (2006). *The Superalloys: Fundamentals and Applications.* Cambridge. doi:10.1017/CBO9780511541285

Reisman, G. (1996). *Capitalism: A Complete and Integrated Understanding of the Nature and Value of Human Economic Life*. Jameson Books.

Rengier, F., Mehndiratta, A., von Tengg-Kobligk, H., Zechmann, C. M., Unterhinninghofen, R., Kauczor, H.-U., & Giesel, F. L. (2010). 3D printing based on imaging data: Review of medical applications. *International Journal of Computer Assisted Radiology and Surgery*, *5*(4), 335–341. doi:10.100711548-010-0476-x PMID:20467825

Renishaw plc. (2016). *Metal 3D printing pushes the boundaries in Moto2™ through defiant innovation.* Retrieved from http://www.renishaw.com/en/metal-3d-printing-pushes-the-boundaries-in-moto2-through-defiant-innovation--38935

Renishaw plc. (2019). *Resource centre - Case study: First metal 3D printed bicycle frame manufactured by Renishaw for Empire cycles.* Retrieved from https://resources.renishaw.com/en/details/case-study-first-metal-3d-printed-bicycle-frame-manufactured-by-renishaw-for-empire-cycles--57197

Revathi, A., Mitun, D., Balla, V. K., Dwaipayan, S., Devika, D., & Manivasagam, G. (2019). Surface properties and cytocompatibility of Ti-6Al-4V fabricated using Laser Engineered Net Shaping. *Materials Science and Engineering C*, *100*, 104–116. doi:10.1016/j.msec.2019.02.099 PMID:30948044

Rickenbacher, L., Etter, T., Hövel, S., & Wegener, K. (2013). High temperature material properties of IN738LC processed by selective laser melting (SLM) technology. *Rapid Prototyping Journal*, *19*(4), 282–290. doi:10.1108/13552541311323281

Risse, J. (2019). *Additive Manufacturing of Nickel-Base Superalloy IN738LC by Laser Powder Bed Fusion* (Doctoral Dissertation). Rheinisch-Westfälischen Technischen Hochschule Aachen, Aachen, Germany.

Robotic Industries Association (RIA). (2018). *Collaborative robot market growing.* Retrieved from https://www.controleng.com/articles/collaborative-robot-market-growing/

Rodrigues, T. A., Duarte, V., Miranda, R. M., Santos, T. G., & Oliveira, J. P. (2019). Current Status and Perspectives on Wire and Arc Additive Manufacturing (WAAM). *Materials (Basel)*, *12*(7), 1121. doi:10.3390/ma12071121 PMID:30987382

Rodriguez, N., Vázquez, L., Huarte, I., Arruti, E., Tabernero, I., & Alvarez, P. (2018). Wire and arc additive manufacturing: A comparison between CMT and TopTIG processes applied to stainless steel. *Welding in the World*, *62*(5), 1083–1096. doi:10.100740194-018-0606-6

Roehling, T. T., Wu, S. S. Q., Khairallah, S. A., Roehling, J. D., Soezeri, S. S., Crumb, M. F., & Matthews, M. J. (2017). Modulating laser intensity pro fi le ellipticity for microstructural control during metal additive manufacturing. *Acta Materialia*, *128*, 197–206. doi:10.1016/j.actamat.2017.02.025

Rojko, A. (2017). Industry 4.0 Concept: Background and Overview. *International Journal of Interactive Mobile Technologies*, *11*(5), 77. doi:10.3991/ijim.v11i5.7072

Romano, J., Ladani, L., Razmi, J., & Sadowski, M. (2015). Temperature distribution and melt geometry in laser and electron-beam melting processes – A comparison among common materials. *Additive Manufacturing*, *8*, 1–11. doi:10.1016/j.addma.2015.07.003

Rosa, B., Mognol, P., & Hascoët, J. Y. (2014). Topography modeling of laser polishing on AISI 316L. *Mechanics & Industry*, *15*(1), 51–61. doi:10.1051/meca/2014003

Ruttert, B., Ramsperger, M., Mujica Roncery, L., Lopez-Galilea, I., Körner, C., & Theisen, W. (2016). Impact of hot isostatic pressing on microstructures of CMSX-4 Ni-base superalloy fabricated by selective electron beam melting. *Materials & Design*, *110*, 720–727. doi:10.1016/j.matdes.2016.08.041

Ryan, E. M., Sabin, T. J., Watts, J. F., & Whiting, M. J. (2018). The influence of build parameters and wire batch on porosity of wire and arc additive manufactured aluminium alloy 2319. *Journal of Materials Processing Technology*, *262*(April), 577–584. doi:10.1016/j.jmatprotec.2018.07.030

Sabol, G. P., & Stickler, R. (1969). Microstructure of Nickel-Based Superalloys. *Physica Status Solidi. B, Basic Research*, *35*(1), 11–52. doi:10.1002/pssb.19690350102

Saboori, A., Gallo, D., Biamino, S., Fino, P., & Lombardi, M. (2017). An Overview of Additive Manufacturing of Titanium Components by Directed Energy Deposition: Microstructure and Mechanical Properties. *Applied Sciences*, *7*(9), 883, 1–23. doi:10.3390/app7090883

Saewe, J., Gayer, C., Vogelpoth, A., & Schleifenbaum, J. H. (2019). Feasability Investigation for Laser Powder Bed Fusion of High-Speed Steel AISI M50 with Base Preheating System. *BHM. Berg- und Hüttenmännische Monatshefte*, *164*(3), 101–107. doi:10.100700501-019-0828-y

Samantaray, M., Sahoo, S., & Thatoi, D. (2018). Computational modeling of heat transfer and sintering behavior during direct metal laser sintering of AlSi10Mg alloy powder. *Comptes Rendus. Mécanique*, *346*(11), 1043–1054. doi:10.1016/j.crme.2018.08.006

Sames, W. J., Unocic, K. A., Dehoff, R. R., Lolla, T., & Babu, S. S. (2014). Thermal effects on microstructural heterogeneity of Inconel 718 materials fabricated by electron beam melting. *Journal of Materials Research*, *29*(17), 1920–1930. doi:10.1557/jmr.2014.140

Sandia National Laboratories. (2017). *Laser Engineered Net Shaping™*. Retrieved from https://www.sandia.gov/mst/technologies/net-shaping.html

Sarkar, S., Kumar, C. S., & Nath, A. K. (2017). Effect of Different Heat Treatments on Mechanical Properties of Laser Sintered Additive Manufactured Parts. *Journal of Manufacturing Science and Engineering*, *139*(11), 1–11. doi:10.1115/1.4037437

Sateesh, N. H., Kumar, G. C. M., & Prasad, K., C.K., S., & Vinod, A. R. (2014). Microstructure and Mechanical Characterization of Laser Sintered Inconel-625 Superalloy. *Procedia Materials Science*, *5*, 772–779. doi:10.1016/j.mspro.2014.07.327

Satpathy, B. B., Nandy, J., & Sahoo, S. (2018). Investigation of consolidation kinetics and microstructure evolution of Al alloys in direct metal laser sintering using phase field simulation. *IOP Conference Series. Materials Science and Engineering*, *338*(1), 012045. doi:10.1088/1757-899X/338/1/012045

Schmutzler, C., Stiehl, T. H., & Zaeh, M. F. (2019). Empirical process model for shrinkage-induced warpage in 3D printing. *Rapid Prototyping Journal*, *25*(4), 721–727. doi:10.1108/RPJ-04-2018-0098

Scipioni Bertoli, U., Wolfer, A. J., Matthews, M. J., Delplanque, J.-P. R., & Schoenung, J. M. (2017). On the limitations of Volumetric Energy Density as a design parameter for Selective Laser Melting. *Materials & Design*, *113*, 331–340. doi:10.1016/j.matdes.2016.10.037

Scott, C. (2016). *China develops new metal 3D printing technology, Combining old and new manufacturing techniques*. Retrieved from https://3dprint.com/143613/china-metal-3d-printing-tech/

Sculpteo. (n.d.). *DMLS Metal Technology for 3D Printing*. Sculpteo.

Seifert, T., Sowade, E., Roscher, F., Wiemer, M., Gessner, T., & Baumann, R. R. (2015). Additive Manufacturing Technologies Compared: Morphology of Deposits of Silver Ink Using Inkjet and Aerosol Jet Printing. *Industrial & Engineering Chemistry Research*, *54*(2), 769–779. doi:10.1021/ie503636c

Sercombe, T. B., Li, X., Sercombe, T. B., & Li, X. (2016). *Selective laser melting of aluminium and aluminium metal matrix composites : review*. doi:10.1080/10667857.2016.1161147

Seufzer, W. J., & Taminger, K. M. (n.d.). *Control Methods for the Electron Beam Free Form Fabrication Process*. Academic Press.

Shahzad, K., Deckers, J., Zhang, Z., Kruth, J.-P., & Vleugels, J. (2014). Additive manufacturing of zirconia parts by indirect selective laser sintering. *Journal of the European Ceramic Society*, *34*(1), 81–89. doi:10.1016/j.jeurceramsoc.2013.07.023

Shaikh, A. S. (2018). *Development of γ'Precipitation Hardening Ni-Base Superalloy for Additive Manufacturing* (Master's Thesis). Chalmers University.

Shamsaei, N., Yadollahi, A., Bian, L., & Thompson, S. M. (2015). An overview of Direct Laser Deposition for additive manufacturing; Part II: Mechanical behavior, process parameter optimization and control. *Additive Manufacturing*, *8*, 12–35. doi:10.1016/j.addma.2015.07.002

Sharma, A., Bandari, V., Ito, K., Kohama, K., & Ramji, M., & BV, H. S. (. (2017). A new process for design and manufacture of tailor-made functionally graded composites through friction stir additive manufacturing. *Journal of Manufacturing Processes*, *26*, 122–130. doi:10.1016/j.jmapro.2017.02.007

Shen, C., Pan, Z., Cuiuri, D., Roberts, J., & Li, H. (2016). Fabrication of Fe-FeAl functionally graded material using the wire-arc additive manufacturing process. *Metallurgical and Materials Transactions. B, Process Metallurgy and Materials Processing Science*, *47*(1), 763–772. doi:10.100711663-015-0509-5

Shen, C., Pan, Z., Ma, Y., Cuiuri, D., & Li, H. (2015). Fabrication of iron-rich Fe – Al intermetallics using the wire-arc additive manufacturing process. *Additive Manufacturing*, *7*, 20–26. doi:10.1016/j.addma.2015.06.001

Shen, J., & Yang, X. (2010). Numerical approximations of allen-cahn and cahn-hilliard equations. *Discrete and Continuous Dynamic Systems*, *28*(4), 1669–1691. doi:10.3934/dcds.2010.28.1669

Shiga, M. (1996). Invar alloys. *Current Opinion in Solid State and Materials Science*, *1*(3), 340–348. doi:10.1016/S1359-0286(96)80023-4

Shi, Q., Gu, D., Xia, M., Cao, S., & Rong, T. (2016). Effects of laser processing parameters on thermal behavior and melting/solidification mechanism during selective laser melting of TiC/Inconel 718 composites. *Optics & Laser Technology*, *84*, 9–22. doi:10.1016/j.optlastec.2016.04.009

Shiva, S., Palani, I. A., Paul, C. P., & Singh, B. (2016). Laser annealing of laser additive– manufactured Ni-Ti structures: An experimental–numerical investigation. *Proceedings of the Institution of Mechanical Engineers. Part B, Journal of Engineering Manufacture*, *232*(6), 1054–1067. doi:10.1177/0954405416661582

Shi, Y., Wang, Y., Chen, J., & Huang, S. (2008). Experimental investigation into the selective laser sintering of high-impact polystyrene. *Journal of Applied Polymer Science*, *108*(1), 535–540. doi:10.1002/app.27686

Shunmugavel, M., Polishetty, A., & Littlefair, G. (2015). Microstructure and Mechanical Properties of Wrought and Additive Manufactured Ti-6Al-4V Cylindrical Bars. *Procedia Technology*, *20*, 231–236. doi:10.1016/j.protcy.2015.07.037

Silva, J. V., & Rezende, R. A. (2013). Additive manufacturing and its future impact in logistics. *IFAC Proceedings Volumes, 46*(24), 277-282.

Silva, F. J. G., Campilho, R. D. S. G., Gouveia, R. M., Pinto, G., & Baptista, A. (2018). A Novel Approach to Optimize the Design of Parts for Additive Manufacturing. *Procedia Manufacturing*, *17*, 53–61. doi:10.1016/j.promfg.2018.10.012

Simchi, A., Petzoldt, F., & Pohl, H. (2003). On the development of direct metal laser sintering for rapid tooling. *Journal of Materials Processing Technology*, *141*(3), 319–328. doi:10.1016/S0924-0136(03)00283-8

Sims, C. T. (1984). *A history of superalloy metallurgy for superalloy metallurgists*. Superalloys. doi:10.7449/1984/Superalloys_1984_399_419

Singare, S., Lian, Q., Wang, W.P., Wang, J., Liu, Y., Li, D., & Lu, B. (2009). Rapid prototyping assisted surgery planning and custom implant design. *Rapid Prototyp*, *19–23*. Advance online publication. doi:10.1108/13552540910925027

Singh, S., & Ramakrishna, S. (2017). Biomedical applications of additive manufacturing: Present and future. *Current Opinion in Biomedical Engineering*, *2*, 105–115. doi:10.1016/j.cobme.2017.05.006

Sing, S. L., An, J., Yeong, W. Y., & Wiria, F. E. (2016). Laser and electron-beam powder-bed additive manufacturing of metallic implants: A review on processes, materials and designs. *Journal of Orthopaedic Research*, *34*(3), 369–385. doi:10.1002/jor.23075 PMID:26488900

Smith, J., Xiong, W., Yan, W., Lin, S., Cheng, P., Kafka, O. L., Wagner, G. J., Cao, J., & Liu, W. K. (2016). Linking process, structure, property, and performance for metal-based additive manufacturing: Computational approaches with experimental support. *Computational Mechanics*, *57*(4), 583–610. doi:10.100700466-015-1240-4

Snyder, J., Rin Son, A., Hamid, Q., & Sun, W. (2016). Fabrication of Microfluidic Manifold by Precision Extrusion Deposition and Replica Molding for Cell-Laden Device. *Journal of Manufacturing Science and Engineering*, *138*(4), 041007. Advance online publication. doi:10.1115/1.4031551

Solutions, S. L. M. (2018). *SLM® machines*. Retrieved from https://slm-solutions.com

Song, B., Dong, S., Coddet, P., Liao, H., & Coddet, C. (2014). Fabrication of NiCr alloy parts by selective laser melting: Columnar microstructure and anisotropic mechanical behavior. *Materials & Design*, *53*, 1–7. doi:10.1016/j.matdes.2013.07.010

Song, S., Gao, Z., Lu, B., Bao, C., Zheng, B., & Wang, L. (2019). Performance optimization of complicated structural SiC/Si composite ceramics prepared by selective laser sintering. *Ceramics International*, *46*(1), 568–575. doi:10.1016/j.ceramint.2019.09.004

Souza, H. (2019). *3D printing: a growing market in Brazil.* Retrieved from https://3dprint.com/236228/3d-printing-a-growing-market-in-brazil/

Spears, T. G., & Gold, S. A. (2016). In-process sensing in selective laser melting (SLM) additive manufacturing. *Integrating Materials and Manufacturing Innovation, 5*(1), 16–40. doi:10.118640192-016-0045-4

Spierings, A. B., Herres, N., & Levy, G. (2011). Influence of the particle size distribution on surface quality and mechanical properties in AM steel parts. *Rapid Prototyping Journal, 17*(3), 195–202. doi:10.1108/13552541111124770

Spierings, A. B., Voegtlin, M., Bauer, T., & Wegener, K. (2016). Powder flowability characterisation methodology for powder-bed-based metal additive manufacturing. *Progress in Additive Manufacturing, 1*(1-2), 9–20. doi:10.100740964-015-0001-4

Spoerk, M., Arbeiter, F., Cajner, H., Sapkota, J., & Holzer, C. (2017). Parametric optimization of intra- and inter-layer strengths in parts produced by extrusion-based additive manufacturing of poly(lactic acid). *Journal of Applied Polymer Science, 134*(41), 45401. doi:10.1002/app.45401

Spowart, J. E., Gupta, N., & Lehmhus, D. (2018). Additive Manufacturing of Composites and Complex Materials. *JOM, 70*(3), 272–274. doi:10.100711837-018-2742-2

Staffa, G., Barbanera, A., Faiola, A., & Fricia, M. (2012). Custom made bio ceramic implants in complex and large cranial re-construction: A two-year follow-up. *Journal of Cranio-Maxillofacial Surgery, 40*(3),65-70. https:// doi:10.1016/j.jcms.2011.04.014

Strauss, J. T., & Stucky, M. J. (2016). Laser Additive Manufacturing Processing of a Mixture of Iron and Nickel Powders. *Proceedings of the 27th Annual International Solid Freeform Fabrication Symposium – An Additive Manufacturing Conference.*

Studart, A. R. (2016). Additive manufacturing of biologically-inspired materials. *Chemical Society Reviews, 45*(2), 359–376. doi:10.1039/C5CS00836K PMID:26750617

Students Aim for Space with 3D-Printed Rocket Engine | Space. (n.d.). Retrieved February 25, 2020, from https://www.space.com/29808-students-3d-printed-rocket-engine.html

Stull, J. A., Hill, M. A., Lienert, T. J., Tokash, J., Bohn, K. R., & Hooks, D. E. (2018). Corrosion characteristics of laser-engineered net shaping additively-manufactured 316L stainless steel. *JOM, 70*(11), 2677–2683. doi:10.100711837-018-3123-6

Sturm, L. D., Williams, C. B., Camelio, J. A., White, J., & Parker, R. (2017). Cyber-physical vulnerabilities in additive manufacturing systems: A case study attack on the. STL file with human subjects. *Journal of Manufacturing Systems, 44*, 154–164. doi:10.1016/j.jmsy.2017.05.007

Su, C., Chen, X., Gao, C., & Wang, Y. (2019). Applied Surface Science E ff ect of heat input on microstructure and mechanical properties of Al-Mg alloys fabricated by WAAM. *Applied Surface Science, 486*(December), 431–440. doi:10.1016/j.apsusc.2019.04.255

Suard, M. (2014). *Scheme of the electron beam melting process.* Download Scientific Diagram.

Sufiiarov, V. S., Popovich, A. A., Borisov, E. V., Polozov, I. A., Masaylo, D. V., & Orlov, A. V. (2017). The Effect of Layer Thickness at Selective Laser Melting. *Procedia Engineering, 174*, 126–134. doi:10.1016/j.proeng.2017.01.179

Sugavaneswaran, M., & Arumaikkannu, G. (2014). Modelling for randomly oriented multi material additive manufacturing component and its fabrication. *Materials & Design, 54*, 779-785. doi:10.1016/j.matdes.2013.08.102

Sui, S., Tan, H., Chen, J., Zhong, C., Li, Z., Fan, W., Gasser, A., & Huang, W. (2019). The influence of Laves phases on the room temperature tensile properties of Inconel 718 fabricated by powder feeding laser additive manufacturing. *Acta Materialia, 164*, 413–427. doi:10.1016/j.actamat.2018.10.032

Sun, R., Li, L., Zhu, Y., Guo, W., Peng, P., & Cong, B. (2018). *Microstructure, residual stress and tensile properties control of wire- arc additive manufactured 2319 aluminum alloy with laser shock peening.* doi:10.1016/j.jallcom.2018.02.353

Sun, C., Kirk, M., Li, M., Hattar, K., Wang, Y., Anderoglu, O., Valdez, J., Uberuaga, B. P., Dickerson, R., & Maloy, S. A. (2015). Microstructure, chemistry and mechanical properties of Ni-based superalloy Rene N4 under irradiation at room temperature. *Acta Materialia, 95*, 357–365. doi:10.1016/j.actamat.2015.04.061

Sun, S.-H., Koizumi, Y., Saito, T., Yamanaka, K., Li, Y.-P., Cui, Y., & Chiba, A. (2018). Electron beam additive manufacturing of Inconel 718 alloy rods: Impact of build direction on microstructure and high-temperature tensile properties. *Additive Manufacturing, 23*(August), 457–470. doi:10.1016/j.addma.2018.08.017

Sun, S., Hagihara, K., & Nakano, T. (2018). Effect of scanning strategy on texture formation in Ni-25 at. % Mo alloys. *Materials & Design, 140*, 307–316. doi:10.1016/j.matdes.2017.11.060

Sun, Y., & Hao, M. (2012). Statistical analysis and optimization of process parameters in Ti6Al4V laser cladding using Nd:YAG laser. *Optics and Lasers in Engineering, 50*(7), 985–995. doi:10.1016/j.optlaseng.2012.01.018

Suryakumar, S., Karunakaran, K. P., Bernard, A., Chandrasekhar, U., Raghavender, N., & Sharma, D. (2011). Computer-Aided Design Weld bead modeling and process optimization in Hybrid Layered Manufacturing. *Computer Aided Design, 43*(4), 331–344. doi:10.1016/j.cad.2011.01.006

Taminger, K. M. B., & Hafley, R. A. (2019). *Electron beam freeform fabrication : A rapid metal deposition process.* Academic Press.

Tammas-Williams, S., Withers, P. J., Todd, I., & Prangnell, P. B. (2016). The Effectiveness of Hot Isostatic Pressing for Closing Porosity in Titanium Parts Manufactured by Selective Electron Beam Melting. *Metallurgical and Materials Transactions. A, Physical Metallurgy and Materials Science, 47*(5), 1939–1946. doi:10.100711661-016-3429-3

Tan, Z. E., Pang, J. H., Kaminski, J., & Pepin, H. (2019). Characterisation of porosity, density, and microstructure of directed energy deposited stainless steel AISI 316L. *Additive Manufacturing, 25*, 286–296. doi:10.1016/j.addma.2018.11.014

Tanzi & Farè. (2019). *Manufacturing Technologies 3.1.* doi:10.1016/B978-0-08-101034-1.00003-7

Tao, F., Tang, Y., Zou, X., & Qi, Q. (2019). A field programmable gate array implemented fibre channel switch for big data communication towards smart manufacturing. *Robotics and Computer-integrated Manufacturing, 57*, 166–181. doi:10.1016/j.rcim.2018.12.005

Tascioglu, E., Kaynak, Y., Poyraz, Ö., Orhangül, A., & Ören, S. (2019, September). The Effect of Finish-Milling Operation on Surface Quality and Wear Resistance of Inconel 625 Produced by Selective Laser Melting Additive Manufacturing. In *Proceeding of International Conference on Advanced Surface Enhancement* (pp. 263-272). Springer.

Technavio. (2020). *Metal additive manufacturing market by application and geography - forecast and analysis 2020-2024*. Retrieved from https://www.technavio.com/report/global-metal-additive-manufacturing-market-analysis

Tepylo, N., Huang, X., & Patnaik, P. C. (2019). Laser-Based Additive Manufacturing Technologies for Aerospace Applications. *Advanced Engineering Materials, 21*(11), 1900617. doi:10.1002/adem.201900617

Teslenko, I. B., Digilina, O. B., & Abdullaev, N. V. (2019). Status and perspectives for the use of additive technologies in various branches of Russian industry. *IOP Conference Series. Materials Science and Engineering, 483*, 012093. doi:10.1088/1757-899X/483/1/012093

Thakar, R. A., & Trivedi, S. V. (2017). An Overview of Dimensional Stability of Invar 36 Material for Space Based Optical Mounting Applications. *International Conference on Ideas, Impact and Innovation in Mechanical Engineering (ICIIME 2017)*, *5*(6), 146-151.

The Fabricator. (2016). *Aerospace 3-D printing market to grow at 55.85 percent CAGR to 2020*. Retrieved from https://www.thefabricator.com/thefabricator/news/shopmanagement/aerospace-3-d-printing-market-to-grow-at-55-85-percent-cagr-to-2020

Thijs, L., Verhaeghe, F., Craeghs, T., Humbeeck, J. V., & Kruth, J.-P. (2010). A study of the microstructural evolution during selective laser melting of Ti–6Al–4V. *Acta Materialia*, *58*(9), 3303–3312. doi:10.1016/j.actamat.2010.02.004

Thomas, M., Baxter, G. J., & Todd, I. (2016). Normalised model-based processing diagrams for additive layer manufacture of engineering alloys. *Acta Materialia*, *108*, 26–35. doi:10.1016/j.actamat.2016.02.025

Thompson, S. M., Bian, L., Shamsaei, N., & Yadollahi, A. (2015). An overview of Direct Laser Deposition for additive manufacturing; Part I: Transport phenomena, modeling and diagnostics. *Additive Manufacturing*, *8*, 36–62. doi:10.1016/j.addma.2015.07.001

Tillmann, W., Schaak, C., Nellesen, J., Schaper, M., Aydinöz, M. E., & Hoyer, K.-P. (2017). Hot isostatic pressing of IN718 components manufactured by selective laser melting. *Additive Manufacturing*, *13*, 93–102. doi:10.1016/j.addma.2016.11.006

Ting, W., Martin, E., Lorenz, F., & Aleksandra, P. (2006). Anatomically constrained deforma-tion for design of cranial implant: methodology and validation. *International Conference on Medical Image Computing and Computer-Assisted Intervention*, 9-16. 565_210.1007/ 11866

Tofail, S. A. M., Koumoulos, E. P., Bandyopadhyay, A., Bose, S., O'Donoghue, L., & Charitidis, C. (2018). Additive manufacturing: Scientific and technological challenges, market uptake and opportunities. *Materials Today*, *21*(1), 22–37. doi:10.1016/j.mattod.2017.07.001

Tolochko, N. K., Mozzharov, S. E., Yadroitsev, I. A., Laoui, T., Froyen, L., Titov, V. I., & Ignatiev, M. B. (2004). Balling processes during selective laser treatment of powders. *Rapid Prototyping Journal*, *10*(2), 78–87. doi:10.1108/13552540410526953

Tomus, D., Rometsch, P. A., Heilmaier, M., & Wu, X. (2017). Effect of minor alloying elements on crack-formation characteristics of Hastelloy-X manufactured by selective laser melting. *Additive Manufacturing*, *16*, 65–72. doi:10.1016/j.addma.2017.05.006

Tortorelli, P. F., & Natesan, K. (1998). Critical factors affecting the high-temperature corrosion performance of iron aluminides. *Materials Science and Engineering A*, *258*(1-2), 115–125. doi:10.1016/S0921-5093(98)00924-1

Toursangsaraki, M. (2018). *A review of multi-material and composite parts production by modified additive manufacturing methods*. arXiv preprint arXiv:1808.01861

Travitzky, N., Bonet, A., Dermeik, B., Fey, T., Filbert-Demut, I., Schlier, L., Schlordt, T., & Greil, P. (2014). Additive Manufacturing of Ceramic-Based Materials. *Advanced Engineering Materials*, *16*(6), 729–754. doi:10.1002/adem.201400097

Tucho, W. M., Cuvillier, P., Sjolyst-Kverneland, A., & Hansen, V. (2017). Microstructure and hardness studies of Inconel 718 manufactured by selective laser melting before and after solution heat treatment. *Materials Science and Engineering A*, *689*, 220–232. doi:10.1016/j.msea.2017.02.062

Tucho, W. M., & Hansen, V. (2019). Characterization of SLM-fabricated Inconel 718 after solid solution and precipitation hardening heat treatments. *Journal of Materials Science*, *54*(1), 823–839. doi:10.100710853-018-2851-x

Tucker, D. (2017). *Additive Manufacturing: Electron Beam Freeform Fabrication*. Tata Technologies PLM Solutions.

Tun, T., & Lwin, K.T. (2008). Optimizing the microstructure and mechanical properties of austempered ductile iron for automobile differential gear. *Journal of Metals Materials and Minerals*, *18*(2), 199–205.

Tuomi, J., Paloheimo, K.-S., Vehviläinen, J., Björkstrand, R., Salmi, M., Huotilainen, E., Kontio, R., Rouse, S., Gibson, I., & Mäkitie, A. A. (2014). A Novel Classification and Online Platform for Planning and Documentation of Medical Applications of Additive Manufacturing. *Surgical Innovation*, *21*(6), 553–559. doi:10.1177/1553350614524838 PMID:24616012

Tytko, D., Choi, P., Klo, J., Kostka, A., Inden, G., & Raabe, D. (2012). Microstructural evolution of a Ni-based superalloy (617B) at 700 ° C studied by electron microscopy and atom probe tomography. *Acta Materialia*, *60*(4), 1731–1740. doi:10.1016/j.actamat.2011.11.020

Udroiu, R., Braga, I. C., & Nedelcu, A. (2019). Evaluating the Quality Surface Performance of Additive Manufacturing Systems: Methodology and a Material Jetting Case Study. *Materials (Basel)*, *12*(6), 995. doi:10.3390/ma12060995 PMID:30917576

Unocic, R. R., & DuPont, J. N. (2004). Process efficiency measurements in the laser engineered net shaping process. *Metallurgical and Materials Transactions. B, Process Metallurgy and Materials Processing Science*, *35*(1), 143–152. doi:10.100711663-004-0104-7

V, C. (2019). *The Complete Guide to Electron Beam Melting (EBM) in 3D Printing*. 3Dnatives.

Vaezi, M., Chianrabutra, S., Mellor, B., & Yang, S. (2013). Multiple material additive manufacturing–Part 1: A review. *Virtual and Physical Prototyping*, *8*(1), 19–50. doi:10.1080/17452759.2013.778175

Vaidya, S., Ambad, P., & Bhosle, S. (2018). Industry 4.0 – A Glimpse. *Procedia Manufacturing*, *20*, 233–238. doi:10.1016/j.promfg.2018.02.034

Van Hooreweder, B., & Kruth, J.-P. (2017). Advanced fatigue analysis of metal lattice structures produced by Selective Laser Melting. *CIRP Annals*, *66*(1), 221–224. doi:10.1016/j.cirp.2017.04.130

Vayre, B., Vignat, F., & Villeneuve, F. (2012). Designing for additive manufacturing. *Procedia CIRP*, *3*, 632–637. doi:10.1016/j.procir.2012.07.108

Vayre, B., Vignat, F., & Villeneuve, F. (2012). Metallic additive manufacturing: State-of-the-art review and prospects. *Mechanics & Industry*, *13*(2), 89–96. doi:10.1051/meca/2012003

Vayssette, B., Saintier, N., Brugger, C., Elmay, M., & Pessard, E. (2018). Surface roughness of Ti-6Al-4V parts obtained by SLM and EBM: Effect on the High Cycle Fatigue life. *Procedia Engineering*, *213*, 89–97. doi:10.1016/j.proeng.2018.02.010

Velu, R., Calais, T., Jayakumar, A., & Raspall, F. (2019). A Comprehensive Review on Bio-Nanomaterials for Medical Implants and Feasibility Studies on Fabrication of Such Implants by Additive Manufacturing Technique. *Materials (Basel)*, *13*(1), 92. doi:10.3390/ma13010092 PMID:31878040

Venturini, G., Montevecchi, F., Scippa, A., & Campatelli, G. (2016). Optimization of WAAM Deposition Patterns for T-crossing Features. *Procedia CIRP*, *55*, 95–100. doi:10.1016/j.procir.2016.08.043

Vieira, A. A., Dias, L. M., Santos, M. Y., Pereira, G. A., & Oliveira, J. A. (2018). Setting an industry 4.0 research and development agenda for simulation - a literature review. *International Journal of Simulation Modelling*, *17*(3), 377–390. doi:10.2507/IJSIMM17(3)429

Vilaro, T., Colin, C., Bartout, J. D., Nazé, L., & Sennour, M. (2012). Microstructural and mechanical approaches of the selective laser melting process applied to a nickel-base superalloy. *Materials Science and Engineering A*, *534*, 446–451. doi:10.1016/j.msea.2011.11.092

Voznesenskaya, A. A., Kochuev, D. A., Chkalov, R. V., Kireev, A. V., & Morozov, V. V. (2020). Research of post-processing approaches for parts obtained by the method of selective laser melting. *Journal of Physics: Conference Series*, *1439*, 012028. doi:10.1088/1742-6596/1439/1/012028

Walker, J., Andani, M. T., Haberland, C., & Elahinia, M. (2014). *Additive Manufacturing of Nitinol Shape Memory Alloys to Overcome Challenges in Conventional Nitinol Fabrication. Volume 2A: Advanced Manufacturing*. American Society of Mechanical Engineers.

Wang, F. (2011). Mechanical property study on rapid additive layer manufacture Hastelloy® X alloy by selective laser melting technology. *International Journal of Advanced Manufacturing Technology*, *58*(5-8), 545–551. doi:10.100700170-011-3423-2

Wang, F., Williams, S., Colegrove, P., & Antonysamy, A. A. (2013). Microstructure and Mechanical Properties of Wire and Arc Additive Manufactured Ti-6Al-4V. *Metallurgical and Materials Transactions. A, Physical Metallurgy and Materials Science*, *44*(2), 968–977. doi:10.100711661-012-1444-6

Wang, J., Pan, Z., Ma, Y., Lu, Y., Shen, C., Cuiuri, D., & Li, H. (2018). Characterization of wire arc additively manufactured titanium aluminide functionally graded material: Microstructure, mechanical properties and oxidation behaviour. *Materials Science and Engineering A*, *734*, 110–119. doi:10.1016/j.msea.2018.07.097

Wang, L., & Felicelli, S. (2006). Analysis of thermal phenomena in LENS™ deposition. *Materials Science and Engineering A*, *435*, 625–631. doi:10.1016/j.msea.2006.07.087

Wang, L., Wang, S., & Hong, X. (2018). Pulsed SLM-manufactured AlSi10Mg alloy: Mechanical properties and microstructural effects of designed laser energy densities. *Journal of Manufacturing Processes*, *35*, 492–499. doi:10.1016/j.jmapro.2018.09.007

Wang, R., Gu, D., Xi, L., Lin, K., Guo, M., & Zhang, H. (2019). Selective laser melted TiB2/Ti6Al4V graded materials and first-principle calculations. *Materials Letters*, *254*, 33–36. doi:10.1016/j.matlet.2019.07.015

Wang, W., Lu, T., Yang, M., Sun, D., Xia, Y., & Wang, T. (2019). Hydrogel 3D printing with the capacitor edge effect. *Science Advances*, *5*, 1–7. PMID:30915393

Wang, X., Jiang, M., Zhou, Z., Gou, J., & Hui, D. (2017). 3D printing of polymer matrix composites: A review and prospective. *Composites. Part B, Engineering*, *110*, 442–458. doi:10.1016/j.compositesb.2016.11.034

Wang, Y., Kamath, C., Voisin, T., & Li, Z. (2018). A processing diagram for high-density Ti-6Al-4V by selective laser melting. *Rapid Prototyping Journal*, *24*(9), 1469–1478. doi:10.1108/RPJ-11-2017-0228

Wasserman, S. (2015, November 19). *3D Printing brings out the full potential of topology optimization.* Retrieved from https://www.engineering.com/ DesignSoftware/Design SoftwareArticles/ArticleID/11008/3D-Printing-Brings-Out-the-Full-Potential-of-TopologyOptimization.aspx?utm_ source=solid Thinking&utm_ campaign = c2b 97 dfaab-solid-Thinking_Newsletter_Nov_2015_11_2015 &utm_medium =email&utm _term=0 _2eedfcce8a -c2b97dfaab-37115393

Wei, H. (2019). *Enabler of digital transformation.* Retrieved from http://global.chinadaily.com.cn/a/201904/15/WS5cb3dda3a3104842260b62af.html

Wei, H. L., Mazumder, J., & DebRoy, T. (2015). Evolution of solidification texture during additive manufacturing OPEN. *Scientific Reports*, *5*(1), 1–7. doi:10.1038rep16446 PMID:26553246

Wei, K., Yang, Q., Ling, B., Yang, X., Xie, H., Qu, Z., & Fang, D. (2020). Mechanical properties of Invar 36 alloy additively manufactured by selective laser melting. *Materials Science and Engineering A, 772*, 138799. doi:10.1016/j.msea.2019.138799

Wiberg, A., Persson, J., & Ölvander, J. (2019). Design for additive manufacturing – a review of available design methods and software. *Rapid Prototyping Journal, 25*(6), 1080–1094. doi:10.1108/RPJ-10-2018-0262

Williams, S. W., Martina, F., Addison, A. C., Ding, J., Pardal, G., & Colegrove, P. (2016). *Wire þ Arc Additive Manufacturing.* doi:10.1179/1743284715Y.0000000073

Williams. L.R., Fan, K.F., & Bentley, R.P. (2015). Custom-made titanium cranioplasty: early and late complications of 151 cranioplasties and review of the literature. *Int J Oral Maxillofac Surg, 44*(5), 599-608. doi:10.1016/j.ijom.2014.09.006

Williams, C. B., Mistree, F., & Rosen, D. W. (2011). A Functional Classification Framework for the Conceptual Design of Additive Manufacturing Technologies. *Journal of Mechanical Design, 133*(12), 121002. Advance online publication. doi:10.1115/1.4005231

Willis, K., Brockmeyer, E., Hudson, S., & Poupyrev, I. (2012). Printed optics. In *Proceedings of the 25th Annual ACM Symposium on User Interface Software and Technology - UIST '12*, 589. New York: ACM Press. 10.1145/2380116.2380190

Wilts, E. M., Pekkanen, A. M., White, B. T., Meenakshisundaram, V., Aduba, D. C., Williams, C. B., & Long, T. E. (2019). Vat photopolymerization of charged monomers: 3D printing with supramolecular interactions. *Polymer Chemistry, 10*(12), 1442–1451. doi:10.1039/C8PY01792A

Winter, K., Hartmann, J., Jeschke, P., & Lahmer, M. (2013). Experimental and numerical investigation of streamwise surface waviness on axial compressor blades. In *ASME Turbo Expo 2013: Turbine Technical Conference and Exposition.* American Society of Mechanical Engineers Digital Collection. 10.1115/GT2013-95983

Wolff, S. J., Wu, H., Parab, N., Zhao, C., Ehmann, K. F., Sun, T., & Cao, J. (2019). In-situ high-speed X-ray imaging of piezo-driven directed energy deposition additive manufacturing. *Scientific Reports, 9*(1), 962. doi:10.103841598-018-36678-5 PMID:30700736

Wu, X. (2015). *The world 1st 3D printed jet engine was on display at this years Paris Airshow.* https://www.monash.edu/mcam/news/articles/paris-le-bourget-airshow

Wu, A. S., Brown, D. W., Kumar, M., Gallegos, G. F., & King, W. E. (2014). An Experimental Investigation into Additive Manufacturing-Induced Residual Stresses in 316L Stainless Steel. *Metallurgical and Materials Transactions. A, Physical Metallurgy and Materials Science, 45*(13), 6260–6270. doi:10.100711661-014-2549-x

Wu, B., Pan, Z., Ding, D., Cuiuri, D., Li, H., Xu, J., & Norrish, J. (2018). A review of the wire arc additive manufacturing of metals: Properties, defects and quality improvement. *Journal of Manufacturing Processes, 35*(August), 127–139. doi:10.1016/j.jmapro.2018.08.001

Wu, H., Liu, W., He, R., Wu, Z., Jiang, Q., Song, X., Chen, Y., Cheng, L., & Wu, S. (2017). Fabrication of dense zirconia-toughened alumina ceramics through a stereolithography-based additive manufacturing. *Ceramics International, 43*(1), 968–972. doi:10.1016/j.ceramint.2016.10.027

Xia, M., Gu, D., Yu, G., Dai, D., Chen, H., & Shi, Q. (2017). Porosity evolution and its thermodynamic mechanism of randomly packed powder-bed during selective laser melting of Inconel 718 alloy. *International Journal of Machine Tools & Manufacture, 116*, 96–106. doi:10.1016/j.ijmachtools.2017.01.005

Xiao, L., Song, W., Hu, M., & Li, P. (2019). Compressive properties and micro-structural characteristics of Ti–6Al–4V fabricated by electron beam melting and selective laser melting. *Materials Science and Engineering A, 764*, 138204. doi:10.1016/j.msea.2019.138204

Xiong, W., Hao, L., Li, Y., Tang, D., Cui, Q., Feng, Z., & Yan, C. (2019). Effect of selective laser melting parameters on morphology, microstructure, densification and mechanical properties of supersaturated silver alloy. *Materials & Design, 170*, 107697. doi:10.1016/j.matdes.2019.107697

Xu, F., Dhokia, V., Colegrove, P., McAndrew, A., Williams, S., Henstridge, A., & Newman, S. T. (2018). Realisation of a multi-sensor framework for process monitoring of the wire arc additive manufacturing in producing Ti-6Al-4V parts. *International Journal of Computer Integrated Manufacturing, 31*(8), 785–798. doi:10.1080/0951192X.2018.1466395

Xu, X. (2012). From cloud computing to cloud manufacturing. *Robotics and Computer-integrated Manufacturing, 28*(1), 75–86. doi:10.1016/j.rcim.2011.07.002

Yadollahi, A., Shamsaei, N., Thompson, S. M., Elwany, A., & Bian, L. (2017). Effects of building orientation and heat treatment on fatigue behavior of selective laser melted 17-4 PH stainless steel. *International Journal of Fatigue, 94*, 218–235. doi:10.1016/j.ijfatigue.2016.03.014

Yakout, M., & Elbestawi, M. A. (2017). *Additive Manufacturing of Composite Materials: An Overview*. Paper presented at the 6th International Conference on Virtual Machining Process Technology (VMPT), Montréal, Canada.

Yakout, M., & Elbestawi, M. A. (2019). *Selective laser melting of graphene-reinforced aluminum matrix composites for electrical batteries*. Paper presented at the HI-AM Annual Conference 2019, Vancouver, Canada.

Yakout, M., Elbestawi, M. A., & Veldhuis, S. C. (2018b). *Process-Structure-Property relationship for selective laser melting of aerospace alloys*. Paper presented at the 7th International Conference on Virtual Machining Process Technology (VMPT), Hamilton, Canada.

Yakout, M., Elbestawi, M., Wang, L., & Muizelaar, R. (2019a). *Selective laser melting of soft magnetic alloys for automotive applications*. Paper presented at the Joint Special Interest Group meeting between euspen and ASPE; Advancing Precision in Additive Manufacturing.

Yakout, M. (2019). *Process-Structure-Property Relationships in Selective Laser Melting of Aerospace Alloys*. McMaster University.

Yakout, M., Cadamuro, A., Elbestawi, M. A., & Veldhuis, S. C. (2017). The selection of process parameters in additive manufacturing for aerospace alloys. *International Journal of Advanced Manufacturing Technology, 92*(5), 2081–2098. doi:10.100700170-017-0280-7

Yakout, M., Elbestawi, M. A., & Veldhuis, S. C. (2018a). On the characterization of stainless steel 316L parts produced by selective laser melting. *International Journal of Advanced Manufacturing Technology, 95*(5), 1953–1974. doi:10.100700170-017-1303-0

Yakout, M., Elbestawi, M. A., & Veldhuis, S. C. (2018c). A Review of Metal Additive Manufacturing Technologies. *Diffusion and Defect Data, Solid State Data. Part B, Solid State Phenomena, 278*, 1–14. doi:10.4028/www.scientific.net/SSP.278.1

Yakout, M., Elbestawi, M. A., & Veldhuis, S. C. (2018d). A study of thermal expansion coefficients and microstructure during selective laser melting of Invar 36 and stainless steel 316L. *Additive Manufacturing, 24*, 405–418. doi:10.1016/j.addma.2018.09.035

Yakout, M., Elbestawi, M. A., & Veldhuis, S. C. (2019b). Density and mechanical properties in selective laser melting of Invar 36 and stainless steel 316L. *Journal of Materials Processing Technology*, *266*, 397–420. doi:10.1016/j.jmatprotec.2018.11.006

Yakout, M., Elbestawi, M. A., & Veldhuis, S. C. (2020a). A study of the relationship between thermal expansion and residual stresses in selective laser melting of Ti-6Al-4V. *Journal of Manufacturing Processes*, *52*, 181–192. doi:10.1016/j.jmapro.2020.01.039

Yakout, M., Elbestawi, M. A., Veldhuis, S. C., & Nangle-Smith, S. (2020b). Influence of thermal properties on residual stresses in SLM of aerospace alloys. *Rapid Prototyping Journal*, *26*(1), 213–222. doi:10.1108/RPJ-03-2019-0065

Yan, L., Chen, Y., & Liou, F. (2019). Additive Manufacturing of Functionally Graded Metallic Materials Using Laser Metal Deposition. *Additive Manufacturing, 100901.*

Yang, Y. P., Jamshidinia, M., Boulware, P., & Kelly, S. M. (2018). Prediction of microstructure, residual stress, and deformation in laser powder bed fusion process. *Computational Mechanics*, *61*(5), 599–615. doi:10.100700466-017-1528-7

Yang, Y., Song, X., Li, X., Chen, Z., Zhou, C., Zhou, Q., & Chen, Y. (2018). Recent Progress in Biomimetic Additive Manufacturing Technology: From Materials to Functional Structures. *Advanced Materials*, *30*(1706539), 1–34. doi:10.1002/adma.201706539 PMID:29920790

Yan, R., Luo, D., Huang, H., Li, R., Yu, N., Liu, C., H, Min., & Rong, Q. (2018). Electron beam melting in the fabrication of three-dimensional mesh titanium mandibular prosthesis scaffold. *Scientific Reports*, *750*. Advance online publication. doi:10.103841598-017-15564-6 PMID:29335485

Yan, W., Lin, S., Kafka, O. L., Yu, C., Liu, Z., Lian, Y., Wolff, S., Cao, J., Wagner, G. J., & Liu, W. K. (2018). Modeling process-structure-property relationships for additive manufacturing. *Frontiers of Mechanical Engineering*, *13*(4), 482–492. doi:10.100711465-018-0505-y

Yasa, E., & Ersoy, K. (2018). Additive Manufacturing of Polymer Matrix Composites. In M. C. Kuşhan (Ed.), *Aircraft Technology* (pp. 147–169). InTech. doi:10.5772/intechopen.75628

Yasa, E., Poyraz, O., Solakoglu, E. U., Akbulut, G., & Oren, S. (2016). A study on the stair stepping effect in direct metal laser sintering of a nickel-based superalloy. *Procedia CIRP*, *45*, 175–178. doi:10.1016/j.procir.2016.02.068

Yin, J., Zhu, H., Ke, L., Lei, W., Dai, C., & Zuo, D. (2012). Simulation of temperature distribution in single metallic powder layer for laser micro-sintering. *Computational Materials Science*, *53*(1), 333–339. doi:10.1016/j.commatsci.2011.09.012

Yin, S., Yan, X., Chen, C., Jenkins, R., Liu, M., & Lupoi, R. (2018). Hybrid additive manufacturing of Al-Ti6Al4V functionally graded materials with selective laser melting and cold spraying. *Journal of Materials Processing Technology*, *255*, 650–655. doi:10.1016/j.jmatprotec.2018.01.015

Yoon, O. S. (2013, September 11). *Issue quest publishes a report on business opportunities and response strategies in the spotlight 3D printing market*. Retrieved from http:// www. epnc. co. kr / news / articleView.html?idxno=10132

Yoon, H.S., Choi, J.S., & Oh, Y.K. (2007). Thermal deformation of aluminum alloy casting materials for tire mold by numerical analysis. *Transactions of Nonferrous Metals Society of China*, *22*(3), 859–863. doi:10.1016/S1003-6326(12)61816-0

Yoo, Y. S. J., Book, T. A., Sangid, M. D., & Kacher, J. (2018). Identifying strain localization and dislocation processes in fatigued Inconel 718 manufactured from selective laser melting. *Materials Science and Engineering A*, *724*, 444–451. doi:10.1016/j.msea.2018.03.127

Yuan, S., Zheng, Y., Chua, C. K., Yan, Q., & Zhou, K. (2018). Electrical and thermal conductivities of MWCNT/polymer composites fabricated by selective laser sintering. *Composites. Part A, Applied Science and Manufacturing*, *105*, 203–213. doi:10.1016/j.compositesa.2017.11.007

Yu, J. Y., Jung, T., Kim, S., & Rhee, S. (2011). Laser welding of cast iron and carburized steel for differential gear. *Journal of Mechanical Science and Technology*, *25*(11), 2887–2893. doi:10.100712206-011-0809-x

Yun, G. B. (2013). Study in weight lightening of differential gear and process development of cold forging for electric vehicle. Graduate School, Gyeongsang National University.

Yu, T., Zhao, Y., Sun, J., Chen, Y., & Qu, W. (2018). Process parameters optimization and mechanical properties of forming parts by direct laser fabrication of YCF101 alloy. *Journal of Materials Processing Technology*, *262*, 75–84. doi:10.1016/j.jmatprotec.2018.06.023

Zaharin, H. A., Abdul Rani, A. M., Ginta, T. L., & Azam, F. I. (2018). Additive Manufacturing Technology for Biomedical Components: A review. *IOP Conference Series. Materials Science and Engineering*, *328*(012003), 1–7. doi:10.1088/1757-899X/328/1/012003

Zander, N. E. (2019). *Recycled Polymer Feedstocks for Material Extrusion Additive Manufacturing*. Academic Press.

Zegard, T., & Paulino, G. H. (2016). Bridging topology optimization and additive manu-facturing. *Structural and Multidisciplinary Optimization*, *53*(1), 175–192. doi:10.100700158-015-1274-4

Zenou, M., & Grainger, L. (2018). Additive manufacturing of metallic materials. In *Additive Manufacturing* (pp. 53–103). Elsevier. doi:10.1016/B978-0-12-812155-9.00003-7

Zhai, Y., Lados, D. A., & LaGoy, J. L. (2014). Additive Manufacturing: Making Imagination the Major Limitation. *JOM*, *66*(5), 808–816. doi:10.100711837-014-0886-2

Zhang, B., Jaiswal, P., Rai, R., & Nelaturi, S. (2018). Additive manufacturing of functionally graded material objects: A review. *Journal of Computing and Information Science in Engineering*, *18*(4), 041002. doi:10.1115/1.4039683

Zhang, C., Chen, F., Huang, Z., Jia, M., Chen, G., Ye, Y., ... Zhang, L. (2019). Additive manufacturing of functionally graded materials: A review. *Materials Science and Engineering A*, *764*, 138209. doi:10.1016/j.msea.2019.138209

Zhang, D., Niu, W., Cao, X., & Liu, Z. (2015). Effect of standard heat treatment on the microstructure and mechanical properties of selective laser melting manufactured Inconel 718 superalloy. *Materials Science and Engineering A*, *644*, 32–40. doi:10.1016/j.msea.2015.06.021

Zhang, F., Levine, L. E., Allen, A. J., Stoudt, M. R., Lindwall, G., Lass, E. A., Williams, M. E., Idell, Y., & Campbell, C. E. (2018). Effect of heat treatment on the microstructural evolution of a nickel- based superalloy additive-manufactured by laser powder bed fusion. *Acta Materialia*, *152*, 200–214. doi:10.1016/j.actamat.2018.03.017 PMID:31080354

Zhang, J., Li, X., Xu, D., & Yang, R. (2019). Recent progress in the simulation of microstructure evolution in titanium alloys. *Progress in Natural Science: Materials International*, *29*(3), 295–304. doi:10.1016/j.pnsc.2019.05.006

Zhang, J., Wang, X., Paddea, S., & Zhang, X. (2016). Fatigue crack propagation behaviour in wire+arc additive manufactured Ti-6Al-4V: Effects of microstructure and residual stress. *Materials & Design*, *90*, 551–561. doi:10.1016/j.matdes.2015.10.141

Zhang, K., Fu, G., Zhang, P., Ma, Z., Mao, Z., & Zhang, D. Z. (2018). Study on the Geometric Design of Supports for Overhanging Structures Fabricated by Selective Laser Melting. *Materials (Basel, Switzerland)*, *12*(1), 27. doi:10.3390/ma12010027 PMID:30577654

Zhang, Y. N., Cao, X., Wanjara, P., & Medraj, M. (2013). Oxide films in laser additive manufactured Inconel 718. *Acta Materialia, 61*(17), 6562–6576. doi:10.1016/j.actamat.2013.07.039

Zhang, Y., Wei, Z., Shi, L., & Xi, M. (2008). Characterization of laser powder deposited Ti–TiC composites and functional gradient materials. *Journal of Materials Processing Technology, 206*(1-3), 438–444. doi:10.1016/j.jmatprotec.2007.12.055

Zhong, H., Qi, B., Cong, B., Qi, Z., & Sun, H. (2019). Microstructure and Mechanical Properties of Wire + Arc Additively Manufactured 2050 Al–Li Alloy Wall Deposits [English Edition]. *Chinese Journal of Mechanical Engineering, 32*(1), 92. Advance online publication. doi:10.118610033-019-0405-z

Zhong, Y., Rännar, L.-E., Wikman, S., Koptyug, A., Liu, L., Cui, D., & Shen, Z. (2017). Additive manufacturing of ITER first wall panel parts by two approaches: Selective laser melting and electron beam melting. *Fusion Engineering and Design, 116*, 24–33. doi:10.1016/j.fusengdes.2017.01.032

Zhou, Z., Tan, Z., He, D., Zhou, Z., Cui, L., Wang, Y., & Shao, W. G. W. (2020). *Fabrication of three-dimensional connected W-Cu10Sn composites by selective laser melting _ Elsevier Enhanced Reader.pdf.* Elsevier.

Zhou, J., Li, H., Yu, Y., Firouzian, K., Qian, Y., & Lin, F. (2020). Characterization of interfacial transition zone of functionally graded materials with graded composition from a single material in electron beam powder bed fusion. *Journal of Alloys and Compounds, 832*, 154774. doi:10.1016/j.jallcom.2020.154774

Zhou, Y., Lin, X., Kang, N., Huang, W., Wang, J., & Wang, Z. (2019). Influence of travel speed on microstructure and mechanical properties of wire + arc additively manufactured 2219 aluminum alloy. *Journal of Materials Science and Technology*. Advance online publication. doi:10.1016/j.jmst.2019.06.016

Zhou, Z., Li, D., Zeng, J., & Zhang, Z. (2007). Rapid fabrication of metal-coated composite stereolithography parts. *Proceedings of the Institution of Mechanical Engineers. Part B, Journal of Engineering Manufacture, 221*(9), 1431–1440. doi:10.1243/09544054JEM827

Zhou, Z., Tan, Z., He, D., Zhou, Z., Cui, L., Wang, Y., Shao, W., & Wang, G. (2020). Fabrication of three-dimensional connected W-Cu10Sn composites by selective laser melting. *Materials Letters, 264*, 127377. doi:10.1016/j.matlet.2020.127377

Zhu, J.H., Zhang, W.H., & Xia, L. (2015). Topology optimization in aircraft and aerospace structures design. *Archives of Computational Methods in Engineering*, 1–28. doi:10.100711831-015-9151-2

Zhu, Z., Dhokia, V. G., Nassehi, A., & Newman, S. T. (2013). A review of hybrid manufacturing processes–state of the art and future perspectives. *International Journal of Computer Integrated Manufacturing, 26*(7), 596–615. doi:10.1080/0951192X.2012.749530

Ziaee, M., & Crane, N. B. (2019). *Binder jetting: A review of process, materials, and methods.* Additive Manufacturing.

Zumofen, L., Beck, C., Kirchheim, A., & Dennig, H.-J. (2018). Quality Related Effects of the Preheating Temperature on Laser Melted High Carbon Content Steels. *Proceedings of Additive Manufacturing in Products and Applications - AMPA2017.*

Zwier, M. P., & Wits, W.W. (2016). Design for additive manufacturing: automated build orientation selection and optimization. *Procedia CIRP, 55*, 128-133.

About the Contributors

Raja A. completed his Ph.D. at the Indian Institute of Technology Roorkee. He is working on the problems related to metal additive manufacturing at the Department of Engineering Design, Indian Institute of Technology Madras as an Institute Post-Doctoral Fellow.

Esther Akinlabi is a Full Professor at the Department of Mechanical Engineering Science, Faculty of Engineering and the Built Environment, University of Johannesburg. Her research interest is in the field of modern and advanced manufacturing processes – Friction stir welding and additive manufacturing. Her research in the field of laser based additive manufacturing include laser material processing and surface engineering. She also conducts research in the field of renewable energy, and biogas production from waste. She is a rated National Research Foundation (NRF) researcher and has demonstrated excellence in all fields of endeavors. Her mentorship and research experience is enviable as she guides her team of postgraduate students through the research journey. She is a recipient of several research grants and has received many awards of recognition to her credit. She is a member of the prestigious South African Young Academy of Science and registered with the Engineering Council of South Africa. Prof Akinlabi has filed two patents, edited one book, published four books and authored/co-authored over 250 peer reviewed publications.

Ajit Behera is working as Assistant Professor in National Institute of Technology, Rourkela. He has published more than 60 publications. He is a active editorial board of two international journal and also acting as a active reviewer of many high impact journals. He already awarded with "Young Faculty award" in 2017, "C. V. Raman Award" in 2019 and "Yuva Rattan National Award" in 2020.

Velmurugan C. has graduated with Bachelor's Degree in Mechanical Engineering from Anna University. He obtained his Master and Doctoral Degree (Ph.D.) from Anna University and National Institute of Technology-Tiruchirappalli respectively in India. His research interest in "Development of Smart materials (Shape Memory Alloys) using advanced manufacturing process" bridges the gap between fundamental research and advanced applications into the industry. He has published his research in reputed international journals and books.

Ravi D. is a Professor in the Department of Mechanical Engineering, IIT Delhi. He obtained his Ph. D. from IIT Madras in 1996 and joined R&D, Tata Steel, Jamshedpur. He joined as a faculty at IIT Delhi in 1998. His research interests include Sheet Metal Forming and FE Analysis.

M. A. Elbestawi is a Professor of Mechanical Engineering at McMaster University. Dr. Elbestawi's research interests are focused on Manufacturing Engineering, specifically additive manufacturing, machining systems, and computer aided manufacturing. He is a Fellow of the Canadian Academy of Engineering (CAE), the International Academy for Production Engineering (CIRP), the Canadian Society for Mechanical Engineering (CSME), the American Society of Mechanical Engineers (ASME), and a member of the Association of Professional Engineers of Ontario (PEO).

Joyjeet Ghose is currently working as Associate Professor, in the Department of Production Engineering, BIT Mesra. His research areas include, Additive manufacturing and Advanced manufacturing processes.

Sanjay Kumar Jha is currently Head of the Department of Production Engineering, BIT Mesra. His research areas include Robotics, Advanced Welding and Additive manufacturing processes.

Arackal N. Jinoop is working as Senior research fellow at Raja Ramanna Centre for Advanced Technology, Indore from 2016 for pursuing PhD from Homi Bhabha National Institute – an aided institute of Department of Atomic Energy, Government of India and a deemed to be university. His research interests are laser additive manufacturing, laser surface treatment and 3D printing. He has published international journals, conference papers and book chapters in the above said area.

Jithin Joseph is a Post-doctoral Research Fellow at Deakin University. He graduated in Mechanical Engineering (University of Kerala) and did a post graduation in Materials Engineering (NIT Surathkal, India). He received his PhD from IFM-Deakin University in 2016, conducting research on the structure-property relationship of direct laser fabricated high entropy alloys from elemental powders. Jithin has worked for IFM-Deakin University since his PhD graduation, particularly on rapid alloy development methods, computational alloy design, additive manufacturing and surface coatings. His research experience includes a wide range of manufacturing processes (additive manufacturing, casting, machining, heat treating), materials characterisation techniques (dilatometry, thermal analysis, X-Ray diffraction and electron microscopy) and mechanical/wear testing methods.

Animesh Mandal is currently working as Associate Professor, in the School of Minerals, Metallurgical and Materials Engineering, IIT Bhubaneswar. His research areas include composite materials, materials characterization and semi-solid processing.

Krishna Mugada completed PhD in the area of friction stir welding. He is currently working in the Department of Mechanical Engineering, IIT Delhi as a Postdoctoral Researcher. His research interests include the welding and joining, wire+arc additive manufacturing and polymer composites.

Fredrick Mwema is a postdoctoral researcher at the University of Johannesburg, South Africa and a lecturer at the Dedan Kimathi University of Technology (DeKUT), Kenya. Dr. Mwema has interests in advanced manufacturing processes and materials engineering and has teaching experience at the undergraduate level for over six (6) years and he has mentored several masters students. He has BSc, MSc, and Ph.D. in Mechanical Engineering, which he acquired in 2011, 2015 and 2019 respectively. He has published several articles in journals, conferences and book chapters.

Saurav K. Nayak is working as Senior research fellow at Raja Ramanna Centre for Advanced Technology, Indore from 2017 for pursuing PhD from Homi Bhabha National Institute – an aided institute of Department of Atomic Energy, Government of India and a deemed to be university. His research interests are laser additive manufacturing and 3D printing. He has published international journals, conference papers and book chapters in the above said area.

Mythreyi O. V. is currently pursuing my doctoral degree in additive manufacturing at iit madras. my fields of interest are manufacturing, additive manufacturing, metrology, quality management, and CAM.

Alini Christ Paul is pursuing her bachelor degree in Mechanical Engineering from Nitte Meenakshi Institute of Technology (NMIT),an autonomous institute having Diamond ranking for Teaching and Learning, Facilities and Accreditations by QS-iGuage, situated at Begaluru, India. She has research interests in Additive Manufacturing, Business Analytics and Intelligence. She has presented one conference paper and attended several workshops to enhance her skills in the area of interests. She did internship at Indian Institute of Technology Indore and Raja Ramanna Centre for Advanced Technology, Indore, India.

C. P. Paul is leading a research group in area of Laser Additive Manufacturing (LAM) as Head, Laser Additive Manufacturing Laboratory at Raja Ramanna Centre for Advanced Technology (RRCAT). He has over 15 years of experience in research and development of LAM, from system development to its applications in engineering and prosthesis. His research programs focus on next generation additive manufacturing system and processes. To this end, his group explores novel techniques to develop advanced materials, innovative products, modeling and simulation tools, and holistic in-situ and ex-situ characterization techniques. He is instrumental in promoting LAM technology in India through collaborations with other national labs and reputed academic institutes by extending the indigenous developed facility for experiments. He has contributed chapters to twelve books and published over 100 research papers. Dr. Paul is also Dean-Student Affairs (RRCAT) at Homi Bhabha National Institute (HBNI) - a deemed to be University under Department of Atomic Energy, Government of India.

Özgür Poyraz holds a PhD and a MSc degree in design and manufacturing engineering and works as an Assistant Professor in Eskisehir Technical University Department of Mechanical Engineering focusing on manufacturing technology development projects for aircraft and engine parts. Besides his university carrier, he has worked for various industrial companies and participated in multi-disciplinary projects. He has authored many articles and proceedings for indexed journals and international conferences. His research interests include additive manufacturing, machining, CAD/CAM/CAE, process modeling and reverse engineering.

Jayaganthan R. is a professor at the Department of Engineering Design, IIT Madras and his research interests are in the area of Additive Manufacturing, SPD Processing of Materials, Material Mechanics, FEA, Surface Engineering, Structural Health Monitoring, and Biomaterials. He has graduated several Ph.D. and MTech students in the various areas of materials engineering covering mechanical behavior of materials, Tribology, SPD processing, FEA, and Surface Engineering. He has successfully executed several sponsored research projects in the various domains of materials engineering. He is an Alexander Von Humboldt Fellow, Germany, an IIM life member, India, IE(I) fellow, Calcutta, India, and AMSI Life member. He is currently an Editor, Transaction of IIM Journal, Springer publishing.

Divakar S. has Graduated with Bachelor's Degree in Mechanical Engineering from Dr. Mahalingam College of Engineering and Technology, and Master's Degree in Manufacturing Technology from Government College of Technology, Coimbatore. He is presently a Research Scholar in Mechanical Engineering Department from National Institute of Technology, Tiruchirappalli. His Research Interests include Additive Manufacturing of Metals and Materials Processing.

Deepak Kumar Sahini is currently working as Assistant Professor, in the Department of Production Engineering, BIT Mesra. His research areas include, SMART Manufacturing, Additive Manufacturing; Thixoforming of Al based composites. His current Research focus is on Manufacturing Process Simulations.

Seshadev Sahoo working as Associate Professor in Mechanical Engineering at Siksha O Anusandhan University, Bhubaneswar. His research expertise includes additive manufacturing, Solidification and Casting, and Microstructure modeling. He has published more than 50 international journal papers.

Aravindan Sivanandam is a Professor in the Department of Mechanical Engineering at Indian Institute of Technology Delhi. His Areas of research include Ceramics, Composites, Welding, and Nano-Manufacturing.

Mostafa Yakout is a Postdoctoral Fellow in Mechanical Engineering at McMaster University. His research interests include metal additive manufacturing, design engineering, and manufacturing engineering. After completing his B.Sc. and M.Sc. at Alexandria University, he received his Ph.D. in Mechanical Engineering from McMaster University. He is a member of the Canadian Society for Mechanical Engineering (CSME), the Ontario Society of Professional Engineers (OSPE), the American Society for Testing and Materials (ASTM), the American Society of Mechanical Engineers (ASME), and the International Institute of Acoustics and Vibrations (IIAV).

Evren Yasa graduated Mechanical Engineering in Istanbul Technical University (Istanbul, Turkey) in 2002 and began her master in the University of British Columbia (Vancouver, BC, Canada) under the supervision of Prof. Yusuf Altintas and she studied Machine Tool Metrology during her master studies. After working as a research engineer at Sabancı University for s short period, in September 2005 she became a member of the research group on production processes in Catholic University of Leuven (Leuven, Belgium) under the supervision of Prof. Jean-Pierre Kruth. She completed her Ph.d. study with her thesis on Combined Process of Selective Laser Melting and Selective Laser Erosion/Laser Re-melting. She won "Emerald Outstanding Doctoral Study-Highly commended" award with her doctoral dissertation. After her graduation and post-doc studies at K.U.Leuven, she started to work as a senior engineer at TEI, a GE-joint venture company specialized in manufacturing aero-engine parts. She led the activities on Additive Manufacturing for 5 years and afterwards she joined Eskisehir Osmangazi University as an assistant professor. She has also been working as an independent expert in laser-based manufacturing processes on behalf of European Commission in FP7 and Horizon2020 projects.

Index

F

G

H

I

L

M

N

O

P

R

S

T

W

www.ingramcontent.com/pod-product-compliance
Lightning Source LLC
LaVergne TN
LVHW062135300926
849161LV00010B/2536
9781799840541